FINITE MATHEMATICS

PRACTICAL APPLICATIONS

CONTENTS

Barbara Beaudin
University of Hartford

Barbara A. Becker
St. Xavier University

W. E. Bonnice
University of New Hampshire

Chris Braun
Coconino County Community College

Eleanor Canter
Wentworth Institute of Technology

Frederick J. Carter
St. Mary's University

Jerry Davis
Johnson State College

Joseph W. Fidler
Triton College

Bennette Harris
University of Wisconsin–Whitewater

Marian Harty
Edgewood College

Yvette Hester
Texas A&M University

Vern H. Klotz
Concordia University

Cen-Tsong Lin
Central Washington University

Lewis Lum
University of Portland

Thomas Ralley
Portland State University

Jane M. Rood
Eastern Illinois University

Arthur Rosenthal
Salem State College

Janet S. Schachtner
San Jacinto College North

Dan Schapiro
Yakima Valley Community College

Linda Schultz
McHenry County College

Ghulam M. Shah
University of Wisconsin–Waukesha

Mary Jane Sterling
Bradley University

We wish to thank our wives for their patience and support. We are also grateful to Margot Hanis, Bob Pirtle, Melissa Duge, Kirk Bomont, Roy Neuhaus, and all the wonderful people at Brooks/Cole who worked on this project, as well as Deann Christianson, Sarah Merz, Mike Rosenborg, and Amy Wasserman.

DAVID JOHNSON
THOMAS MOWRY

- Exercises that have an answer in the back of the book but no solution in the Student Solutions Manual.
- Exercises that have neither an answer in the back of the book nor a solution in the Student Solutions Manual.
- Exercises that require the student to check his or her answer.

ANCILLARIES

Instructor's Resource Manual (ISBN 0-534-93407-2), by David Johnson, Thomas Mowry, and Michael Rosenborg, contains answers to all even-numbered exercises, chapter summaries, and suggestions for teaching from the text.

Student Solutions Manual (ISBN 0-534-93408-0) provides the solutions to every other odd-numbered text exercise.

Student Tutorial CD (ISBN 0-534-36425-X) provides the student with tutorial items and practice problems.

Amortrix (Macintosh ISBN 0-534-35597-8; Windows 95/NT ISBN 0-534-35699-0; Windows 3.x ISBN 0-534-35695-8; Java http://www.brookscole.com/math/amortrix/) This software accompanies Chapter 2 (Systems of Linear Equations and Inequalities), Chapter 3 (Linear Programming: The Simplex Method), and Chapter 4 (Matrix Equations). The software executes matrix row operations and creates amortization schedules. It shows students the value of using a computer in computationally intensive areas. It will run on a network, independent computer, or over the Web, and is free to adopters of the text.

Printed Test Items (ISBN 0-534-35849-7) contains printed test forms, with answers, for instructors.

Thomson World Class Testing Tools (Macintosh ISBN 0-534-35859-4 and 0-534-36284-2; Windows ISBN 0-534-35848-9 and 0-534-36285-0) This fully-integrated suite of test creation, delivery, and classroom management tools includes World Class Test, Test Online, and World Class Management software. World Class Testing Tools allows professors to deliver tests via print, floppy, hard drive, LAN, or Internet. With these tools, professors can create cross-platform exam files from publisher files or existing WESTest 3.2 test banks, create and edit questions, and provide their own feedback to objective test questions—enabling the system to work as a tutorial or an examination. In addition, professors can generate questions algorithmically, creating tests that include multiple-choice, true/false, and matching questions. Professors can also track the progress of an entire class or an individual student. Testing and tutorial results can be integrated into the class management tool, which offers scoring, gradebook, and reporting capabilities.

Acknowledgments

Special thanks go to the reviewers who evaluated the manuscript:

Alfred Bachman
Cal Poly State University

Gary G. Bard
Taylor University

Ronald Barnes
University of Houston, Downtown

Carol A. Barnett
St. Louis Community College at Meramec

Subsections that make use of the Amortrix software or spreadsheets are identified by a computer icon. Some technology subsections provide support for both graphing calculators and software, and are identified by both icons.

History

The history of the subject matter is interwoven throughout most chapters. In addition, Historical Notes give in-depth biographies of the prominent people involved. It is our hope that students will see that there is a human aspect to mathematics. After all, mathematics was invented by real people for real purposes and is a part of our culture. Interesting research topics are given, and writing assignments are suggested. Short-answer historical questions are also included; they are intended to focus and reinforce the students' understanding of the historical material. They also serve to warn students that history questions may appear on exams.

Exercises

The exercises vary in difficulty. Some are exactly like the examples, and others expect more of the students. The exercises are not explicitly graded into A, B, and C categories, nor are any marked "optional" (students in this audience tend to react negatively if asked to do anything labeled in this manner). The more difficult exercises are indicated in the Instructor's Resource Manual.

Applications are stressed, and the student is usually given real or realistic data. Furthermore, the student is usually given information at a realistic level. For example, in Chapter 1 on Linear Equations, the student is not given cost and revenue functions, since it is not realistic to assume that this level of information would be available. Instead, he or she is given data and is asked to compute the cost and revenue functions, as well as the break-even point.

Critical thinking is also stressed. For example, the student is frequently asked to interpret a quantitative answer, give advice based on a quantitative answer, discuss assumptions, or make a prediction. Writing exercises are common, as are exercises that could be used in a group situation. Essay questions are also common; they can be used as an integral part of the students' grades, as a background for classroom discussion, or as extra credit work. Many are research topics and are kept as open-ended as possible.

Throughout the text, there is emphasis on the importance of checking one's answers. Thus, students learn to evaluate the reasonableness of their answers rather than accepting them at face value.

Answers

Answers to the odd-numbered exercises are given in the back of the book, with two exceptions:

- Answers to historical questions, interpretive questions, essay questions, and other open-ended questions are not given.
- Answers are not given when the exercises instruct the students to check the answers themselves.

The Students' Solutions Manual contains solutions to every other odd exercise. Thus, the instructor has access to four different types of exercises:

- Exercises that have an answer in the back of the book and a solution in the Student Solutions Manual.

Technology

Calculators and computers are useful and powerful tools that have become an integral part of the classroom and workplace. However, many students are unable to use their calculators effectively and have no mathematical experience using computers. Therefore, instructions for graphing calculator, scientific calculator, and computer use are included.

Detailed instructions for both scientific and graphing calculator use are given in calculator boxes throughout the text:

 Scientific calculator instructions are identified with this scientific calculator icon.

 Graphing calculator instructions are identified with this graphing calculator icon.

Furthermore, a number of optional technology subsections address some of the more advanced capabilities of graphing calculators and computers. These subsections allow instructors to incorporate technology into their classes if and when they desire, but they are entirely optional, and the text is in no way technology dependent. The subsections, identified with italics in the table of contents, are clearly identified in the text with an icon at the beginning of the subsection, and with a colored bar at the edge of the page as in this portion of the preface. The subsections are always preceded by technology-free discussion and exercises.

 The technology subsections that focus on graphing calculators were specifically written for Texas Instruments models TI-82, TI-83, TI-85, and TI-86; however, they frequently apply to other brands as well. They are identified by the graphing calculator icon. See the table of contents for a complete listing.

 The text also features Amortrix, a computer software supplement written specifically to accompany this text. It is available for Macintosh and Windows-based computers; it is also accessible on the World Wide Web (see Ancillaries section, below). The software shows students the value of using a computer in computationally intensive areas, without relieving the student of decision-making responsibilities. Amortrix has two capabilities:

- It will execute specific matrix row operations. After inputting a matrix, the student can instruct the computer to multiply row 2 by 3, and add the result to row 1. However, Amortrix will not "take over" and do a problem for the student—the student must decide where and how to pivot, and the computer will perform only the calculations.
- It will create an amortization schedule. However, Amortrix will not compute the last line correctly; instead, it uses the same algorithm on *all* lines of the schedule, forcing the student to correct the last line so that there is a zero balance.

The use of Amortrix is addressed in optional technology subsections in Chapter 2 (Systems of Linear Equations and Inequalities), Chapter 3 (Linear Programming: The Simplex Method), and Chapter 10 (Finance). The software is *not* an integral part of this book; the topics can be covered quite reasonably without any computer use.

Finally, an optional technology subsection gives instructions on the use of computerized spreadsheets (such as Microsoft Excel and Lotus 1-2-3) in creating amortization schedules.

PREFACE

Typically, a student takes a finite mathematics course to satisfy either a major or graduation requirement. Thus, the course can be populated with students ranging from business majors to biology majors to liberal arts majors. The goal of *Finite Mathematics: Practical Applications* is to familiarize students with the mathematics used in their major fields of study and to expose liberal arts students to topics in mathematics that are usable and relevant to any educated person. It is our hope that each student will encounter several topics that will prove useful over the course of his or her life. In addition, we hope that students will see that mathematics is relevant to their education and that there is a human aspect to mathematics.

TOUR OF THE BOOK

Ease of Use

This book is user-friendly. The examples don't skip steps; key points are boxed for emphasis; step-by-step procedures are given; and there is an abundance of explanation. The Instructor's Manual includes a "prerequisite map" so that the instructor can easily tell which earlier topics must be covered.

Course Prerequisite

Finite Mathematics: Practical Applications is written for the student who has successfully completed a course in algebra. A background in beginning algebra may be sufficient; however, the authors have found that the student who has a background in intermediate algebra is significantly better prepared for the course.

Algebra Review

Where appropriate, algebraic topics are reviewed, but in a very selective and focused manner. Only those topics that are used in the book are covered. There is no "Review of Algebra" chapter; instead, the reviews are placed as close as possible to the topics that utilize them, usually in a Section 0 at the beginning of the chapter. These sections are direct and to the point. They do not attempt to provide a thorough treatment of the algebra in question but rather focus on the algebra that will be used in the sections that follow. Typically, they do not cover applications of the algebra, which are covered in the following sections.

Algebra courses vary significantly from school to school. Among the Section 0 topics are some that students may not have seen before, such as matrix arithmetic and the elimination method. In these cases, the reviews are more detailed and assume less prior knowledge.

Sponsoring Editor: *Margot Hanis*
Signing Representative: *Dwayne Coy*
Marketing Manager: *Caroline Croley*
Editorial Assistants: *Kimberly Raburn,*
    Jennifer Wilkinson, Melissa Duge
Advertising Communications: *Christine Davis*
Production Editor: *Kirk Bomont*
Manuscript Editor: *Barbara Kimmel*
Permissions Editor: *Sue C. Howard*

Interior Design: *John Edeen*
Interior Illustration: *Dartmouth Publishing, Inc.*
Cover Design: *Roy R. Neuhaus*
Cover Illustration: *Amy L. Wasserman*
Art Editor: *Lisa Torri*
Photo Editors: *Kathleen Olson, Terry Powell*
Typesetting: *Carlisle Communications, Ltd.*
Printing and Binding: *R. R. Donnelley*
    and Sons, Inc.

For more information, contact:

BROOKS/COLE PUBLISHING COMPANY
511 Forest Lodge Road
Pacific Grove, CA 93950
USA

International Thomson Editores
Seneca 53
Col. Polanco
11560 México, D. F., México

International Thomson Publishing Europe
Berkshire House 168-173
High Holborn
London WC1V 7AA
England

International Thomson Publishing GmbH
Königswinterer Strasse 418
53227 Bonn
Germany

Thomas Nelson Australia
102 Dodds Street
South Melbourne, 3205
Victoria, Australia

International Thomson Publishing Asia
60 Albert Street
#15-10 Albert Complex
Singapore 189969

Nelson Canada
1120 Birchmount Road
Scarborough, Ontario
Canada M1K 5G4

International Thomson Publishing Japan
Hirakawacho Kyowa Building, 3F
2-2-1 Hirakawacho
Chiyoda-ku, Tokyo 102
Japan

Printed in the United States of America

10 9 8 7 6 5 4 3 2 1

Library of Congress Cataloging-in-Publication Data

Johnson, David B. (David Bruce), [date]
    Finite mathematics : practical applications / David B. Johnson,
  Thomas A. Mowry.
       p.   cm.
    Includes bibliographical references (p.   -) and index.
    ISBN 0-534-94782-4 (alk. paper)
    1. Mathematics.   I. Mowry, Thomas A.   II. Title.
  QA39.2.J59   1999
510—dc21
                                               98-24798
                                                 CIP

FINITE
MATHEMATICS
PRACTICAL APPLICATIONS

David B. Johnson

Diablo Valley College

Thomas A. Mowry

Diablo Valley College

BROOKS/COLE PUBLISHING COMPANY

 ITP® *An International Thomson Publishing Company*

Pacific Grove • Albany • Belmont • Bonn • Boston • Cincinnati• Detroit • Johannesburg
London • Madrid • Melbourne • Mexico City • New York • Paris • Singapore
Tokyo • Toronto • Washington

LINEAR EQUATIONS 1

You have studied three different branches of mathematics: arithmetic, algebra, and geometry. Algebra involves equations and inequalities, while geometry involves shapes and graphs. **Analytic geometry** is a bridge between algebra and geometry. It emphasizes a correspondence between equations and graphs; every equation has a graph, and most graphs have equations. This correspondence is a fruitful one; it allows algebra problems to be attacked with the tools of geometry in addition to the tools of algebra.

Business analysts and economists use a great deal of analytic geometry in their fields.

In this chapter we will investigate some of the mathematics used in business and economics. In particular, we will investigate the Central State University's Business Club and its attempt to make money selling T-shirts at football games. This enterprise involves the careful determination of the quantity of shirts to order as well as of the shirts' sales price. Ordering too many shirts or charging too much would result in the club's buying shirts they can't sell, and ordering too few or charging too little would result in not having enough to sell; in either event, the club would lose revenue.

1.0

LINES AND THEIR EQUATIONS

Cartesian Coordinates

Stuart sells sunglasses from a stand at Venice Beach. After experimenting with prices, he discovered (not surprisingly) that the more he charges, the less he sells. For several days Stuart charged $6 per pair. He kept records on the number of pairs sold and found that he sold an average of 40 pairs per day at that price. This and similar data are given in Figure 1.1.

FIGURE 1.1

Price	Number sold
$6	40
$7	38
$8	36

FIGURE 1.2

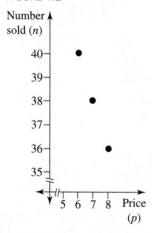

Stuart's data can be illustrated graphically by drawing two perpendicular number lines, where the horizontal number line represents price and the vertical number line represents number sold, as shown in Figure 1.2. Notice that both number lines have breaks; the price is always above 5, and the number sold is always above 35.

The upper-left point in Figure 1.2 is directly above 6 on the horizontal number line, so it corresponds to a price of $6. It is also directly across from 40 on the vertical number line, so it corresponds to 40 pairs sold. If we let p refer to price and n refer to number sold, the upper-left point could be labeled $p = 6$, $n = 40$. A more traditional way of labeling this point is to write $(p, n) = (6, 40)$. This is called an **ordered pair,** because it is a pair of numbers written in a certain order. The order is important; if we write (40, 6), we get the incorrect statement that at a price of $40, 6 pairs are sold.

The number 6 is called the ***p*-coordinate** of the ordered pair, and the number 40 is called the ***n*-coordinate.** The system of graphing is called **Cartesian coordinates,** in honor of René Descartes, a mathematician and philosopher. (Oddly, while Descartes explored the relationship between algebra and geometry, he neither invented nor utilized the system that bears his name.)

There are two different traditions of the use of letters in this type of situation:

- Use x and y
- Use letters that refer to the quantity being measured, as p refers to price and n refers to number sold above

Descartes started the former tradition; he used letters from the end of the alphabet to represent variables. This tradition is usually adhered to in algebra classes. However, in an application like this one, the latter tradition can serve as a valuable memory aid.

When the "x and y" tradition is followed, we have **x-coordinates** and **y-coordinates** rather than p-coordinates and n-coordinates. The horizontal axis is called the **x-axis,** and the vertical axis is called the **y-axis.** In the above discussion, we used p and n rather than x and y, respectively, so the horizontal axis is the p-axis, and the vertical axis is the n-axis. The two axes meet where both p and n are 0; this point, (0, 0), is called the **origin.**

In Figure 1.3, we show x- and y-axes that include both positive and negative values (the negative values are on the left end of the x-axis and on the lower end of the y-axis). The upper-left point corresponds to the ordered pair $(x, y) = (-2, 3)$, since it is above -2 on the x-axis and across from 3 on the y-axis.

FIGURE 1.3

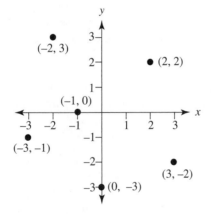

Slope

A line's steepness is measured by its slope. **Slope** (usually denoted by the letter m) is the ratio of the **rise** (the change in y) to the **run** (the change in x):

$$\text{slope} = m = \frac{\text{rise}}{\text{run}} = \frac{\text{change in } y}{\text{change in } x}$$

EXAMPLE 1
a. Calculate the slope of the line between the ordered pairs $(p, n) = (6, 40)$ and $(p, n) = (7, 38)$ from Stuart's sales data.
b. Calculate the slope of the line between the ordered pairs $(p, n) = (7, 38)$ and $(p, n) = (8, 36)$ from Stuart's sales data.
c. Determine what these slopes measure in the context of the problem.
d. Use the slope to predict the number of sunglasses that will sell at $9. What is this prediction based on?

Solution Here our ordered pairs are (p, n), rather than (x, y). Thus,

$$\text{slope} = \frac{\text{rise}}{\text{run}} = \frac{\text{change in } n}{\text{change in } p}$$

a. In moving from the point $(p, n) = (6, 40)$ to the point $(p, n) = (7, 38)$, p increases from 6 to 7, so p changes by $7 - 6 = 1$. Similarly, n decreases from 40 to 38, so n changes by $38 - 40 = -2$. Thus, the slope is

$$\text{slope} = \frac{\text{rise}}{\text{run}} = \frac{\text{change in } n}{\text{change in } p} = \frac{-2}{1} = -2$$

b. In moving from the point $(p, n) = (7, 38)$ to the point $(p, n) = (8, 36)$, p increases from 7 to 8, so p changes by 1. Similarly, n decreases from 38 to 36, so n changes by -2. Thus, the slope is

$$\text{slope} = \frac{\text{rise}}{\text{run}} = \frac{\text{change in } n}{\text{change in } p} = \frac{-2}{1} = -2$$

c. In the context of the problem, the 1 in the denominator represents a price increase of $1, and the -2 in the numerator indicates that the number sold decreased by 2. The fact that the slopes are the same in parts (a) and (b) indicates that a price increase of $1 *consistently* corresponds to a sales decrease of 2 pairs of sunglasses. In the context of the graph, equal slopes means that the steepness doesn't change; that is, the 3 points lie on a line.

d. A charge of $9 per pair of sunglasses is a $1 increase above an $8 price. Each $1 increase consistently corresponded to a sales decrease of 2 pairs, so sales should decrease to 34 pairs per day, if future sales are consistent with past sales. ●

Since a change in y is calculated by subtracting y-values, and a change in x is calculated by subtracting x-values, we have the following formula.

Slope Formula

The **slope** m of the line passing through the points (x_1, y_1) and (x_2, y_2) is

$$m = \frac{y_2 - y_1}{x_2 - x_1}$$

In the preceding formula, each of the symbols is meant to be a memory aid. For example, y_2 means the y-coordinate of the second point, and x_1 means the x-coordinate of the first point.

EXAMPLE 2 Find and interpret the slope of the line passing through the points $(-1, -3)$ and $(4, 7)$. Graph the points and the line, and show the rise and the run.

Solution Select $(-1, -3)$ as the first point [so $(x_1, y_1) = (-1, -3)$] and $(4, 7)$ as the second point [so $(x_2, y_2) = (4, 7)$], and substitute into the slope definition.

$$m = \frac{y_2 - y_1}{x_2 - x_1} \qquad \text{Slope definition}$$

$$= \frac{7 - -3}{4 - -1} \qquad \text{Substituting}$$

$$= \frac{10}{5} = \frac{2}{1} = 2$$

This means that in moving from $(-1, -3)$ to $(4, 7)$, the y-value increases by 10 while the x-value increases by 5—in other words, the line rises 10 while it runs 5. And since 10/5 reduces to 2/1, the line rises 2 while it runs 1. The nonreduced slope (10/5) and the reduced slope (2/1) are illustrated in Figure 1.4.

FIGURE 1.4

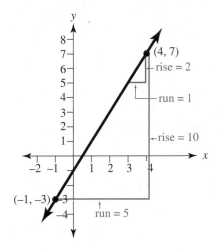

In Figure 1.5, we show two lines with different rises but the same runs. The steeper line has the larger rise and therefore the larger slope; the less steep line has the smaller rise and therefore the smaller slope.

If the slope of a line is positive (as in Figure 1.5), the line *rises* from left to right and the value of y *increases* as the value of x increases. If the slope of a line is negative (as in

FIGURE 1.5

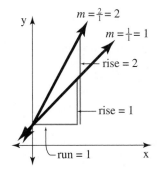

The steeper line has the larger slope; the less steep line has the smaller slope.

Example 1), the line *falls* from left to right and the value of *y* *decreases* as the value of *x* increases. These differences are illustrated in Figure 1.6.

FIGURE 1.6

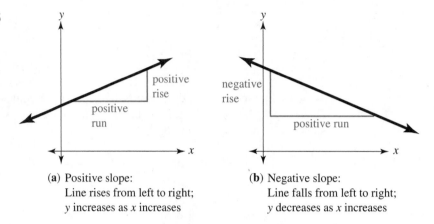

(a) Positive slope:
Line rises from left to right;
y increases as *x* increases

(b) Negative slope:
Line falls from left to right;
y decreases as *x* increases

There are two cases in which the slope of a line is something other than a positive or a negative number. A horizontal line has no rise, so its slope is $\frac{\text{rise}}{\text{run}} = \frac{0}{\text{run}} = 0$. A vertical line has no run, so its slope is $\frac{\text{rise}}{\text{run}} = \frac{\text{rise}}{0}$, which is undefined. This is shown in Figure 1.7.

FIGURE 1.7

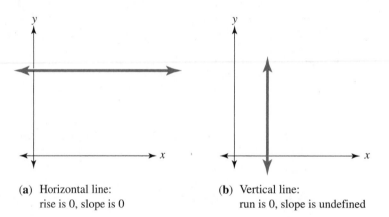

(a) Horizontal line:
rise is 0, slope is 0

(b) Vertical line:
run is 0, slope is undefined

Finding the Equation of a Line

The Point-Slope Formula

As we saw in Example 1, different points on the same line yield the same slope; a line doesn't change its steepness. If (x_1, y_1) is one particular point on a line, then for *all other* points (x, y) on that line, the slope calculation yields the same result:

$$m = \frac{y - y_1}{x - x_1}$$

If we multiply both sides of this equation by $x - x_1$, we get an important formula.

$$m(x - x_1) = \frac{y - y_1}{x - x_1}(x - x_1)$$

$$m(x - x_1) = y - y_1 \qquad \text{Simplifying}$$

This is called the **point-slope formula,** because it is used to find the equation of a line through a point (x_1, y_1) with slope *m*.

> **Point-Slope Formula for a Line**
>
> A line through a point (x_1, y_1) with slope m has the equation
>
> $$y - y_1 = m(x - x_1)$$

EXAMPLE 3   Find the equation of the line in Example 2.

Solution   We found that the line through $(-1, -3)$ and $(4, 7)$ has slope 2. Substitute $(-1, -3)$ for (x_1, y_1) and 2 for m into the point-slope formula.

$y - y_1 = m(x - x_1)$	Point-slope formula
$y - -3 = 2(x - -1)$	Substituting
$y + 3 = 2x + 2$	Distributing
$y = 2x - 1$	Simplifying

> ✔   We can check our answer by verifying that the points $(-1, -3)$ and $(4, 7)$ satisfy our equation.
> Substituting $(-1, -3)$ for (x, y):
>
> | | |
> |---|---|
> | $y = 2x - 1$ | The line's equation |
> | $-3 = 2(-1) - 1$ | Substituting |
> | $-3 = -2 - 1$ | A true statement ✔ |
>
> Substituting $(4, 7)$ for (x, y):
>
> | | |
> |---|---|
> | $y = 2x - 1$ | The line's equation |
> | $7 = 2 \cdot 4 - 1$ | Substituting |
> | $7 = 8 - 1$ | A true statement ✔ |

The Slope-Intercept Formula

FIGURE 1.8

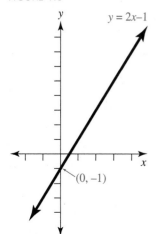

In Example 3, the simplified equation of the line is $y = 2x - 1$. This is called a **linear equation,** because its graph is a line. The 2 in that equation represents the slope; to find out what the -1 represents, substitute 0 for x:

$y = 2x - 1$	The line's equation
$y = 2 \cdot 0 - 1$	Substituting 0 for x
$y = -1$	Simplifying

This means that the line goes through the point $(0, -1)$. This point is on the y-axis, so it is called the line's **y-intercept** (or just **intercept**), as illustrated in Figure 1.8.

If a line's equation is found with the point-slope formula, that equation can always be simplified so that it is in the form $y = mx + b$. The equation $y = mx + b$ is called the **slope-intercept formula,** because it specifies the slope m and the intercept b. In the equation $y = 2x - 1$ [or $y = 2x + (-1)$], the slope is $m = 2$ and the intercept is $b = -1$.

Historical Note

The Inventors of Analytic Geometry

Around 220 B.C., the Greek mathematician Apollonius of Perga wrote an eight-volume work in which he investigated various curves, including circles, ellipses, and parabolas. He used a very early form of algebra to analyze these curves, and his research marks the origin of analytic geometry. Apollonius's work was the standard reference on these curves for almost 2000 years, even though several volumes were lost.

Around 830 A.D., the Arab mathematician Mohammed ibn Musa al-Khowarizmi wrote *Al-Jabr w'al Muqabalah,* in which he discussed linear and quadratic equations. Much of the mathematical knowledge of medieval Europe was derived from the Latin translation of

Apollonius of Perga

Title pages of a translation of al-Khowarizmi's *Al-Jabr w'al Muqabalah*

Slope-Intercept Formula for a Line

A line with slope m and y-intercept b has equation

$$y = mx + b$$

Graphing a Line from Its Equation

The slope-intercept form of a line's equation is especially useful in graphing the line.

EXAMPLE 4

a. Rewrite the equation $x + 2y = 8$ in slope-intercept form.

b. Use part (a) to determine the line's slope and y-intercept.

c. Sketch the line's graph, showing the slope and y-intercept.

René Descartes

philosopher René Descartes wrote a book on his philosophy of science. This book contained an appendix, *La géométrie,* in which he explored the relationship between algebra and geometry. Descartes's analytic geometry did not especially resemble our modern analytic geometry, which consists of ordered pairs, x- and y-axes, and a correspondence between algebraic equations and their graphs. Descartes used an x-axis, but he did not have a y-axis. Although he knew that an equation in two unknowns determines a curve, he had very little interest in sketching curves. Descartes's algebra, on the other hand, was more modern than that of any of his contemporaries. Algebra had been advancing steadily since the Renaissance, and it found its culmination in Descartes's *La géométrie.*

In 1629, eight years before Descartes wrote *La géométrie,* the French lawyer and amateur mathematician Pierre de Fermat attempted to recreate one of the lost works of Apollonius, using references to that work made by

other Greek mathematicians. The restoration of lost works of antiquity was a popular pastime of the upper class at that time. Fermat applied al-Khowarizmi's algebra to Apollonius's analytic geometry, and in so doing created modern analytic geometry. Unfortunately, Fermat never published his work; it was released only after his death, almost 50 years after it was written.

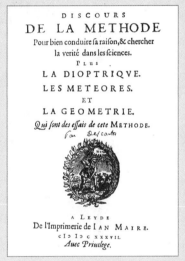

Title page of Descartes's *La géométrie*

al-Khowarizmi's works. In fact, the word *algebra* comes from the title.

Around 1360, Parisian scholar Nicole Oresme drew graphs of the speed of a falling object. In his graphs he used vertical and horizontal lines (somewhat similar to our x- and y-axes). He called these lines *longitudo* and *latitudo,* which indicate the concept's origins in mapmaking. Oresme's work is one of the earliest appearances of the idea of x- and y-axes and the graphing of a variable quantity.

In 1637, the French mathematician, scientist, and

Solution   **a.** To rewrite the equation in slope-intercept form, solve the equation for y.

$$x + 2y = 8$$
$$2y = -x + 8 \qquad \text{Subtracting } x$$
$$y = \frac{1}{2}(-x + 8) \qquad \text{Multiplying by 1/2}$$
$$y = \frac{-1}{2}x + 4 \qquad \text{Distributing}$$

The equation is now in slope-intercept form.

b. By comparing the slope-intercept formula with the above equation, we can find our line's slope and y-intercept:

$$y = mx + b$$

$$y = \frac{-1}{2}x + 4$$

The slope is $m = -1/2$, and the y-intercept is $b = 4$.

c. To sketch the graph of the line, first use the fact that the y-intercept is $b = 4$. This means that the line intercepts the y-axis at 4 [i.e., at the point $(0, 4)$]. Next, use the fact that the slope is $m = -1/2$.

$$m = \frac{-1}{2}$$

$$\frac{\text{rise}}{\text{run}} = \frac{-1}{2}$$

From the point $(0, 4)$, rise -1 (go 1 down) and run 2 (go 2 to the right). The graph is shown in Figure 1.9.

FIGURE 1.9

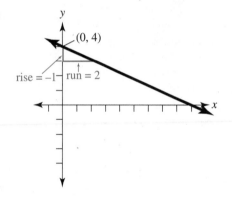

1.0

EXERCISES

1. a. Find the rise, the run, and the slope of the line between $(x_1, y_1) = (3, 7)$ and $(x_2, y_2) = (5, 11)$.

  b. Find the rise, the run, and the slope of the line between $(x_1, y_1) = (5, 11)$ and $(x_2, y_2) = (3, 7)$.

  c. What conclusion can you make regarding the computation of the slope of a line between two given points?

  d. Draw the line between $(3, 7)$ and $(5, 11)$. Show the rise and run.

2. Find the slope between the given points, as illustrated in Figure 1.10.

  a. The line through points A and B.

  b. The line through points B and C.

  c. The line through points A and C.

FIGURE 1.10

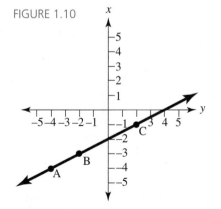

In Exercises 3–4:

    a. Plot the points corresponding to the two given ordered pairs.
    b. Find the slope of the line joining these two points.
    c. Graph the line.
    d. On your graph, show the rise and the run from the nonreduced slope, as well as the rise and the run from the reduced slope, as done in Example 2.

3. $(1, 4)$ and $(3, 10)$

4. $(1, -1)$ and $(4, -7)$

In Exercises 5–10:

    a. Plot the points corresponding to the three given ordered pairs.
    b. Find the slope of the line joining the points given by the first and second ordered pairs.
    c. Find the slope of the line joining the points given by the second and third ordered pairs.
    d. Are the three points in a line? Why or why not?

5. $(1, 3), (2, 5), (3, 7)$

6. $(-2, 1), (0, 6), (2, 11)$

7. $(-5, -6), (-1, -9), (3, -11)$

8. $(-7, 5), (-3, 3), (0, 1)$

9. $(-8, -11), (-4, -8), (4, -2)$

10. $(3, 1), (-5, -7), (-9, -11)$

In Exercises 11–14:

    a. Find the slope of the line joining the points given by the two ordered pairs.
    b. Use the slope to find a third point on that same line.
    c. Draw the line. Show the two given points as well as the point found in part (b).
    NOTE: A line consists of many points, so there are many different correct answers to part (b). For this reason, answers to part (b) are not given at the back of the book. Instead, you will check these answers in Exercises 19–22.

11. $(1, 5)$ and $(3, -3)$ **12.** $(2, -8)$ and $(-3, 4)$

13. $(-6, -3)$ and $(2, 4)$ **14.** $(-8, -4)$ and $(-2, -9)$

15. Jim and Barry's, a nationwide chain of ice cream stores, has found that it sells an average of 42,000 cones per summer day when it charges $1.30 each. When it increased the price to $1.40, average sales dropped to 41,000.
    a. Express these data as two ordered pairs (p, n).
    b. Find the slope of the line between these two ordered pairs.
    c. Determine what this slope measures in the context of the problem.
    d. Use the slope to predict the number of cones that will sell at $1.50.

    e. Use the slope to predict the number of cones that will sell at $1.35.

16. The owners of Blondie's Pizzas found that they sell an average of 120 slices of pizza per day when they charge $1.10 each. When they decreased the price to $1.00, average sales increased to 140.
    a. Express these data as two ordered pairs (p, n).
    b. Find the slope of the line between these two ordered pairs.
    c. Determine what this slope measures in the context of the problem.
    d. Use the slope to predict the number of slices that will sell at $.90.
    e. Use the slope to predict the number of slices that will sell at $1.05.

17. Why are the results of Exercises 15 and/or 16 parts (d) and (e) merely predictions and not guaranteed occurrences? What assumptions are these predictions based on?

18. **a.** Find the equation of the line in Exercise 1, using $(3, 7)$ for (x_1, y_1).
    b. Check your answer as shown in Example 3.
    c. Find the equation of the line in Exercise 1, using $(5, 11)$ for (x_1, y_1).
    d. What conclusion can you make regarding the computation of the equation of a line between two given points?

In Exercises 19–22:

    a. Find the equation of the line through the two given points.
    b. Check your answer to part (a) by verifying that the two given points satisfy the equation, as shown in Example 3.
    c. Check your answers to part (b) of Exercises 11–14 by verifying that the points you found there satisfy the equations.
    NOTE: Answers are not given in the back of the book.

19. $(1, 5)$ and $(3, -3)$ *(see Exercise 11)*

20. $(2, -8)$ and $(-3, 4)$ *(see Exercise 12)*

21. $(-6, -3)$ and $(2, 4)$ *(see Exercise 13)*

22. $(-8, -4)$ and $(-2, -9)$ *(see Exercise 14)*

In Exercises 23–26:

    a. Find the slope of the line through the two given points.
    b. Find the equation of the line through the two given points. What is the line's intercept?
    c. Check your answers to parts (a) and (b) by verifying that the two given points and the intercept satisfy the equation.
    d. Graph the line whose equation was found in part (c). Show the two given points, the slope, and the intercept.
    NOTE: Answers are not given in the back of the book.

23. $(3, -5)$ and $(8, -2)$ **24.** $(-5, -7)$ and $(1, 1)$

25. $(8, 11)$ and $(12, 11)$ **26.** $(3, 9)$ and $(-2, 4)$

27. Use the information given in Exercise 15 to do the following.
  a. Find the equation of the line through the two given points, using the letters p and n rather than x and y, respectively.
  b. Predict the number of cones that will sell at $1.45 each.
  c. Graph the line for $p \geq 0$ and $n \geq 0$. Why must both p and n be greater than 0? (Discuss why n can be 0 as well as why n cannot be negative.)

28. Use the information given in Exercise 16 to do the following.
  a. Find the equation of the line through the two given points, using the letters p and n rather than x and y, respectively.
  b. Predict the number of pizzas that will sell at $1.25 each.
  c. Graph the line for $p \geq 0$ and $n \geq 0$. Why must both p and n be greater than 0? (Discuss why n can be 0 as well as why n cannot be negative.)

29. Jim and Barry's, the chain of ice cream stores in Exercise 15, has found that the cost of making and selling 1000 cones per day at a typical store is $900, and the cost of making and selling 800 cones per day is $780.
  a. Express these data as two ordered pairs (n, c), where n is the number sold and c is the corresponding cost of production.
  b. Find the slope, and interpret it in the context of the problem.
  c. Find the equation of the line through these two points, using the letters n and c rather than x and y, respectively.
  d. Graph the line, for $n \geq 0$.
  e. Find the intercept, and interpret it in the context of the problem.

30. Blondie's Pizzas, the pizza store in Exercise 16, has found that the cost of making and selling 1100 slices per day is $1100, and the cost of making and selling 1000 slices per day is $1030.
  a. Express these data as two ordered pairs (n, c), where n is the number sold and c is the corresponding cost.
  b. Find the slope, and interpret it in the context of the problem.
  c. Find the equation of the line through these two points, using the letters n and c rather than x and y, respectively.
  d. Graph the line, for $n \geq 0$.
  e. Find the intercept, and interpret it in the context of the problem.

31. The average life expectancy of a newborn female in the United States in 1970 was 74.7 years; in 1990, the average was 78.8 years. (*Source:* National Center for Health Statistics)
  a. Express these data as two ordered pairs (t, e), where t is the year and e is the average life expectancy.
  b. Find the slope, and interpret it in the context of the problem.

  c. Find the equation of the line through these two points, using the letters t and e rather than x and y, respectively.
  d. Graph the line, for $t \geq 1900$.
  e. Predict when the average life expectancy will be 85 years.
  f. Predict the average life expectancy in the year 2000.
  g. What assumptions are the calculations in (b) through (f) based on?

32. The average life expectancy of a newborn male in the United States in 1970 was 67.1 years; in 1990, the average was 72.0 years. (*Source:* National Center for Health Statistics)
  a. Express these data as two ordered pairs (t, e), where t is the year and e is the average life expectancy.
  b. Find the slope, and interpret it in the context of the problem.
  c. Find the equation of the line through these two points, using the letters t and e rather than x and y, respectively.
  d. Graph the line, for $t \geq 1900$.
  e. Predict when the average life expectancy will be 85 years.
  f. Predict the average life expectancy in the year 2000.
  g. What assumptions are the calculations in (b) through (f) based on?

In Exercises 33–36:

  a. Rewrite the equation in slope-intercept form.
  b. Use part (a) to determine the line's slope and y-intercept.
  c. Sketch the line's graph.

33. $3x + y = 9$ **34.** $4x - 2y = 16$

35. $5x + 7y = 35$ **36.** $4x - 6y = 12$

37. a. Sketch the line through the points $(1, 3)$ and $(2, 3)$.
  b. Find the slope of the line.
  c. Find the equation of the line.
  d. List three other points on the line.
  e. How could you have found the equation of the line without using any formulas?

38. a. Sketch the line through the points $(2, 1)$ and $(2, 3)$.
  b. Find the slope of the line.
  c. List three other points on the line.
  d. Find the equation of the line.
  HINT: See Exercise 37(e).

In Exercises 39–43, find (a) the slope and (b) the equation of the line through the two given points.

39. $(5, 7)$ and $(8, 7)$ **40.** $(6, 2)$ and $(6, 3)$

41. $(-3, 1)$ and $(-3, 2)$ **42.** $(1, -3)$ and $(-2, -3)$

43. $(2, 11)$ and $(2, 14)$

44. a. On one set of axes, draw lines with the following slopes (where each line goes through the origin): 3, 2, 1, 1/2, 1/3, 0, $-1/3$, $-1/2$, -1, -2, -3. For each line, show the rise and the run.

b. Which line(s) is/are the steepest?

c. Which line(s) is/are the least steep?

d. If you were given the slopes of two different lines, how could you tell which line is steepest? How could you tell whether either line rises from left to right or falls from left to right?

45. Draw a line with the given slope and intercept, and give the equation of that line.

   a. positive slope and positive intercept

   b. positive slope and zero intercept

   c. positive slope and negative intercept

   d. zero slope and positive intercept

   e. zero slope and zero intercept

   f. zero slope and negative intercept

   g. negative slope and positive intercept

   h. negative slope and zero intercept

   i. negative slope and negative intercept

   j. undefined slope

     HINT: See Exercises 38–43

46. Answer the following using complete sentences. Why are Cartesian coordinates named in honor of René Descartes? In honor of which mathematician would it be more accurate to name them?

GRAPHING LINES ON A GRAPHING CALCULATOR

As their name implies, graphing calculators can graph lines. They follow the "x and y" tradition discussed earlier; they don't accept letters that refer to the quantity being measured, such as p for price and n for number sold. Furthermore, the equation must be given in slope-intercept form (or some unsimplified versions of the slope-intercept form).

The Graphing Buttons on a TI Graphing Calculator

The graphing buttons on a TI graphing calculator are all at the top of the keypad, directly under the screen. The labels on these buttons vary a little from model to model, but their uses are the same. The button labels and their uses are listed in Figure 1.11.

FIGURE 1.11

	Use this button to tell the calculator:				
Calculator Model	What to graph	What part of the graph to draw	To zoom in or out	To give the coordinates of a highlighted point	To draw the graph
TI-82/83:	Y=	WINDOW	ZOOM	TRACE	GRAPH
TI-85/86: Labels on buttons	M1 F1	M2 F2	M3 F3	M4 F4	M5 F5
TI-85/86: Labels on screen in graphing mode*	y(x) =	RANGE or WIND	ZOOM	TRACE	GRAPH

*__TI-85/86 users:__ Your calculator is different from the other TI models in that its graphing buttons are labeled "F1" through "F5" ("M1" through "M5" when preceded by the [2nd] button). The use of these buttons varies, depending on what you're doing with the calculator. When these buttons are active, their uses are displayed at the bottom of the screen.

Steps for Graphing a Line on a TI Graphing Calculator

To graph $y = 2x - 1$ on a TI graphing calculator, follow these steps.

Step 1 **(For TI-85/86 calculators only!)** *Put the calculator into graphing mode* by pressing the $\boxed{\text{GRAPH}}$ button. This activates the "F" buttons and puts labels at the bottom of the screen, as shown in Figure 1.12.

FIGURE 1.12
The screen labels for a TI-85's "F" buttons

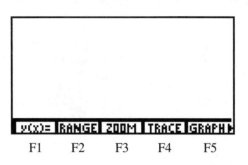

F1   F2   F3   F4   F5

Step 2 *Set the calculator up for instructions on what to graph* by pressing $\boxed{\text{Y=}}$ ($\boxed{\text{y(x)=}}$ or $\boxed{\text{F1}}$ on a TI-85/86). This produces a screen similar to that shown in Figure 1.13. If your screen has things written after the equals symbols, use the $\boxed{\uparrow}$ and $\boxed{\downarrow}$ buttons along with the $\boxed{\text{CLEAR}}$ button to erase them.

FIGURE 1.13
A TI-82's "Y=" screen (other models' screens are similar)

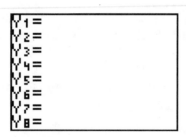

Step 3 *Tell the calculator what to graph* by typing "$2x - 1$" where the screen reads "$Y_1 =$." To type the x symbol:

TI-82:	Press $\boxed{\text{X,T,}\Theta}$
TI-83:	Press $\boxed{\text{X,T,}\Theta\text{,n}}$
TI-85/86:	Press either $\boxed{\text{x-VAR}}$ or $\boxed{\text{F1}}$ (which is now labeled "x" at the bottom of the screen)

After typing "$2x - 1$," press the $\boxed{\text{ENTER}}$ button.

Step 4 *Set the calculator up for instructions on what part of the graph to draw.*

TI-82/83:	Press $\boxed{\text{WINDOW}}$
TI-85:	Press $\boxed{\text{RANGE}}$ (i.e., $\boxed{\text{2nd}}$ $\boxed{\text{M2}}$)*
TI-86:	Press $\boxed{\text{WIND}}$ (i.e., $\boxed{\text{2nd}}$ $\boxed{\text{M2}}$)*

Step 5 *Tell the calculator what part of the graph to draw* by entering the values shown in Figure 1.14 (if necessary, use the $\boxed{\uparrow}$ and $\boxed{\downarrow}$ buttons to move from line to line).

- "xmin" and "xmax" refer to the left and right boundaries of the graph, respectively
- "ymin" and "ymax" refer to the lower and upper boundaries of the graph, respectively
- "xscl" and "yscl" refer to the scales on the *x*- and *y*-axes (i.e., to the location of the tick marks on the axes)

FIGURE 1.14
A TI-82's "WINDOW" screen (other models' screens are similar)

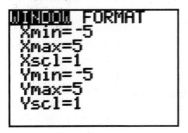

Step 6 *Tell the calculator to draw a graph* by pressing the $\boxed{\text{GRAPH}}$ button.** This produces the screen shown in Figure 1.15.

FIGURE 1.15
A calculator's graph of $y = 2x - 1$

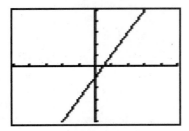

Step 7 Discontinue graphing by pressing $\boxed{\text{2nd}}$ $\boxed{\text{QUIT}}$.

---

*TI-85/86 users: Your calculator now has a double row of labels at the bottom of the screen. The bottom row of labels refers to the current use of the $\boxed{\text{F1}}$ through $\boxed{\text{F5}}$ buttons, and the top row of labels refers to the current use of the $\boxed{\text{M1}}$ through $\boxed{\text{M5}}$ buttons (which require the use of the $\boxed{\text{2nd}}$ button, as indicated by the orange lettering). Pressing the $\boxed{\text{EXIT}}$ button removes one row of labels.

**TI-85/86 users: Your calculator has two different $\boxed{\text{GRAPH}}$ buttons. One is labeled "GRAPH" on the button itself; this button puts the calculator in graphing mode, as discussed in Step 1. The other is labeled "F5" on the button itself, and "GRAPH" on the screen, when the calculator is in graphing mode; this tells the calculator to draw a graph, as discussed in Step 6.

EXERCISES

In the following exercises, you will explore some of your calculator's graphing capabilities. Answers for these exercises are not given in the back of the book.

47. *Exploring the "Zoom Standard" command.* Use your calculator to graph $y = 2x - 1$ as discussed in this section. When that graph is on the screen, select the "Zoom Standard" command from the "Zoom menu" by doing the following:

TI-82/83:   Press ZOOM and then select option 6 "ZStandard" by either:
- using the down-arrow to scroll down to that option and pressing ENTER , or
- typing the number "6"

TI-85/86:   Press ZOOM (i.e., F3) and then press ZSTD (i.e., F4)

---

a. What is the result of the "Zoom Standard" command?
b. How else could you accomplish the same thing, without using any zoom commands?

48. *Exploring the "RANGE" or "WINDOW" screen.* Use your calculator to graph $y = 2x - 1$ as discussed in this section. When that graph is on the screen, use the "RANGE" or "WINDOW" screen, described in Steps 4 and 5 of this section, to reset:
- xmin to 1
- xmax to 20
- ymin to 1
- ymax to 20

a. Why are there no axes shown?
b. Why does the graph start exactly in the lower-left corner of the screen?

49. *Exploring the "RANGE" or "WINDOW" screen.* Use your calculator to graph $y = 2x - 1$ as discussed in this section. When that graph is on the screen, use the "RANGE" or "WINDOW" screen, described in Steps 4 and 5 of this section, to reset:
- xmin to −1
- xmax to −20
- ymin to −1
- ymax to −20

Why did the calculator respond with an error message?

50. *Exploring the "RANGE" or "WINDOW" screen.* Use your calculator to graph $y = 2x - 1$ as discussed in this section. When that graph is on the screen, use the "RANGE" or "WINDOW" screen, described in Steps 4 and 5 of this section, to reset:
- xmin to −5
- xmax to 5
- ymin to 10
- ymax to 25

a. Why was only one axis shown?
b. Why was no graph shown?

51. *Exploring the "Trace" command.* Use your calculator to graph $y = 2x - 1$ as discussed in this section, with the viewing window set as shown in Figure 1.14. When that graph is on the screen, press TRACE . This causes two things to happen:
- A mark appears at a point on the line. This mark can be moved with the left-arrow and right-arrow buttons.
- The corresponding ordered pair is printed out at the bottom of the screen.

a. Use the "TRACE" feature to locate the line's x-intercept, the point at which the line hits the x-axis. (You may need to approximate it.)
b. Use algebra, rather than the graphing calculator, to find the x-intercept.
c. Use the "TRACE" feature to locate another ordered pair on the line.
d. Use substitution to check that the ordered pair found in part (c) is in fact a point on the line.

52. *Exploring the "Zoom Box" command.* Use your calculator to graph $y = 2x - 1$ as discussed in this section. When that graph is on your screen, press ZOOM and select option 1, "BOX" or "ZBOX," in the manner described in Exercise 47. This seems to have the same result as TRACE , except the mark does not have to be a point on the line. Use the four arrow buttons to move to a point of your choice (which may be on or off the line). Press ENTER . Use the arrow buttons to move to a different point, so that the resulting box encloses a part of the line. Press ENTER again. What is the result of using the "Zoom Box" command?

53. *Exploring the "Zoom In" command.* Use your calculator to graph $y = 2x - 1$ as discussed in this section. When that graph is on your screen, press ZOOM and select "Zoom In" or "ZIN," in the manner described in Exercise 47. This causes a mark to appear on the screen. Use the four arrow buttons to move the mark to a point of your choice, which is either on or near the line. Press $\boxed{\text{ENTER}}$.

a. What is the result of the "Zoom In" command?

b. How else could you accomplish the same thing, without using any zoom commands?

c. The "Zoom Out" command is listed right next to the "Zoom In" command? What does it do?

54. *Zooming in on x-intercepts.* Use the "Zoom In" command described in Exercise 53 and the "Trace" command described in Exercise 52 to approximate the location of the *x*-intercept of $y = 2x - 1$ as accurately as possible. You may need to use these commands more than once.

a. Describe the procedure you used to generate this answer.

b. According to the calculator, what is the *x*-intercept?

c. Is this answer the same as that of Exercise 51(b)? Why or why not?

55. *Calculating x-intercepts.* The calculator will calculate the *x*-intercept (also called a "root" or "zero") without using the "Zoom In" and "Trace" commands. First, use your calculator to graph $y = 2x - 1$ as discussed in this section. When that graph is on the screen, do the following.

On a TI-82/83:

- Press $\boxed{\text{2nd}}$ $\boxed{\text{CALC}}$ and select option 2, "root" or "zero."

- The calculator responds by asking "Lower Bound?" or "Left Bound?" Use the left- and right-arrow buttons to move the mark to a point slightly to the left of the *x*-intercept and press $\boxed{\text{ENTER}}$.

- When the calculator asks "Upper Bound?" or "Right Bound?" move the mark to a point slightly to the right of the *x*-intercept and press $\boxed{\text{ENTER}}$.

- When the calculator asks "Guess?" move the mark to a point close to the *x*-intercept, and press $\boxed{\text{ENTER}}$. The calculator will then display the location of the *x*-intercept.

On a TI-85:

- Press $\boxed{\text{MORE}}$ until the "MATH" label appears above $\boxed{\text{F1}}$ button, and then press that button.

- Press $\boxed{\text{ROOT}}$. This causes an ordered pair to appear on the screen and a mark to appear at the corresponding point on the line. The calculator is asking you if it is on the correct line. However, there is only one line in the exercise, so press $\boxed{\text{ENTER}}$. (If there were graphs of several different lines on the screen, you could use the up-arrow and down-arrow buttons to select the correct line.) The calculator will then display the location of the *x*-intercept.

On a TI-86:

- Press $\boxed{\text{MORE}}$ until the "MATH" label appears above $\boxed{\text{F1}}$ button, and then press that button.

- Press $\boxed{\text{ROOT}}$.

- The calculator responds by asking "Left Bound?" Use the left- and right-arrow buttons to move the mark to a point slightly to the left of the *x*-intercept, and press $\boxed{\text{ENTER}}$.

- When the calculator asks "Right Bound?" move the mark to a point slightly to the right of the *x*-intercept and press $\boxed{\text{ENTER}}$.

- When the calculator asks "Guess?" move the mark to a point close to the *x*-intercept and press $\boxed{\text{ENTER}}$. The calculator will then display the location of the *x*-intercept.

a. According to the calculator, what is the *x*-intercept of the line $y = 2x - 1$?

b. Is this answer the same as that of Exercise 51(b)? Why or why not?

1.1

FUNCTIONS

Suppose you asked your instructor, "When is our next exam?" and he responded, "In two weeks." If we call your question the "input" and your instructor's response the "output," then we could say that in answering your question your instructor assigned a unique output to your input (unique in that your instructor gave only one answer to your question).

A **function** is a mathematical procedure for assigning a unique output to any acceptable input. Functions can be represented numerically, graphically, symbolically, or verbally, as the following four examples show.

EXAMPLE 1 **A Function Represented Numerically** Figure 1.16 gives the world population at various times throughout history (*Source:* Population Division, United Nations Secretariat). Why does this represent a function, if the input is the year and the output is the population?

FIGURE 1.16

Year	0	1000	1500	1750	1800	1850	1900	1950	1980	1990
Population (in millions)	300	310	500	790	980	1260	1650	2520	4450	5300

Solution   For any acceptable input (such as the year 1900), there is an output (a world population of 1650 million people). And each of these outputs is unique—there is only one population for a given year. Thus, the data in Figure 1.16 are a function; in particular, world population is a function of time.

When a relationship is a function, **functional notation** can be used to describe that relationship. This involves giving the function a name. Usually, the name is either a letter that refers to the meaning of the function, the letter f (which stands for "function"), or some other letter near f in the alphabet. For a function named f, functional notation consists of a statement of the form

$$f(\text{input}) = \text{output}$$

If we name the function in Example 1 p for population, we could use functional notation and write $p(1900) = 1650$. More generally, we could write $p(\text{year}) = $ world population in that year.

The notation "$p(1900)$" looks like multiplication, but it is not. It is read "p of 1900," rather than "p times 1900"; the equation "$p(1900) = 1650$" means "the population of the world in 1900 was 1650 million people."

EXAMPLE 2   **A Function Represented Graphically**   A certain relationship between the input variable x and the output variable y is given by the graph in Figure 1.17. Why does this represent a function? Use functional notation to describe this relationship.

FIGURE 1.17

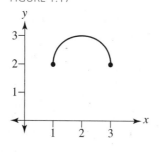

FIGURE 1.18

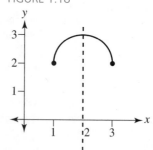

All acceptable vertical lines
hit the graph once.

Solution   Consider a typical input such as $x = 2$. This input has only one output, $y = 3$. One way to see this is to draw the vertical line $x = 2$, as shown in Figure 1.18. That line hits the graph once, at $y = 3$, so the input $x = 2$ has $y = 3$ as a unique output. All other vertical lines that hit the graph hit it once; that is, all other acceptable values of x have one corresponding value of y. Some vertical lines, such as $x = 4$, do not hit the graph. This means that $x = 4$ is not an acceptable input. Thus, each acceptable input has one output, and the graph is that of a function. In particular, y is a function of x.

If we name this function f, we could use functional notation and write $f(2) = 3$. More generally, we could write $f(x) = y$.   •

EXAMPLE 3   **A Function Represented Symbolically**   A certain relationship between the input variable x and the output variable y is given by the equation $y = x^2$. Why does this equation represent a function? Use functional notation to describe this relationship.

Solution   Any acceptable input (such as $x = 3$) has a corresponding output (in this case, $y = 3^2 = 9$). And this output is unique—for an input of 3 (or any other number), there is only one output. Thus, the equation represents a function. In particular, y is a function of x.

If we name this function g, we could use functional notation and write $g(3) = 3^2 = 9$. More generally, we could write $g(x) = x^2$.   •

EXAMPLE 4   **A Function Represented Verbally**   Why does the relationship between U.S. presidents and their political parties (Democrat, Republican, etc.) represent a function, if we consider the input to be the president and the output to be the president's party? Use functional notation to describe this relationship.

Solution   Each president belongs to one unique party, so this relationship represents a function, and the party is a function of the candidate. If we name this function p for "party," we could use functional notation and write $p(\text{Clinton}) = \text{Democrat}$. More generally, we could write $p(\text{president}) = $ the president's party.   •

The Vertical Line Test

In Example 2, we drew vertical lines to determine if a graph was that of a function. This is called the **vertical line test.** Drawing a vertical line is choosing an input. Counting the number of places where that vertical line hits the graph is counting the number of outputs assigned to that input. If every vertical line that hits a graph hits the graph once, then each acceptable input has one unique output, and the graph is that of a function. If there is one (or more) vertical line that hits the graph more than once, then one (or more) input has more than one output, and the graph is not that of a function.

EXAMPLE 5   A certain relationship between the input variable x and the output variable y is given by the graph in Figure 1.19. Use the vertical line test to determine if this graph represents a function.

FIGURE 1.19

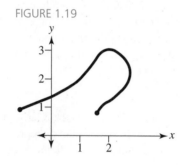

FIGURE 1.20

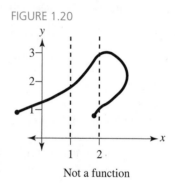

Not a function

Solution   If we draw the vertical line $x = 1$, that line hits the graph once, at $y = 1.7$, so the input $x = 1$ has a unique output. However, a vertical line at $x = 2$ hits the graph at $y = 1$ and at $y = 2.7$, as shown in Figure 1.20. Thus, some inputs have a unique output, and some inputs do not; the graph is not that of a function.

●

Why Must the Outputs Be Unique?

A function is a procedure for assigning a *unique* output to any acceptable input. Why must the outputs be unique?

One of the reasons is notational. When a relationship is a function, functional notation can be used to describe that relationship. The relationship in Example 5 is not a function because the input $x = 2$ has two outputs, $y = 1$ and $y = 3$. If we ignore this fact and use functional notation anyway, we would write $f(2) = 1$ and $f(2) = 3$. Clearly, $f(2)$ must equal itself—anything equals itself. This leads to the following statement:

$$f(2) = 1$$
$$\|$$
$$f(2) = 2.7$$

The statement is false; $1 \neq 2.7$. If we use ordered pair notation rather than functional notation, we merely state that the graph goes through the points $(2, 1)$ and $(2, 2.7)$, and there is no false implication that $1 = 2.7$.

Ordered pair notation can be used with both functions and nonfunctions; functional notation can be used only with functions. However, functional notation can be more useful and informative than ordered pair notation. In Example 1, we wrote $p(1900) = 1650$. The presence of the letter p reminds us that the function's output is population, and we can interpret the statement as "the population in the year 1900 was 1650." If we used ordered pair notation, we would write (1900, 1650) and there would be no reminder that population is being described. When functional notation can be used, functional notation is preferred over ordered pair notation.

There is another reason the outputs must be unique. Many questions have only one answer. If you asked your instructor for the date of your next exam, you would be rather perplexed if you were given more than one answer. You would also be perplexed if you were given more than one answer to the question, "What was the world's population in the year 1990?" Saying that this question has only one answer is saying that world population is a function of time.

Independent and Dependent Variables

In each of the preceding examples, the output depends on the input. In Example 4 the political party depends on the president, and in Example 1 the population depends on the year. For this reason, the output is called the **dependent variable,** and the input is called the **independent variable.** In Example 4 the political party is the dependent variable and the president is the independent variable (because the party depends on the president), and in Example 1 the population is the dependent variable and the year is the independent variable (because the population depends on the year).

Domain and Range

A function's **domain** is the set of all acceptable inputs; that is, the domain is the set of all acceptable values of the independent variable. Its **range** is the set of all values of the dependent variable.

EXAMPLE 6   Find the domain and range of the function p in Example 4.

Solution   As discussed above, the president is the independent variable, so the domain of the function p is the set of all presidents:

{Washington, . . . , Bush, Clinton}.

The political party is the dependent variable, so the range of p is the set of all political parties:

{Democratic, Republican, Bull Moose, . . .}.

By definition, the domain and range are sets; one way to describe a set is to list its elements (in any order) inside of set brackets (the symbols { and }). Sets will be discussed in greater detail in Chapter 5.   •

EXAMPLE 7   Find the domain and range of the function $y = x^2$ in Example 3.

Solution   The function does something to x to get y (in particular, it squares x), so the value of y depends on the value of x; y is the dependent variable, and x is the independent variable.

The domain is then the set of all values of the independent variable x. Any number can be squared, so the domain is the set of all real numbers. The range is the set of all values of the dependent variable y. Since y is the result of squaring something, y cannot be negative, and the range is

$\{y \mid y \geq 0\}$.

This is read as "the set of all y such that $y \geq 0$."

In Example 6 we described the domain and range of p by listing input and output values inside of set brackets. In Example 7 we described the domain with the phrase "the set of all real numbers," and we described the range by enclosing a rule ($y \geq 0$) inside set brackets. These are examples of the three ways of describing a set:

1. Enclosing a list in set brackets.
2. Enclosing a rule in set brackets.
3. Giving a verbal description that begins with the phrase "the set of all."

EXAMPLE 8   Find the domain and range of the function $f(x)$ in Example 2. Sketch the function's graph, and show the domain and range.

Solution   The domain is the set of all values of the independent variable x. In Example 2 we decided that a vertical line at $x = 4$ does not hit the graph, so $x = 4$ is not an acceptable input. Vertical lines to the left of $x = 1$ or to the right of $x = 3$ do not hit the graph. Thus, the domain is

$$\{x \mid 1 \leq x \leq 3\}$$

In other words, there is no graph to the left of $x = 1$ or to the right of $x = 3$. The range is the set of all values of the dependent variable y. We can see that there is no graph below $y = 2$ or above $y = 3$. Thus, the range is

$$\{y \mid 2 \leq y \leq 3\}$$

The function's domain and range are shown in Figure 1.21.

FIGURE 1.21

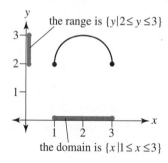
the range is $\{y \mid 2 \leq y \leq 3\}$
the domain is $\{x \mid 1 \leq x \leq 3\}$

EXAMPLE 9   Find the range of the function $f(x) = 2x + 6$, if the domain is given as $\{x \mid 1 \leq x \leq 2\}$.

Solution   To find the range, start with the inequality $1 \leq x \leq 2$, and use algebra to convert it to an inequality involving $f(x) = 2x + 6$.

$1 \leq x \leq 2$	The domain
$2 \leq 2x \leq 4$	Multiplying by 2
$8 \leq 2x + 6 \leq 10$	Adding 6
$8 \leq y \leq 10$	Replacing $2x + 6$ with y

The range is $\{y \mid 8 \leq y \leq 10\}$.

**Evaluating
Functions**

In Example 3 we decided that the equation $y = x^2$ is a function, and we used functional notation to rewrite that equation as $g(x) = x^2$. The instructions "evaluate the function g at $x = 3$"
or "find $g(3)$" mean "substitute 3 for x in the function named g." The process of substituting
a value for x looks different when functional notation is used, as shown in Figure 1.22.

FIGURE 1.22

	With functional notation	**Without functional notation**
Instructions	Evaluate g at $x = 3$ *or* find $g(3)$	Find y if $x = 3$
Computation	$g(x) = x^2$ $g(3) = 3^2 = 9$	$y = x^2$ $y = 3^2 = 9$
Result	$g(3) = 9$	$(3, 9)$

Notice that the statement "if $x = 3$, then $y = 9$" can be written as "$g(3) = 9$" or as
"$(3, 9)$"; the only difference is that the former uses functional notation and the latter doesn't.

EXAMPLE 10   Given the functions $g(x) = x^2$ and $h(x) = 2x + 1$, find the following:

a. $g(-5)$
b. $-g(5)$
c. $h(x + 1)$
d. $h(x) + 1$
e. $h(g(x))$

Solution   a. To find $g(-5)$, substitute -5 for x in the function g:

$$g(-5) = (-5)^2 = 25$$

b. To find $-g(5)$, first find $g(5)$ by substituting 5 for x in the function g:

$$g(5) = 5^2 = 25$$

To find $-g(5)$, put a negative symbol in front of $g(5)$:

$$-g(5) = -25$$

c. To find $h(x + 1)$, substitute $x + 1$ for x in the function h:

$$
\begin{aligned}
h(x + 1) &= 2(x + 1) + 1 &&\text{Substituting} \\
&= 2x + 2 + 1 &&\text{Distributing} \\
&= 2x + 3
\end{aligned}
$$

d. To find $h(x) + 1$, don't substitute anything. Instead, add 1 to $h(x)$:

$$
\begin{aligned}
h(x) + 1 &= 2x + 1 + 1 \\
&= 2x + 2
\end{aligned}
$$

e. To find $h(g(x))$, first substitute x^2 for $g(x)$:

$$h(g(x)) = h(x^2)$$

Next, substitute x^2 for x in the function h:

$$h(g(x)) = h(x^2)$$
$$= 2x^2 + 1$$

Functional notation is very sensitive to position. Notice that the instructions for parts (a) and (b) of Example 10 are identical *except for the position of the negative symbol.* Their solutions, however, are different, in spite of the similarity of the instructions. Similarly, the instructions for parts (c) and (d) are identical *except for the position of the "+1,"* whereas their solutions are quite different.

The instructions "find f(something)" always mean "substitute that thing for x in the function labeled f." Thus, we solved parts (a), (b), (c), and (e) of Example 10 by substituting the thing inside of the parentheses for x. Part (d), however, has an x inside of the parentheses, so there was no substituting to do. Instead, the instructions meant "add 1 to $h(x)$."

1.1

EXERCISES

1. a. Find $p(1000)$, using the function p given in Example 1.
   b. Write a sentence in which you explain the meaning of the equation "$p(1000) = $ (the answer to part a)."

2. a. Find $f(1)$, using the function f given in Example 2.
   b. Write a sentence in which you explain the meaning of the equation "$f(1) = $ (the answer to part a)."

3. a. Find $g(-2)$, using the function g given in Example 3.
   b. Write a sentence in which you explain the meaning of the equation "$g(-2) = $ (the answer to part a)."

4. a. Find $p(\text{Bush})$, using the function p given in Example 4.
   b. Write a sentence in which you explain the meaning of the equation "$p(\text{Bush}) = $ (the answer to part a)."

5. a. Find $f(3)$, using the function f given in Example 2.
   b. Write a sentence in which you explain the meaning of the equation "$f(3) = $ (the answer to part a)."

6. a. Find $p(1750)$, using the function p given in Example 1.
   b. Write a sentence in which you explain the meaning of the equation "$p(1750) = $ (the answer to part a)."

7. a. Find $p(\text{Nixon})$, using the function p given in Example 4.
   b. Write a sentence in which you explain the meaning of the equation "$p(\text{Nixon}) = $ (the answer to part a)."

8. a. Find $g(-5)$, using the function g given in Example 3.
   b. Write a sentence in which you explain the meaning of the equation "$g(-5) = $ (the answer to part a)."

In Exercises 9–14, justify each negative answer by giving an input that has more than one output.

9. Does the relationship between the 50 states and their capitals represent a function if:
   a. the input is the state and the output is its capital?
   b. the input is the capital and the output is its state?
   HINT: No two capitals share the same name.

10. Does the relationship between a state and the date that state achieved statehood represent a function if:
   a. the input is the state and the output is its statehood date?
   b. the input is the statehood date and the output is its state?
   HINT: Think of the first 13 states.

11. Do the data in Example 4 represent a function if the input is the political party and the output is the president?

12. Do the data in Example 1 represent a function if the input is the population and the output is the year?

13. Do the data in Figure 1.23 represent a function if:
   a. the input is the instructor and the output is that instructor's course?
   b. the input is the course and the output is the instructor?

FIGURE 1.23

Instructor	Mr. Kersting	Mr. Freidman	Mr. Feinman	Ms. Landre	Ms. Davis
Course	Statistics	Economics	Physics	Statistics	Sociology

14. Do the data in Figure 1.24 represent a function if:
  a. the input is the person and the output is that person's height?
  b. the input is the height and the output is the person?

FIGURE 1.24

Person	David	Gail	Lauren	Peter	Eric
Height	6′4″	5′6″	4′	3′2″	6′4″

In Exercises 15–18:

  a. Determine whether the given equation is a function if the input is x and the output is y. Justify each negative answer by giving an input that has more than one output.
  b. If the equation is a function, determine the domain and the range.

15. $y = 2x + 3$ **16.** $y = \sqrt{x}$

17. $y^2 = x$ **18.** $x^2 = y^2$

In Exercises 19–24:

  a. Determine whether the given graph represents a function if the input is x and the output is y. Justify each negative answer by giving an input that has more than one output.
  b. If the graph represents a function, determine the domain and the range.

19.

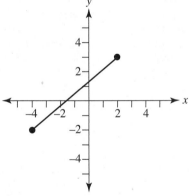

20.

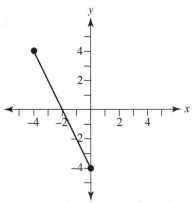

21.

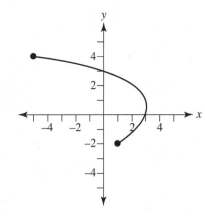

22.

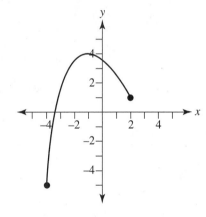

23.

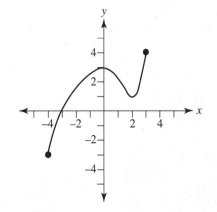

24.

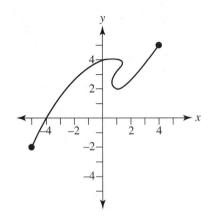

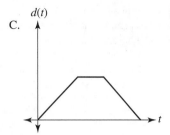

C.

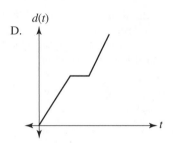

D.

25. Match each of the following activities with one of the given graphs, where t = time and $d(t)$ = distance from home. Justify your answers.

   a. You drove from your house to the gas station, filled your car up with gas, and returned.

   b. You drove from your house to school, stopping along the way at the gas station, where you filled your car up with gas.

   c. You drove one mile from your house, and then walked an additional mile.

   d. You walked one mile from your house to your car, and then drove an additional mile.

A.

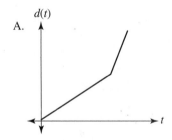

B.

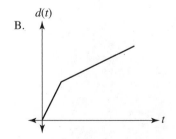

26. a. The day gradually warmed up until the sun went down, when it started to cool off. Sketch a graph of temperature (T) as a function of time (t).

   b. As soon as the sun rose it started to get hot. At noon it was 100°. At 2 P.M. a thunderstorm came through the area and the temperature dropped. At dusk it was 82°. Sketch a graph of temperature (T) as a function of time (t).

   c. Joan had no income during her four years of college. As soon as she graduated, she got a job that paid her $35,000 per year. After one year, she got a 20% raise. Sketch a graph of Joan's salary (S) as a function of time (t).

27. If the domain of $f(x) = 2x - 1$ is given as $\{x \mid x \geq 0\}$, find the range. Sketch the function's graph, and show the domain and range.

28. If the domain of $f(x) = 3x + 2$ is given as $\{x \mid x \geq 0\}$, find the range. Sketch the function's graph, and show the domain and range.

29. If the domain of $f(x) = 2x + 4$ is given as $\{x \mid 0 \leq x \leq 10\}$, find the range. Sketch the function's graph, and show the domain and range.

30. If the domain of $f(x) = 3x - 1$ is given as $\{x \mid 1 \leq x \leq 8\}$, find the range. Sketch the function's graph, and show the domain and range.

In Exercises 31–38, use the following functions:

$$f(x) = 2x \qquad g(x) = 3x - 1 \qquad h(x) = x^2$$

31. Find:

   a. $f(-3)$ **b.** $-f(3)$ **c.** $-f(-3)$

32. Find:
   a. $g(-2)$ **b.** $-g(2)$ **c.** $-g(-2)$

33. Find:
   a. $h(-4)$ **b.** $-h(4)$ **c.** $-h(-4)$

34. Find:
   a. $f(2x-1)$ **b.** $f(2x)-1$ **c.** $2f(x)-1$

35. Find:
   a. $g(4x+2)$ **b.** $4g(x+2)$ **c.** $4g(x)+2$

36. Find:
   a. $h(x+2)$ **b.** $h(x)+2$ **c.** $h(x+2)+2$

37. Find:
   a. $f(g(x))$ **b.** $g(f(x))$

38. Find:
   a. $g(h(x))$ **b.** $h(g(x))$

39. In Section 1.0, Exercise 27, you developed an equation relating p, the price of an ice cream cone, and n, the number that Jim and Barry's sells at that price.
   a. Explain why that equation is a function if the input is p and the output is n.
   b. Rewrite the equation in functional notation, using N as the function's name.

40. In Section 1.0, Exercise 28, you developed an equation relating p, the price of a pizza slice, and n, the number that Blondie's sells at that price.
   a. Explain why that equation is a function if the input is p and the output is n.
   b. Rewrite the equation in functional notation, using N as the function's name.

41. In Section 1.0, Exercise 29, you developed an equation relating n, the number of cones that Jim and Barry's sells, and c, the corresponding cost of production.
   a. Explain why that equation is a function if the input is n and the output is c.
   b. Rewrite the equation in functional notation, using C as the function's name.

42. In Section 1.0, Exercise 30, you developed an equation relating n, the number of pizza slices that Blondie's sells, and c, the corresponding cost of production.
   a. Explain why that equation is a function if the input is n and the output is c.
   b. Rewrite the equation in functional notation, using C as the function's name.

43. Find $C(N(p))$, where C is the function you developed in Exercise 41 and N is the function you developed in Exercise 39. Describe the input and output of this function.

44. Find $C(N(p))$, where N is the function you developed in Exercise 40 and C is the function you developed in Exercise 42. Describe the input and output of this function.

Exercises 45–48 refer to linear functions and linear relationships. If $y = mx + b$ [or $f(x) = mx + b$], we say that y is a linear function of x and that there is a linear relationship between x and y.

45. The air above us exerts a pressure on us. At sea level that pressure is 14.7 pounds per square inch, and at higher elevations it is less. Under the ocean the pressure increases as a linear function of depth, due to the weight of the water. At a depth of 33 feet, the pressure is double what it is at sea level.
   a. Express the above information as two ordered pairs (d, p), where d is depth and p is pressure.
   b. Find the slope, and interpret it in the context of the problem.
   c. Find the equation of the line through these two points, using the letters d and p rather than x and y, respectively.
   d. Find pressure as a linear function of depth using functional notation.
   e. Find the domain and range of this function.
   f. Beginning scuba divers may go to a depth of 60 feet. What pressure is on a diver at that depth? How many times the pressure at sea level is that?
   g. Experienced sport scuba divers may go to a depth of 100 feet. What pressure is on a diver at that depth? How many times the pressure at sea level is that?
   h. The average depth of the world's oceans is 12,540 feet. What is the pressure at that depth? How many times the pressure at sea level is that?
   i. Sketch the graph of the pressure function, showing its domain and range, as well as the solutions to (f), (g), and (h).

46. If an object falls for t seconds, then the distance (in feet) it falls is given by the function $D(t) = 16t^2$. (This function has limited accuracy. After a certain amount of time, depending on the shape of the object, air resistance becomes so great that the object stops accelerating and the function D no longer applies.)
   a. Is this a linear function? Why or why not?
   b. A parachutist falls for 10 seconds before she opens her parachute. How far does she fall in those 10 seconds?
   c. A rock is dropped from the top of a building. It falls for 2 seconds. What is the height of the building?

47. The speed of a falling object is a linear function of its time in motion. If an object is dropped, its speed will be 32 feet/second, after 1 second in motion. (This speed has the same limitation as the function D in Exercise 46.)
   a. Express the above information as an ordered pair (t, s) where t is time and s is speed.
   b. Your personal experience with falling objects should tell you the speed 0 seconds after an object is dropped. Express this information as a second ordered pair (t, s),

and find the slope of the line joining the two points.

c. Find the equation of the line through these two points, using the letters t and s rather than x and y, respectively.

d. Find speed as a linear function of time, using functional notation.

e. How fast is the parachutist in Exercise 46 going when she opens her parachute?

f. How fast is the rock in Exercise 46 going when it hits the ground?

48. A sketch pad is 15 inches by 12 inches; equivalently, it is 38.1 centimeters by 30.48 centimeters. The relationship between inches and centimeters is linear.

a. Express the given information as two ordered pairs (i, c), where i is length in inches and c is length in centimeters.

b. Find the slope of the line through the two points, and interpret it in the context of the problem.

c. Find the equation of the line through these two points, using the letters i and c rather than x and y, respectively.

d. Find length in centimeters as a linear function of length in inches, using functional notation.

e. Give the domain and range of this function.

f. Convert 36 inches (or 1 yard) to centimeters.

g. Convert 100 cm (or 1 meter) to inches.

h. Sketch the graph of the function found in (d), showing its domain and range, as well as the solutions to (f) and (g).

49. The Intergovernmental Panel on Climate Change predicted that, in the absence of effective international emissions control, the average global temperature would rise 0.3° centigrade every ten years (*Source: Science News,* June 30, 1990, p. 391). In 1970, the average global temperature was 15° C.

a. Find a linear function whose input is the number of years after 1970 and whose output is the predicted average global temperature.

b. Explain why the panel's information must result in a linear function.

c. It has been predicted that the sea level will rise by 65 cm if the average global temperature rises to 19° C. Why would this happen? When would it happen, according to the intergovernmental panel's prediction?

50. One aspect of sensory perception studied by psychologists is the amount by which a certain stimulus (a light or sound, for example) must be increased in order for a person to notice a difference. This amount, called the **just-noticeable difference** (or JND), is known to vary with the level of stimulus; that is, a louder sound would have to be increased by more than a softer sound would in order for the difference to be noticeable. In the mid-1800s, German physiologist E. H. Weber formulated Weber's law, which states that the JND is a linear function of the stimulus (with an intercept of 0), for any given stimulus.

a. A conversational voice has an intensity of $1 \times 10^{-10} = 0.0000000001$ watts/cm^2. A sound with this level of intensity is played for a subject, and then increased in intensity until the subject hears a difference. This does not occur until the experimenter increases its intensity to $1.25 \times 10^{-10} = 0.000000000125$ watts/cm^2. Find the JND for these data.

b. Use Weber's law and the answer to (a) to find the JND as a function of the intensity of the sound.

c. A bus interior has an intensity of $1 \times 10^{-7} = 0.0000001$ watts/cm^2. At what intensity will the subject hear a difference?

FUNCTIONS ON A GRAPHING CALCULATOR

Evaluating a Function on a TI Graphing Calculator

Graphing calculators can evaluate functions. To evaluate the function $f(x) = x^2$ at $x = 3$ [i.e., to find $f(3)$] on a TI graphing calculator, first store the function $f(x) = x^2$ as Y_1 (as discussed in Section 1.0), erase any other functions, and erase the "Y=" screen by pressing 2nd QUIT . Then use one of the following methods.

Method 1: Evaluating a Function by Storing a Value for x

The "variable x" button is labeled "X,T,Θ" on a TI-82, "X,T,Θ,n" on a TI-83, and "x-VAR" on a TI-85/86. For convenience, we will use $\boxed{\text{X,T}}$ to refer to each of these differently labeled but similarly used buttons.

TI graphing calculators have a memory for each letter of the alphabet. We can let x equal 3 by storing that number in the x memory.

Step 1   Store the number 3 in the x memory by typing 3 $\boxed{\text{STO}\rightarrow}$ $\boxed{\text{X,T}}$ $\boxed{\text{ENTER}}$.

Step 2   Evaluate the function at $x = 3$ by making the screen read "Y_1." To do this:

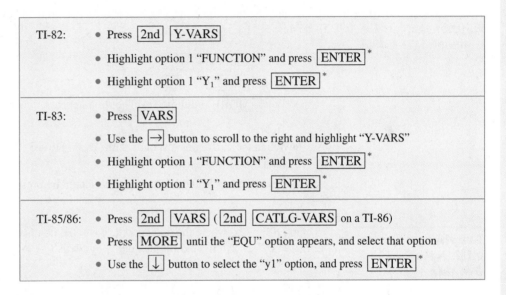

TI-82:	• Press $\boxed{\text{2nd}}$ $\boxed{\text{Y-VARS}}$
	• Highlight option 1 "FUNCTION" and press $\boxed{\text{ENTER}}$*
	• Highlight option 1 "Y_1" and press $\boxed{\text{ENTER}}$*
TI-83:	• Press $\boxed{\text{VARS}}$
	• Use the $\boxed{\rightarrow}$ button to scroll to the right and highlight "Y-VARS"
	• Highlight option 1 "FUNCTION" and press $\boxed{\text{ENTER}}$*
	• Highlight option 1 "Y_1" and press $\boxed{\text{ENTER}}$*
TI-85/86:	• Press $\boxed{\text{2nd}}$ $\boxed{\text{VARS}}$ ($\boxed{\text{2nd}}$ $\boxed{\text{CATLG-VARS}}$ on a TI-86)
	• Press $\boxed{\text{MORE}}$ until the "EQU" option appears, and select that option
	• Use the $\boxed{\downarrow}$ button to select the "y1" option, and press $\boxed{\text{ENTER}}$*

Once "Y_1" is on the screen, press $\boxed{\text{ENTER}}$ and the display will read "9," since $f(3) = 3^2 = 9$.

Method 2: Evaluating a Function While the Graph Is on the Screen

Step 1   Graph $y = x^2$ using the "Zoom Standard" feature discussed in Section 1.0.

Step 2   To evaluate the graphed function:

| TI-82/83: | Press $\boxed{\text{2nd}}$ $\boxed{\text{CALC}}$ and select option 1, "value" |
| TI-85/86: | Press $\boxed{\text{MORE}}$ until the "EVAL" option appears, and select that option by pressing $\boxed{\text{F1}}$ |

Step 3   When the calculator responds with "Eval X=" or "X=" type "3" and press $\boxed{\text{ENTER}}$. The calculator will respond by printing an ordered pair at the bottom of the screen and putting a mark on the corresponding point.

---

*Option 1 is automatically highlighted. If we were selecting some other option, we would use the $\boxed{\uparrow}$ and $\boxed{\downarrow}$ buttons to highlight it.

Method 3: Evaluating a Function Using Functional Notation (TI-82/83/86)

First, erase the graph by pressing CLEAR. To evaluate the function Y_1 at $x = 3$, make the screen read "$Y_1(3)$" and press ENTER.

- Generate the "Y_1" part of this in the manner discussed above in Method 1.
- Generate the "(3)" part by using the parentheses on the keyboard.

This form of functional notation reads $Y_1(3)$ rather than the more traditional $f(3)$. Unfortunately, TI-85s do not use functional notation, and "$Y_1(3)$" is interpreted as multiplication.

Method 4: Evaluating a Function Using the EVAL Command (TI-85/86)

To evaluate the function Y_1 at $x = 3$, make the screen read "eval 3" and press ENTER. To do this:

- Press 2nd MATH.
- Press MISC (i.e., F5).
- Press MORE until the "eval" option appears, and select that option by pressing F5.
- When the calculator responds with "eval" press 3 ENTER and the display will read "{9}," since $f(3) = 3^2 = 9$.
- Erase the labels at the bottom of the screen by pressing EXIT twice.

Graphing a Function with a Restricted Domain

To graph $f(x) = 2x$, where the domain is $\{x \mid x \geq 0\}$, do not enter "$2x$" as Y_1. Instead, enter "$(2x)(x \geq 0)$" as Y_1. To find the inequality symbols ($<$, $>$, $\leq$, $\geq$), press 2nd TEST.

EXERCISES

In the following exercises, you will explore some of your calculator's graphing capabilities. Answers for these exercises are not given in the back of the book.

51. *Exploring the limitations of the various methods of evaluating functions.* If $f(x) = 3x - 2$, do the following:
   a. Find $f(11)$ without using a calculator.
   b. Find $f(11)$ using Method 1.
   c. Graph $f(x)$ using the "Zoom Standard" command, and find $f(11)$ using Method 2.
   d. Find $f(11)$ using Methods 3 and/or 4, if your calculator has that capability.
   e. Do the different methods give the same result? Explain any differences.

52. *Exploring the simultaneous graphing of multiple functions.* If $f(x) = x$ and $g(x) = 2x$, do the following:

   a. Find $f(2)$ and $g(2)$ without using a calculator.
   b. Enter x for Y_1 and $2x$ for Y_2. Then find $f(2)$ [i.e., $Y_1(2)$] and $g(2)$ [i.e., $Y_2(2)$] using Method 1.
   c. Graph $f(x)$ and $g(x)$ using the "Zoom Standard" command. Then find $f(2)$ and $g(2)$ using Method 2. Use the up-arrow and down-arrow keys to switch between $f(2)$ and $g(2)$. Notice the "1" or "2" in the upper-right corner of the screen, indicating whether function Y_1 or Y_2 is utilized.
   d. Find $f(2)$ and $g(2)$ using Method 3 and/or Method 4, if your calculator has that capability.
   e. Explain how to use each method to evaluate two different functions.

53. Discuss the advantages and disadvantages of Methods 1, 2, 3, and 4. Take into account the results of Exercises 51 and/or 52.

54. *Exploring the limitations of graphing functions with restricted domains.* If $f(x) = x + 1$ with a domain of $\{x \mid x \geq 2\}$, do the following:
a. Graph $f(x)$ without using a calculator.
b. Graph $f(x)$ on your graphing calculator, and sketch the result.
c. Change the graphing mode from "connected" to "dot" as discussed below. Then sketch the graph that results.

TI-82/83:	Press $\boxed{\text{MODE}}$, scroll down to the line that says "Connected Dot," press $\boxed{\rightarrow}$ to highlight "Dot," and press $\boxed{\text{ENTER}}$.

TI-85/86:	While in graphing mode, press $\boxed{\text{MORE}}$ until the "FORMT" option appears, and select that option. Scroll down to the line that says "DrawLine DrawDot," press $\boxed{\rightarrow}$ to highlight "DrawDot," and press $\boxed{\text{ENTER}}$.

d. A graphing calculator graphs by computing and plotting a large number of ordered pairs. The corresponding points (or dots) are then either connected or not connected, depending on the calculator's mode. With this in mind, discuss the different graphs obtained in parts (a), (b), and (c).

55. *Exploring the graphing of functions with restricted domains.*
a. Graph $f(x) = 3x - 2$, $x \geq 2$ on your graphing calculator.
b. Use the "TRACE" feature to find f's range.

56. *Exploring the graphing of functions with restricted domains.*
a. Graph $g(x) = -x + 3$, $-2 \leq x \leq 2$ on your graphing calculator. This involves entering "$(-x + 3)(x \geq -2)$ $(x \leq 2)$" for Y_1.
b. Use the "TRACE" feature to find g's range.

57. *Exploring the graphing of functions with restricted domains.*
a. Graph $h(x) = x^2 - 3x + 4$, $-2 \leq x \leq 3$ on your graphing calculator.
b. Use the "TRACE" feature to find h's range.

1.2

LINEAR MODELS IN BUSINESS AND ECONOMICS

A **mathematical model** is a mathematical description of a real-world situation. A mathematical model that utilizes a linear function is called a **linear model.** In the exercises of Section 1.1 we explored several linear models, including underwater pressure and the speed of a falling object. These linear models follow the laws of physics and are quite accurate.

In this section we explore linear models in business and economics. These linear models involve human behavior rather than the laws of physics, and thus tend to be somewhat inaccurate. Furthermore, they can involve simplistic descriptions of the situations being modeled, but they can also provide powerful insights. Economists use linear models to analyze the relationship between supply and demand, as well as the relationship between cost and revenue.

Cost Functions

A **cost function** is a function whose input is the number of items produced and whose output is the cost of producing those items. All costs are classified as either fixed costs or variable costs. **Variable costs** are those costs that vary with the level of production, and **fixed costs** are those that do not vary with the level of production.

EXAMPLE 1   Every year, approximately 2000 students and alumni of Central State University attend the "Big Game" between Central State and its archrival, Western State University. One year, the CSU Business Club decides to sell Big Game T-shirts. A Central State alumnus owns Prints Alive, the local silk-screen studio, and she agrees to make the T-shirts and sell them to the business club at her cost. Production of the T-shirts will require one silk-screen master at $50. In addition, production of each individual shirt will use $4.50 worth of blank T-shirts and ink.

a. Determine which costs are fixed and which are variable.
b. Find the cost function $C(x)$ for Big Game T-shirts.
c. Find the domain and range of C.
d. Sketch the graph of the cost function C, showing the domain and range.

Solution   a. One silk-screen master will be used, regardless of the number of T-shirts produced, so the cost of the master is a fixed cost. The cost of blank T-shirts and ink will vary with the number of shirts produced, so these costs are variable costs.

b. To find the cost function $C(x)$, compute the cost of producing various numbers of shirts until a pattern appears.

To produce:	The cost will be:
1 shirt	$50 + 1 \cdot $4.50 = $54.50
2 shirts	$50 + 2 \cdot $4.50 = $59.00
3 shirts	$50 + 3 \cdot $4.50 = $63.50

The pattern is clear. To produce x T-shirts, the cost will be $50 + x \cdot $4.50. Thus, the cost function is $C(x) = 4.50x + 50$.

c. The domain of $C(x) = 4.50x + 50$ would seem to be the set of all real numbers, because any number can be substituted for x in the function C. However, $C(x)$ is a cost function, and x measures the number of shirts produced. Thus, x cannot be negative. Furthermore, it is reasonable to assume that the club could not sell more than 2000 shirts, since 2000 CSU fans attend the game. Thus, the domain is $\{x \mid 0 \le x \le 2000\}$. (You could certainly argue that x could be 1 or 2 but not 1.5, since x counts T-shirts. However, this domain would result in a graph that is a series of disconnected dots, rather than a line. Economists prefer the latter type of graph and thus allow fractions and decimals.)

To find the range, start with the inequality $0 \le x \le 2000$ and use algebra to convert it to an inequality involving $y = 4.50x + 50$.

$0 \le x \le 2000$	The domain
$0 \le 4.50x \le 9000$	Multiplying by 4.50
$50 \le 4.50x + 50 \le 9050$	Adding 50
$50 \le y \le 9050$	Replacing $4.50x + 50$ with y

The range is $\{y \mid 50 \le y \le 9050\}$.

d. The graph of $C(x) = 4.50x + 50$ is a line with slope 4.5 and intercept 50. Its graph is shown in Figure 1.25. Notice that only the part of the line that fits with the domain and range is shown.

FIGURE 1.25

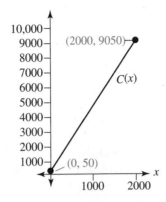

In Example 1, the fixed costs are $50. The only other cost is $4.50 per shirt. This $4.50 is called the **unit variable cost** because it is the variable cost for one shirt. The *variable cost* refers to the total cost of all the T-shirts and ink; it is 4.50x (the cost per shirt times the number of shirts).

Notice that the slope of the cost function $C(x)$ is $m = 4.50$, and the unit variable cost is $4.50 per T-shirt. Furthermore, the intercept of the cost function $C(x)$ is $b = 50$, and the fixed costs are $50. This always happens with a linear cost function—the slope is the unit variable cost, and the intercept is the fixed cost.

Marginality

Marginal cost is the change in cost when one more unit is produced. **Marginal revenue** is the change in revenue (or income) when one more unit is sold. **Marginality** refers to an increase of production by one. The idea of marginality—of looking at the potential costs and benefits of a decision to increase production—is one of the most important ideas in economics. Economists determine whether or not an increase in production is financially justifiable by comparing marginal cost and marginal revenue.

EXAMPLE 2   Find the marginal cost of the Business Club's T-shirt enterprise, at the following production levels:

a. 100 shirts
b. 200 shirts

Solution   a. If 100 shirts are produced, the cost of production is

$$C(100) = 4.50 \cdot 100 + 50 = 500$$

If production is increased by 1, then the cost of production would be

$$C(101) = 4.50 \cdot 101 + 50 = 504.50$$

The marginal cost is the change in cost when one more shirt is produced:

$$C(101) - C(100) = 504.50 - 500 = 4.50$$

b. If 200 shirts are produced, the cost of production is

$$C(200) = 4.50 \cdot 200 + 50 = 950$$

If production is increased by 1, then the cost of production would be

$$C(201) = 4.50 \cdot 201 + 50 = 954.50$$

The marginal cost is the change in cost:

$$C(201) - C(200) = 954.50 - 950 = 4.50$$

•

In Example 2, the marginal cost was the same for a production level of 100 shirts as it was for a production level of 200 shirts. This should not surprise you. If production were increased by 1, the only extra cost would be $4.50 for one more shirt and one more unit of ink, at any level of production. Thus, the marginal cost should be $4.50 at any level of production, not just at the production levels used in Example 2.

EXAMPLE 3 Use algebra to show that the marginal cost of the Business Club's T-shirt enterprise is $4.50, regardless of production levels.

Solution To do this, we must compute the marginal cost as we did in Example 2, but for an arbitrary production level x rather than the specific production levels used in Example 2.
If x shirts are produced, the cost of production is

$$C(x) = 4.50x + 50$$

If production is increased by 1, then the production level would be $x + 1$, and the cost of production would be $C(x + 1)$. To find $C(x + 1)$, we substitute $x + 1$ for x in the function named C.

$$\begin{aligned} C(x + 1) &= 4.50(x + 1) + 50 \\ &= 4.50x + 4.50 + 50 \\ &= 4.50x + 54.50 \end{aligned}$$

The marginal cost is the change in cost when one more shirt is produced:

$$\begin{aligned} C(x + 1) - C(x) &= (4.50x + 54.50) - (4.50x + 50) \\ &= 4.50x + 54.50 - 4.50x - 50 \\ &= 4.50 \end{aligned}$$

•

In Examples 2 and 3, the slope of the cost function $C(x)$ is $m = 4.50$, and the marginal cost is $4.50. This always happens with a linear cost function—the slope and the marginal cost are the same.

Revenue Functions

A **revenue function** is a function whose input is the number of items produced and whose output is the **revenue** or income generated by selling those items.

EXAMPLE 4 a. If the Business Club sells Big Game T-shirts for $8 each, find the revenue function $R(x)$.
b. Find the domain and range of R.
c. Sketch the graph of R, showing the domain and range.

Solution a. To find the revenue function $R(x)$, compute the revenue from selling various numbers of shirts until a pattern appears.

Sell:	And the revenue will be:
1 shirt	$1 \cdot \$8 = \$8$
2 shirts	$2 \cdot \$8 = \$16$
3 shirts	$3 \cdot \$8 = \$24$

The pattern is clear. If x T-shirts are sold, the revenue will be $x \cdot \$8$. Thus, the revenue function is $R(x) = 8x$.

b. The club can't sell a negative number of shirts, and we assumed in Example 1 that it can't sell more than 2000 shirts, so the domain is $\{x \mid 0 \leq x \leq 2000\}$.

To find the range, start with the inequality $0 \leq x \leq 2000$ and use algebra to convert it to an inequality involving $8x$.

$0 \leq x \leq 2000$ The domain

$0 \leq 8x \leq 16,000$ Multiplying by 8

$0 \leq y \leq 16,000$ Replacing $8x$ with y

The range is $\{y \mid 0 \leq y \leq 16,000\}$.

c. The graph of $R(x) = 8x = 8x + 0$ is a line with slope 8 and intercept 0. Its graph is shown in Figure 1.26. Notice that only the part of the line that fits with the domain and range is shown.

FIGURE 1.26

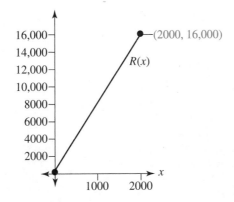

A company would neither make a profit nor sustain a loss if its revenue matched its costs. The ordered pair (x, y) at which this matching occurs is called the **break-even point.** The value of x at which this occurs is called the **break-even quantity;** the corresponding revenue is called the **break-even revenue.**

Break-Even Analysis

EXAMPLE 5   a. Use the information in Examples 1 and 4 to find the Business Club's break-even point. Find and interpret its break-even quantity and break-even revenue.

b. Graph $R(x)$ and $C(x)$ on the same axes, and show the break-even point.

Solution **a.** We are asked to find the point at which revenue equals costs.

$$R(x) = C(x)$$

$$8x = 4.5x + 50 \qquad \text{From Examples 1 and 4}$$

$$3.5x = 50 \qquad \text{Subtracting } 4.5x$$

$$x = 50/3.5 = 14.28 \ldots$$

This is the break-even quantity.

The corresponding revenue is

$$R(14.28 \ldots) = R(50/3.5) = 8 \cdot 50/3.5 = 114.285 \ldots \approx \$114.29.$$

This is the break-even revenue. If the Business Club produces and sells 14.28 . . . T-shirts, its revenue will be \$114.285 . . . , its costs will be \$114.285 . . . , and the club will break even. (In this example it is not possible to exactly break even, because it is not possible to sell exactly 14.28 . . . T-shirts.) The break-even point is the ordered pair (14.28 . . . , 114.285 . . .).

b. The graphs from Examples 1 and 4 are combined in Figure 1.27.

FIGURE 1.27

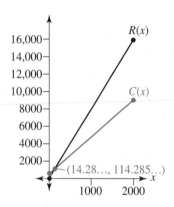

We can check our answer to Example 5(a) by verifying that the costs equal the revenue. The corresponding cost is

$$C(14.28 \ldots) = C(50/3.5) = 4.5 \cdot 50/3.5 + 50 = 114.285 \ldots \approx \$114.29.$$

Thus, $R(14.28 \ldots) = C(14.28 \ldots)$. ✔

Notice that 50/3.5, not 14.28, is substituted for x in Example 5(a). If 14.28 were used, the revenue would be different:

$$R(14.28) = 8 \cdot 14.28 = 114.24$$

This is only 4.5 . . . cents off from the true break-even revenue of \$114.285 . . . . However, in a big business situation the units could well have been millions of dollars rather than dollars; in that case, the error could have been more than your salary. To be accurate, do not round off until the end of the calculation.

Demand

A **consumer** is anyone who buys a good or service. Consumers react strongly to price. An increase in price would cause consumers to be less interested in purchasing an item, and thus would decrease the demand for the item. **Demand** is the number of units a group of consumers will buy at a given price (within a given time period).

EXAMPLE 6   To determine the best sales price for their Big Game T-shirts, Business Club members show a number of students, faculty, and alumni a sample T-shirt and ask them if they would pay $8 or $9 for the shirt. Of the 40 people asked, 30 said that they would pay $8, and 20 said that they would pay $9.

    a. Use these data to project the number of shirts that the CSU Business Club would sell at these prices if 2000 CSU fans attend the game. Write these projections as ordered pairs (p, d), where p is price and d is demand.

    b. Find the linear demand function $D(p)$ for Big Game T-shirts.

    c. Find the domain and range of D, assuming that no one buys more than one shirt.

    d. Sketch the graph of D, showing the domain and range.

Solution   **a.** At $8, 30 of 40 people (or 75%) would buy the shirt. This implies that 75% of the 2000 game attendees (or 1500 people) would buy the shirt at that price. This gives us the ordered pair $(p, d) = (8, 1500)$.

        At $9, 20 of 40 people (or 50%) would buy the shirt. Thus, 1000 people (50% of 2000) are projected to buy the shirt. This gives us the ordered pair $(9, 1000)$.

    b. The slope of the line between these two points is

$$m = \frac{d_2 - d_1}{p_2 - p_1} = \frac{1000 - 1500}{9 - 8} = \frac{-500}{1} = -500$$

The equation of the line is

$$d - d_1 = m(p - p_1)$$
$$d - 1500 = -500(p - 8)$$
$$d - 1500 = -500p + 4000$$
$$d = -500p + 5500$$

Switching to functional notation, we have the demand function.

$$D(p) = -500p + 5500$$

This demand function has an inherent weakness—it assumes that a linear model fits the situation. We will explore this issue in Section 1.3.

    c. If we followed the procedure used in Examples 1 and 4, we would figure that price can't be negative, so the domain is $\{p \mid p \geq 0\}$. However, this would include $p = 1$ in the domain, and at a price of $1 the demand would be

$$D(1) = -500 \cdot 1 + 5500 = 5000$$

Thus, the demand function projects sales of 5000 shirts at this price, even though there are only 2000 potential buyers! Clearly, $p = 1$ is not in the domain, and the domain is limited by the range.

The range is the set of all possible values of the output d. If 2000 students and alumni attend the game, then the demand must be between 0 and 2000 (assuming no one buys more than one shirt).

The range is $\{d \mid 0 \le d \le 2000\}$.

To find the domain, start with the inequality $0 \le d \le 2000$ and use algebra to convert it to an inequality involving p.

$$0 \le d \le 2000$$

$0 \le -500p + 5500 \le 2000$ Replacing d with $-500p + 5500$

$-5500 \le -500p \le -3500$ Subtracting 5500

$\dfrac{-5500}{-500} \ge \dfrac{-500p}{-500} \ge \dfrac{-3500}{-500}$ Multiplying or dividing by a negative reverses the direction of an inequality

$11 \ge p \ge 7$ Simplifying

The domain is $\{p \mid 7 \le p \le 11\}$.

d. The graph of $D(p) = -500p + 5500$ is shown in Figure 1.28. Notice that only the part its with the domain and range is shown.

FIGURE 1.28

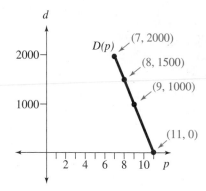

A **supplier** is anyone who produces or sells a good or service. Suppliers react to price, as do consumers. An increase in price (the supplier's price to the consumer) would motivate a supplier to increase the supply of items available for sale. **Supply** is the number of units a group of suppliers of a product will offer for sale at a given price.

Supply

EXAMPLE 7   Members of the CSU Business Club will be selling the T-shirt at booths at the Big Game. This means that the number of shirts they can supply to consumers depends on how many members are willing to staff the booths. The members agree that the club needs the income, but they have limited leisure time to invest in this enterprise. (After all, they are business majors.) At $8 per shirt, the members are willing to put in enough time to sell 600 shirts, but at $9 per shirt, more members would be willing to put in their time and 700 shirts could be sold.

a. Use these data to find the linear supply function $S(p)$ for Big Game T-shirts.

b. Find the domain and range of S, assuming that no one buys more than one shirt.

c. Sketch the graph of S, showing the domain and range.

Solution   a. Using ordered pairs (p, s), where p is price and s is supply, we have the ordered pairs $(8, 600)$ and $(9, 700)$. The slope of the line between these two points is

$$m = \frac{s_2 - s_1}{p_2 - p_1} = \frac{700 - 600}{9 - 8} = \frac{100}{1} = 100$$

The equation of the line is

$$s - s_1 = m(p - p_1)$$
$$s - 600 = 100(p - 8)$$
$$s - 600 = 100p - 800$$
$$s = 100p - 200$$

Switching to functional notation, we have the supply function.

$$S(p) = 100p - 200$$

b. As in Example 6, we will find the range first and the domain second. The range is the set of all possible values of the output s. If 2000 students and alumni attend the game, then the supply must be between 0 and 2000 (assuming no one buys more than one shirt). The range is $\{s \mid 0 \le s \le 2000\}$.

To find the domain, start with the inequality $0 \le s \le 2000$, and use algebra to convert it to an inequality involving p.

$$0 \le s \le 2000$$
$$0 \le 100p - 200 \le 2000$$
$$200 \le 100p \le 2200$$
$$\frac{200}{100} \le \frac{100p}{100} \le \frac{2200}{100}$$
$$2 \le p \le 22$$

The domain is $\{p \mid 2 \le p \le 22\}$.

c. The graph of S is shown in Figure 1.29.

FIGURE 1.29

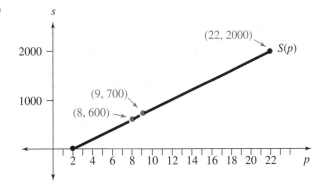

Supply and demand are both functions of price. The **theory of supply and demand** maintains that, in some situations, if a product's price is too *high:*

The Theory of Supply and Demand

- demand will be low (few consumers will be interested in purchasing the product)
- supply will be high (many suppliers will be interested in supplying a product at a high price)
- supply will exceed demand (see Figure 1.30)

FIGURE 1.30

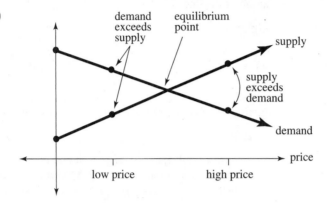

As a result, suppliers will accumulate unsold product and will tend to lower the product's price.

If a product's price is too *low:*

- demand will be high (many consumers will be interested in purchasing the product)
- supply will be low (few suppliers will be interested in supplying a product at a low price)
- demand will exceed supply (see Figure 1.30)

As a result, suppliers will tend to increase the product's price.

Thus, a price that is too high will tend to decrease, and a price that is too low will tend to increase. In either event, the price will converge to the point where supply and demand are equal. That point is called the **equilibrium point.**

EXAMPLE 8   a. Use the information in Examples 6 and 7 to find and interpret the equilibrium price.
b. Graph $S(p)$ and $D(p)$ on the same axes, and show the equilibrium point.

Solution   a. The equilibrium price is the price at which supply equals demand.

$$S(p) = D(p)$$
$$100p - 200 = -500p + 5500$$
$$600p = 5700$$
$$p = 5700/600 = 9.5$$

The equilibrium price is $9.50.

This means that if the Business Club charges $9.50 per T-shirt, supply will equal demand. That is, members won't have extra unsold shirts, and they won't lack shirts to sell.
b. The graphs from Examples 6 and 7 are combined in Figure 1.31.

FIGURE 1.31

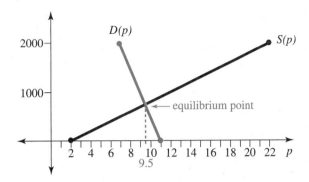

Depreciation

When a business purchases a piece of equipment (such as a copier, drill press, or oven), it expects the asset to lose value from year to year; daily wear and tear and technological obsolescence are anticipated. This anticipated loss of value is a cost of doing business; in many cases, it can be claimed as a deduction on an income tax return. When a piece of equipment has lost value over time, we say that the value has **depreciated.** Depreciation can be calculated in several ways; the easiest and most common method is straight-line depreciation, which is described in Example 9.

EXAMPLE 9   The manager of Pine Town Furniture purchased a new copier for $28,000. She plans to keep it for four years, at which point she expects to trade the copier in and receive a $6000 credit toward the purchase of a new machine. She is going to use straight-line depreciation on the copier.

a. Find the slope, and determine what it measures in the context of the problem.
b. Express the value of the copier as a function $V(t)$ of the number of years after purchase.
c. Find the value three years after purchase.
d. Find the domain and range of $V(t)$.

Solution   a. We know the value of the copier at two different times: $28,000 at the point of purchase, and $6000 four years later. If we let v be the value of the copier and t the number of years after the copier's purchase, we have two ordered pairs (t, v): (0, 28,000) and (4, 6000). To not only find the slope but also interpret what it measures, we will include units in calculating the slope.

$$m = \frac{v_2 - v_1}{t_2 - t_1}$$

$$= \frac{\$6000 - \$28,000}{4 \text{ years} - 0 \text{ years}}$$

$$= \frac{-\$22,000}{4 \text{ years}}$$

$$= \frac{-\$5500}{1 \text{ year}}$$

$$= -\$5500/\text{year}.$$

This means that a change in value of $-\$5500$ corresponds to a change in time of one year. In other words, the copier loses value at a rate of $\$5500$ per year.

b. The equation of the line is:

$$v - v_1 = m(t - t_1)$$
$$v - 28,000 = -5500(t - 0)$$
$$v = -5500t + 28,000$$

Switching to functional notation, we have $V(t) = -5500t + 28,000$.

c. The value after three years is:

$$V(3) = -5500 \cdot 3 + 28,000 = \$11,500.$$

d. The value function is valid only from 0 to 4 years after purchase, so the domain is $\{t \mid 0 \le t \le 4\}$. We are told that the value of the copier ranges between $\$28,000$ (at the point of purchase) to $\$6000$ (four years later), so the range is $\{v \mid 6000 \le v \le 28,000\}$. •

In Example 9, the intercept of $V(t) = -5500t + 28,000$ is $b = 28,000$, and the initial value of the copier is $\$28,000$. Does this always happen? To answer the question, consider an arbitrary value function $V(t) = mt + b$ (using t rather than x as the independent variable, where t measures the number of years of use). The initial value is the value after 0 years of use; it is

$$V(0) = m \cdot 0 + b = b$$

Thus, the initial value is always the same as the intercept.

1.2

EXERCISES

1. In Example 8, we found that the equilibrium price for the Business Club's T-shirts is $\$9.50$. Accordingly, the club members decided to charge $\$9.50$ per shirt, rather than $\$8$ as they were planning earlier. They used the supply and demand functions from Examples 6 and 7 to determine the number of shirts to order. They sold all of their shirts, and they could have sold more.

   a. How many shirts did the club order? Use the information in Examples 6 and 7 to find out.

   b. Find the revenue function that corresponds to a sales price of $\$9.50$. How much revenue did the club take in?

   c. How much did the club pay for the shirts ordered? Use the cost function from Example 1 to find out.

   d. Did the club lose money or earn a profit? How much did the club lose or earn?

2. Find the marginal revenue of the Business Club's T-shirt enterprise, using the information in Exercise 1, at the following sales levels:

   a. 100 T-shirts
   b. 200 T-shirts
   c. x T-shirts (use the method of Example 3)
   d. Compare the Business Club's marginal cost and marginal revenue at an arbitrary production level. Discuss the implications of this comparison regarding an appropriate level of production.

3. Profit is the difference between revenue and cost. If revenue $-$ cost is positive, then profit is positive and money is made. If revenue $-$ cost is negative, then profit is negative and money is lost. Use the information in Example 1 and Exercise 1 to:

   a. Find the profit if the Business Club produces 750 T-shirts and sells 500.

   b. Find the profit if the Business Club produces 1000 T-shirts and sells them all.

   c. Find the **profit function**—that is, the profit if the Business Club produces and sells x T-shirts.

d. What assumptions are made about the number of T-shirts produced and the number sold in both Example 5 and in applying the profit function?

4. Use the information in Example 1 and Exercises 1 and 3 to:
   a. Find the break-even point.
   b. Find the cost, the revenue, and the profit if the Business Club produces and sells the break-even quantity.
   c. Find the cost, the revenue, and the profit if the Business Club produces and sells twice the break-even quantity.
   d. Find the cost, the revenue, and the profit if the Business Club produces and sells half the break-even quantity.
   e. Sketch the graphs of the cost, revenue, and profit functions, and show the answers to parts (b), (c), and (d) on the graph.

5. The Zephyr Bicycle Company's plant in Gualala makes one model of bicycle, the Zephyr Zipper.
   a. Classify the following costs as either fixed or variable:
   metal tubing for the bicycle frames
   bearings for the pedals and wheels
   wages for the men and women who work on the assembly line
   wages for managers
   fire insurance
   worker's compensation insurance (covers workers' on-the-job accidents)
   b. The Gualala plant has a monthly fixed cost of $125,418 and unit variable costs of $57.19 per bike. Zephyr sells its Zippers for $118.53 each. Find the cost function, the revenue function, and the profit function and their domains and ranges.
   c. Find the break-even point, the break-even quantity, and the break-even revenue. Check your answer, using the method shown in Example 5. (The answer is not given in the back of the book.)
   d. Find the cost, revenue, and profit at a production level that's 25% above the break-even quantity (rounded to the nearest whole number of bikes).
   e. Find the cost, revenue, and profit at a production level that's 25% below the break-even quantity (rounded to the nearest whole number of bikes).
   f. Graph the cost, revenue, and profit functions on one set of axes, using the appropriate domains and ranges. Show the answers to (c), (d), and (e) on the graph.
   g. Find the marginal revenue, the marginal cost, and the marginal profit at an arbitrary level of production. Discuss their implications regarding an appropriate level of production.

6. The Dull Calculator Corp. makes graphing calculators at its plant in Hayfork.
   a. Classify the following costs as either fixed or variable:

electronic chips
printed circuit boards
wages for the men and women who work on the assembly line
wages for managers
gas and electricity
   b. The Hayfork plant has a monthly fixed cost of $98,387 and unit variable costs of $22.12 per calculator. Dull sells its calculators for $52.19 each. Find the cost function, the revenue function, and the profit function and their domains and ranges.
   c. Find the break-even point, the break-even quantity, and the break-even revenue. Check your answer, using the method shown in Example 5. (The answer is not given in the back of the book.)
   d. Find the cost, revenue, and profit at a production level that's 25% above the break-even quantity (rounded to the nearest whole number of calculators).
   e. Find the cost, revenue, and profit at a production level that's 25% below the break-even quantity (rounded to the nearest whole number of calculators).
   f. Graph the cost, revenue, and profit functions on one set of axes, using the appropriate domains and ranges. Show the answers to (c), (d), and (e) on the graph.
   g. Find the marginal revenue, the marginal cost, and the marginal profit at an arbitrary level of production. Discuss their implications regarding an appropriate level of production.

7. Use the information in Examples 6, 7, and 8 to do the following.
   a. Find the supply and the demand if the Business Club sells T-shirts for $9 each.
   b. Find the supply and the demand if the Business Club sells T-shirts for $10 each.
   c. Sketch the graphs of the supply and demand functions. Show the solutions to (a) and (b) on the graph.
   d. Discuss how the answers to (a) and (b) pertain to the theory of supply and demand.

8. Crunchy Crisps is trying to determine the appropriate price for a new product: a low-fat, lime salsa tortilla chip. At several supermarkets consumers are offered free tastes of the new chips and are asked to fill out a brief questionnaire. Forty-two percent of the respondents indicated that they would buy the chip the next time they bought chips if the price were $2.09 per bag; 28% indicated that they would buy it at $2.49 per bag.
   a. Use these data to project the number of bags that Crunchy Crisps would sell at these prices in a region that has 15,000 potential customers. Write these projections as ordered pairs (p, d), where p is price and d is demand.

b. Find the linear demand function $D(p)$ for the new chips.

c. Find the domain and range of D.

d. Sketch the graph of D, using the appropriate domains and ranges.

9. The Petrolia Oil Co. is trying to determine the appropriate price for a new regular gasoline that will lower exhaust emissions. A random sampling of Petrolia credit card holders are mailed a questionnaire regarding this new gasoline. Sixty-three percent of the respondents indicated that they would buy the gas if the price were 5¢ higher than regular unleaded; 34% indicated that they would buy it if it were 10¢ higher.

a. Use these data to project the volume that Petrolia would sell at these price increases if it currently sells 648 million gallons of regular gas per month. Write these projections as ordered pairs (p, d), where p is price increase and d is demand.

b. Find the linear demand function $D(p)$ for the new gas.

c. Find the domain and range of D.

d. Sketch the graph of D, using the appropriate domains and ranges.

10. After analyzing cost, revenue, and profit functions, the Crunchy Crisps analysts (from Exercise 8) have determined that they could produce 5000 bags of the new chips if they sold for $2.09 per bag, and they could produce 6000 bags if they sold for $2.49 per bag.

a. Use these data to find the linear supply function $S(p)$.

b. Find the domain and range of S.

c. Find the equilibrium price for the new chips (round off to the nearest penny).

d. At what level should Crunchy Crisps produce its new chips?

e. Sketch the graphs of S and D (from Exercise 8) using the appropriate domains and ranges and showing the equilibrium point.

f. According to the theory of supply and demand, what would happen if Crunchy Crisps charged significantly less than the equilibrium price? What would happen if it charged significantly more?

11. After analyzing cost, revenue, and profit functions, the Petrolia analysts (from Exercise 9) have determined that they could produce 200 million gallons per month if the price were 20¢ higher than regular unleaded, and they could produce 300 million gallons if the price were 40¢ higher.

a. Use these data to find the linear supply function $S(p)$.

b. Find the domain and range of S.

c. Find the equilibrium price for the new gas (round off to the nearest penny).

d. At what level should Petrolia produce its new gas?

e. Sketch the graphs of S and D (from Exercise 9) using the appropriate domains and ranges and showing the equilibrium point.

f. According to the theory of supply and demand, what would happen if Petrolia charged significantly less than the equilibrium price? What would happen if it charged significantly more?

12. An author purchased a new computer system for $3400. The IRS allows it to be depreciated over seven years, with straight-line depreciation and zero value at the end of those seven years.

a. Express the value of the computer as a function of the number of years after purchase.

b. Find its value after one year.

c. Find its value after two years.

d. How much value is lost each year?

e. How much can be deducted from income taxes each year?

13. A machine shop buys a new drill press for $13,500. For tax purposes, the IRS allows it to be depreciated over seven years, with straight-line depreciation and zero value at the end of those seven years.

a. Express the value of the drill press as a function of the number of years after purchase.

b. Find its value after one year.

c. Find its value after two years.

d. How much value is lost each year?

e. How much can be deducted from income taxes each year?

14. The bookkeeper of the machine shop in Exercise 13 must create annual profit and loss statements for the machine shop's owners. For that analysis, the bookkeeper depreciates the drill press over ten years, and assumes that the press can be sold at that time for $1000.

a. Express the value of the drill press as a function of the number of years after purchase.

b. Find its value after one year.

c. Write a paragraph in which you compare the depreciation for the IRS and the bookkeeper's depreciation. In particular, compare the time intervals and the annual loss in value. Also discuss why the two different depreciations would be used.

*Exercises 15 and 16 involve cost functions that are **quadratic** rather than linear; that is, they are of the form $C(x) = ax^2 + bx + c$, rather than $C(x) = mx + b$.*

15. The cost function for a certain product is $C(x) = 0.0667x^2 + 10.09x + 2542.88$, where x is the number of thousands of units produced.

a. The product sells for $45 per thousand units. Find the revenue function.

b. Find the profit function.

c. Find the marginal profit at an arbitrary level of production.

d. If the company produces and sells 271,000 of these items, find the corresponding marginal profit. Should it increase or decrease production? By how much? Why?

e. If the company produces and sells 80,000 of these items, find the corresponding marginal profit. Should it increase or decrease production? By how much? Why?

16. The cost function for a certain product is $C(x) = 0.0333x^2 + 10.29x + 2542.88$, where x is the number of thousands of units produced.

a. The product sells for $35 per thousand units. Find the revenue function.

b. Find the profit function.

c. Find the marginal profit at an arbitrary level of production.

d. If the company produces and sells 371,000 of these items, find the corresponding marginal profit. Should it increase or decrease production? By how much? Why?

e. If the company produces and sells 352,000 of these items, find the corresponding marginal profit. Should it increase or decrease production? By how much? Why?

Almost all scientists follow the mathematicians' tradition of assigning the independent variable to the horizontal axis and the dependent variable to the vertical axis. We followed this tradition when we assigned price to the horizontal axis and demand (in Example 6) and supply (in Example 7) to the vertical axis. Economists, however, plot price on the vertical axis, and supply

and demand on the horizontal axis, following a tradition initiated by English economist Alfred Marshall (1842–1924).

17. Sketch the graph of the demand function from Example 6, following the economists' tradition. Show the slope in your sketch.

18. Sketch the graph of the supply function from Example 7, following the economists' tradition. Show the slope in your sketch.

19. Sketch the graph of the supply and demand functions from Exercises 9 and 11 following the economists' tradition. Show the slopes in your sketch.

20. Sketch the graph of the supply and demand functions from Exercises 8 and 10, following the economists' tradition. Show the slopes in your sketch.

21. Both break-even points and equilibrium points occur where two functions intersect. Discuss the differences between these two types of points of intersection. Include a discussion of the implications if the independent variable is to the left of the point of intersection, for either type of point, and a discussion of the implications if the independent variable is to the right of the point of intersection.

22. What does it mean if marginal cost is positive? Negative?

23. What does it mean if marginal revenue is positive? Negative?

24. What does it mean if marginal profit is positive? Negative?

LINEAR MODELS ON A GRAPHING CALCULATOR

Using the Domain and Range to Appropriately Place the Screen

One big problem with using a graphing calculator to graph a function is determining the values of xmin, xmax, ymin, and ymax. Using the wrong values can result in a very limited view of the function's graph, or even a view that includes none of the function's graph. If, however, the domain and range of the function are known, it is easy to appropriately place the screen and get a good view of the function's graph.

In Example 1 of this section, we found that the cost function for the Business Club's T-shirt enterprise is $C(x) = 4.50x + 50$. If we graph this on a graphing calculator using the "Zoom Standard" feature, we get an unsatisfactory graph. However, we also found the function's domain and range, and if we use that information to determine the values of xmin, xmax, ymin, and ymax, we get a good graph.

EXAMPLE 10   Graph $C(x) = 4.50x + 50$, where the domain is $\{x \mid 0 \leq x \leq 2000\}$ and the range is $\{y \mid 50 \leq y \leq 9050\}$, as found in Example 1.

Solution   It's a good idea to set xmin and xmax so that the screen shows a bit more than the domain, so that it's clear the line doesn't go on forever. Thus, we'll set xmin to -500 rather than 0, and xmax to 2500 rather than 2000. Also, an xscl of 500 reasonably spreads the tick marks. Similarly, we'll set ymin to -500, ymax to 9500, and yscl to 500. Notice that the actual domain is given in the way we enter the function on the "Y=" screen, as shown in Figure 1.32(a). The graph is shown in Figure 1.32(b).

FIGURE 1.32

 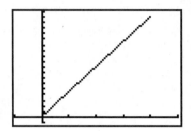

If your graph doesn't look like this, check to see if you're in "Connected" or "Dot" mode ("DrawLine" or "DrawDot" on a TI-85/86), as discussed in Section 1.1.   •

Finding the Intersection of Two Functions

A break-even point is a point at which revenue and cost are equal. That is, it is the point at which the graphs of the revenue and cost functions intersect. Similarly, an equilibrium point is a point at which the graphs of the supply and demand functions intersect. The graphing calculator can be used to find these points of intersection.

The functions $Y_1(x) = x + 3$ and $Y_2(x) = -x + 9$ intersect on the standard viewing screen. To find their point of intersection with a TI graphing calculator, first store the two functions (as Y_1 and Y_2), erase any other functions, and erase the "Y=" screen by pressing [2nd] [QUIT]. Then proceed according to the following instructions.

On a TI-82/83:

- Graph the two functions.
- Press [2nd] [CALC] and select option 5, "intersect."
- When the calculator responds with "First curve?" and a mark on the first function's graph, press [ENTER].
- When the calculator responds with "Second curve?" and a mark on the second function's graph, press [ENTER].
- When the calculator responds with "Guess?" use the left- and right-arrow buttons to place the mark near the point of intersection, and press [ENTER].
- Check your answer, as discussed below.

On a TI-85/86:

- Graph the two functions.
- Press [MORE] until the "MATH" option appears, and select that option.

- Press ⟨MORE⟩ until the "ISECT" option appears, and select that option.
- When the calculator responds with a mark on the first function's graph, press ⟨ENTER⟩.
- When the calculator responds with a mark on the second function's graph, press ⟨ENTER⟩.
- **(TI-86 only)**   When the calculator responds with "Guess?" use the arrow buttons to move the mark near the point of intersection and press ⟨ENTER⟩.
- Check your answer, as discussed below.

✔   In some circumstances, calculators will give incorrect points of intersection. Always check your answer by substituting the ordered pair into each of the two functions. If it is incorrect, either regraph without restricting the domains, or zoom in on the point of intersection before calculating the point of intersection.

Figure 1.33 shows the results of computing a break-even point after entering the following information from Examples 1, 4, 5, and 10 into a TI-82:

- the cost and revenue functions, from Examples 1 and 4
- their domains and ranges, from Examples 1 and 4
- appropriate values of xmin, xmax, ymin, and ymax, as discussed in Example 10

Notice that the computed break-even point agrees with that found in Example 5.

FIGURE 1.33

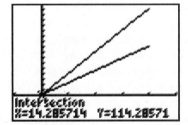

Intersection
X=14.285714   Y=114.28571

EXERCISES

In the following exercises, you will explore some of your calculator's graphing capabilities. Answers to these exercises are not given in the back of the book.

25. *Exploring the placement of the screen.* Determine the values of xmin, xmax, xscl, ymin, ymax, and yscl that would result in a graph like that in Figure 1.33. Discuss how you determined these values.

26. *Exploring the placement of the screen.* Graph the supply and demand functions found in Examples 6 and 7, using their domains and ranges to appropriately place the screen as discussed in Example 10. Determine the values of xmin,

xmax, xscl, ymin, ymax, and yscl that would result in an appropriate graph, and discuss how you determined these values. Also specify exactly how the two functions should be entered in order to restrict their domains.

27. *Exploring finding the intersection of two functions.* Use the graphing calculator and the result of Exercise 26 to locate the equilibrium point. Compare your answer with that of Example 8.

Exploring finding the intersection of two functions. In Exercises 28–33, use the graphing calculator to find the point(s) of intersection of the given functions. Check your solutions.

28. $f(x) = 3x + 2$ and $g(x) = 5x + 5$

29. $f(x) = 2x - 6$ and $g(x) = 3x + 4$

30. $f(x) = 8x - 14$ and $g(x) = 11x + 23$

31. $f(x) = -7x + 12$ and $g(x) = -12x - 71$

32. $f(x) = x^2 - 2x + 3$ and $g(x) = -x^2 - 3x + 12$ (There are two points of intersection—find both.)

33. $f(x) = 8x^2 - 3x - 7$ and $g(x) = 2x + 4$ (There are two points of intersection—find both.)

Exploring finding the maximum value of a function. Exercises 34 and 35 involve finding the maximum value of a function. This is done in a manner similar to that used to find the intersection of two functions, except: (1) graph one function, not two, and (2) choose "maximum" rather than "intersect" on a TI-82/83, "FMAX" rather than "ISECT" on a TI-85/86.

34. Graph the profit function found in Exercise 15. Determine the point at which profit is maximum. What advice would you give the manufacturing firm?

35. Graph the profit function found in Exercise 16. Determine the point at which profit is maximum. What advice would you give the manufacturing firm?

1.3

LINEAR REGRESSION

In Section 1.2 the Business Club determined the relationship between price and demand for their Big Game T-shirts (and thus an appropriate sales price) by showing a certain number of students, faculty, and alumni a sample T-shirt and asking them if they would pay $8 or $9 for the shirt. This generated two ordered pairs, which in turn generated the equation of the line through the two ordered pairs. That linear equation was the demand function.

This method of generating a demand function has an inherent weakness—it assumes that a linear model fits the situation, without any evidence to support that assumption. In Section 1.2 we found the Business Club's cost and revenue functions without assuming that these functions were linear; we just computed the cost and the revenue and the results turned out to be linear functions. If we assume that a relationship is linear when it is not, we'll get an unrealistic result.

After a successful experience selling Big Game T-shirts, members of the Business Club reflected on the fact that they could have sold more shirts if they had had more to sell; that is, demand had exceeded supply. They ordered the number of shirts dictated by their supply and demand functions, so they should have been able to meet demand. Club members concluded that their demand function did a poor job of predicting demand.

The next year, club members used a different method to create a demand function. They asked 100 students, faculty, and alumni if they would pay $8, $9, $10, or $11 for a Big Game T-shirt, and they generated four ordered pairs rather than two. The results are given in Figure 1.34.

FIGURE 1.34

Price	$8	$9	$10	$11
Number of those queried who would buy	80	64	44	17

At $8, 80 of 100 people (or 80%) would buy the shirt. This implies that 80% of the 2000 CSU fans that generally attend the game (or 1600 people) would buy the shirt at that price. Similar computations yield the data in Figure 1.35.

FIGURE 1.35

Price p	$8	$9	$10	$11
Projected demand d	1600	1280	880	340
(p, d)	(8, 1600)	(9, 1280)	(10, 880)	(11, 340)

The Business Club has done a number of things to obtain a more realistic demand function. Members asked 100 people how much they would pay for a shirt, rather than the 40 they asked the year before. And they generated four ordered pairs, rather than two. This allows them to check to see if the relationship between price and demand is linear, rather than assuming that it is. When these four ordered pairs are graphed, they appear to lie on a line. However, the slopes between consecutive pairs of points are significantly different, as shown in Figure 1.36; so the points do not actually lie on a line and the data do not exactly fit into a linear relationship. In this situation, we say that the data fit into an **approximate linear relationship.**

FIGURE 1.36

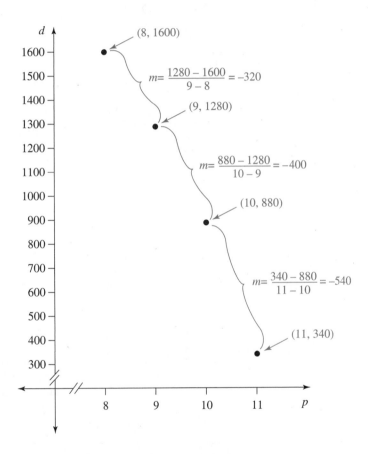

Mathematicians and statisticians frequently find this type of situation, where real-world data generate points that fit an approximate linear relationship. In our situation, a sample of 100 students and alumni was used to approximate the population of 2000 Big Game attendees. If the club used a different sample of students, they could well get different answers. Thus, errors can occur in collecting data. For this and other reasons, real-world data rarely fit into a perfect linear relationship. However, real-world data frequently fit into an approximate linear relationship, as do the Business Club's demand data.

Line of Best Fit

The line that best fits a collection of points is called either the **line of best fit,** the **regression line,** or the **least squares line,** and the points themselves are called **data points.** A graph of the data points is called a **scatter diagram.** The scatter diagram and the line of best fit for the Business Club's demand data are shown in Figure 1.37.

FIGURE 1.37

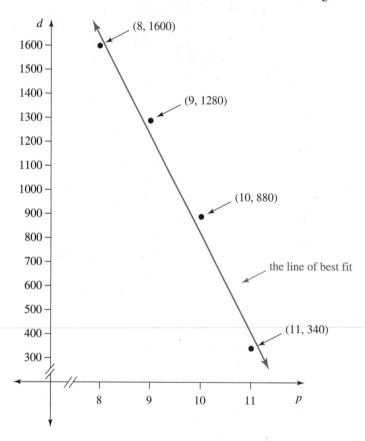

Formulas have been developed for the slope and the intercept of the line of best fit, using mathematics beyond the scope of this text. The process of using these formulas to find the equation of the line of best fit is called **linear regression.** The formulas use the Greek letter sigma (Σ). When used in a formula, Σ means "the sum of," because Σ is the Greek letter most similar to S, which stands for "sum."

Line of Best Fit Formulas

Given n data points (x, y), the line of best fit has slope

$$m = \frac{n(\Sigma xy) - (\Sigma x)(\Sigma y)}{n(\Sigma x^2) - (\Sigma x)^2}$$

and y-intercept

$$b = \frac{(\Sigma y) - m(\Sigma x)}{n}$$

These formulas involve a lot of notation. Let's look at them piece by piece.

Notation	Meaning
n	The number of ordered pairs
Σxy	Multiply each x-coordinate by its corresponding y-coordinate (i.e., find xy), and add the results (Σxy)
Σx	The sum of all the x-coordinates
Σy	The sum of all the y-coordinates
(Σx^2)	Square each x-coordinate (i.e., find x^2), and add the results (Σx^2)
$(\Sigma x)^2$	Add all the x-coordinates (i.e., find Σx), and square the result ($\Sigma x)^2$

We'll start with an example with minimal computations; we'll return to the Business Club later.

EXAMPLE 1   Consider the three ordered pairs (1, 1), (5, 3), and (10, 5).

    a. Find the slope and the y-intercept of the line of best fit.
    b. Find the equation of the line of best fit.
    c. Draw a scatter diagram, and sketch the graph of the line of best fit on the same set of axes. Does the line fit the points well?

Solution   **a.** To find the slope, organize the data in a table and compute the appropriate sums.

(x, y)	x	x^2	y	xy
(1, 1)	1	$1^2 = 1$	1	$1 \cdot 1 = 1$
(5, 3)	5	$5^2 = 25$	3	$5 \cdot 3 = 15$
(10, 5)	10	$10^2 = 100$	5	$10 \cdot 5 = 50$
Total	$\Sigma x = 16$	$\Sigma x^2 = 126$	$\Sigma y = 9$	$\Sigma xy = 66$

$$m = \frac{n(\Sigma xy) - (\Sigma x)(\Sigma y)}{n(\Sigma x^2) - (\Sigma x)^2}$$

$$= \frac{3(66) - (16)(9)}{3(126) - (16)^2}$$

$$= 0.4426\ldots$$

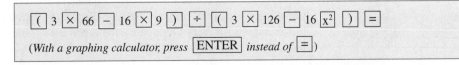

(*With a graphing calculator, press* ENTER *instead of* =)

Once m has been calculated, we store it in the memory of our calculator. We need it to calculate b, the y-intercept. (See Appendix 1 regarding the use of the memory.)

$$b = \frac{(\Sigma y) - m(\Sigma x)}{n}$$

$$= \frac{9 - (0.4426\ldots)(16)}{3}$$

$$= 0.6393\ldots$$

b. The equation of the line of best fit is

$$\hat{y} = mx + b$$

$$= (0.4426\ldots)x + 0.6393\ldots$$

$$\approx 0.443x + 0.639$$

The "hat" over the y indicates that the calculated value of y is a prediction based on linear regression.

c. The scatter diagram and the line of best fit are graphed in Figure 1.38. The line fits the points quite well.

FIGURE 1.38
A good fit

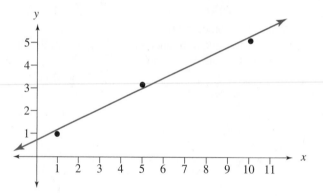

Best Fit or Good Fit?

A line found in this manner is called "the line of *best* fit," because it fits the given points better than would any other line. Unfortunately, this does not ensure that the line provides a *good* fit. Consider the following collection of ordered pairs: (2, 3), (3, 4), (4, 5), (5, 4), and (6, 3). These ordered pairs are graphed in Figure 1.39, and they do not appear to lie on a line. Nevertheless, the line of best fit formulas can still be applied to them; the resulting line (with equation $y = 0x + 3.8$) does not fit the points well.

By looking at the scatter diagrams, it is easy to observe that the data in Example 1 follow a strong linear trend and that the data in Figure 1.39 do not follow a linear trend at all. In either case, the line of best fit can be found, but in one case that line doesn't fit the data very well. How can this be evaluated when the data are not so obviously linear or nonlinear? One way is to compare the actual y-values with the y-values that are predicted by the line of best fit. When doing this, the predicted y-values are denoted by $\hat{y}$, to distinguish them from the actual y-values.

FIGURE 1.39
A bad fit

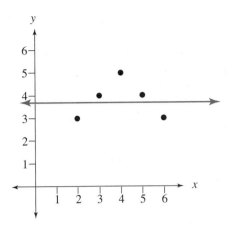

EXAMPLE 2 Make a chart in which you compare the actual y-values with the predicted y-values for the data in Example 1. Use $\hat{y} = 0.443x + 0.639$ as the equation of the line of best fit.

Solution

x	y	$\hat{y} =$ 0.443x + 0.639
1	1	1.082
5	3	2.854
10	5	5.069

In Example 2 the predicted y-values are all quite close to the actual y-values. For example, the difference between the two (or **error**) is 0.069, if $x = 10$. The corresponding **relative error** (i.e., the error compared to the actual y-value) is $0.069/5 = 0.0138 = 1.38\%$. In other words, the predicted y-value is slightly more than 1% off of the actual y-value, if $x = 10$. If the relative error is this small with the other given points, then this line of best fit fits the points well.

The Correlation Coefficient

We've discussed two ways to measure the strength of a linear trend exhibited by a collection of ordered pairs. One way is to inspect the graph of the ordered pairs and determine whether they seem to lie on a line. A second way is to calculate the relative error for each data point. A third way is to calculate the **correlation coefficient.**

Correlation Coefficient Formula

Given n ordered pairs (x, y), the correlation coefficient is

$$r = \frac{n(\Sigma xy) - (\Sigma x)(\Sigma y)}{\sqrt{n(\Sigma x^2) - (\Sigma x)^2} \sqrt{n(\Sigma y^2) - (\Sigma y)^2}}$$

The value of the correlation coefficient r is always between -1 and 1, inclusive; that is, $-1 \le r \le 1$. If the given ordered pairs lie perfectly on a line whose slope is *positive,* then the calculated value of r will equal 1 (think 100% correlation with positive slope). In this case, both variables have the same behavior: if one increases (or decreases), so will the other. (See Figure 1.40.) On the other hand, if the data points fall perfectly on a line whose slope is *negative,* the calculated value of r will equal -1 (think 100% correlation with a negative slope). In this case, the variables have opposite behavior: if one increases, the other decreases. (See Figure 1.41.)

FIGURE 1.40

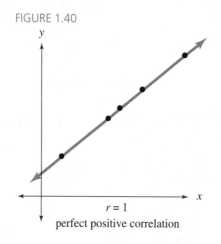

$r = 1$
perfect positive correlation

FIGURE 1.41

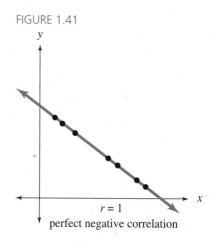

$r = 1$
perfect negative correlation

If the value of r is close to 0, there is little or no *linear* relation between the variables. This does not mean that the variables are not related; they might be related in some non-linear fashion.

EXAMPLE 3 Find the coefficient of linear correlation for the data given in Figure 1.39, and interpret it.

Solution

(x, y)	x	x^2	y	y^2	xy
$(2, 3)$	2	4	3	9	6
$(3, 4)$	3	9	4	16	12
$(4, 5)$	4	16	5	25	20
$(5, 4)$	5	25	4	16	20
$(6, 3)$	6	36	3	9	18
Total	$\Sigma x = 20$	$\Sigma x^2 = 90$	$\Sigma y = 19$	$\Sigma y^2 = 75$	$\Sigma xy = 76$

$$
\begin{aligned}
r &= \frac{n(\Sigma xy) - (\Sigma x)(\Sigma y)}{\sqrt{n(\Sigma x^2) - (\Sigma x)^2}\ \sqrt{n(\Sigma y^2) - (\Sigma y)^2}} \\[2mm]
&= \frac{5(76) - (20)(19)}{\sqrt{5(90) - (20)^2}\ \sqrt{5(75) - (19)^2}} \\[2mm]
&= 0
\end{aligned}
$$

This means that there is no linear relationship between the variables. The fact that r is not positive mirrors the fact that y does not predictably increase as x increases, and the fact that r is not negative mirrors the fact that y does not predictably decrease as x increases.   ●

1.3

EXERCISES

1. Consider the following three ordered pairs: (1, 10), (5, 4), and (9, 1). (Answers are not given in the back of the book, except as indicated.)
   a. Make a scatter diagram carefully and accurately on graph paper.
   b. Use a straight edge to draw a line where you think the line of best fit should be. Do this without making any calculations.
   c. By looking at the graph, determine your line's y-values at $x = 1$, $x = 5$, and $x = 9$, as accurately as you can.
   d. Compute your line's relative errors at $x = 1$, $x = 5$, and $x = 9$.
   e. Use the line of best fit formulas to determine the equation of the actual line of best fit.
      NOTE: Only the slope is given in the back of the book.
   f. Compute the line of best fit's relative errors at $x = 1$, $x = 5$, and $x = 9$.
      NOTE: Only the relative error at $x = 1$ is given in the back of the book.
   g. Compare the accuracy of your line with that of the line of best fit.
   h. Without computing the correlation coefficient, predict whether it is positive or negative. Also predict whether it is closer to 0, 1, or −1. Explain your predictions.

2. Consider the following three ordered pairs: (1, 3), (3, 4), and (5, 6).
   a. Make a scatter diagram carefully and accurately on graph paper.
   b. Use a straightedge to draw a line where you think the line of best fit should be. Do this without making any calculations.
   c. By looking at the graph, determine your line's y-values at $x = 1$, $x = 3$, and $x = 5$, as accurately as you can.
   d. Compute your line's relative errors at $x = 1$, $x = 3$, and $x = 5$.
   e. Use the line of best fit formulas to determine the equation of the actual line of best fit.
   f. Compute the line of best fit's relative errors at $x = 1$, $x = 3$, and $x = 5$.
   g. Compare the accuracy of your line with that of the line of best fit.
   h. Without computing the correlation coefficient, predict whether it is positive or negative. Also predict whether it is closer to 0, 1, or −1. Explain your predictions.

3. Consider the following three ordered pairs: (1, 3), (2, 2), and (2, 4). (Answers are not given in the back of the book.)
   a. Make a scatter diagram carefully and accurately on graph paper.
   b. Without computing the correlation coefficient, predict whether it is positive or negative. Also predict whether it is close to 0, 1, or −1. Explain your predictions.

4. Consider the following two ordered pairs: (2, 7) and (9, 3).
   a. Use the point-slope formula to find the equation of the line between them.
   b. Use the line of best fit formulas to find the equation of the line of best fit.
   c. Discuss the relationship between the solutions of (a) and (b).

5. Consider the three ordered pairs given in Example 2.
   a. Compute the relative error for each of the given data points.
   b. Compute the correlation coefficient.
   c. Compare the results of Example 1(c) with parts (a) and (b) of this exercise. Do these three methods of measuring the strength of a linear trend agree? Discuss.

6. Consider the following five ordered pairs: (2, 3), (3, 4), (4, 5), (5, 4), and (6, 3). These ordered pairs are graphed in Figure 1.39.
   a. Compute the relative error for each of the given data points.
   b. Compute the correlation coefficient.
   c. Compare the bad fit observed in Figure 1.39 with parts (a) and (b) of this exercise. Do these three methods of measuring the strength of a linear trend agree? Discuss.

7. In Example 3, for which values of x does y increase while x increases? For which values of x does y decrease while x increases? Use these results to justify the last sentence in the solution of Example 3.

8. Write a paper in which you compare and contrast the three methods of measuring the strength of a linear trend: graphing the data points and observing whether they seem to fit on a line, computing relative errors for each of the data points, and computing the correlation coefficient. Discuss the strengths and weaknesses of each method.

REGRESSION ON A GRAPHING CALCULATOR

In Example 1 of this section, we computed the slope and y-intercept of the line of best fit for the three ordered pairs $(1, 1)$, $(5, 3)$, and $(10, 5)$. These calculations can be tedious when done by hand, even with only three data points. In the real world, there are usually a large number of data points, and the calculations are usually done with the aid of technology.

Graphing calculators can compute the slope and y-intercept of the line of best fit, graph the line, and create a scatter diagram.

Entering the Data

On a TI-82/83/86:

• Put the calculator into statistics mode by pressing STAT (TI-86: 2nd STAT).
• Set the calculator up for entering the data by selecting "Edit" from the "EDIT" menu (TI-86: Select "Edit" by pressing F2), and the "list screen" appears, as shown in Figure 1.42. (A TI-86's list screen says "xStat," "yStat," and "fStat" instead of "L_1," "L_2," and "L_3," respectively.) If data already appear in a list (as they do in list L_1 in Figure 1.42) and you want to clear it, use the arrow buttons to highlight the name of the list (i.e., "L_1" or "xStat") and press CLEAR ENTER .
• Use the arrow buttons and the ENTER button to enter the x-coordinates from Example 1 in list L_1 and the corresponding y-coordinates in list L_2. (TI-86: Enter the x-coordinates in list xStat and the corresponding y-coordinates in list yStat. Also enter a frequency of 1 in list fStat, for each of the three entries in list xStat.) When done, your screen should look like Figure 1.42. (TI-86 screens should also have a list of three 1's in the last column.)

FIGURE 1.42

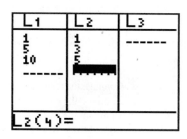

On a TI-85:

• Put the calculator into statistics mode by pressing STAT .
• Set the calculator up for entering the data by pressing EDIT (i.e., F2).
• The calculator will respond by automatically naming the list of x-coordinates "xStat" and the list of y-coordinates "yStat," as shown in Figure 1.43. (You could rename these lists by typing over the calculator-selected names.) Press ENTER twice to indicate that these two lists are named to your satisfaction.

FIGURE 1.43

- Enter 1 for x_1, 1 for y_1, 5 for x_2, etc.

Finding the Equation of the Line of Best Fit

Once the data have been entered, it's easy to find the equation.

On a TI-82/83:

- Press ⎡STAT⎤, scroll to the right and select the "CALC" menu, and select "LinReg(ax+b)" from the "CALC" menu.
- When "LinReg(ax+b)" appears on the screen, press ⎡ENTER⎤.
- The slope (labeled "a" rather than "m"), the y-intercept and the correlation coefficient will appear on the screen.

Note: The TI-83 does not display the correlation coefficient unless you tell it to. To do so, press ⎡2nd⎤ ⎡CATALOG⎤, scroll down and select "DiagnosticOn," and press ⎡ENTER⎤. When this is done, follow the above procedure to find the correlation coefficient.

On a TI-85:

- Press ⎡CALC⎤ (i.e., ⎡2nd⎤ ⎡M1⎤).
- The calculator will respond with the list names you selected above. Press ⎡ENTER⎤ twice to indicate that these two names are correct, or change them if they are incorrect.
- Select linear regression by pressing ⎡LINR⎤ (i.e., ⎡F2⎤).
- The slope (confusingly labeled "b" rather than "m"), the y-intercept (confusingly labeled "a" rather than "b"), the correlation coefficient (labeled "corr") and the number of data points (labeled "n") will appear on the screen.

On a TI-86:

- Press ⎡EXIT⎤.
- Press ⎡2nd⎤ ⎡STAT⎤ and select "CALC" by pressing ⎡F1⎤.
- Select linear regression by pressing ⎡LINR⎤ (i.e., ⎡F3⎤).
- When "LinR" appears on the screen, press ⎡ENTER⎤.
- The slope (confusingly labeled "b" rather than "m"), the y-intercept (confusingly labeled "a" rather than "b"), the correlation coefficient (labeled "corr") and the number of data points (labeled "n") will appear on the screen. (Use the ⎡↓⎤ button to see it all.)

Drawing a Scatter Diagram and the Line of Best Fit

Once the equation has been found, the line can be graphed.

On a TI-82/83:

- Press $\boxed{\text{Y=}}$ and clear any functions that may appear.
- Set the calculator up to draw a scatter diagram by pressing $\boxed{\text{2nd}}$ $\boxed{\text{STAT PLOT}}$ and selecting "Plot 1." Turn the plot on and select the scatter icon.
- Tell the calculator to put the data entered in list L_1 on the x-axis by selecting "L_1" for "Xlist," and to put the data entered in list L_2 on the y-axis by selecting "L_2" for "Ylist."
- Automatically set the range of the window and obtain a scatter diagram by pressing $\boxed{\text{ZOOM}}$ and selecting option 9: "ZoomStat."
- If you don't want the data points displayed on the line of best fit, press $\boxed{\text{2nd}}$ $\boxed{\text{STAT PLOT}}$ and turn plot 1 off.
- Quit the statistics mode by pressing $\boxed{\text{2nd}}$ $\boxed{\text{QUIT}}$.
- Enter the equation of the line of best fit by pressing $\boxed{\text{Y=}}$ $\boxed{\text{VARS}}$, selecting "Statistics," scrolling to the right to select the "EQ" menu, and selecting "RegEQ" for "Regression Equation."
- Automatically set the range of the window and obtain a scatter diagram by pressing $\boxed{\text{ZOOM}}$ and selecting option 9: "ZoomStat."
- Press $\boxed{\text{TRACE}}$ to read off the data points as well as points on the line of best fit. Use the up- and down-arrow buttons to switch between them.
- Press $\boxed{\text{2nd}}$ $\boxed{\text{STAT PLOT}}$ and turn plot 1 off, so that the scatter diagram does not appear on future graphs.

On a TI-85/86:

- Quit the statistics mode by pressing $\boxed{\text{2nd}}$ $\boxed{\text{QUIT}}$.
- Put the calculator into graphing mode, press $\boxed{\text{y(x)=}}$, and clear any functions that may appear. Press $\boxed{\text{RANGE}}$ or $\boxed{\text{WIND}}$ and enter appropriate values. For this problem, let x and y range from -1 to 10. Quit the graphing mode by pressing $\boxed{\text{2nd}}$ $\boxed{\text{QUIT}}$.
- Put the calculator into statistics mode by pressing $\boxed{\text{STAT}}$ (TI-86: $\boxed{\text{2nd}}$ $\boxed{\text{STAT}}$).
- Press $\boxed{\text{DRAW}}$. If this immediately produces a drawing, clear it by pressing $\boxed{\text{CLDRW}}$. (You may need to press $\boxed{\text{MORE}}$ to find $\boxed{\text{CLDRW}}$.)
- Press either $\boxed{\text{SCAT}}$ (for a scatter diagram) or $\boxed{\text{DRREG}}$ (for the regression line) or both. (You may need to press $\boxed{\text{MORE}}$ to find $\boxed{\text{DRREG}}$.) If the button labels obscure the line, press $\boxed{\text{EXIT}}$ once.
- Press $\boxed{\text{CLDRW}}$ or else the scatter diagram will appear on future graphs.

EXERCISES

FIGURE 1.44
Record times for the mile run

Year	1911	1923	1933	1942	1945	1954	1964	1967	1975	1980	1985
Time (min: sec)	4:15.4	4:10.4	4:07.6	4:04.6	4:01.4	3:59.4	3:54.1	3:51.1	3:49.4	3:48.8	3:46.3

9. Throughout the 20th century, the record time for the mile run has steadily decreased, from 4 minutes 15.4 seconds in 1911 to 3 minutes 46.3 seconds in 1985. Some of the record times for the mile are given in Figure 1.44.
   a. Convert the given data to ordered pairs (x, y), where x is the number of years since 1900 and y is the time in seconds.
   b. Use your calculator to find the slope and y-intercept of the line of best fit, as well as the correlation coefficient.
   c. Find the equation of the line of best fit.
   d. Predict the record time for the mile run in the year 2000.
   e. Predict when the record time for the mile run will reach the 3½ -minute mark.
   f. What assumption are these predictions based on? If this assumption is correct, how accurate are these predictions?

10. The consumption of beef in the United States is given in Figure 1.45.

FIGURE 1.45
Beef consumption in the U.S.

Year	1950	1960	1970	1980	1990
Consumption (billions of pounds)	9.529	15.465	23.391	23.513	24.031

Source: U.S. Department of Agriculture

   a. Plot the data points. Do they exhibit a linear trend?
   b. Find the equation of the line of best fit.
   c. Graph the line of best fit on the scatter diagram from (a).
   d. Predict the consumption of beef in 1999.
   e. Predict when 30 billion pounds of beef will be consumed.
   f. Find and interpret the correlation coefficient.
   g. What assumption are these predictions based on? If this assumption is correct, how accurate are these predictions?

11. The number of U.S. residents (in millions) whose job is classified as a "farm occupation" is given in Figure 1.46.
    a. Plot the data points. Do they exhibit a linear trend?

   b. Find the equation of the line of best fit.
   c. Graph the line of best fit on the scatter diagram from (a).
   d. Predict the number of people with farm occupations in the year 2000.
   e. Predict when there will be 125 million people with farm occupations.
   f. Find and interpret the correlation coefficient.
   g. What assumption are these predictions based on? If this assumption is correct, how accurate are these predictions?

FIGURE 1.46
People with farm occupations

Year	1940	1950	1960	1970	1980	1990
Millions of people	51.724	59.230	67.990	79.802	104.058	117.491

Source: U.S. Department of Agriculture

12. In each of the above problems, find the value of x that results in a y-value of 0. Interpret this ordered pair in the context of the problem. Discuss the impact of this on the validity of regression-based predictions.

13. The number of marriages and divorces (in millions) in the United States is given in Figure 1.47.
    a. Letting x be the number of marriages and y the number of divorces, plot the data points. Do they exhibit a linear trend?

FIGURE 1.47
Marriages and divorces

Year	1965	1970	1975	1980	1985	1990
Millions of marriages	1.800	2.158	2.152	2.413	2.425	2.448
Millions of divorces	0.479	0.708	1.036	1.182	1.187	1.175

Source: National Center for Health Statistics

b. Find the equation of the line of best fit.

c. Graph the line of best fit on the scatter diagram from (a).

d. Predict the number of divorces when there are 2,750,000 marriages.

e. Predict the number of marriages when there are 2,750,000 divorces.

f. Find and interpret the correlation coefficient.

g. What assumption are these predictions based on? If this assumption is correct, how accurate are these predictions?

14. Consider the Business Club's four ordered pairs (p, d) given in Figure 1.35.

a. Use your calculator to find the slope and y-intercept of the line of best fit, as well as the correlation coefficient. Interpret the correlation coefficient.

b. Find the equation of the line of best fit.

c. Use the demand function found in (b) and the supply function in Example 7 in Section 1.2 to find and interpret the equilibrium price.

d. Graph $S(p)$ and $D(p)$ on the same axes, and show the break-even point.

15. The Metropolis Transit Authority is attempting to increase its revenue. Although the subway system operates at capacity during the commute hours, it operates far below capacity during the late morning and early afternoon. The MTA plans to increase demand by offering a reduced fare during these hours. Reducing the fare will have two conflicting effects: it will cause more people to use the subway (and therefore increase the revenue), but it will also cause the people who use the subway to pay less (and therefore decrease the revenue). The MTA needs to know

the fare that will result in maximum off-peak revenue. To this end, they have sent out a questionnaire to a random sample of city residents; the results are given in Figure 1.48. They estimate that the city has 120,000 residents who could ride the subway on a regular basis.

FIGURE 1.48

Fare in cents	75	60	50	40	25
Percent who would ride at this fare	9.1	21.2	33.7	34.6	37.5

a. Convert the given data into 5 data points (p, d), where p is price in cents and d is demand in thousands of riders.

b. Use your calculator to find the slope and y-intercept of the line of best fit, as well as the correlation coefficient. Interpret the correlation coefficient.

c. Find the equation of the line of best fit.

d. Revenue is the fare p times the number of riders d. Use the result of part (c) to find the revenue as a function of p. Write this function using functional notation.

e. Determine appropriate values of xmin, xmax, xscl, ymin, ymax, and yscl. Discuss how you determined those values.

f. Graph the revenue function.

g. Determine the fare that would yield maximum revenue and the corresponding revenue. (Finding maximum points was discussed in Section 1.2, Exercises 34 and 35.)

h. Determine the fare that the MTA should charge. Discuss why it is different than the fare found in (g).

CHAPTER 1 REVIEW

Terms

analytic geometry	demand	linear equation	range	theory of supply and
approximate linear	dependent variable	linear function	regression line	demand
relationship	depreciation	linear model	relative error	unit variable cost
break-even point	domain	linear regression	revenue	variable costs
break-even quantity	equilibrium point	linear relationship	revenue function	vertical line test
break-even revenue	fixed costs	marginal cost	rise	x-axis
Cartesian coordinates	function	marginal revenue	run	x-coordinate
consumer	functional notation	marginality	scatter diagram	x-intercept
correlation	independent variable	mathematical model	slope	y-axis
coefficient	intercept	ordered pair	supplier	y-coordinate
cost function	least squares line	origin	supply	y-intercept
data points	line of best fit	profit function		

Review Exercises

In Exercises 1–4:

  a. Find the slope of the line through the two given points.
  b. Graph the line, showing the two given points, the rise, and the run.

1. $(1, 7)$ and $(2, -1)$ **2.** $(3, -2)$ and $(5, -2)$

3. $(2, 6)$ and $(2, -1)$ **4.** $(-1, 5)$ and $(3, 12)$

In Exercises 5–8:

  a. Find the equation, in slope-intercept form, of the line through the two given points.
  b. Check your answers to part (a) by verifying that the two given points satisfy the equation.
  c. Graph the line, showing the two given points and the y-intercept.
  [NOTE: Answers are not given at the back of the book.]

5. $(3, 9)$ and $(5, 5)$

6. $(-1, 7)$ and $(-1, 3)$ (Find the equation in *any* form.)

7. $(2, 6)$ and $(8, 10)$

8. $(-3, 9)$ and $(12, 9)$

In Exercises 9 and 10:

  a. Rewrite the equation in slope-intercept form.
  b. Use part (a) to determine the line's slope and y-intercept.
  c. Sketch the line's graph.

9. $2x - 6y = 9$

10. $4x + 2y = 12$

11. a. Must the relationship between the classes taught at your school and the number of students enrolled in those classes represent a function, if the input is the class and the output is its enrollment?
  b. If that relationship represents a function, use functional notation to describe it. If it is not a function, explain why.
  [NOTE: Math 181 is a course, not a class. Math 181, Section 1 is a class.]

12. a. Must the relationship between the classes taught at your school and the number of students enrolled in those classes represent a function, if the input is the enrollment and the output is the class?
  b. If that relationship represents a function, use functional notation to describe it. If it is not a function, explain why.
  [NOTE: Math 181 is a course, not a class. Math 181, Section 1 is a class.]

In Exercises 13–15:

  a. Determine whether the given equation is a function, if the input is x and the output is y. Justify each negative answer by giving an input that has more than one output.
  b. If the equation is a function, determine its domain and the range.

13. $y = 2x + 3$

14. $y = x^2$

15. $x = y^2$

In Exercises 16 and 17:

  a. Determine whether the given graph represents a function, if the input is x and the output is y.
  b. If the graph represents a function, determine the domain and the range.

16.

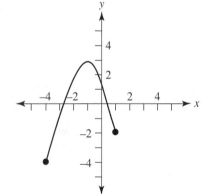

17.

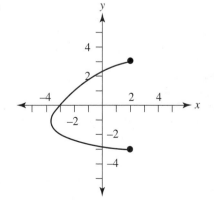

18. If the domain of $f(x) = 3x - 7$ is given as $\{x \mid 0 \le x \le 10\}$, find the range. Sketch the function's graph, and show the domain and range.

In Exercises 19–21, use the following functions:

$$f(x) = 2x + 1 \qquad g(x) = x^2$$

19. Find:
    a. $f(-3)$ **b.** $-f(3)$

20. Find:
    a. $g(2x - 1)$ **b.** $g(2x) - 1$
    c. $2g(x) - 1$

21. a. $f(g(x))$ **b.** $g(f(x))$

22. Temperature can be measured using different scales; the two most common are Fahrenheit and centigrade (also called Celsius). The relationship between these two scales is linear. Water boils at 212° F, which is equivalent to 100° C; water freezes at 32° F, which is equivalent to 0° C.
    a. Express the given information as two ordered pairs (F, C).
    b. Find the slope of the line between those two points and interpret it in the context of the problem.
    c. Find the equation of the line through these two points, using the letters F and C rather than x and y, respectively.
    d. Find centigrade temperature as a linear function of Fahrenheit temperature, using functional notation.
    e. Convert 0° F to centigrade.
    f. Convert 98.6° F to Celsius.
    g. Find the function whose input is centigrade temperature and whose output is Fahrenheit temperature.
    h. Convert 25° C to Fahrenheit.

23. The Down Depot makes down comforters at its plant in Osh Kosh.
    a. Classify the following costs as either fixed or variable:
        down
        cotton fabric
        thread
        sewing machines repair
        wages for the men and women who work on the assembly line
        wages for managers
        gas and electricity
    b. The plant has a monthly fixed cost of $111,412, and unit variable costs of $38 per comforter. The Down Depot sells its comforters for $349 each. Find the cost function, the revenue function, and the profit function and their domains and ranges.
    c. Find the break-even point, the break-even quantity, and the break-even revenue. Check your answer. (The answer is not given in the back of the book.)
    d. Find the cost, revenue, and profit at a production level that's 10% above the break-even quantity (rounded to the nearest whole number of comforters).
    e. Find the cost, revenue, and profit at a production level that's 10% below the break-even quantity (rounded to the nearest whole number of comforters).
    f. Graph the cost, revenue, and profit functions on one set of axes, using the appropriate domains and ranges. Show the answers to (c), (d), and (e) on the graph.
    g. Find the marginal revenue and the marginal cost at an arbitrary level of production. Discuss the implications of this comparison regarding an appropriate level of production.

24. Describe the theory of supply and demand.

25. An architect purchased a new computer system for $7200. The IRS allows it to be depreciated over seven years, with straight-line depreciation and zero value at the end of those seven years.
    a. Express the value of the computer as a function of the number of years after purchase.
    b. Find its value after three years.
    c. How much value is lost each year?
    d. How much can be deducted from income taxes each year?

26. The number of milk cows in the United States is given in Figure 1.49.
    a. Plot the data points. Do they exhibit a linear trend?
    b. Find the equation of the line of best fit.
    c. Graph the line of best fit on the scatter diagram from (a).
    d. Predict the number of cows in the year 2010.
    e. Predict when there will be 9 million milk cows.
    f. Find and interpret the correlation coefficient.
    g. What assumption are these predictions based on? If this assumption is correct, how accurate are these predictions?

FIGURE 1.49
Milk cows in the United States

(Source: U.S. Department of Agriculture)

Year	1970	1975	1980	1985	1990
Millions of cows	12.091	11.220	10.758	10.777	10.153

SYSTEMS OF LINEAR EQUATIONS AND INEQUALITIES

2

A system of linear equations is a set of more than one linear equation. Systems of linear equations have a number of widely varying applications, one of which is break-even analysis, which we covered in Section 1.2. There, we had a system of two linear equations—a cost equation and a revenue equation. We found the break-even point by solving the system of equations.

Systems of linear equations play a major role in half of the chapters of this book. Perhaps because of their abundant applications, there are numerous methods for solving them, each of which has its own advantages and disadvantages. In this chapter we will study the elimination method and the Gauss-Jordan method.

The elimination method is appropriate for problems that have a small number of equations, as did those in Section 1.2 on break-even analysis. However, some applications have so many equations that the elimination method is not practical. In these circumstances, the Gauss-Jordan method is more appropriate. Furthermore, the Gauss-Jordan method is amenable to the use of technology, whereas the elimination method is not.

A system of linear inequalities is a set of more than one linear inequality. The most important application of systems of linear inequalities is linear programming, which we will introduce at the end of this chapter and expand on in Chapter 3.

2.0
THE ELIMINATION METHOD

2.1
INTRODUCTION TO MATRICES AND THE GAUSS-JORDAN METHOD

2.2
MORE ON THE GAUSS-JORDAN METHOD

2.3
LINEAR INEQUALITIES

2.4
THE GEOMETRY OF LINEAR PROGRAMMING

2.0

THE ELIMINATION METHOD

Linear Equations

Because the graph of an equation of the form $ax + by = c$ is always a line, such an equation is called a linear equation. To graph an equation of the form $ax + by + cz = d$, we would need a z-axis in addition to the usual x- and y-axes; the graph of such an equation is always a plane. Despite its geometrical appearance, however, such an equation is called a linear equation because of its algebraic similarity to $ax + by = c$. A **linear equation** is an equation that can be written in the form $ax + by = c$, the form $ax + by + cz = d$, or a similar form with more unknowns.

Systems of Equations and Their Solutions

A **system of equations** is a set of more than one equation. Solving a system of equations means finding all ordered pairs (x, y) [or ordered triples (x, y, z) and so on] that will satisfy each equation in the system. Geometrically, this means finding all points that are on the graph of each equation in the system. The system

$$x + y = 3$$
$$2x + 3y = 8$$

is a **system of linear equations** because it is a set of two equations and each equation is of the form $ax + by = c$. The ordered pair $(1, 2)$ is a solution to the system because $(1, 2)$ satisfies each equation:

$$x + y = 1 + 2 = 3$$
$$2x + 3y = 2 \cdot 1 + 3 \cdot 2 = 8$$

Does this system have any other solutions? If we solve each equation for y, we get

$$y = -1x + 3$$
$$y = \frac{-2}{3}x + \frac{8}{3}$$

FIGURE 2.1

The first equation is the equation of a line with slope -1 and y-intercept 3; the second is the equation of a line with slope $\frac{-2}{3}$ and y-intercept $\frac{8}{3}$. Because the slopes are different, the lines are not parallel and intersect in only one point. Thus, $(1, 2)$ is the only solution to the system; it is the point at which the two lines intersect. This system is illustrated in Figure 2.1.

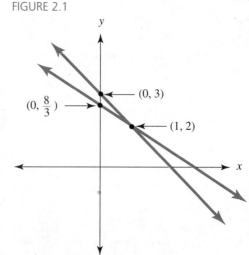

Not all systems have only one solution. A system with two linear equations in two unknowns describes a pair of lines. The two lines can be parallel and not intersect (in which case the system has no solutions), or the two lines can be the same (in which case the system has an infinite number of solutions), as illustrated in Figure 2.2.

FIGURE 2.2

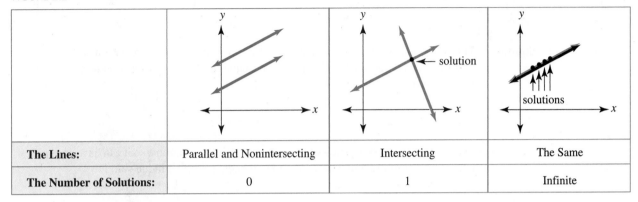

The Lines:	Parallel and Nonintersecting	Intersecting	The Same
The Number of Solutions:	0	1	Infinite

A system of two linear equations in *three* unknowns (x, y, and z) describes a set of two planes. Such a system can have no solutions or an infinite number of solutions, but it cannot have only one solution, as illustrated in Figure 2.3. In fact, *any system of linear equations that has fewer equations than unknowns cannot have a unique solution.*

FIGURE 2.3

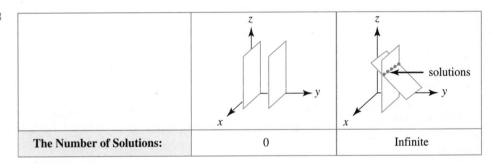

	The Number of Solutions:	0	Infinite

A system of *three* linear equations in three unknowns describes a set of three planes. Such a system can have no solutions, only one solution, or an infinite number of solutions, as illustrated in Figure 2.4. In fact, *any system of linear equations that has as many equations as unknowns can have either no solutions, one unique solution, or an infinite number of solutions.*

FIGURE 2.4

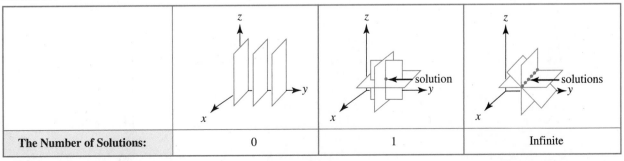

	The Number of Solutions:	0	1	Infinite

EXAMPLE 1   How many solutions could the following system have?

$$3x - y + z = 5$$
$$2x + y - 3z = 1$$
$$20x + 10y - 30z = 10$$

Solution   This system seems to have as many equations as unknowns; however, this is not the case. The third equation is 10 times the second equation, so its presence is unnecessary. The above system is equivalent to the system

$$3x - y + z = 5$$
$$2x + y - 3z = 1$$

This system has three unknowns but only two *unique* equations, so it can have either no solutions or an infinite number of solutions. It cannot have a unique solution.   ●

FIGURE 2.5

Number of Solutions of a System of Linear Equations	
A system with:	**Can have:**
Fewer equations than unknowns	No solutions or an infinite number of solutions
As many unique equations as unknowns	No solutions, only one solution, or an infinite number of solutions

Solving Systems: The Elimination Method

Many methods are used to solve systems of equations. Perhaps the easiest method for solving a small system is the **elimination method** (also called the **addition method**). With this method, you add together equations or multiples of equations in such a way that a variable is eliminated. You may have studied this method in an algebra class.

EXAMPLE 2   Use the elimination method to solve the following system:

$$3x - 2y = 5$$
$$2x + 3y = 4$$

Solution   We'll eliminate x if we multiply the first equation by 2 and the second equation by -3 and add the results:

$$
\begin{array}{ll}
6x - 4y = 10 & \text{2 times equation 1} \\
\underline{+\ -6x - 9y = -12} & \text{-3 times equation 2} \\
\quad\ -13y = -2 & \\
\end{array}
$$

$$y = \frac{2}{13}$$

We can find x by substituting $y = \frac{2}{13}$ into either of the original equations:

$$3x - 2y = 5$$

$$3x - 2 \cdot \frac{2}{13} = 5$$

$$3x - \frac{4}{13} = \frac{65}{13} \qquad \text{Rewriting 5 with a common denominator}$$

$$3x = \frac{69}{13}$$

$$x = \frac{23}{13}$$

Thus, the solution to the system of equations is the ordered pair $\left(\frac{23}{13}, \frac{2}{13}\right)$.

✔ We can easily check our work by substituting the solution back in the original equations.

$$3x - 2y = 3 \cdot \frac{23}{13} - 2 \cdot \frac{2}{13} = \frac{69}{13} - \frac{4}{13} = \frac{65}{13} = 5 \quad ✔$$

$$2x + 3y = 2 \cdot \frac{23}{13} + 3 \cdot \frac{2}{13} = \frac{46}{13} + \frac{6}{13} = \frac{52}{13} = 4 \quad ✔$$

Naturally, these calculations can be performed on a calculator.

EXAMPLE 3 Use the elimination method to solve the following system:

$$12x - 18y = 48$$
$$-10x + 15y = -40$$

Solution We'll eliminate the x if we multiply the first equation by 5 and the second by 6 and add the results:

$$
\begin{array}{ll}
60x - 90y = 240 & \text{5 times equation 1} \\
\underline{-60x + 90y = -240} & \text{6 times equation 2} \\
\quad 0 + 0 \;\;= 0 &
\end{array}
$$

This result seems to be useless—after all, we already know that $0 + 0 = 0$. What can this result mean?

Let's look at the preceding elimination work from a slightly different perspective. If we don't simplify the left side quite as much as we did, we get

$$
\begin{array}{l}
60x - 90y = 240 \\
\underline{-60x + 90y = -240} \\
\;\; 0x \;+\; 0y = 0
\end{array}
$$

This equation ($0x + 0y = 0$) is true for an infinite number of ordered pairs (x, y), since zero times anything plus zero times anything else adds to zero. Thus, the system has an infinite number of solutions. (The same conclusion could be reached by comparing the two equations in the above elimination work—they clearly describe the same lines.)

We can obtain the solutions to the system if we solve either of the original equations for y.

$$12x - 18y = 48 \qquad \text{The first original equation}$$

$$-18y = -12x + 48$$

$$\frac{-18y}{-18} = \frac{-12x + 48}{-18}$$

$$y = \frac{-12}{-18}x - \frac{48}{18} \qquad \text{Distributing the } -18$$

$$y = \frac{2}{3}x - \frac{8}{3}$$

This means that any ordered pair (x, y) where $y = \frac{2}{3}x - \frac{8}{3}$ is a solution to the system. In other words, all ordered pairs of the form $(x, \frac{2}{3}x - \frac{8}{3})$ are solutions to the system. This ordered pair $(x, \frac{2}{3}x - \frac{8}{3})$ is called the **general solution** to the system, and the variable x is called **a parameter** when it is used in this fashion.

Replacing x with a number results in a **particular solution** to the system. Some particular solutions to this system are:

$x = 0$	$x = 1$	$x = 2$
$(x, \frac{2}{3}x - \frac{8}{3})$	$(x, \frac{2}{3}x - \frac{8}{3})$	$(x, \frac{2}{3}x - \frac{8}{3})$
$= (0, \frac{2}{3} \cdot 0 - \frac{8}{3})$	$= (1, \frac{2}{3} \cdot 1 - \frac{8}{3})$	$= (2, \frac{2}{3} \cdot 2 - \frac{8}{3})$
$= (0, -8/3)$	$= (1, -2)$	$= (2, -4/3)$

Each of these particular solutions can be checked in the manner used in Example 2. ●

EXAMPLE 4 Use the elimination method to solve the following system:

$$3x - 2y = 5$$
$$6x - 4y = 7$$

Solution We'll eliminate the x if we multiply the first equation by -2 and add the result to the second equation:

$$-6x + 4y = -10$$
$$\underline{6x - 4y = 7}$$
$$0x + 0y = -3$$

This equation ($0x + 0y = -3$) is false for any ordered pair (x, y), since zero times anything plus zero times anything else never adds to -3. Thus, the system has no solutions.   ●

2.0

EXERCISES

In Exercises 1–6, determine whether the given ordered pair or ordered triple solves the given system of equations.

1. $(4, 1)$
$3x - 5y = 7$
$2x + 2y = 10$

2. $(7, -2)$
$2x - 4y = 30$
$x + y = 5$

3. $(-5, 3)$
$3x + y = 4$
$10x - 4y = -62$

4. $(-1, -2)$
$2x + 2y = -5$
$3x - 4y = 2$

5. $(4, -1, 2)$
$2x + 3y - z = 3$
$x + y + z = 5$
$10x - 2y = 3$

6. $(0, 5, -1)$
$3x - 2y + z = -11$
$2x + 4y + z = 19$
$x - z = 1$

In Exercises 7–12, determine whether the given system has no solutions, one solution, or an infinite number of solutions by finding the slopes and y-intercepts. Do not actually solve the system.

7. $5x + 2y = 4$
$6x - 19y = 27$

8. $3x + 132y = 19$
$45x + 17y = 4$

9. $4x + 3y = 12$
$8x + 6y = 24$

10. $19x - 22y = 1$
$190x - 220y = 10$

11. $x + y = 7$
$3x + 2y = 8$
$2x + 2y = 14$

12. $3x - y = 12$
$2x + 3y = 5$
$5x + 2y = 17$

In Exercises 13–18, determine whether the given system could have a single solution. Do not actually solve the system.

13. $3x - 2y + 5z = 1$
$2x + y = 2$
$5x + 7y - z = 0$

14. $8x - 4y + 2z = 10$
$3x + y + z = 1$
$4x - 2y + z = 5$

15. $x + y + z = 1$
$2x + 2y + 2z = 2$
$3x - y + 10z = 45$

16. $x + y + z = 34$
$5x - y + 2z = 9$
$3x + 3y + 3z = 102$

17. $x + 2y + 3z = 4$
$5x - y = 2$

18. $9x - 21y = 476$
$x + 3y + z = 12$

Solve the systems in Exercises 19–28. If there is more than one solution, give the general solution and three particular solutions. Check your answers by substituting them back in. (Answers are not given at the back of the book.)

19. $2x + 3y = 5$
$4x - 2y = 2$

20. $5x + 3y = 11$
$2x + 7y = 16$

21. $3x - 7y = 27$
$4x - 5y = 23$

22. $5x - 2y = -23$
$x + 2y = 5$

23. $5x - 9y = -12$
$3x + 7y = -1$

24. $5x - 12y = 9$
$3x + 3y = 2$

25. $8x - 4y = 16$
$6x - 3y = 12$

26. $12x + 28y = 20$
$15x + 35y = 25$

27. $9x - 21y = 15$
$6x - 14y = 12$

28. $5x + 3y = 8$
$10x + 6y = 9$

In Exercises 29–32, you are asked to solve a system of three equations in three unknowns using the elimination method. To do that, first combine two equations so that one variable is eliminated. Then combine two other equations so that the same one variable is eliminated. This results in two equations in two unknowns. Solve that system of two equations with the elimination method.

29. $5x + y - z = 17$
$2x + 5y + 2z = 0$
$3x + y + z = 11$

30. $9x - 2y + 4z = 29$
$2x + 3y - 4z = 3$
$x + y + z = 1$

31. $x + y + z = 14$
$\quad 3x - 2y + z = 3$
$\quad 5x + y + 2z = 29$

32. $2x + y - z = 4$
$\quad 3x + 2y - 7z = 0$
$\quad 5x - 3y + 2z = 20$

 Answer the following questions using complete sentences.

33. By sketching lines, determine whether a system of three equations in two unknowns could have no solutions, one solution, or an infinite number of solutions. What if the three equations were unique?

34. By sketching planes, determine whether a system of four equations in three unknowns could have no solutions, one solution, or an infinite number of solutions. What if the four equations were unique?

GRAPHING SYSTEMS OF EQUATIONS ON A GRAPHING CALCULATOR

Systems of two linear equations in two unknowns can be solved on a graphing calculator by graphing the two lines and finding their intersection, as we did in Section 1.2 when we found break-even points. First, solve each equation for y. Compare the results and determine if the lines are the same or are parallel and nonintersecting. If neither, enter and graph the two equations. If your viewing window doesn't allow you to see the intersection point, alter it. Finally, find the intersection point.

Finding the intersection point with a graphing calculator was thoroughly covered in Section 1.2; it is summarized in the following table.

TI-82/83:	Select option 5 "intersect" from the "CALC" menu
TI-85/86:	Select "MATH" from the "GRAPH" menu, and then "ISECT" from the "GRAPH MATH" menu

EXERCISES

Use a graphing calculator to solve the following exercises. Check your answers by substituting them back in. (Answers are not given at the back of the book.)

35. $8x - 7y = 24$
$\quad 3x + 3y = 19$

36. $11x + 19y = 35$
$\quad 12x - 7y = 53$

37. $91x - 32y = 4$
$\quad 2x + 17y = 9$

38. $33x - 52y = 1$
$\quad 59x - 11y = 2$

2.1

INTRODUCTION TO MATRICES AND THE GAUSS-JORDAN METHOD

Matrix Terminology

A **matrix** (plural *matrices*) is a rectangular arrangement of numbers enclosed by brackets. For example,

$$\begin{bmatrix} 3 & -2 & 0 \\ 1 & -27 & 5 \end{bmatrix}$$

is a matrix. Each number in the matrix is called an **element** or an **entry** of the matrix; 3 is an element of the above matrix, as is -27. The horizontal listings of elements are called **rows;** the vertical listings are called **columns.** The first row of the matrix

$$\begin{bmatrix} 3 & -2 & 0 \\ 1 & -27 & 5 \end{bmatrix} \quad \leftarrow \text{First row}$$

$$\uparrow$$

Last column

is "3 -2 0," and the last column is "0 5." The **dimensions** of a matrix with m rows and n columns are $m \times n$ (read "m by n"). The matrix above has two rows and three columns, so its dimensions are 2×3; that is, it is a 2×3 matrix.

The dimensions of a matrix are like the dimensions of a room. If a room is 7 feet on one side and 10 feet on the other, then the dimensions of that room are 7×10. We do not actually multiply 7 times 10 to determine the dimension (although we would to find the area). Likewise, the dimensions of the above matrix are 2×3; we do not actually multiply 2 times 3 and get 6. If we did, we would have the *number of elements* in the matrix rather than the *dimensions* of the matrix.

A matrix with only one row is called a **row matrix.** A matrix with only one column is called a **column matrix.** A matrix is **square** if it has as many rows as columns. For example:

$\begin{bmatrix} 3 & -2 & 0 \end{bmatrix}$ is a row matrix.

$\begin{bmatrix} 26 \\ 89 \end{bmatrix}$ is a column matrix.

$\begin{bmatrix} 0 & -1 \\ 5 & 3 \end{bmatrix}$ is a square matrix.

$\begin{bmatrix} 3 & -4 & 5 \\ 0 & 18 & 2 \end{bmatrix}$ is neither a row matrix, a column matrix, nor a square matrix.

A matrix is usually labeled with a capital letter, and the entries of that matrix are labeled with the same letter in lowercase with a double subscript. The first subscript refers to the entry's row, the second subscript refers to its column. For example, b_{23} refers to the element of matrix B that is in the second row and the third column. Similarly, if

$$A = \begin{bmatrix} 5 & 7 \\ 9 & 11 \end{bmatrix}$$

then $a_{11} = 5$, $a_{12} = 7$, $a_{21} = 9$, and $a_{22} = 11$.

Solving Systems: The Gauss-Jordan Method

The elimination method described in Section 2.0 is fine for solving small systems of equations—for instance, systems with two equations and two unknowns—but the method becomes too complicated with larger systems. A more formalized version of elimination works much better; it is the Gauss-Jordan method, a systematic approach that uses matrices to abbreviate the amount of writing necessary. We will illustrate the Gauss-Jordan method with a small system, but keep in mind that the Gauss-Jordan method is really more appropriate for use with larger systems. A small system simply provides an easy introduction to the method.

EXAMPLE 1   Use the Gauss-Jordan method to solve the following system:

$$x + y = 3$$
$$2x + 3y = 8$$

Solution   **Step 1**   *Write the system in such a way that all the variables are on the left side of the equations.* Write in all the coefficients, including coefficients of 1.

$$1x + 1y = 3$$
$$2x + 3y = 8$$

Step 2   *Rewrite the system in matrix form.* This is done by eliminating the letters, addition symbols, and equal symbols and putting the remaining numbers into a matrix.

$$\begin{matrix} x & y & \\ \begin{bmatrix} 1 & 1 & 3 \\ 2 & 3 & 8 \end{bmatrix} \end{matrix}$$

Notice that we've labeled the first two columns x and y. This is an optional way of keeping track of the origins of the numbers.

The goal of the remaining steps is to rewrite the matrix in the following form:

$$\begin{matrix} x & y & \\ \begin{bmatrix} 1 & 0 & ? \\ 0 & 1 & ? \end{bmatrix} \end{matrix}$$

This is called **reduced row echelon form.** When this goal is achieved, we will have our solution, because the form above is the matrix equivalent of the system.

$$1x + 0y = ?$$
$$0x + 1y = ?$$

which simplifies to the system

$$x = ?$$
$$y = ?$$

In order to achieve this goal, we use the following **row operations,** each of which is the matrix equivalent of a tool used in the elimination method.

Row Operations

1. *Multiply or divide a row by any number (except zero).* This is equivalent to multiplying or dividing an equation by a number.
2. *Add one row to another row or add a multiple of one row to another row.* This is equivalent to adding one equation to another equation or adding a multiple of an equation to another equation.
3. *Interchange any two rows.* This is equivalent to interchanging two equations.

Note: Each of the following steps in the Gauss-Jordan method is accompanied by the equivalent step in the elimination method. Following the Gauss-Jordan steps is easier if you see them as the equivalent of the elimination method steps with many of the symbols left out.

Gauss-Jordan Method   **Elimination Method**

$$\begin{matrix} x & y & \\ \begin{bmatrix} 1 & 1 & 3 \\ 2 & 3 & 8 \end{bmatrix} \end{matrix}$$

$1x + 1y = 3$
$2x + 3y = 8$

Step 3   *Use the row operations to rewrite the matrix so its first columns is "1 0."* In other words, we want the matrix to become

$$\begin{bmatrix} 1 & ? & ? \\ 0 & ? & ? \end{bmatrix}$$

The first column already has 1 as the first entry; what remains is to use the row operations to change 2 to 0. We will use row operation 2 to accomplish this, adding -2 times the first row to the second row (or "$-2R1 + R2$").

$$
\begin{array}{ccc}
-2 \cdot 1 & -2 \cdot 1 & -2 \cdot 3 \\
+ \quad 2 & + \quad 3 & + \quad 8 \\
\hline
0 & 1 & 2
\end{array}
\qquad
\begin{array}{cc}
-2 \cdot 1x + -2 \cdot 1y = -2 \cdot 3 \\
+ \quad 2x + \quad 3y = + \quad 8 \\
\hline
0x + \quad 1y = \quad 2
\end{array}
$$

If we replace the original second row with this result, then the resulting matrix will have the desired column.

$$\begin{matrix} x & y & \\ \begin{bmatrix} 1 & 1 & 3 \\ 0 & 1 & 2 \end{bmatrix} \end{matrix}$$

$1x + 1y = 3$
$0x + 1y = 2$

Step 4   *Use the row operations to rewrite the matrix so its second column is "0 1."* In other words, we want the matrix to become

$$\begin{bmatrix} 1 & 0 & ? \\ 0 & 1 & ? \end{bmatrix}$$

The second column already has 1 as the second entry; what remains is to use the row operations to change the first entry (also 1) to 0. We will accomplish this by adding the first row to -1 times the second row (or "R1 $+ -$1R2").

$$
\begin{array}{ccc}
1 & 1 & 3 \\
\underline{+ -1 \cdot 0} & \underline{+ -1 \cdot 1} & \underline{+ -1 \cdot 2} \\
1 & 0 & 1
\end{array}
\qquad
\begin{array}{ccc}
1x + & 1y = & 3 \\
\underline{+ -1 \cdot 0x +} & \underline{-1 \cdot 1y =} & \underline{+ -1 \cdot 2} \\
1x + & 0y = & 1
\end{array}
$$

If we replace the first row with this result, then the resulting matrix will have the desired column.

$$
\begin{array}{cc}
x & y
\end{array}
$$
$$
\begin{bmatrix} 1 & 0 & 1 \\ 0 & 1 & 2 \end{bmatrix}
\qquad
\begin{array}{l}
1x + 0y = 1 \\
0x + 1y = 2
\end{array}
$$

Step 5   *Read off the solution.* The matrix

$$
\begin{bmatrix} 1 & 0 & 1 \\ 0 & 1 & 2 \end{bmatrix}
$$

is just shorthand for the system

$$1x + 0y = 1$$
$$0x + 1y = 2$$

This system simplifies to

$$x = 1$$
$$y = 2$$

The solution is the ordered pair (1, 2).

Alternatively, the solution can be read from the matrix without reverting to the system of equations. We can find the value of the variable heading each column by reading down the column, turning at 1, and stopping at the end of the row.

$$
\begin{array}{cc}
x & y
\end{array}
$$
$$
\begin{bmatrix} 1 & 0 & 1 \\ 0 & 1 & 2 \end{bmatrix}
$$

Step 6   ✔   *Check the solution by substituting it into the original system.*

$$x + y = 1 + 2 = 3$$
$$2x + 3y = 2 \cdot 1 + 3 \cdot 2 = 8$$

Notice that in Example 1, Steps 3 and 4 are really the same steps applied to different columns. In each of these two steps, we rewrote a column as an identity matrix column; this process is called **pivoting.** Also, in the execution of Steps 3 and 4, the columns already had 1 in the desired location; if that had not been the case, more work would have been necessary, as shown in the next example. The following procedure (a more complete version of the steps used above) will be used on all problems.

Gauss-Jordan Steps

1. *Write the system with all the variables on the left side of the equations and with all coefficients showing.*
2. *Rewrite the system in matrix form.* Eliminate the letters, addition symbols, and equal symbols. Keep all signs for negative numbers.
3. *Use the row operations to rewrite the matrix in reduced row echelon form.* (This is called *pivoting.*) Decimals can be used rather than fractions as long as you round *each* calculation off to four or more decimal places, to keep some accuracy.
4. *Read off the solution.*
5. *Check the solution by substituting it into the original system.*

There are many different ways to execute Step 3. Some people prefer to do each problem with a routine procedure that always works, whereas others prefer to do each problem differently, finding clever shortcuts that minimize the calculations. If you are among the former, the following procedure is recommended.

Routine Pivoting Procedure

1. *Change the appropriate entry of the first column to 1 by multiplying or dividing the row containing that entry by a number.*
2. *Change each of the other entries in the column to 0 by adding a multiple of the row with 1 in it to the row in which you wish to have a zero.*
3. *Repeat the above steps with the next column.*

EXAMPLE 2 Use the Gauss-Jordan method to solve the following system:

$$2x + z = y + 5$$
$$x + 2y + z = 3$$
$$4x - 3z = 0$$

Solution **Step 1** *Write the system with all the variables on the left side of the equations and with all coefficients showing.*

$$2x - 1y + 1z = 5$$
$$1x + 2y + 1z = 3$$
$$4x + 0y - 3z = 0$$

Step 2 *Rewrite the system in matrix form.* Eliminate the letters, addition symbols, and equal symbols.

$$\begin{array}{ccc} x & y & z \end{array}$$
$$\begin{bmatrix} 2 & -1 & 1 & 5 \\ 1 & 2 & 1 & 3 \\ 4 & 0 & -3 & 0 \end{bmatrix}$$

Step 3 *Use the row operations to rewrite the matrix in reduced row echelon form.* We rewrite the matrix so that the first column is "1 0 0." The routine pivoting procedure specifies that we start this by dividing the first row by 2, but it's easier to use row operation 3 and interchange the first two rows. The routine pivoting procedure isn't necessarily the most efficient procedure.

Pivot on 1
$$\begin{array}{ccc} x & y & z \end{array}$$
$$\begin{bmatrix} 1 & 2 & 1 & 3 \\ 2 & -1 & 1 & 5 \\ 4 & 0 & -3 & 0 \end{bmatrix}$$

$-2R1 + R2 : R2$
$$\begin{array}{ccc} x & y & z \end{array}$$
$$\begin{bmatrix} 1 & 2 & 1 & 3 \\ 0 & -5 & -1 & -1 \\ 4 & 0 & -3 & 0 \end{bmatrix}$$

$-4R1 + R3 : R3$
$$\begin{array}{ccc} x & y & z \end{array}$$
$$\begin{bmatrix} 1 & 2 & 1 & 3 \\ 0 & -5 & -1 & -1 \\ 0 & -8 & -7 & -12 \end{bmatrix}$$

Now that we're finished with the first column, we move on to the second column and execute the same steps.

Pivot on -5
$$\begin{array}{ccc} x & y & z \end{array}$$
$$\begin{bmatrix} 1 & 2 & 1 & 3 \\ 0 & -5 & -1 & -1 \\ 0 & -8 & -7 & -12 \end{bmatrix}$$

$R2 \div -5$
$$\begin{array}{ccc} x & y & z \end{array}$$
$$\begin{bmatrix} 1 & 2 & 1 & 3 \\ 0 & 1 & 0.2 & 0.2 \\ 0 & -8 & -7 & -12 \end{bmatrix}$$

$R1 - 2R2 : R1$
$8R2 + R3 : R3$
$$\begin{array}{ccc} x & y & z \end{array}$$
$$\begin{bmatrix} 1 & 0 & 0.6 & 2.6 \\ 0 & 1 & 0.2 & 0.2 \\ 0 & 0 & -5.4 & -10.4 \end{bmatrix}$$

Now that we're finished with the second column, we move on to the third column and execute the same steps.

Pivot on −5.4

$$\begin{array}{ccc} x & y & z \end{array}$$
$$\begin{bmatrix} 1 & 0 & 0.6 & 2.6 \\ 0 & 1 & 0.2 & 0.2 \\ 0 & 0 & -5.4 & -10.4 \end{bmatrix}$$

R3 ÷ −5.4

$$\begin{array}{ccc} x & y & z \end{array}$$
$$\begin{bmatrix} 1 & 0 & 0.6 & 2.6 \\ 0 & 1 & 0.2 & 0.2 \\ 0 & 0 & 1 & 1.9259 \end{bmatrix}$$ Rounding to four decimal places

R1 − 0.6R3 : R1
R2 − 0.2R3 : R2

$$\begin{array}{ccc} x & y & z \end{array}$$
$$\begin{bmatrix} 1 & 0 & 0 & 1.4445 \\ 0 & 1 & 0 & -0.1852 \\ 0 & 0 & 1 & 1.9259 \end{bmatrix}$$ Rounding to four decimal places

We have finished Step 3; we've used the row operations to rewrite the matrix in reduced row echelon form.

Step 4 *Read off the solution.* The above matrix is just shorthand for the system

$$1x + 0y + 0z = 1.4445$$
$$0x + 1y + 0z = -0.1852$$
$$0x + 0y + 1z = 1.9259$$

This system simplifies to

$$x = 1.4445$$
$$y = -0.1852$$
$$z = 1.9259$$

The (approximate) solution is the ordered triple (1.4445, −0.1852, 1.9259). Alternatively, the solution can be read from the matrix:

$$\begin{array}{ccc} x & y & z \end{array}$$
$$\begin{bmatrix} 1 & 0 & 0 & 1.4445 \\ 0 & 1 & 0 & -0.1852 \\ 0 & 0 & 1 & 1.9259 \end{bmatrix}$$

Step 5 *Check the solution by substituting it into the original system.*

$$2x - y + z = 2(1.4445) - (-0.1852) + (1.9259) = 5.0001$$
$$x + 2y + z = (1.4445) + 2(-0.1852) + (1.9259) = 3$$
$$4x - 3z = 4(1.4445) - 3(1.9259) = 0.0003$$

Note that our solution doesn't check perfectly; we should get 5, 3, and 0 when we substitute the solution into the three equations. This discrepancy is a result of rounding off. •

2.1

EXERCISES

In Exercises 1–10, (a) find the dimensions of the given matrix and (b) determine whether the matrix is a row matrix, a column matrix, a square matrix, or none of these.

1. $A = \begin{bmatrix} 5 & 0 \\ 22 & -3 \\ 18 & 9 \end{bmatrix}$

2. $B = \begin{bmatrix} 1 & 13 & 207 \\ -4 & 8 & 100 \\ 0 & 1 & 5 \end{bmatrix}$

3. $C = \begin{bmatrix} 23 \\ 41 \end{bmatrix}$

4. $D = \begin{bmatrix} 2 & 0 & 19 & -3 \\ 62 & 13 & 44 & 1 \\ 5 & 5 & 30 & 12 \\ 0 & 0 & 0 & 0 \end{bmatrix}$

5. $E = \begin{bmatrix} 3 & 0 \end{bmatrix}$

6. $F = \begin{bmatrix} -2 & 10 \\ 4 & -3 \end{bmatrix}$

7. $G = \begin{bmatrix} 12 & -11 & 5 \\ -9 & 4 & 0 \\ 1 & 9 & 5 \end{bmatrix}$

8. $H = \begin{bmatrix} 1 & 0 \\ 0 & -1 \end{bmatrix}$

9. $J = \begin{bmatrix} 5 \\ -3 \\ 11 \end{bmatrix}$

10. $K = \begin{bmatrix} 2 & -5 & 13 & 0 \\ -1 & 4 & 3 & 6 \\ 8 & -10 & 4 & 0 \end{bmatrix}$

In Exercises 11–20, find the indicated elements of the matrices given in Exercises 1–10.

11. a_{21}
HINT: You're asked to find the entry in row two, column one of the matrix A, which is given in Exercise 1.

12. b_{23} **13.** c_{21} **14.** d_{34}

15. e_{11} **16.** f_{22} **17.** g_{12}

18. h_{21} **19.** j_{21} **20.** k_{23}

Rewrite the systems of equations in Exercises 21–28 in matrix form.

21. $2x + 7y = 11$
$3x - 2y = 15$

22. $4x + 5y = 6$
$3x + 2y = 9$

23. $2x = y + 5$
$3x - y = 41$

24. $2x - 19y = 55$
$3y = 2x + 8$

25. $2x + 3y - 7z = 53$
$5x - 2y + 12z = 19$
$x + y + z = 55$

26. $3x - y + z = 22$
$5x + y + 10z = 38$
$18x + 10y - 121z = 15$

27. $5x + z = 2$
$3x - y = 15$
$2x + 2y + 2z = 53$

28. $8x + 2y = 43$
$2x - z = 15$
$5x + y + 3z = 0$

In Exercises 29–34, use the row operations to change the indicated column so that it is in reduced row echelon form.

29. $\begin{bmatrix} 2 & 6 & 10 \\ -2 & 1 & 4 \end{bmatrix}$ Change the first column.

30. $\begin{bmatrix} 4 & 12 & 40 \\ -2 & 1 & 4 \end{bmatrix}$ Change the first column.

31. $\begin{bmatrix} 1 & 3 & 12 \\ 0 & 5 & 7 \end{bmatrix}$ Change the second column.

32. $\begin{bmatrix} 1 & 4 & 3 \\ 0 & 2 & -4 \end{bmatrix}$ Change the second column.

33. $\begin{bmatrix} 2 & 2 & 4 & 12 \\ 2 & -1 & 4 & 3 \\ 1 & 2 & -9 & 2 \end{bmatrix}$ Change the first column.

34. $\begin{bmatrix} 4 & 8 & 12 & 4 \\ 2 & 2 & 3 & 4 \\ -1 & 4 & -2 & 2 \end{bmatrix}$ Change the first column.

In Exercises 35–38, find the solution that corresponds to the given matrix.

35. $\begin{bmatrix} 1 & 0 & 9 \\ 0 & 1 & 2 \end{bmatrix}$

36. $\begin{bmatrix} 1 & 0 & 0.7654 \\ 0 & 1 & -3.7990 \end{bmatrix}$

37. $\begin{bmatrix} 1 & 0 & 0 & 0.5281 \\ 0 & 1 & 0 & -0.6205 \\ 0 & 0 & 1 & 0 \end{bmatrix}$ **38.** $\begin{bmatrix} 1 & 0 & 0 & 3.4076 \\ 0 & 1 & 0 & 1.2066 \\ 0 & 0 & 1 & -0.9934 \end{bmatrix}$

Use the Gauss-Jordan method to solve the systems in Exercises 39–50. If you use decimals, round off each calculation to four decimal places. Check your answers by substituting them back in. (Answers are not given at the back of the book.)

39. $x + y = 3$
$\quad 2x - 3y = -4$

40. $x - y = 2$
$\quad 5x + 2y = 17$

41. $2x + 4y = 12$
$\quad 3x - y = 4$

42. $3x - 9y = 15$
$\quad 2x + 11y = 10$

43. $5x + 2y = 19$
$\quad 3x - 4y = -25$

44. $4x + 9y = 35$
$\quad 3x + 2y = 12$

45. $5x + y - z = 17$
$\quad 2x + 5y + 2z = 0$
$\quad 3x + y + z = 11$

46. $9x - 2y + 4z = 29$
$\quad 2x + 3y - 4z = 3$
$\quad x + y + z = 1$

47. $x + y + z = 14$
$\quad 3x - 2y + z = 3$
$\quad 5x + y + 2z = 29$

48. $2x + y - z = 4$
$\quad 3x + 2y - 7z = 0$
$\quad 5x - 3y + 2z = 20$

49. $x - y + 4z = -13$
$\quad 2x - z = 12$
$\quad 3x + y = 25$

50. $2x + 2y + z = 5$
$\quad 3x - y = 5$
$\quad 2y + 7z = -9$

 Answer the following question using complete sentences.

51. Compare and contrast the elimination method and the Gauss-Jordan method. How are they similar? How are they different? What advantages does the elimination method have? What advantages does the Gauss-Jordan method have?

TECHNOLOGY AND THE ROW OPERATIONS

One advantage of the Gauss-Jordan method over the elimination method is that it will solve larger systems of equations in a more systematic and efficient manner. Another advantage is that it is amenable to the use of technology and thus allows you to relieve yourself of the task of making tedious calculations. Some graphing calculators can perform the row operations on a matrix, as can a computer combined with the Amortrix computer program available with this book. A computer provides an easier interface than does a graphing calculator; however, a graphing calculator may be more available to you. Each is discussed below.

The Row Operations and Amortrix

The Amortrix computer program available with this book will enable you to easily and quickly perform the row operations on a matrix. Amortrix is available for both Macintosh and Windows-based computers.

We will discuss how to use Amortrix to solve the following system:

$$3.1x + 2.7y - 4.9z = 5.3$$
$$6.9x + 4.2y + 3.7z = 9.2$$
$$4.1x - 1.7y + 9.3z = 7.7$$

This system can be rewritten as the following matrix:

$$\begin{bmatrix} 3.1 & 2.7 & -4.9 & 5.3 \\ 6.9 & 4.2 & 3.7 & 9.2 \\ 4.1 & -1.7 & 9.3 & 7.7 \end{bmatrix}$$

When you start Amortrix, a main menu appears. Use the mouse to click on the "Matrix Row Operations" option. Once you're into the "Matrix Row Operations" part of the program, press the "Create a New Matrix" button. The program will ask for the dimensions of the matrix you want to work on and will tell you how to make a response. Once you've responded, the program will display a matrix of the appropriate dimensions. Use the mouse to click on the "Edit Values" button and follow the instructions that appear on the screen to enter the numbers.

In order to pivot on 3.1, first divide row 1 by 3.1. To do this, use the mouse to click on the "ROW 1" label to the left of the matrix. This causes two boxes to appear: one that allows you to multiply the row by a number, and one that allows you to divide the row by a number. Use the mouse to press the division box, click on the white box, type in "3.1," and press "Return" or "Enter." When you press the "OK" box, the computer will execute the row operation. See Figure 2.6.

FIGURE 2.6

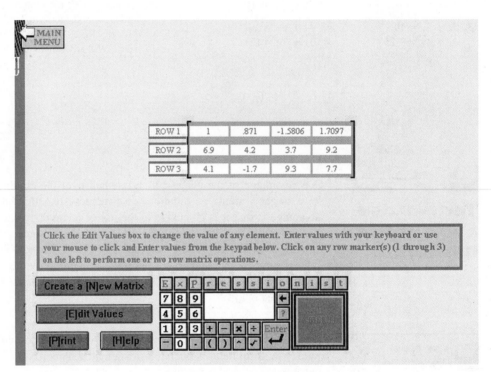

In order to multiply row 1 by −6.9 and add the result to row 2, use the mouse to click first on the "ROW 1" label and then on the "ROW 2" label to the left of the matrix. *When adding a multiple of one row to another row, always click first on the row to be multiplied.* This causes two boxes to appear: one that allows you to add two rows, and one that allows you to swap two rows. Use the mouse to press the add box, click on the first white box, type in "−6.9," and press "Return" or "Enter." When you press the "OK" box, the computer will execute the row operation. In a similar manner, multiply row 1 by −4.1 and add the result to row 3. This completes the first pivot. See Figure 2.7.

At this point, you're ready to move on to the second column. The next step is to divide row 2 by the entry in row 2, column 2. The computer shows that entry as

FIGURE 2.7

MAIN MENU				

ROW 1	1	.871	-1.5806	1.7097
ROW 2	0	-1.8097	14.6065	-2.5968
ROW 3	0	-5.271	15.7806	.6093

Click the Edit Values box to change the value of any element. Enter values with your keyboard or use your mouse to click and Enter values from the keypad below. Click on any row marker(s) (1 through 3) on the left to perform one or two row matrix operations.

Create a [N]ew Matrix

[E]dit Values

[P]rint   [H]elp

EXPressionist

7 8 9 ←
4 5 6 ?
1 2 3 + − × ÷ Enter
− 0 . ( ) ^ √ ↵

Amortrix

−1.8097; however, this number is rounded off, and if you type −1.8097 in the white box, the row operation will result in a 1.0054 in row 2, column 2, rather than the desired 1. To avoid this, use the mouse to click on the −1.8097 in row 2, column 2; the computer will divide by the *nonrounded* version of this number, and the row operation will result in a 1 in row 2, column 2.

Continue in a similar manner until you obtain the solution.

Row Operations on a Graphing Calculator

Some graphing calculators (including the TI-82, TI-83, TI-85, and TI-86) can perform the row operations.

First, we discuss how to enter the matrix

$$A = \begin{bmatrix} 1 & 2 \\ 3 & 4 \end{bmatrix}$$

and how to perform the row operations on that matrix. The TI-82 and TI-83 require that you name this matrix "[A]" with brackets; the TI-85 and TI-86 require that you name it "A" without brackets.

TI-85/86 Users When your calculator is in "ALPHA" mode, some buttons will generate a letter of the alphabet. If your calculator has a flashing "A" on the screen, then it is in "ALPHA" mode. One way to put your calculator in "ALPHA" mode is to press the ALPHA button. However, some other commands will automatically put your calculator in "ALPHA" mode. Don't confuse the ⌷2nd⌷ and ⌷ALPHA⌷ buttons. The ⌷2nd⌷ button refers to the mathematical labels above and to the *left* of the buttons.

Entering a Matrix

To enter matrix *A*, select "EDIT" from the "MATRIX" menu and then matrix *A* from the "MATRX EDIT" menu:

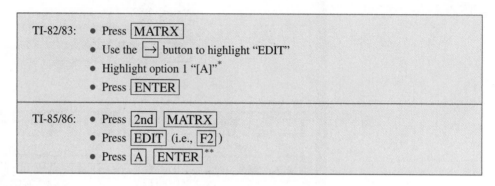

Then enter the dimensions of our 2×2 matrix *A* by typing

2 ⌷ENTER⌷ 2 ⌷ENTER⌷

To enter the elements, type

1 ⌷ENTER⌷ 2 ⌷ENTER⌷ 3 ⌷ENTER⌷ 4 ⌷ENTER⌷

Notice that the calculator uses double subscript notation. When we entered the "4" above, the calculator did its best to show that this was entered as a_{22}.

When you're done entering, type ⌷2nd⌷ ⌷QUIT⌷.

Viewing a Matrix

To view matrix A:

TI-82/83:	• Press ⌷MATRX⌷
	• Highlight option 1: "[A]"
	• Press ⌷ENTER⌷ and "[A]" will appear on the screen
	• Press ⌷ENTER⌷ and the matrix itself will appear on the screen, as shown in Figure 2.8
TI-85/86:	Type ⌷ALPHA⌷ ⌷A⌷ ⌷ENTER⌷ and matrix A will appear on the screen, as shown in Figure 2.8

---

[*]Option 1 is automatically highlighted. If we were selecting some other option, we would use the ⌷↑⌷ and ⌷↓⌷ buttons to highlight it.

[**]The ⌷LOG⌷ button becomes the ⌷A⌷ button when the calculator is in "ALPHA" mode. Usually this requires preceding the ⌷A⌷ button with the ⌷ALPHA⌷ button. However, the TI-85/86 was automatically placed in "ALPHA" mode when you pressed ⌷EDIT⌷, as shown by the flashing "A" on the screen.

FIGURE 2.8
Matrix A on a TI-82/83

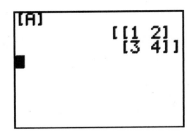

Interchanging Two Rows

The "RowSwap" command interchanges two rows. This and other row operations are available under the "MATRX" menu. To swap rows 1 and 2 of matrix *A*, make the screen read "RowSwap([A],1,2)" ["rSwap(A,1,2)" on a TI-85/86] by doing the following.

TI-82/83:	• Press MATRX • Scroll to "MATH" • Highlight the "rowSwap(" option (it's down the page) • Press ENTER	• Press MATRX • Highlight option 1: "[A]" • Press ENTER	Press , 1 , 2) ENTER
TI-85/86:	• Press 2nd MATRX, • Press OPS (i.e., F4) • Select "rSwap" by pressing MORE F2	Press ALPHA A	Press , 1 , 2) ENTER
	This generates "RowSwap(" *or "rSwap("*	*This generates "[A]"* *or "A"*	*This generates ",1,2)"*

The result should be

$$\begin{bmatrix} 3 & 4 \\ 1 & 2 \end{bmatrix}$$

Adding Two Rows

To add rows 1 and 2 of matrix *A* and place the result in row 2, make the screen read "Row+([A],1,2)" [or "rAdd(A,1,2)" on a TI-85/86] by doing the following.

TI-82/83:	• Press MATRX • Scroll to "MATH" • Highlight "row+(" • Press ENTER	• Press MATRX • Highlight option 1: "[A]" • Press ENTER	, 1 , 2) ENTER
TI-85/86:	• Press 2nd MATRX • Select "OPS" (i.e., F4) • Select "rAdd" (i.e., MORE F3)	Press ALPHA A	, 1 , 2) ENTER
	This generates "Row+(" *or "rAdd("*	*This generates "[A]"* *or "A"*	*This generates ",1,2)"*

The result should be

$$\begin{bmatrix} 1 & 2 \\ 4 & 6 \end{bmatrix}$$

Multiplying or Dividing a Row by a Number

To multiply row 1 of matrix A by 5, make the screen read "*Row(5,[A],1)" ["multR(5,A,1)" on a TI-85/86]. The result should be

$$\begin{bmatrix} 5 & 10 \\ 3 & 4 \end{bmatrix}$$

To divide a row by a number, multiply by the number's reciprocal. For example, divide a row by 5 by multiplying it by 1/5.

Adding a Multiple of One Row to Another Row

To multiply row 1 by 5, add the result to row 2 and place the result in row 2—make the screen read "*Row+(5,[A],1,2)" [or "mRAdd(5,A,1,2)" on a TI-85/86]. The result should be

$$\begin{bmatrix} 1 & 2 \\ 8 & 14 \end{bmatrix}$$

Solving a System on a Graphing Calculator

With this experience, you should be able to use your graphing calculator to solve the following system:

$$3.1x + 2.7y - 4.9z = 5.3$$
$$6.9x + 4.2y + 3.7z = 9.2$$
$$4.1x - 1.7y + 9.3z = 7.7$$

This system can be rewritten as the following matrix:

$$\begin{bmatrix} 3.1 & 2.7 & -4.9 & 5.3 \\ 6.9 & 4.2 & 3.7 & 9.2 \\ 4.1 & -1.7 & 9.3 & 7.7 \end{bmatrix}$$

Enter this matrix into your calculator. Check to make sure that you made no errors by viewing the matrix. (Use the $\boxed{\rightarrow}$ button to view the right end of the matrix.)

In order to pivot on 3.1, select the row operation "*Row(" [TI-85/86: "multR("]. Use that command to divide row 1 by 3.1 (i.e., multiply it by 1/3.1).

After performing this or any row operation, *you must store the resulting matrix or it will be lost!* To store the result as matrix *B:*

TI-82/83:	Type $\boxed{\text{STO}\rightarrow}$ $\boxed{\text{MATRX}}$ 2 $\boxed{\text{ENTER}}$
TI-85/86:	Type $\boxed{\text{STO}\rightarrow}$ $\boxed{\text{B}}$ $\boxed{\text{ENTER}}$

Store the result of the next row operation as matrix *C,* and then alternate between *B* and *C* in storing later results. This keeps the previous matrix available, in case you made any errors. It also keeps the original matrix available, as matrix *A.*

It is important to maintain accuracy by not rounding off until the end of the problem. For example, partway through solving the above system, one obtains the following matrix:

```
B
[[1 .870967741935    -...
 [0 -1.80967741935  1...
 [0 -5.27096774194  1...
```

It would be tempting to divide row 2 by -1.8096; unfortunately, doing so results in an inaccurate final solution. Instead, divide row 2 by -1.80967741935; this maintains as much accuracy as possible.

A similar issue arises a little further along in solving the above system, when one obtains the following matrix:

```
B
[[1 -1.87E-12       5...
 [0 1               -...
 [0 -5.27096774194  1...
```

The "$-1.87E - 12$" in row 1, column 2 is the calculator's version of -1.87×10^{-12}. This number is not the desired 0; however, it is awfully close to 0. This discrepancy occurs because the calculator has of necessity rounded off the results of various previous calculations; it is unavoidable.

Reduced Row Echelon Form

The TI-83, TI-85, and TI-86 have a "rref" command, which automatically computes the reduced row echelon form of a matrix. To compute the reduced row echelon form of matrix A, make the screen read "rref [A]" on a TI-83, or "rrefA" on a TI-85/86, by typing:

TI-83:	MATRX , scroll to "MATH" and select "rref("	MATRX 1 ENTER
TI-85/86:	2nd MATRX , select "OPS" (i.e., F4) and then "rref " (i.e., F5)	ALPHA A ENTER
	This generates "rref(" or "rref"	*This generates "[A]" or "A"*

EXERCISES

52. Use the course software or graphing calculator to finish solving the system discussed in this section. Check your answer with your calculator; it is not given in the back of the book. (Your answer should check, but not perfectly, since the program rounded off its calculations.)

Use the course software or graphing calculator to solve the systems in Exercises 53–64. Check your solutions with your calculator; they are not given in the back of the book. Your answer should consist of a list of the row operations used in the order in which you used them, in addition to the final solution of the problem.

53. $4.6x - 7.2y + 9.8z = 16.36$
$3.2x + 6.2y + 5.3z = 105.7$
$9.5x - 3.6y - 2.5z = 5.58$

54. $5.1x - 3.2y + 9.8z = 15$
$7.3x - 4.6y - 1.2z = 5$
$8.0x + y + 2z = 14$

55. $18x - 22y + 53z = 9$
$-3x + 28y - 10z = 0$
$5x + y - 10z = 33$

56. $x = 2y - z$
$3y + 2z - 15x = 0$
$9x + 2y + 7z = 29$

57. $y - z = 20x$
$8x + 2y - 5z = 19$
$37x + 22y - 55z = 39$

58. $1.3x + 5.3y - 8.9z + 5.2w = 2.8$
$4.7x - 5.5y - 3.8z - 7.3w = 5.0$
$5.3x - 1.0y - 3.3z + 8.9w = 8.3$
$7.4x - 3.2y + 9.9z + 5.7w = 82.9$

59. $38x + 39y + 18z + 32w = 449$
$-57x + 48y + 29z - 12w = 39$
$-38x + 30y - 94z + 42w = 93$
$42x + 15y - 34z + 12w = 38$

60. $3.9x_1 + 4.9x_2 - 3.9x_3 - 4.2x_4 = 4.2$
$4.7x_1 + 3.7x_2 + 3.8x_3 + 9.7x_4 = 1.1$
$7.0x_1 - 7.2x_2 + 3.9x_3 + 3.6x_4 = 5$
$8.5x_1 + 6.3x_2 - 9.0x_3 - 3.7x_4 = 0$

61. $63x_1 + 95x_2 - 36x_3 - 99x_4 = 148$
$74x_1 - 77x_2 + 38x_3 + 96x_4 = 95$
$12x_1 - 6x_2 + x_3 = 5$
$84x_1 + 96x_2 - 46x_3 - 145x_4 = 80$

62. $6.3x_1 + 5.9x_2 - 3.9x_3 - 4.7x_4 + 9.1x_5 = 71.8$
$6.4x_1 + 9.2x_2 + 5.1x_3 + 2.2x_4 - 7.6x_5 = 81.1$
$3.4x_1 - 7.0x_2 + 2.9x_3 + 3.5x_4 + 4.2x_5 = 15$
$5.7x_1 + 3.6x_2 - 9.0x_3 - 2.4x_4 + 3.3x_5 = 100$
$3.6x_1 + 4.3x_2 - 5.7x_3 - 6.4x_4 - 2.8x_5 = 53$

63. $63x_1 + 85x_2 - 26x_3 - 89x_4 + 72x_5 = 148$
$32x_1 - 54x_2 + 95x_3 + 146x_4 - 32x_5 = 95$
$12x_1 - 6x_2 + x_3 + 12x_4 = 5$
$84x_1 + 96x_2 - 46x_3 - 145x_4 = 80$
$472x_1 - 59x_2 + 98x_3 + 16x_4 - 72x_5 = 83$

64. $6.3x_1 + 5.9x_2 - 3.9x_3 - 4.7x_4 + 9.1x_5 = 71.8$
$6.4x_1 + 9.2x_2 + 5.1x_3 + 2.2x_4 - 7.6x_5 = 81.1$
$3.4x_1 - 7.0x_2 + 2.9x_3 + 3.5x_4 + 4.2x_5 = 15$
$5.7x_1 + 3.6x_2 - 9.0x_3 - 2.4x_4 + 3.3x_5 = 100$
$3.6x_1 + 4.3x_2 - 5.7x_3 - 6.4x_4 - 2.8x_5 = 53$

2.2

MORE ON THE GAUSS-JORDAN METHOD

Applications

EXAMPLE 1 A craftsman produces two products: coffee tables and end tables. Production of one coffee table requires 6 hours of his labor, and the materials cost him $200. Production of one end table requires 5 hours of labor, and the materials cost $100. How many coffee tables and end tables will the craftsman make in one week if he works 40 hours and spends $1000 on materials?

Solution   Our two unknowns are the number of coffee tables made and the number of end tables made. We will let x and y, respectively, measure these two quantities.

We have two different sets of information: one about the craftsman's time, and the other about his money. Each of these sets yields an equation.

The Money Equation:

money spent = $1000

(coffee table money)   +   (end table money)   = 1000

$\left(\begin{array}{c}\$200 \text{ per} \\ \text{coffee table}\end{array}\right) \cdot \left(\begin{array}{c}\text{number of} \\ \text{coffee tables}\end{array}\right)$ + $\left(\begin{array}{c}\$100 \text{ per} \\ \text{end table}\end{array}\right) \cdot \left(\begin{array}{c}\text{number of} \\ \text{end tables}\end{array}\right)$ = 1000

$\qquad\quad$ 200 $\qquad\qquad\quad$ x $\qquad$ + $\qquad$ 100 $\qquad\qquad\quad$ y $\qquad$ = 1000

The Time Equation:

hours worked = 40

(coffee table hours)   +   (end table hours)   = 40

$\left(\begin{array}{c}6 \text{ hours per} \\ \text{coffee table}\end{array}\right) \cdot \left(\begin{array}{c}\text{number of} \\ \text{coffee tables}\end{array}\right)$ + $\left(\begin{array}{c}5 \text{ hours per} \\ \text{end table}\end{array}\right) \cdot \left(\begin{array}{c}\text{number of} \\ \text{end tables}\end{array}\right)$ = 40

$\qquad\quad$ 6 $\qquad\qquad\quad$ x $\qquad$ + $\qquad$ 5 $\qquad\qquad\quad$ y $\qquad$ = 40

Our two equations:

$$200x + 100y = 1000 \qquad \text{The money equation}$$
$$6x + 5y = 40 \qquad\qquad \text{The time equation}$$

form the following matrix:

$$\begin{bmatrix} 200 & 100 & 1000 \\ 6 & 5 & 40 \end{bmatrix}$$

Pivot on 200
$$\begin{bmatrix} 200 & 100 & 1000 \\ 6 & 5 & 40 \end{bmatrix}$$

R1 ÷ 200
$$\begin{bmatrix} 1 & 0.5 & 5 \\ 6 & 5 & 40 \end{bmatrix}$$

−6 R1 + R2 : R2
$$\begin{bmatrix} 1 & 0.5 & 5 \\ 0 & 2 & 10 \end{bmatrix}$$

Pivot on 2
$$\begin{bmatrix} 1 & 0.5 & 5 \\ 0 & 2 & 10 \end{bmatrix}$$

R2 ÷ 2
$$\begin{bmatrix} 1 & 0.5 & 5 \\ 0 & 1 & 5 \end{bmatrix}$$

R1 − 0.5 R2 : R1
$$\begin{bmatrix} 1 & 0 & 2.5 \\ 0 & 1 & 5 \end{bmatrix}$$

The solution is (2.5, 5). This means that the craftsman will make $2\frac{1}{2}$ coffee tables and 5 end tables if he works exactly 40 hours and spends exactly $1000 on materials.

✔ Check your work by substituting your solution into the original equations:

$200x + 100y = 200 \cdot 2.5 + 100 \cdot 5 = 500 + 500 = 1000$

$6x + 5y = 6 \cdot 2.5 + 5 \cdot 5 = 15 + 25 = 40$

Systems with No Solutions

EXAMPLE 2   The matrix

$$\begin{bmatrix} 1 & 0 & 5 & 0 \\ 0 & 1 & 4 & 0 \\ 0 & 0 & 0 & 1 \end{bmatrix}$$

is the result of using the Gauss-Jordan method to solve a system of three equations in three unknowns, x, y, and z. Rewrite this matrix as a system of equations and solve that system.

Solution   Notice that the normal Gauss-Jordan procedure could not continue because there is no way to make the third column into

$$\begin{bmatrix} 0 \\ 0 \\ 1 \end{bmatrix}$$

The matrix corresponds to the system

$1x + 0y + 5z = 0$
$0x + 1y + 4z = 0$
$0x + 0y + 0z = 1$

The third equation ($0x + 0y + 0z = 1$) is false for any ordered triple (x, y, z); there is nothing that we can substitute for (x, y, z) that will make the left side add up to 1. Thus, the system has no solutions.

Systems with More Than One Solution

EXAMPLE 3   The matrix

$$\begin{bmatrix} 1 & 0 & 7 & 1 \\ 0 & 1 & 3 & 2 \\ 0 & 0 & 0 & 0 \end{bmatrix}$$

is the result of using the Gauss-Jordan method to solve a system of three equations in three unknowns, *x*, *y*, and *z*. Rewrite this matrix as a system of equations and solve that system.

Solution Again, the normal Gauss-Jordan procedure could not continue because there is no way to make the third column into

$$\begin{bmatrix} 0 \\ 0 \\ 1 \end{bmatrix}$$

The matrix corresponds to the system

$$1x + 0y + 7z = 1$$
$$0x + 1y + 3z = 2$$
$$0x + 0y + 0z = 0$$

The third equation ($0x + 0y + 0z = 0$) is true for an infinite number of ordered triples (x, y, z); the left side will always add up to 0, regardless of what we substitute for (x, y, z). Thus, the system has an infinite number of solutions.

We can obtain the solutions to the system if we solve the first equation for *x* and the second equation for *y*.

$$1x + 0y + 7z = 1 \rightarrow x = 1 - 7z$$
$$0x + 1y + 3z = 2 \rightarrow y = 2 - 3z$$

This means that any ordered triple (x, y, z), where $x = 1 - 7z$ and $y = 2 - 3z$, is a solution to the system. In other words, all ordered triples of the form $(1 - 7z, 2 - 3z, z)$ are solutions to the system. This ordered triple $(1 - 7z, 2 - 3z, z)$ is the general solution to the system, and the variable *z* is the parameter. Replacing *z* with a number results in a particular solution to the system. Some particular solutions to this system are:

$z = 0$	$z = 1$	$z = 2$
$(1 - 7z, 2 - 3z, z)$	$(1 - 7z, 2 - 3z, z)$	$(1 - 7z, 2 - 3z, z)$
$= (1 - 7 \cdot 0, 2 - 3 \cdot 0, 0)$	$= (1 - 7 \cdot 1, 2 - 3 \cdot 1, 1)$	$= (1 - 7 \cdot 2, 2 - 3 \cdot 2, 2)$
$= (1, 2, 0)$	$= (-6, -1, 1)$	$= (-13, -4, 2)$

2.2

EXERCISES

Use the Gauss-Jordan method to solve Exercises 1–4.

1. The Wet Pleasure Kayak Co. makes two types of kayaks: the river model and the oceangoing model. Each river model requires 40 minutes in the cutting department and 20 minutes in the assembly department, and each oceangoing model requires 50 minutes in the cutting department and 30 minutes in the assembly department. The cutting department has a maximum of 460 hours available each week, and the assembly department has 240 hours available each week. How many kayaks should be produced each week for the plant to operate at full capacity?

2. The Eternal Slumber Casket Co. makes two types of caskets: the luxury model and the pine box. Each luxury model requires 120 minutes in the carpentry department and 180 minutes in the finishing department, and each pine box requires 30 minutes in the carpentry department and 30 minutes in the finishing department. The carpentry department has a maximum of 22 hours available each week, and the finishing department has 32 hours available each week. How many caskets should be produced each week for the plant to operate at full capacity?

3. A hospital patient is on a strict diet that is to include exactly 740 calories and 28 grams of fat. The hospital dietician plans to accomplish this by serving the patient only beef liver and Velveeta surprise. Each ounce of beef liver has 40 calories and 1.4 grams of fat, and each cup of Velveeta surprise has 340 calories and 14 grams of fat. What will the patient be served each day?

4. Dogs at a kennel are to be fed exactly 15 IUs of vitamin E and 60 mg of vitamin C each day. Each Pup-O-Vit tablet contains 5 IUs of vitamin E and 20 mg of vitamin C. Each Dogamins tablet contains 10 IUs of vitamin E and 40 mg of vitamin C. How many tablets should a dog be fed each day?

5. A farmer can buy three different types of fertilizer, each of which contains different amounts of nitrogen, phosphorus, and potassium. A barrel of Green Beauty contains 50 lb of nitrogen, 30 lb of phosphorus, and 20 lb of potassium. A barrel of NoMoorMan's contains 30 lb of nitrogen, 30 lb of phosphorus, and 30 lb of potassium. A barrel of Purismal contains 50 lb of phosphorus and 50 lb of potassium. Soil tests indicate that a certain field needs 900 lb of nitrogen, 800 lb of phosphorus, and 700 lb of potassium. How many barrels of each type of fertilizer should the farmer mix together to supply the necessary nutrients to the field?

6. A farmer can buy three different types of fertilizer, each of which contains different amounts of nitrogen, phosphorus, and potassium. Green Beauty is 10% nitrogen, 8% phosphorus, and 6% potassium. NoMoorMan's is 10% nitrogen, 10% phosphorus, and 10% potassium. Purismal is 10% phosphorus and 10% potassium. Soil tests indicate that a certain field needs 1370 lb of nitrogen, 1270 lb of phosphorus, and 1060 lb of potassium. How many pounds of each type of fertilizer should the farmer mix together to supply the necessary nutrients to the field?

In Exercises 7–10, the given matrix is the result of using the Gauss-Jordan method to solve a system of three equations in three unknowns, x, y, and z. Rewrite this matrix as a system of equations and solve that system. If there is more than one solution, give the general solution and three particular solutions.

7. $\begin{bmatrix} 1 & 0 & 6 & 2 \\ 0 & 1 & 4 & 7 \\ 0 & 0 & 0 & 1 \end{bmatrix}$

8. $\begin{bmatrix} 1 & 0 & 3 & 9 \\ 0 & 1 & 2 & 6 \\ 0 & 0 & 0 & 0 \end{bmatrix}$

9. $\begin{bmatrix} 1 & 0 & 5 & -3 \\ 0 & 1 & -7 & 8 \\ 0 & 0 & 0 & 0 \end{bmatrix}$

10. $\begin{bmatrix} 1 & 0 & 9 & -2 \\ 0 & 1 & -3 & -8 \\ 0 & 0 & 0 & 15 \end{bmatrix}$

Use the Gauss-Jordan method to solve Exercises 11–14. If there is more than one solution, give the general solution and three particular solutions. Check your solutions by substituting them back in. (Answers are not given at the back of the book.)

11. $3x - 2y + 7z = -22$
    $4x + 5y - 2z = 20$
    $10x + y + 12z = -24$

12. $5x - 6y + 12z = 18$
    $2x + y + z = 9$
    $16x - 9y + 27z = 52$

13. $8x + 2y - z = 21$
    $3x + 2y + 4z = 10$
    $-7x + 2y + 14z = 20$

14. $x - y - z = 12$
    $x + y + z = 2$
    $2x - 4y - 4z = 34$

15. The average rush-hour traffic flow for 4 one-way streets is shown in Figure 2.9. A Street, for example, is a north-to-south one-way street; a traffic meter has shown that an average of 2000 cars per hour approach the intersection of A Street and

FIGURE 2.9

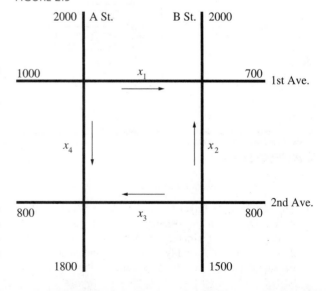

1st Avenue via A Street. City planners are analyzing the rush-hour traffic flow in this area. They wish to determine the traffic flow through the remaining portions of the 4 illustrated streets, but they need to minimize the number of traffic meters they use.

a. The number of cars approaching an intersection must equal the number departing that intersection. For example, 3000 cars approach the intersection of A Street and 1st Avenue (1000 on 1st Ave. and 2000 on A St.), and $x_1 + x_4$ depart that intersection (x_1 on 1st Ave. and x_4 on A St.), so $x_1 + x_4 = 3000$. Find similar equations for the other three intersections.

b. Solve the system of equations found in part (a).

c. Are any more traffic meters needed? If so, how many? Why?

d. If a new traffic meter measures x_4 as 1900 cars per hour, determine the traffic flow through the remaining portions of the other three streets.

16. Refer to Exercise 15. An accident stops all traffic on 2nd Avenue between A and B Streets. Assume that all of the cars that would have gone west on 2nd Avenue now attempt to proceed two blocks north on B Street and turn left.

a. If the streets are all capable of absorbing the excess flow, which of the previously measured flow rates [including that in part (d)] could reasonably be assumed to remain unchanged? Which flow rates would have to change, and what predictions could be made about them?

b. If B Street can handle at most 2500 cars per hour, where would an accident have the most profound effect?

The systems in Exercises 17–20 have more equations than unknowns. Use the Gauss-Jordan method to solve the given system.

HINT: Temporarily ignore any one of the given equations and solve the resulting system. Then substitute the resulting solution into the ignored equation.

17. $3x - 5y + z = 12$
    $2x + y + z = 3$
    $5x - 4y + z = 0$
    $x + y + z = 4$

18. $x + 2y - 3z = 4$
    $5x + 2y + z = 12$
    $3x - y = 7$
    $4x + z = 12$

19. $x + y = 4$
    $3x + z = 5$
    $2x - y = 8$
    $x + 2y + 3z = -17$

20. $5x - 4y + 2z = 8$
    $3x + 2y + z = 1$
    $4x - y - z = 12$
    $8x + 10y + 2z = 18$

If a system has fewer equations than unknowns, it may have an infinite number of solutions. In Exercises 21–24, use the Gauss-Jordan method to find the general solution and three particular solutions. Check your solutions by substituting them back in. (Answers are not given at the back of the book.)

21. $x + y + z = 3$
    $2x - y + z = 2$

22. $3x + y + 5z = 12$
    $4x - 12y = 9$

23. $6x + 10y = 22$
    $3x - 2z = 5$

24. $x + 2y + 3z = 5$
    $4x - 3y - z = 2$

2.3

LINEAR INEQUALITIES

In Example 1 of Section 2.2 we studied a craftsman who makes coffee tables and end tables. We determined the number of tables he needs to make each week in order to work exactly 40 hours and spend exactly $1000 on materials each week. However, in determining how best to allocate his resources of time and money, the craftsman's goal is not to use up all the resources; rather, it is to maximize profit. In some cases, profit is greater when resources are not all exhausted. Perhaps the craftsman profits more from selling a coffee table than from selling an end table. If so, and if more resources were put into coffee tables and less into end tables, the craftsman might increase his profit without exhausting all resources.

Thus, in determining how to allocate his resources, we should require that the craftsman work *no more than* 40 hours and spend *at most* $1000 on materials each week. These restrictions, and others that involve language like "no more than" and "at most," are described mathematically with inequalities. To analyze the craftsman's restricted resources and the effect of these restrictions on profit, we must be able to graph the inequalities that describe the restrictions.

Recall that a linear equation in two variables x and y is an equation that can be written in the form $ax + by = c$, where a, b, and c are constants. Such an equation is referred to as a *linear* equation because its graph is a line. A **linear inequality** in two variables x and y is an inequality that can be written in the form $ax + by < c$ (or with $>$, $\leq$, or $\geq$ instead of $<$). In other words, a linear inequality is the result of replacing a linear equation's equals symbol with an inequality symbol.

We can graph a linear inequality by solving the inequality for y, graphing the line described by the associated equation, and then shading the region to one side of that line.

EXAMPLE 1 Graph the linear inequality $2x + y \le 6$.

Solution **Step 1** *Solve the linear inequality for y.*

$$2x + y \le 6 \rightarrow y \le -2x + 6$$

Step 2 *Graph the line.* The equation associated with the inequality is $y = -2x + 6$, which is a line with slope $m = -2$ and y-intercept $b = 6$. Because "$=$" is a part of "$\le$," any point on the line $y = -2x + 6$ must also be a point on the graph of the inequality $y \le -2x + 6$. We show this by using a solid line, as in Figure 2.10. (If our inequality were $y < -2x + 6$, then a point on the line would *not* be a point on the graph of the inequality. We would show this by using a dashed line.)

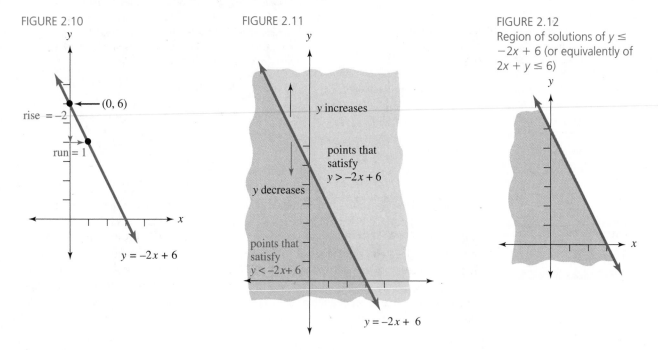

FIGURE 2.10

$y = -2x + 6$

FIGURE 2.11

$y = -2x + 6$

FIGURE 2.12
Region of solutions of $y \le -2x + 6$ (or equivalently of $2x + y \le 6$)

Step 3 *Shade in one side of the line.* Two types of points satisfy the inequality $y \le -2x + 6$: points that satisfy $y = -2x + 6$ and points that satisfy $y < -2x + 6$.

Points that satisfy $y = -2x + 6$ were graphed in Step 2, when we graphed the line. Points that satisfy y *is less than* $-2x + 6$ are the points *below* the line, because values of y decrease if we move down and increase if we move up, as shown in Figure 2.11.

Thus, to graph the inequality $y \le -2x + 6$, we make the line solid and shade in the region *below* the line, as shown in Figure 2.12. The solution of $y \le -2x + 6$ is the set of all points on or below the line; any point on or below the line will successfully substitute into the inequality, and any point above the line will not. This region is called the **region of solutions** of the inequality.   •

EXAMPLE 2   Graph $3x - 2y < 12$.

Solution   **Step 1**   *Solve the linear inequality for y.*

$$3x - 2y < 12$$
$$-2y < -3x + 12$$
$$\frac{-2y}{-2} > \frac{-3x + 12}{-2} \qquad \text{Multiplying or dividing by a negative reverses the direction of an inequality}$$
$$y > \frac{-3x}{-2} + \frac{12}{-2} \qquad \text{Distributing } -2$$
$$y > \frac{3}{2}x - 6$$

FIGURE 2.13
Region of solutions of
$y > \frac{3}{2}x - 6$ (or equivalently
of $3x - 2y < 12$)

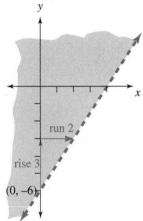

Step 2   *Graph the line.* The associated equation is $y = \frac{3}{2}x - 6$, which is a line with slope $m = \frac{3}{2}$ and y-intercept $b = -6$. Because "=" is *not* part of ">," a point on the line $y = \frac{3}{2}x - 6$ is *not* a point on the graph of the inequality $y > \frac{3}{2}x - 6$. We show this by using a dashed line.

Step 3   *Shade in one side of the line.* Because values of *y* increase if we move upward and because we want to graph where *y is greater* than $\frac{3}{2}x - 6$, we shade in the region above the dashed line. The region of solutions of the inequality is the set of all points above (but not on) the line, as shown in Figure 2.13.   •

EXAMPLE 3   Graph $x \geq 3$.

Solution   **Step 1**   *Solve the linear inequality for y.* This can't be done, since there is no *y* in the inequality $x \geq 3$.

FIGURE 2.14
Region of solutions of $x \geq 3$

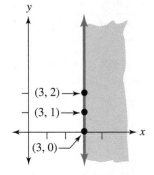

Step 2   *Graph the line.* The association equation is $x = 3$. Because any point with an x-coordinate of 3 satisfies this equation, the graph of $x = 3$ is a vertical line through $(3, 0)$ and $(3, 1)$ and $(3, 2)$. Also, "=" is part of "≥," so any point on the line $x = 3$ is a point on the graph of the inequality $x \geq 3$. We show this by using a solid line.

Step 3   *Shade in one side of the line.* Values of *x* increase if we move to the right and decrease if we move to the left. The points on the line are the points where *x* equals 3, and the points to the right of the line are the points where *x* is greater than 3. Thus, the region of solutions of the inequality is the set of all points on or to the right of the line, as shown in Figure 2.14.   •

Graphing the Region of Solutions of a Linear Inequality

1. *Solve the linear inequality for y.* This puts the inequality in slope-intercept form. If the inequality has no *y*, then solve the inequality for *x*.
2. *Graph the line.* The line is described by the equation associated with the inequality. If the inequality is $\leq$ or $\geq$, then the line is part of the region of solutions; indicate this with a solid line. If the inequality is $<$ or $>$, then the line is not part of the region of solutions; indicate this with a dashed line.
3. *Shade in one side of the line.*

   y increases as you move up, and decreases as you move down

   x increases as you move to the right, and decreases as you move to the left

---

Systems of Linear Inequalities

Frequently, a problem involves more than one restriction and therefore more than one inequality. For example, the craftsperson has two restrictions (and thus two inequalities): to work no more than 40 hours and spend no more than $1000 on materials each week. A **system of linear inequalities** is a set of more than one linear inequality. The **region of solutions** of a system of linear inequalities is the set of all points that simultaneously satisfy each inequality in the system.

To graph the region of solutions of a system of linear inequalities, we graph each inequality on the same axes and shade in the intersection of their solutions.

EXAMPLE 4   Graph the following system of inequalities:

$$x + y \geq 3$$
$$-x + 2y \leq 0$$

Solution   With the aid of the following chart, we can graph the two inequalities on the same axes, as shown in Figure 2.15.

FIGURE 2.15
$x + y \geq 3$ and $-x + 2y \leq 0$
graphed on the same axes

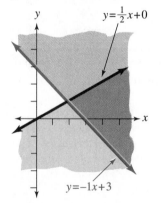

Original Inequality	Slope-Intercept Form	Associated Equation	The Graph of the Inequality
$x + y \geq 3$	$y \geq -x + 3$	$y = -1x + 3$	All points on or above the line with slope -1 and intercept 3
$-x + 2y \leq 0$	$y \leq \dfrac{1}{2}x$	$y = \dfrac{1}{2}x + 0$	All points on or below the line with slope $\frac{1}{2}$ and intercept 0

The region of solutions of the system is the set of all points that satisfy both the first *and* the second inequality; that is, it is the intersection of the graphs of the two inequalities, shown in Figure 2.16.

FIGURE 2.16
Region of solutions of the system
$x + y \geq 3$
$-x + 2y \leq 0$

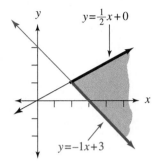

EXAMPLE 5   Graph the following system of inequalities:

$$2x + y \leq 8$$
$$x + 2y \leq 10$$
$$x \geq 0$$
$$y \geq 0$$

Solution   With the aid of the following chart, we can graph the two inequalities on the same axes.

Original Inequality	Slope-Intercept Form	Associated Equation	The Graph of the Inequality
$2x + y \leq 8$	$y \leq -2x + 8$	$y = -2x + 8$	All points on or below the line with slope -2 and intercept 8
$x + 2y \leq 10$	$y \leq \dfrac{-1}{2}x + 5$	$y = \dfrac{-1}{2}x + 5$	All points on or below the line with slope $\dfrac{-1}{2}$ and intercept 5
$x \geq 0$	(not applicable)	$x = 0$	All points on or to the right of the y-axis
$y \geq 0$	$y \geq 0$	$y = 0$	All points on or above the x-axis

The inequalities $x \geq 0$ and $y \geq 0$ appear frequently in linear programming problems. They tell us that our graph is in the first quadrant.

The region of solutions of the system is the intersection of the graph of each individual inequality, shown in Figure 2.17 on page 96.

FIGURE 2.17
Region of solutions of the
system
$2x + y \leq 8$
$x + 2y \leq 10$
$x \geq 0$
$y \geq 0$

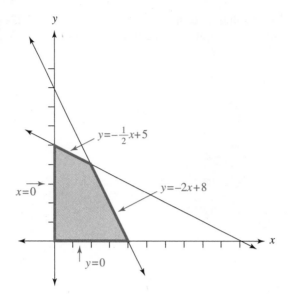

The region of solutions in Example 4 is different from that of Example 5 in that the former is not totally enclosed. For that reason, it is called an **unbounded region.** Regions that are totally enclosed, like that in Example 5, are called **bounded regions.** See Figure 2.18. When we graph a system of inequalities as part of a linear programming problem, we must analyze unbounded regions differently than we do bounded regions.

FIGURE 2.18

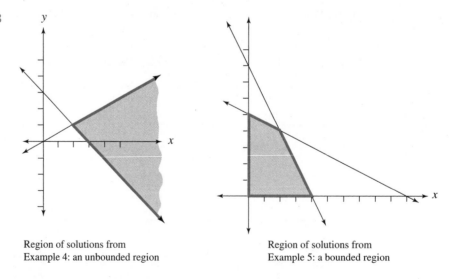

Region of solutions from
Example 4: an unbounded region

Region of solutions from
Example 5: a bounded region

Finding Corner Points

Our work in linear programming will involve graphing the region of solutions of a system of linear inequalities, as we did in Example 5. It will also involve finding the region's corner points. A **corner point** is a point that is at a corner of the region of solutions.

EXAMPLE 6   Find the corner points of the region of solutions in Example 5.

   Solution   The region has four corner points, as shown in Figure 2.19.

FIGURE 2.19

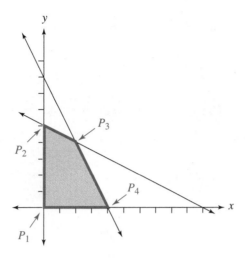

- P_1, at the origin
- P_2, where $y = -\frac{1}{2}x + 5$ intersects $x = 0$ (the y-axis)
- P_3, where $y = -\frac{1}{2}x + 5$ intersects $y = -2x + 8$
- P_4, where $y = -2x + 8$ intersects $y = 0$ (the x-axis)

There is no work to do to find points P_1 and P_2. Point P_1 is clearly $(0, 0)$. Point P_2 is the y-intercept of $y = -\frac{1}{2}x + 5$; that y-intercept is $b = 5$, so P_2 is $(0, 5)$.
We can find P_3 by solving the system of equations

$$y = -\frac{1}{2}x + 5$$

$$y = -2x + 8$$

We can solve this system with the elimination method or the Gauss-Jordan method, from Section 2.0, or with the substitution method you might have encountered in intermediate algebra. To proceed with the elimination method, multiply the second equation by -1 and add the results.

$$y = -\frac{1}{2}x + 5$$

$$\underline{+\ -y = 2x - 8} \qquad \text{-1 times the second equation}$$

$$0 = \frac{3}{2}x - 3$$

$$\frac{3}{2}x = 3$$

$$\frac{2}{3} \cdot \frac{3}{2}x = \frac{2}{3} \cdot 3$$

$$x = 2$$

We can find y by substituting $x = 2$ into either of the original equations.

$$y = -2x + 8$$

$$y = -2 \cdot 2 + 8 = 4$$

Thus, P_3 is (2, 4).

We can find P_4 by solving the system of equations

$$y = -2x + 8$$

$$y = 0 \qquad \qquad \text{The equation of the } x\text{-axis}$$

To proceed with the elimination method, multiply the second equation by -1 and add the results.

$$
\begin{array}{ll}
y = -2x + 8 & \\
\underline{+ \ -y = 0} & \quad -1 \text{ times the second equation} \\
0 = -2x + 8 & \\
2x = 8 & \\
x = 4 &
\end{array}
$$

Since P_4 is on the x-axis, its y-coordinate must be 0. Thus, P_4 is (4, 0). The region of solutions' four corner points therefore are

- P_1 at (0, 0)
- P_2 at (0, 5)
- P_3 at (2, 4)
- P_4 at (4, 0)

2.3

EXERCISES

In Exercises 1–8, graph the region of solutions of the given linear inequality.

1. $3x + y < 4$

2. $8x + y > 2$

3. $4x - 3y \le 9$

4. $5x - 2y \ge 6$

5. $x \ge 4$

6. $x \le -3$

7. $y \le -4$

8. $y \ge -2$

In Exercises 9–21:

a. Graph the region of solutions of the given system of inequalities.

b. Determine whether the region of solutions is bounded or unbounded.

c. Find all of the region's corner points.

9. $y > 2x + 1$
$y \le -x + 4$

10. $y < -2x + 6$
$y \ge -x + 7$

11. $2x + 3y < 17$
$3x - y \ge -2$

12. $5x - y \ge 7$
$2x - 3y < -5$

13. $x + 2y \le 4$
$3x - 2y \le -12$
$x - y < -7$

14. $x - y + 1 \ge 0$
$3x + 2y + 8 \ge 0$
$3x - y < 6$

15. $2x + 5y \le 70$
$5x + y \le 60$
$x \ge 0$
$y \ge 0$

16. $x + 20y \le 460$
$21x + y \le 861$
$x \ge 0$
$y \ge 0$

17. $15x + 22y \le 510$
$35x + 12y \le 600$
$x + y > 10$
$x \ge 0$
$y \ge 0$

18. $3x + 20y \le 2200$
$19x + 9y \le 2755$
$2x + y \le 120$
$x \ge 0$
$y \ge 0$

19. $0.50x + 1.30y \le 2.21$
$6x + y \le 9$
$0.7x + 0.6y < 3.00$
$x \ge 0$
$y \ge 0$

20. $3.70x + 0.30y \le 1.17$
$0.10x + 2.20y \le 0.47$
$x \ge 0$
$y \ge 0$

21. $x - 2y + 16 \ge 0$
$3x + y \le 30$
$x + y \le 14$
$x \ge 0$
$y \ge 0$

Answer the following question using complete sentences.

22. Why do we describe the solution of a system of linear inequalities with a graph, rather than a list of points?

GRAPHING LINEAR INEQUALITIES ON A GRAPHING CALCULATOR

A Texas Instruments graphing calculator can perform all of the specific tasks that are part of graphing the region of solutions of a system of linear inequalities: it can graph the lines associated with the system, shade in the appropriate sides of those lines (with the "Shade" command), and locate corner points. However, the "Shade" command is cumbersome to use—it is much easier to do the shading by hand and use the calculator to both graph the lines associated with the inequalities and find the corner points. If you wish to explore the use of the "Shade" command, consult your calculator's operating manual.

Graphing a Linear Inequality

EXAMPLE 7   Graph $3x - 2y < 12$.

Solution   **Step 1**   *Solve the linear inequality for y.*

$$3x - 2y < 12 \rightarrow -2y < -3x + 12 \rightarrow y > \frac{3}{2}x - 6$$

Step 2   *Graph the line.* The line associated with the inequality is $y = \frac{3}{2}x - 6$. Enter $3/2*x - 6$ for Y_1. (Be sure to include the "*." The calculator would interpret $3/2x - 6$ as $\frac{3}{2x} - 6$.) Be sure that no other equations are selected. Use the standard viewing window. The line's graph is shown in Figure 2.20 on page 100.

Step 3 *Copy the line's graph onto paper.* The inequality is a "> inequality" so use a dashed line.

Step 4 *By hand, shade in one side of the line.* To graph $y > \frac{3}{2}x - 6$, shade in the region above the line $y = \frac{3}{2}x - 6$. The region's graph is shown in Figure 2.20.

FIGURE 2.20

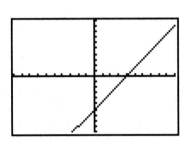

The graph of $y = \frac{3}{2}x - 6$

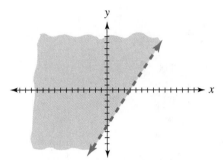

The graph of $y > \frac{3}{2}x - 6$

Graphing a System of Linear Inequalities

EXAMPLE 8 Graph the system of inequalities:

$$x + y \geq 3$$
$$-x + 2y \leq 0$$

Solution **Step 1** *Solve the linear inequalities for y.*

$$x + y \geq 3 \rightarrow y \geq -x + 3$$
$$-x + 2y \leq 0 \rightarrow y \leq \frac{1}{2}x$$

Step 2 *Graph the lines.* Enter $-x + 3$ for Y_1 and 1/2*x for Y_2. Be sure that no other equations are selected. Use the standard viewing window.

Step 3 *Copy the lines' graphs onto paper.* The inequalities are both "≤ or ≥ inequalities," so use solid lines.

Step 4 *By hand, shade in the region of solutions.* The region of solutions is above the line $Y_1 = -x + 3$, because the associated inequality is $y \geq -x + 3$, and y increases as you move up. The region of solutions is also below the line $Y_2 = \frac{1}{2}x$, because the associated inequality is $y \le \frac{1}{2}x$, and y decreases as you move down. The region of solutions is the region that is both above the line $Y_1 = -x + 3$ and below the line $Y_2 = \frac{1}{2}x$. It is shown in Figure 2.21.

FIGURE 2.21

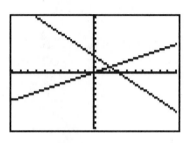

The graphs of $x + y = 3$
and $-x + 2y = 0$

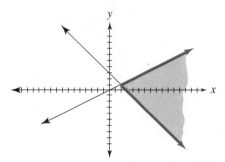

The graph of $x + y \ge 3$
and $-x + 2y \le 0$

**Nonstandard
Viewing Windows**

The previous examples were chosen so that the standard viewing window would be appropriate. In the following example, we must determine an appropriate viewing window.

EXAMPLE 9   Graph the system of inequalities:

$$2x + y \le 8$$
$$x + 2y \le 10$$
$$x \ge 0$$
$$y \ge 0$$

Solution $2x + y \le 8 \rightarrow y \le -2x + 8$

$x + 2y \le 10 \rightarrow 2y \le -x + 10 \rightarrow y \le -\frac{1}{2}x + 5$

$x \ge 0$ cannot be solved for y, and $y \ge 0$ is already solved for y.

Enter $-2x + 8$ for Y_1 and $-1/2*x + 5$ for Y_2. Be sure that no other equations are selected. The last pair of inequalities tell us that the region of solutions is in the first quadrant; if we set Xmin and Ymin equal to -1, we'll leave ourselves a little extra room. By inspecting the two lines' equations, we can tell that the largest y-intercept is 8, so we'll set Ymax equal to 8. Substituting 0 for y in each equation gives x-intercepts of $(4, 0)$ and $(10, 0)$, so we'll set Xmax equal to 10.

The region of solutions is below the line $Y_1 = -2x + 8$, because the associated inequality is $y \le -2x + 8$, and y decreases as you move down. The region of solutions is also below the line $Y_2 = -\frac{1}{2}x + 5$, because the associated inequality is $y \le -\frac{1}{2}x +$ 5. The region of soutions is the part of the first quadrant that is both on or below the line $Y_1 = -2x + 8$ and on or below the line $Y_2 = \frac{1}{2}x + 5$. It is shown in Figure 2.22.

FIGURE 2.22

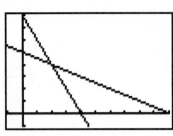

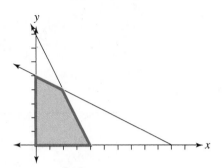

The graphs of $2x + y = 8$
and $x + 2y = 10$

The system's region of solutions

●

Finding Corner Points

After using a graphing calculator to graph a system's region of solutions, you can use the calculator to find the corner points as well. Use the method discussed in Section 1.0, Exercises 54 and 55, to find x-intercepts and the method discussed in Section 1.2 to find intersection points.

EXAMPLE 10   Find the corner points for the region graphed in Example 8.

Solution   This region has four corner points. One is clearly located at $(0, 0)$. A second is the y-intercept of the line $y = -\frac{1}{2}x + 5$; by inspecting the equation, we can tell that it is $(0, 5)$.

A third corner point is the intersection of the two lines; it can be found using the procedure discussed in Section 1.2.

TI-82/83:	Select "intersect" from the "CALC" menu
TI-85/86:	Select "MATH" from the "GRAPH" menu, and then "ISECT" from the "GRAPH MATH" menu

This third corner point is $(2, 4)$, as shown in Figure 2.23.

FIGURE 2.23

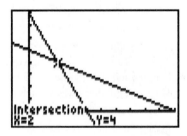

The fourth corner point is the x-intercept or root of the line $y = -2x + 8$; it can be found using the procedure discussed in Section 1.0.

TI-82/83:	Select "root" or "zero" from the "CALC" menu
TI-85/86:	Select "MATH" from the "GRAPH" menu, and then "ROOT" from the "GRAPH MATH" menu

This fourth corner point is (4, 0), as shown in Figure 2.24.

FIGURE 2.24

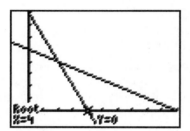

EXERCISES

In Exercises 23–30, use a graphing calculator to graph the region of solutions of the inequality given earlier in this section, if possible.

23. Exercise 1

24. Exercise 2

25. Exercise 3

26. Exercise 4

27. Exercise 5

28. Exercise 6

29. Exercise 7

30. Exercise 8

In Exercises 31–43, use a graphing calculator to (a) graph the region of solutions of the system of inequalities given earlier in this section and (b) find the corner points.

31. Exercise 9

32. Exercise 10

33. Exercise 11

34. Exercise 12

35. Exercise 13

36. Exercise 14

37. Exercise 15

38. Exercise 16

39. Exercise 17

40. Exercise 18

41. Exercise 19

42. Exercise 20

43. Exercise 21

2.4

THE GEOMETRY OF LINEAR PROGRAMMING

Linear programming is a method of solving problems that involve a quantity to be maximized or minimized when that quantity is subject to certain restrictions. Linear programming was invented in the 1940s by George Dantzig as a result of an air force research project concerned with computing the most efficient and economical way to distribute men, weapons, and supplies to the various fronts during World War II. The word *programming* in the name "linear programming" means creating a plan or procedure that solves a problem; it is not a reference to computer programming.

The following is a typical linear programming problem. As in our previous example of a craftsman, suppose a craftsman produces two products: coffee tables and end tables.

Historical Note

George Dantzig

1914–

George Bernard Dantzig's father was both a writer and a mathematician. He had hoped that his first son would be a writer and his second a mathematician, so he named his first son after the playwright and critic George Bernard Shaw and his second after Henri Poincaré, a famous French mathematician under whom the father had studied. As it

happened, both sons became mathematicians.

In 1939, George Dantzig was a graduate student at the University of California at Berkeley. One day he arrived late at a statistics class. He copied the two problems that were written on the blackboard, assuming that they were homework problems. About six weeks after turning the homework in, he was awakened early one Sunday morning by his excited professor, who wanted to send off Dantzig's work right away for publication. The two problems were not homework but famous unsolved problems in statistics. Their solutions became Dantzig's Ph.D. thesis in mathematics.

During World War II, Dantzig was hired by the U.S. Air Force to find practical ways to distribute men, weapons, and supplies to the various fronts. Shortly after the end of the

war, he became mathematics advisor to the U.S. Air Force Comptroller at the Pentagon, where he was responsible for finding a way to mechanize this planning process. The result was linear programming. The procedure was quickly applied to a wide variety of business, economic, and environmental topics. Tjalling Koopmans of the United States and Leonid Kantorovich of the Soviet Union received the 1975 Nobel Prize in economics for their use of linear programming in developing the theory of allocation of resources. Surprisingly, Dantzig himself was not honored. George Dantzig is now a mathematics professor at Stanford University.

Source: Donald J. Albers and Constance Reid, "An Interview with George B. Dantzig. The Father of Linear Programming," *College Mathematics Journal,* vol. 17, no. 4, September 1986.

How should a craftsman allocate his time and money in order to maximize profit?

Production of one coffee table requires 6 hours of his labor, and the materials cost him $200. Production of one end table requires 5 hours of labor, and the materials cost him $100. The craftsman wants to work no more than 40 hours each week, and his financial resources allow him to pay no more than $1000 for materials each week. If he can sell as many tables as he can make and if his profit is $240 per coffee table and $160 per end table, how many coffee tables and how many end tables should he make each week to maximize weekly profit?

Any linear programming problem has three features: *variables,* an *objective,* and *constraints.* In the problem above, the **variables** (or quantities that can vary) are the following:

- the number of coffee tables made each week
- the number of end tables made each week
- the number of hours the craftsman works each week

- the amount of money he spends on materials each week
- the weekly profit

The last three variables depend on the first two, so they are called the **dependent variables,** whereas the first two are called the **independent variables.**

The craftsman's objective is to maximize profit. The **objective function** is a function that mathematically describes the profit.

The **constraints** (or restrictions) are as follows:

- the craftsman's weekly hours ≤ 40
- the craftsman's weekly expenses $\leq \$1000$

The constraints form a system of inequalities. In order to analyze the effect of these constraints on the craftsman's profit, we must graph the system of inequalities. The resulting graph is called the **region of possible solutions,** because it contains all the points that could *possibly* solve the craftsman's problem.

Creating a Model

A **model** is a mathematical description of a real-world situation. In this section, we discuss how to model a linear programming problem (that is, how to translate it into mathematical terms), how to find and graph the region of possible solutions, and how to analyze the effect of the constraints on the objective and solve the problem.

EXAMPLE 1   Model the linear programming problem from the beginning of this section and graph the region of possible solutions. The problem is summarized as follows:

A craftsman produces two products: coffee tables and end tables. Production data are given in Figure 2.25. If the craftsman wants to work no more than 40 hours each week and if his financial resources allow him to pay no more than $1000 for materials each week, how many coffee tables and how many end tables should he make each week to maximize weekly profit?

FIGURE 2.25

	Labor (per table)	**Cost of Materials (per table)**	**Profit (per table)**
Coffee Tables	6 hours	$200	$240
End Tables	5 hours	$100	$160

Solution   **Step 1**   *List the independent variables.* We have already done this. If we call them x and y, the independent variables are

x = number of coffee tables made each week

y = number of end tables made each week

Step 2   *List the constraints and translate them into linear inequalities.* We have already determined that the constraints (or restrictions) are as follows:

the craftsman's weekly hours ≤ 40

the craftsman's weekly expenses $\leq \$1000$

We need to translate these constraints into linear inequalities. First, let's translate the time constraint:

hours ≤ 40

(coffee table hours) $\quad + \quad$ (end table hours) $\quad \leq 40$

$\left(\begin{matrix} 6 \text{ hours per} \\ \text{coffee table} \end{matrix}\right) \cdot \left(\begin{matrix} \text{number of} \\ \text{coffee tables} \end{matrix}\right) + \left(\begin{matrix} 5 \text{ hours per} \\ \text{end table} \end{matrix}\right) \cdot \left(\begin{matrix} \text{number of} \\ \text{end tables} \end{matrix}\right) \leq 40$

$\quad\quad 6 \quad\quad\quad\quad x \quad\quad + \quad\quad 5 \quad\quad\quad\quad y \quad\quad \leq 40$

Next we'll translate the money constraint:

money spent ≤ 1000

(coffee table money) $\quad + \quad$ (end table money) $\quad \leq 1000$

$\left(\begin{matrix} \$200 \text{ per} \\ \text{coffee table} \end{matrix}\right) \cdot \left(\begin{matrix} \text{number of} \\ \text{coffee tables} \end{matrix}\right) + \left(\begin{matrix} \$100 \text{ per} \\ \text{end table} \end{matrix}\right) \cdot \left(\begin{matrix} \text{number of} \\ \text{end tables} \end{matrix}\right) \leq 1000$

$\quad\quad 200 \quad\quad\quad\quad x \quad\quad + \quad\quad 100 \quad\quad\quad\quad y \quad\quad \leq 1000$

There are two more constraints. Both x and y count things (the number of tables), so neither can be negative; therefore,

$$x \geq 0 \quad \text{and} \quad y \geq 0$$

Our constraints are

$$6x + 5y \leq 40$$
$$200x + 100y \leq 1{,}000$$
$$x \geq 0 \quad \text{and} \quad y \geq 0$$

Step 3 *Find the objective and translate it into a linear equation.* The objective is to maximize profit. If we let $z = $ profit, we get

$z = $ (coffee table profit) + (end table profit)

$\quad = (\$240 \text{ per coffee table})(\text{number of coffee tables})$
$\quad\quad + (\$160 \text{ per end table})(\text{number of end tables})$

$\quad = \$240x + 160y$

This equation is our objective function.

Steps 1 through 3 yield the model, or mathematical description, of the problem:

Independent Variables

$x = $ number of coffee tables

$y = $ number of end tables

Constraints

$6x + 5y \leq 40$ $\quad\quad\quad\quad$ The time constraint

$200x + 100y \leq 1000$ $\quad\quad\quad$ The money constraint

$x \geq 0 \quad \text{and} \quad y \geq 0$

Objective Function

$$z = 240x + 160y \qquad \text{\textit{z} measures profit}$$

Step 4 *Graph the region of possible solutions.* With the aid of the following chart, we can graph the two inequalities on the same axes.

Original Inequality	Slope-Intercept Form	Associated Equation	The Graph of the Inequality
$6x + 5y \le 40$	$5y \le -6x + 40 \rightarrow$ $y \le -\dfrac{6}{5}x + 8$	$y = -\dfrac{6}{5}x + 8$	All points on or below the line with slope $-6/5$ and y-intercept 8
$200x + 100y \le 1000$	$100y \le -200x + 1000$ $\rightarrow$ $y \le -2x + 10$	$y = -2x + 10$	All points on or below the line with slope -2 and y intercept 10
$x \ge 0$	(not applicable)	$x = 0$	All points on or to the right of the y-axis
$y \ge 0$	$y \ge 0$	$y = 0$	All points on or above the x-axis

The last two constraints tell us that the region of possible solutions is in the first quadrant. The region of possible solutions is shown in Figure 2.26.

FIGURE 2.26

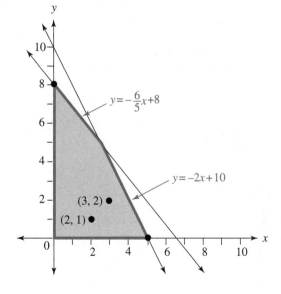

Analyzing the Model

The region of possible solutions graphed in Example 1 consists of all the points that satisfy the constraints, that is, all the points at which the craftsman's weekly hours are no more than 40 and his weekly expenses are no more than $1000. Let's arbitrarily select two points in the region, verify that the constraints are satisfied, and find the craftsman's profit at those points. By inspecting the graph in Figure 2.26, we can see that (2, 1) and (3, 2) are clearly in the region.

Recall that x measures the number of coffee tables to be made each week, y measures the number of end tables, and z measures weekly profit. The first constraint is that the time used, $6x + 5y$, be no more than 40 hours per week. The second constraint is that the money used, $200x + 100y$, be no more than $1000 each week. The weekly profit is $z = 240x + 160y$. Figure 2.27 gives the time used, the money spent, and the profit at our two points.

FIGURE 2.27

Point	(2, 1)	(3, 2)
Time Used	$6 \cdot 2 + 5 \cdot 1 = 17$	$6 \cdot 3 + 5 \cdot 2 = 28$
Money Spent	$200 \cdot 2 + 100 \cdot 1 = 500$	$200 \cdot 3 + 100 \cdot 2 = 800$
Profit	$240 \cdot 2 + 160 \cdot 1 = 640$	$240 \cdot 3 + 160 \cdot 2 = 1040$

The table in Figure 2.27 shows that if the craftsman makes two coffee tables and one end table each week, he will use 17 hours (of 40 available hours), spend $500 (of the $1000 available), and profit $640. If he makes three coffee tables and two end tables each week, he will use 28 hours, spend $800, and profit $1040. Each of these points represents a *possible* solution to the craftsman's problem, because each satisfies the time constraint and the money constraint. Neither represents the *actual* solution, because neither maximizes his profit—he has both money and time left over, so he should be able to increase his profit by building more tables. This is why the region in Figure 2.26 is called the region of *possible* solutions.

Common sense tells us that in order to maximize profit, our craftsman must make more tables. To make more tables means to increase the value of x and/or y, which implies that we should choose points on the boundary of the region of feasible solutions. There are quite a few points on the boundary—too many to find and substitute into the equation for profit. Fortunately, the **corner principle** comes to our rescue. (We discuss why the corner principle is true later in this section.)

Corner Principle

The maximum and minimum values of an objective function occur at corner points of the region of possible solutions if that region is bounded.

Our region is a bounded region, so the corner principle applies. The region has four corner points (see Figure 2.28):

- P_1, at the origin
- P_2, where $y = -\frac{6}{5}x + 8$ intersects the y-axis
- P_3, where $y = -\frac{6}{5}x + 8$ intersects $y = -2x + 10$
- P_4, where $y = -2x + 10$ intersects the x-axis

We can find each of these points by solving a system of equations. For example, P_3 can be found by solving the following system:

FIGURE 2.28

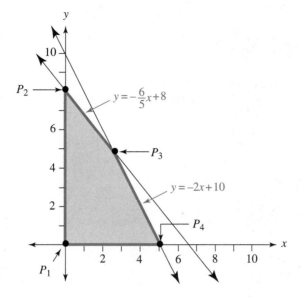

$$y = -\frac{6}{5}x + 8$$
$$y = -2x + 10$$

P_4 can be found by solving the following system:

$$y = -2x + 10$$
$$y = 0$$

The corner points, shown in Figure 2.28, are as follows:

$P_1\,(0, 0)$
$P_2\,(0, 8)$
$P_3\left(\frac{5}{2}, 5\right)$
$P_4\,(5, 0)$

Let's verify that the constraints are satisfied at each of these points and, more important, find the profit at each point.

Point	$P_1\,(0, 0)$	$P_2\,(0, 8)$	$P_3\left(\frac{5}{2}, 5\right)$	$P_4\,(5, 0)$
Time Used	$6 \cdot 0 + 5 \cdot 0 = 0$	$6 \cdot 0 + 5 \cdot 8 = 40$	$6 \cdot \frac{5}{2} + 5 \cdot 5 = 40$	$6 \cdot 5 + 5 \cdot 0 = 30$
Money Spent	$200 \cdot 0 + 100 \cdot 0 = 0$	$200 \cdot 0 + 100 \cdot 8 = 800$	$200 \cdot \frac{5}{2} + 100 \cdot 5 = 1000$	$200 \cdot 5 + 100 \cdot 0 = 1000$
$z =$ **Profit**	$240 \cdot 0 + 160 \cdot 0 = 0$	$240 \cdot 0 + 160 \cdot 8 = 1280$	$240 \cdot \frac{5}{2} + 160 \cdot 5 = 1400$	$240 \cdot 5 + 160 \cdot 0 = 1200$

At each corner point, each constraint is satisfied: the time used is at most 40 hours, and the money spent is at most $1000. The highest profit, $1400, occurs at $P_4(\frac{5}{2}, 5)$. The corner principle tells us that this is the point in the region of possible solutions at which the highest profit occurs. The craftsman will maximize his profit if he makes $2\frac{1}{2}$ coffee tables and 5 end tables each week (finishing that third coffee table during the next week), working his entire 40 hours per week and spending his entire $1000 per week on materials.

Why the Corner Principle Works

Each point in the region of possible solutions has a value of z associated with it. For example, we found that $P_4(\frac{5}{2}, 5)$ has a z-value of 1400. Think of each point in the region as being a light bulb, with the brightness of the light given by the z-value. The point at which the craftsman's profit is maximized is the point with the brightest light.

Some points are just as bright as others. For example, all points that satisfy the equation $320 = 240x + 160y$ have a brightness of $z = 320$. If we solve this equation for y, we get

$$320 = 240x + 160y$$
$$-160y = 240x - 320$$
$$y = \frac{-3}{2}x + 2$$

This is a line with slope $\frac{-3}{2}$ and y-intercept 2. Points satisfying the equation $640 = 240x + 160y$ have a brightness of $z = 640$. Solving this equation for y gives $y = (\frac{-3}{2})x + 4$. Points that satisfy the equation $960 = 240x + 160y$ have a brightness of 960; solving for y gives $y = (\frac{-3}{2})x + 6$. These three lines have the same slope, so they are parallel, as shown in Figure 2.29.

The light bulbs that fill the region of possible solutions form parallel rows, and all bulbs in any one row are equally bright. Rows closer to the upper-right corner of the region are brighter, and rows closer to the lower-left corner are dimmer. The brightest bulb is at corner point $P_3(\frac{5}{2}, 5)$, and the dimmest is at corner point $P_1(0, 0)$.

FIGURE 2.29

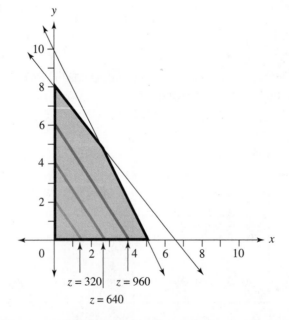

Any linear programming problem (including our problem about the craftsman) must have constraints that are expressible as linear inequalities and an objective function that is expressible as a linear equation. This means that when a problem has two independent variables, the region of possible solutions is bounded by lines, and each value of z will correspond to a line. Thus, the graph must always look something like one of the three possibilities shown in Figure 2.30.

FIGURE 2.30

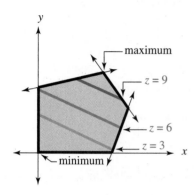

Bounded region: maximum and minimum at corner points

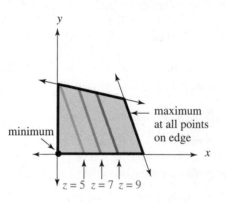

Bounded region: maximum at all points on edge, minimum at corner point

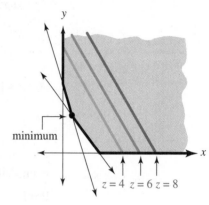

Unbounded region: maximum at corner point, no maximum

If the region is bounded, the maximum and minimum must be at corner points or at all points on an edge. Therefore, we can find all corner points, substitute them into the objective function, and choose the biggest or smallest. If two corners yield the same maximum z-value, we know that the maximum occurs at all points on the boundary line between those corners.

An unbounded region does not necessarily have a maximum or a minimum. Unbounded regions are explored further in the exercises.

Linear Programming Steps

1. *List the independent variables.*
2. *List the constraints and translate them into linear inequalities.*
3. *Find the objective and translate it into a linear equation.* This equation is called the objective function.
4. *Graph the region of possible solutions.* This is the region described by the constraints. Graph each line carefully, using the x-intercept as well as the slope and y-intercept, if you are graphing without the aid of technology.
5. *Find all corner points and the z-values associated with these points.*
   ✔ Check your corner point computations by verifying that a point's computed location fits with the graph, if you are graphing without the aid of technology.
6. *Find the maximum/minimum.* For a bounded region, the maximum occurs at the corner with the largest z-value, and the minimum occurs at the corner with the smallest z-value. If two corners give the same maximum (or minimum) value, then the maximum (or minimum) occurs at all points on the boundary line between those corners.

In Example 1, the craftsman maximized profit by exhausting all resources. In Example 2, we will find that maximizing profit does not necessarily entail exhausting all resources.

EXAMPLE 2   Pete's Coffees sells two blends of coffee beans, Rich Blend and Hawaiian Blend. Rich Blend is one-half Colombian beans and one-half Kona beans, and Hawaiian Blend is one-quarter Colombian beans and three-quarters Kona beans. Profit on the Rich Blend is $2 per pound, while profit on the Hawaiian Blend is $3 per pound. Each day, the shop can obtain 200 pounds of Colombian beans and 60 pounds of Kona beans, and it uses that coffee only in the two blends. If the shop can sell all that it makes, how many pounds of Rich Blend and of Hawaiian Blend should Pete's Coffees prepare each day in order to maximize profit?

Solution   Step 1   *List the independent variables.* The variables are as follows:

- the amount of Rich Blend to be prepared each day
- the amount of Hawaiian Blend to be prepared each day
- the daily profit

Profit depends on the amount of the two blends prepared, so profit is the dependent variable, and the amounts of Rich Blend and Hawaiian Blend are the independent variables. If we call them x and y, then

x = pounds of Rich Blend to be prepared each day

y = pounds of Hawaiian Blend to be prepared each day

Step 2   *List the constraints and translate them into linear inequalities.* The constraints, or restrictions, are that there are 200 pounds of Colombian beans available each day and only 60 pounds of Kona beans. In the blends, no more than the amount available can be used. First, let's translate the Colombian bean constraint:

How should Pete's Coffees blend its beans in order to maximize its profit?

Colombian beans used ≤ 200

(Colombian in Rich Blend) + (Colombian in Hawaiian Blend) ≤ 200

(one-half of Rich Blend)   + (one-fourth of Hawaiian Blend) ≤ 200

$$\frac{1}{2}x \qquad + \qquad \frac{1}{4}y \qquad\qquad \leq 200$$

Next we'll translate the Kona bean constraint:

Kona beans used ≤ 60

(Kona in Rich Blend)   + (Kona in Hawaiian Blend) ≤ 60

(one-half of Rich Blend) + (three-quarters of Hawaiian Blend) ≤ 60

$$\frac{1}{2}x \qquad + \qquad \frac{3}{4}y \qquad\qquad \leq 60$$

Also, x and y count things (pounds of coffee), so neither can be negative.

$$x \geq 0 \qquad \text{and} \qquad y \geq 0$$

There is another implied constraint in this problem. Pete's sells its coffee in 1-pound bags, so x and y must be whole numbers. There are special methods available for handling such constraints, but these methods are beyond the scope of this text. We will ignore such constraints and accept fractional answers should they occur.

Step 3   *Find the objective and translate it into a linear equation.* The objective is to maximize profit. If we let z = profit, we get

z = (Rich Blend profit) + (Hawaiian Blend profit)

  = ($2 per pound)(pounds of Rich Blend)
    + ($3 per pound)(pounds of Hawaiian Blend)

  = $2x + 3y$

Steps 1 through 3 yield the mathematical model:

Independent Variables

x = pounds of Rich Blend to be prepared each day

y = pounds of Hawaiian Blend to be prepared each day

Constraints

$$\frac{1}{2}x + \frac{1}{4}y \leq 200 \qquad\qquad \text{the Colombian bean constraint}$$

$$\frac{1}{2}x + \frac{3}{4}y \leq 60 \qquad\qquad \text{the Kona bean constraint}$$

$$x \geq 0 \qquad \text{and} \qquad y \geq 0$$

Objective Function

$z = 2x + 3y$

Step 4 *Graph the region of possible solutions.* With the aid of the following chart, we can graph the two inequalities on the same axes.

Original Inequality	Slope-Intercept Form	Associated Equation	The Graph of the Inequality
$\frac{1}{2}x + \frac{1}{4}y \leq 200$	$\frac{1}{4}y \leq -\frac{1}{2}x + 200 \rightarrow$ $y \leq -2x + 800$	$y = -2x + 800$	All points on or below the line with slope -2 and y-intercept 800
$\frac{1}{2}x + \frac{3}{4}y \leq 60$	$\frac{3}{4}y \leq -\frac{1}{2}x + 60 \rightarrow$ $y \leq -\frac{2}{3}x + 80$	$y = -\frac{2}{3}x + 80$	All points on or below the line with slope $-2/3$ and y-intercept 80
$x \geq 0$	(not applicable)	$x = 0$	All points on or to the right of the y-axis
$y \geq 0$	$y \geq 0$	$y = 0$	All points on or above the x-axis

The last two constraints tell us that the region of possible solutions is in the first quadrant.

If you are graphing the region without the aid of technology, it's easier to get an accurate graph if you find the x-intercept of each line and use it (as well as the slope and y-intercept) to graph the line. To find the x-intercept of $y = -\frac{2}{3}x + 80$, solve the system

$$y = -\frac{2}{3}x + 80$$

$$y = 0 \qquad \text{The equation of the } x\text{-axis}$$

The x-intercept is the point (120, 0).

The region of possible solutions, shown in Figure 2.31, is a bounded region. Notice that $y = -2x + 800$ is not a boundary of the region of possible solutions; if we were not given the first constraint, we would have the same region of possible solutions. That constraint, which describes the limited amount of Colombian beans available to Pete's Coffees, is not really a limitation.

Step 5 *Find all corner points and the z-values associated with these points.* Clearly, $P_1 = (0, 0)$. P_2 and P_3 have already been found; they are the y- and x-intercepts of $y = -\frac{2}{3}x + 80$. P_2 is (0, 80) and P_3 is (120, 0).

Step 6 *Find the maximum.* The result of substituting the corner points into the objective function $z = 2x + 3y$ is given in Figure 2.32.

FIGURE 2.31

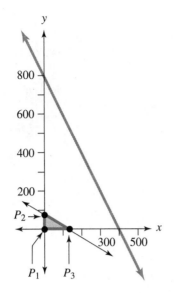

FIGURE 2.32

Point	Value of $z = 2x + 3y$
$P_1 (0, 0)$	$z = 2 \cdot 0 + 3 \cdot 0 = 0$
$P_2 (0, 80)$	$z = 2 \cdot 0 + 3 \cdot 80 = 240$
$P_3 (120, 0)$	$z = 2 \cdot 120 + 3 \cdot 0 = 240$

The two corner points P_2 and P_3 give the same maximum value, $z = 240$. Thus, the maximum occurs at P_2 and P_3 and at all points between them on the boundary line $y = -(2/3)x + 80$, as shown in Figure 2.33. This line includes points such as $(30, 60)$, $(60, 40)$, and $(90, 20)$. (Find points like these by substituting appropriate values of x in the equation.) The meaning of these points is given in Figure 2.34 on page 116. Pete's Coffees can choose to produce its Rich Blend and Hawaiian Blend in any of the amounts given in Figure 2.34 or any other amount given by a point between the corner points P_2 and P_3. Because each of these choices will maximize Pete's profit at \$240 per day, the choice must be made using criteria other than profit. Perhaps the Hawaiian Blend tends to sell out earlier in the day than the Rich Blend. In this case, Pete might choose to produce 30 pounds of Rich Blend and 60 pounds of Hawaiian Blend.

FIGURE 2.33

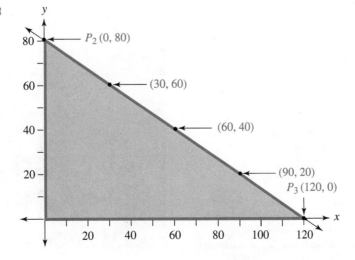

FIGURE 2.34

Point	Interpretation
P_2 (0, 80)	Each day, prepare no Rich Blend and 80 pounds of Hawaiian Blend for a profit of $240.
P_3 (120, 0)	Each day, prepare 120 pounds of Rich Blend and no Hawaiian Blend for a profit of $240.
(30, 60)	Each day, prepare 30 pounds of Rich Blend and 60 pounds of Hawaiian Blend for a profit of $240.
(60, 40)	Each day, prepare 60 pounds of Rich Blend and 40 pounds of Hawaiian Blend for a profit of $240.
(90, 20)	Each day, prepare 90 pounds of Rich Blend and 20 pounds of Hawaiian Blend for a profit of $240.

2.4

EXERCISES

In Exercises 1–6, convert the information to a linear inequality. Give the meaning of each variable used.

1. A landscape architect wants his project to use no more than 100 gallons of water per day. Each shrub requires 1 gallon of water per day, and each tree requires 3 gallons of water per day.

2. A shopper wishes to spend no more than $150. Each pair of pants costs $25 and each shirt costs $21.

3. A bookstore owner wishes to generate at least $5000 in profit this month. Each hardback book generates $4.50 in profit, and each paperback generates $1.25 in profit.

4. Dick Rudd wants to take at least 1000 mg of vitamin C. Each tablet of Megavite has 30 mg of vitamin C, and each tablet of Healthoboy has 45 mg of vitamin C.

5. A warehouse has 1650 cubic feet of unused storage space. Refrigerators take up 63 cubic feet each, and dishwashers take up 41 cubic feet each.

6. A coffee shop owner has 1000 pounds of Java beans. In addition to selling pure Java beans, the shop also sells Blend Number 202, which is 32% Java beans.

In Exercises 7–16, use the method of linear programming to solve the problem.

7. A craftswoman produces two products, floor lamps and table lamps. Production of one floor lamp requires 75 minutes of labor and materials that cost $25. Production of one table lamp requires 50 minutes of labor, and the materials cost $20. The craftswoman wishes to work no more than 40 hours each week, and her financial resources allow her to pay no more than $900 for materials each week. If she can sell as many lamps as she can make and if her profit is $39 per floor lamp and $33 per table lamp, how many floor lamps and how many table lamps should she make each week in order to maximize her weekly profit? What is that maximum profit?

8. Five friends, each of whom is an experienced baker, form a company that will make bread and cakes and sell them to local restaurants and specialty stores. Each loaf of bread requires 50 minutes of labor and ingredients costing $0.90 and can be sold

for $1.20 profit. Each cake requires 30 minutes of labor and ingredients costing $1.50 and can be sold for $4.00 profit. The partners agree that no one will work more than 8 hours a day. Their financial resources do not allow them to spend more than $190 per day on ingredients. How many loaves of bread and how many cakes should they make each day in order to maximize their profit? What is that maximum profit?

9. Pete's Coffees sells two blends of coffee beans, Morning Blend and South American Blend. Morning Blend is one-third Mexican beans and two-thirds Colombian beans, and South American Blend is two-thirds Mexican beans and one-third Colombian beans. Profit on the morning Blend is $3 per pound, while profit on the South American Blend is $2.50 per pound. Each day, the shop can obtain 100 pounds of Mexican beans and 80 pounds of Colombian beans, and it uses that coffee only in the two blends. If the shop can sell all that it makes, how many pounds of Morning Blend and of South American Blend should Pete's Coffees prepare each day in order to maximize profit? What is that maximum profit?

10. Pete's Coffees sells two blends of coffee beans, Yusip Blend and Exotic Blend. Yusip Blend is one-half Costa Rican beans and one-half Ethiopian beans, and Exotic Blend is one-quarter Costa Rican beans and three-quarters Ethiopian beans. Profit on the Yusip Blend is $3.50 per pound, while profit on the Exotic Blend is $4 per pound. Each day, the shop can obtain 200 pounds of Costa Rican beans and 330 pounds of Ethiopian beans, and it uses that coffee only in the two blends. If the shop can sell all that it makes, how many pounds of Yusip Blend and of Exotic Blend should Pete's Coffees prepare each day in order to maximize profit? What is that maximum profit?

11. Bake-Em-Fresh sells its bread to supermarkets. Shopgood Stores needs at least 15,000 loaves each week, and Rollie's Markets needs at least 20,000 loaves each week. Bake-Em-Fresh can ship at most 45,000 loaves to these two stores each week if it wishes to satisfy its other customers' needs. If shipping costs an average of 8¢ per loaf to Shopgood Stores and 9¢ per loaf to Rollie's Markets, how many loaves should Bake-Em-Fresh allot to Shopgood and to Rollie's each week? What shipping costs would this entail?

12. Notel Chips manufactures computer chips. Its two main customers, HAL Computers and Peach Computers, just submitted orders that must be filled immediately. HAL needs at least 130 cases of chips and Peach needs at least 150 cases. Due to a limited supply of silicon, Notel cannot send more than a total of 300 cases. If shipping costs $100 per case for shipments to HAL and $90 per case for shipments to Peach, how many cases should Notel send to each customer in order to minimize shipping costs? What shipping costs would this entail?

13. Global Air Lines has contracted with a tour group to transport a minimum of 1600 first-class passengers and 4800 economy-class passengers from New York to London during a 6-month time period. Global Air has two types of airplanes, the Orville 606 and the Wilbur W-1112. The Orville 606 carries 20 first-class passengers and 80 economy-class passengers and costs $12,000 to operate. The Wilbur W-1112 carries 80 first-class passengers and 120 economy-class passengers and costs $18,000 to operate. During the time period involved, Global Air can schedule no more than 52 flights on Orville 606s and no more than 30 flights on Wilbur W-1112s. How should Global Air Lines schedule its flights? What operating costs would this schedule entail?

14. Compucraft sells personal computers and printers made by Peach Computers. The computers come in 12-cubic-foot boxes, and the printers come in 8-cubic-foot boxes. Compucraft's owner estimates that at least 30 computers can be sold each month and that the number of computers sold will be at least 50% more than the number of printers. The computers cost Compucraft $1000 each and can be sold at a $1000 profit, while the printers cost $300 each and can be sold for a $350 profit. Compucraft has 1000 cubic feet of storage available for the Peach personal computers and printers and sufficient financing to spend $70,000 each month on computers and printers. How many computers and printers should Compucraft order from Peach each month in order to maximize profit? What is that maximum profit?

15. The Appliance Barn has 2400 cubic feet of storage space for refrigerators. The larger refrigerators come in 60-cubic-foot packing crates, and the smaller ones come in 40-cubic-foot crates. The larger refrigerators can be sold for a $250 profit, while the smaller ones can be sold for a $150 profit.
   a. If the manager is required to sell at least 50 refrigerators each month, how many large refrigerators and how many small refrigerators should he order each month in order to maximize profit?
   b. If the manager is required to sell at least 40 refrigerators each month, how many large refrigerators and how many small refrigerators should he order each month in order to maximize profit?
   c. Should the Appliance Barn owner require his manager to sell 40 or 50 refrigerators per month?

16. City Electronics Distributors handles two lines of televisions, the Packard and the Bell. It purchases up to $57,000 worth of television sets from the manufacturers each month and stores them in a 9000-cubic-foot warehouse. The Packards come in 36-cubic-foot packing crates, and the Bells come in 30-cubic-foot crates. The Packards cost City Electronics $200 each and can be sold

to a retailer for a $200 profit, while the Bells cost $250 each and can be sold for a $260 profit. City Electronics must stock enough sets to meet its regular customers' standing orders.

a. If City Electronics has standing orders for 250 sets in addition to orders from other retailers, how many sets should City Electronics order each month in order to maximize profit?

b. If City Electronics' standing orders increase to 260 sets, how many sets should City Electronics order each month in order to maximize profit?

17. How much of their available Mexican beans would be unused if Pete's Coffees of Exercise 9 maximizes its profit? How much of the available Colombian beans would be unused?

18. How much of their available time would be unused if the five friends of Exercise 8 maximize their profit? How much of their available money would be unused?

19. How much of her available time would be unused if the craftswoman of Exercise 7 maximizes profit? How much of the available money would be unused?

20. How much of the available Costa Rican beans would be unused if Pete's Coffees of Exercise 10 maximizes its profit? How much of the available Ethiopian beans would be unused?

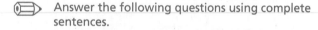 Answer the following questions using complete sentences.

21. Who was George Dantzig named after?

22. How did Dantzig choose the topic of his Ph.D. thesis?

23. What did Dantzig do during World War II?

24. What was surprising about the awarding of a Nobel Prize to two economists for their use of linear programming?

25. Is it possible for an unbounded region to have both a maximum and a minimum? Draw a number of graphs (similar to those in Figure 2.30) to see. Justify your answer with graphs.

26. Discuss why it is that, when two corners give the same maximum (or minimum) value, the maximum (or minimum) occurs at all points on the boundary line between those corners.

27. In Example 2, Pete's Coffees mixes Colombian beans and Kona beans to make Rich Blend and Hawaiian Blend. It may be that Pete's could make slight variations in the proportions of Colombian and Kona beans in these two blends without affecting the taste of the coffee. Why would Pete's consider doing this? What effect would it have on the objective function, the constraints, the region of possible solutions, and the solution to the example?

28. In an open market, the selling price of a product is determined by the supply of the product and the demand for the product. In particular, the price falls if a larger supply of the product becomes available, and the price increases if there is suddenly a larger demand for the product. Any change in the selling price could affect the profit. However, when modeling a linear programming problem, we assume a constant profit. How could linear programming be adjusted to allow for the fact that prices change with supply and demand?

In Exercises 29–32, the region of possible solutions is not bounded; thus, there may not be both a maximum and a minimum. After graphing the region of possible solutions and finding each corner point, you can determine if both a maximum and a minimum exist by choosing two arbitrary values of z and graphing the corresponding lines, as discussed in the section "Why the Corner Principle Works."

29. The objective function $z = 2x + 3y$ is subject to the constraints

$$3x + y \geq 12$$
$$x + y \geq 6$$
$$x \geq 0 \quad \text{and} \quad y \geq 0$$

Find the following.

a. the point at which the maximum occurs (if there is such a point)

b. the maximum value

c. the point at which the minimum occurs (if there is such a point)

d. the minimum value

30. The objective function $z = 3x + 4y$ is subject to the constraints

$$2x + y \geq 10$$
$$3x + y \geq 12$$
$$x \geq 0 \quad \text{and} \quad y \geq 0$$

Find the following:

a. the point at which the maximum occurs (if there is such a point)

b. the maximum value

c. the point at which the minimum occurs (if there is such a point)

d. the minimum value

31. U.S. Motors manufactures quarter-ton, half-ton, and three-quarter-ton panel trucks. United Delivery Service has placed an order for at least 300 quarter-ton, 450 half-ton, and 450 three-quarter-ton panel trucks. U.S. Motors builds the trucks at two plants, one in Detroit and one in Los Angeles. The Detroit plant produces 30 quarter-ton trucks, 60 half-ton trucks, and 90 three-quarter-ton trucks each week, at a total cost of

$540,000. The Los Angeles plant produces 60 quarter-ton trucks, 45 half-ton trucks, and 30 three-quarter-ton trucks each week, at a total cost of $360,000. How should U.S. Motors schedule its two plants so that it can fill this order at minimum cost? What is that minimum cost?

32. Eaton's Chocolates produces semisweet chocolate chips and milk chocolate chips at its plants in Bay City and Estancia. The Bay City plant produces 3000 pounds of semisweet chips and 2000 pounds of milk chocolate chips each day at a cost of $1000, while the Estancia plant produces 1000 pounds of semisweet chips and 6000 pounds of milk chocolate chips each day at a cost of $1500. Eaton's has an order from SafeNShop Supermarkets for at least 30,000 pounds of semisweet chips and 60,000 pounds of milk chocolate chips. How should it schedule its production so it can fill the order at minimum cost? What is that minimum cost?

CHAPTER 2 REVIEW

Terms

addition method	dependent and	linear inequality	region of solutions
bounded and unbounded	independent variables	linear programming	row
regions	dimensions	matrix	row matrix
column	element	model	row operations
column matrix	elimination method	objective function	square matrix
constraint	entry	parameter	system of linear equations
corner point	general and particular	pivoting	system of linear
corner principle	solutions	reduced row echelon form	inequalities
	linear equation	region of possible solutions	

Review Exercises

In Exercises 1–3, determine the number of solutions to the system, and describe the geometric relationship between the lines determined by the given equations. Do not actually solve the given system.

1. $5x - 7y = 2$
$3x + 6y = 9$
$x + 2y = 3$

2. $8x - 4y = 12$
$2x + 3y = 5$
$2x - y = 2$

3. $3x + 2y = 8$
$6x + 4y = 16$

Solve the systems in Exercises 4–7 with the elimination method. If the system has an infinite number of solutions, give the general solution and three particular solutions.

4. $5x - 2y = 17$
$3x - 4y = 13$

5. $8x + 4y = 16$
$6x + 3y = 12$

6. $9x - 6y = 24$
$6x - 4y = 9$

7. $8x - 2y + z = 19$
$3x + y - 4z = 9$
$5x - 3y + 3z = 13$

In Exercises 8–11, determine whether the given matrix is a row matrix, a column matrix, a square matrix, or none of these; also, give the dimensions of the matrix.

8. $\begin{bmatrix} 5 & 7 \\ 3 & 9 \\ 0 & 1 \end{bmatrix}$

9. $\begin{bmatrix} 5 & 7 & 3 \\ 9 & 0 & 1 \\ 5 & 2 & 1 \end{bmatrix}$

10. $\begin{bmatrix} 2 \\ 6 \\ 31 \end{bmatrix}$

11. $\begin{bmatrix} 5 & 7 \end{bmatrix}$

Use the Gauss-Jordan method to solve Exercises 12–14. If the system has an infinite number of solutions, give the general solution and three particular solutions. Check your solutions by substituting them back in. (Answers are not given at the back of the book.)

12. $5x - 6y + 2z = 6$
$4x + 5y - 2z = 11$
$6x + y + 3z = 16$

13. $3x - y + 2z = 18$
$2x + y + z = 9$
$10x + 6z = 54$

14. $8x + 2y - 4z = 2$
$3x + 2y + 4z = 10$
$14x + 6y + 4z = 12$

15. Solve the following with the Gauss-Jordan method. Dogs at a kennel are to be fed exactly 40 IUs of vitamin E and 37 mg of vitamin D each day. Each DoggYums tablet contains 4 IUs of vitamin E and 5 mg of vitamin D. Each Arf-O-Vite tablet contains 5 IUs of vitamin E and 3 mg of vitamin D. How many tablets should a dog be fed each day?

Graph the region of solutions of the inequalities in Exercises 16–19.

16. $4x - 5y > 7$

17. $3x + 4y < 10$

18. $3x + 6y \leq 9$

19. $6x - 8y \geq 12$

Graph the region of solutions of the systems of inequalities in Exercises 20–25, and find the corner points.

20. $8x - 4y < 10$
$3x + 5y \geq 7$

21. $x - 5y < 7$
$3x + 2y > 6$

22. $5x - y > 8$
$x \geq -3$

23. $6x + 4y \leq 7$
$y < 4$

24. $x - y \geq 7$
$5x + 3y \leq 9$
$x \geq 0$
$y \geq 0$

25. $3x - 2y \leq 12$
$5x + 3y \geq 15$
$x \geq 0$
$y \geq 0$

Solve the linear programming problems in Exercises 26 and 27.

26. Mowson Audio Co. makes stereo speaker assemblies. It purchases speakers from a speaker manufacturing firm and installs them in its own cabinets. Mowson's model 110 speaker assembly, which sells for $200, has a tweeter and a mid-range speaker. The model 330 assembly, which sells for $350, has two tweeters, a mid-range speaker, and a woofer. Mowson currently has in stock 90 tweeters, 60 mid-range speakers, and 44 woofers. How many speaker assemblies should Mowson make in order to maximize its income? What is that maximum income?

27. The Stereo Guys store sells two lines of personal stereos, the Sunny and the Iwa. The Sunny comes in a 12-cubic-foot box and can be sold for a $220 profit, and the Iwa comes in an 8-cubic-foot box and can be sold for a $200 profit. The Stereo Guys marketing department estimates that at least 600 personal stereos can be sold each month and that, due to Sunny's reputation for quality, the demand for the Sunny unit is at least twice that for the Iwa. If the Stereo Guys warehouse has 12,000 cubic feet of space available for personal stereos, how many Sunnys and Iwas should the company stock each month to maximize profit? What is that maximum profit?

LINEAR PROGRAMMING

THE SIMPLEX METHOD

3

In Section 2.4 we discussed how to solve a linear programming problem by graphing the region of possible solutions, finding the corner points, and selecting the optimal corner point. This is the **geometric method.** While an excellent method for solving problems that have two independent variables and only a handful of constraints (such as those in Section 2.4), it is not practical for solving more complex problems.

Usually, a real-life problem has many variables. In Section 2.4 we discussed a craftsman who produced coffee tables and end tables, for a total of two independent variables. A more realistic example would have the craftsperson producing two types of coffee tables, two types of end

tables, and three different types of chairs, for a total of seven independent variables.

After developing the graphical method, George Dantzig invented the **simplex method,** a method of linear programming for two *or more* variables that does not involve graphing the region of possible solutions and finding all the corner points. Instead, the method uses matrices and a procedure very similar to the Gauss-Jordan method of solving systems of equations. The simplex method works well with more complex problems that have a large number of variables and constraints. It is also amenable to computer use.

From its beginning, the simplex method has been used successfully by a number of industries and government agencies. Refineries use the simplex method to blend gasoline, decide what crude oil to buy, and determine what products to produce. The steel industry uses it to evaluate ores and to determine what products to produce and when to build new furnaces. Airlines use it to minimize costs related to the scheduling of flights, subject to constraints such as the amount of time a pilot or a crew may fly. The federal energy authorities use linear programming to explore policy alternatives. Various businesses and government agencies use it to determine the best way to control water and air pollution, assign personnel to jobs, and achieve racial balance in schools. Supermarket chains use it to determine which warehouses should ship which products to the stores. Investment companies use it to create portfolios with the best mix of stocks and bonds.

In large corporations, managers must decide how to allocate their limited resources (such as raw materials, labor, and machinery) in order to maximize their profit and meet other objectives. You might assume that such a corporation's resources are not really limited—that a manager could just determine a production level for each of her products and obtain the resources needed to meet those production levels. However, this is frequently not the case. Sufficient resources may not be available, or they may be limited by their location or their cost. Resources are usually limited, and their allocation is a difficult decision. Linear programming analyzes the effect of such limitations and helps corporate managers make informed decisions.

Corporate managers have so many different products and resources to consider that the necessary calculations must be done by computer. It is estimated that linear programming accounts for as much computer time as does payroll or inventory control. And some experts claim that linear programming is the most widely used form of modern mathematics.

The method of linear programming resulted from 1940s research to determine the least expensive way to distribute men and equipment to the various fronts of World War II.

3.1

INTRODUCTION TO THE SIMPLEX METHOD

The simplex method of linear programming involves a number of steps. First, the problem is modeled in exactly the same way that problems are modeled in the geometric method. Next, the constraint inequalities are converted to equations, and the model is converted to a matrix problem, as we will show in this section. Finally, the matrix problem is solved using the row operations and a Gauss-Jordan–like procedure, as we will show in Section 3.2.

We'll use the craftsman problem from Section 2.4, summarized below, to introduce the procedure and explain why it works.

A craftsman produces two products, coffee tables and end tables. Production data are given in Figure 3.1. If the craftsman wants to work no more than 40 hours each week and if his financial resources allow him to pay no more than $1000 for materials each week, how many coffee tables and how many end tables should he make each week to maximize his weekly profit?

FIGURE 3.1

	Labor (per table)	Cost of Materials (per table)	Profit (per table)
Coffee Tables	6 hours	$200	$240
End Tables	5 hours	$100	$160

Step 1 *Set up the model.* This work was done in Section 2.4. Because there are frequently more than two variables, the simplex method uses x_1, x_2, and so on rather than x and y to represent the variables.

Independent Variables

x_1 = number of coffee tables

x_2 = number of end tables

Constraints

$C_1: 6x_1 + 5x_2 \leq 40$ The time constraint

$C_2: 200x_1 + 100x_2 \leq 1000$ The money constraint

$C_3: x_1 \geq 0$

$C_4: x_2 \geq 0$

(Notice that we've labeled the constraints C_1 through C_4, for easy reference.)

Objective Function

$z = 240x_1 + 160x_2$ z measures profit

Step 2 *Convert the constraint inequalities to equations.* Recall that the constraint

$C_1: 6x_1 + 5x_2 \leq 40$

was derived from the data

(coffee table hours) + (end table hours) ≤ 40

If the total number of hours is less than 40, then some "slack" or unused hours are left, and we could say (coffee table hours) + (end table hours) + (unused hours) = 40

C_1 can be rewritten as

$$C_1: 6x_1 + 5x_2 + s_1 = 40$$

where s_1 measures unused hours and $s_1 \geq 0$.

The variable s_1 is called a **slack variable** because it "takes up the slack" between the hours used (which can be less than 40) and the hours available (which are exactly 40). The introduction of slack variables converts the constraint inequalities to equations, allowing us to use matrices and a Gauss-Jordan–like procedure to solve the problem.

The constraint

$$C_2: 200x_1 + 100x_2 \leq 1000$$

was derived from the idea

(coffee table money) + (end table money) $\leq$ 1000

which could be rephrased as

(coffee table money) + (end table money) + (unused money) = 1000

Thus, C_2 can be rewritten as

$$C_2: 200x_1 + 100x_2 + s_2 = 1000$$

where s_2 is a slack variable that measures unused money and $s_2 \geq 0$. This new variable "takes up the slack" between the money used (which can be less than \$1000) and the money available (which is exactly \$1000).

Constraints $C_3: x_1 \geq 0$ and $C_4: x_2 \geq 0$ remind us that variables x_1 and x_2 count things that cannot be negative (coffee tables and end tables, respectively); there is no need to convert these constraints to equations.

Step 3 *Rewrite the objective function with all variables on the left side.* The objective function

$$z = 240x_1 + 160x_2$$

becomes

$$-240x_1 - 160x_2 + z = 0$$

or, using *all* the variables,

$$-240x_1 - 160x_2 + 0s_1 + 0s_2 + 1z = 0$$

We have rewritten our model:

Independent Variables	**Slack Variables**
x_1 = number of coffee tables	s_1 = unused hours
x_2 = number of end tables	s_2 = unused money
$(x_1, x_2, s_1, s_2 \geq 0)$	

Constraints

$C_1: 6x_1 + 5x_2 + s_1 = 40$	The time constraint
$C_2: 200x_1 + 100x_2 + s_2 = 1000$	The money constraint

Objective Function

$$-240x_1 - 160x_2 + 0s_1 + 0s_2 + 1z = 0 \qquad \textit{z measures profit}$$

Step 4 *Make a matrix out of the rewritten constraints and the rewritten objective function.* Basically, we copy down the constraints and the objective function without the variables, as with the Gauss-Jordan method. The constraints go in the first rows, and the objective function goes in the last row.

The first constraint is

$$C_1: 6x_1 + 5x_2 + s_1 = 40$$

or, using *all* the variables,

$$C_1: 6x_1 + 5x_2 + 1s_1 + 0s_2 + 0z = 40$$

This equation becomes the first row of our matrix:

$$6 \quad 5 \quad 1 \quad 0 \quad 0 \quad 40$$

Similar alterations of the second constraint and the objective function yield the following matrix.

$$
\begin{array}{cccccc}
x_1 & x_2 & s_1 & s_2 & z & \\
\end{array}
$$

$$
\begin{bmatrix}
6 & 5 & 1 & 0 & 0 & 40 \\
200 & 100 & 0 & 1 & 0 & 1000 \\
-240 & -160 & 0 & 0 & 1 & 0
\end{bmatrix}
\begin{array}{l}
\leftarrow C_1 \\
\leftarrow C_2 \\
\leftarrow \text{Objective function}
\end{array}
$$

This matrix is called the **first simplex matrix.** The simplex method requires us to explore a series of matrices, just as the geometric method requires us to explore a series of corner points. Each simplex matrix will provide us with a corner point of the region of possible solutions, *without our actually graphing that region.* The last simplex matrix will provide us with the optimal corner point, that is, the point that solves the problem.

Step 5 *Determine the possible solution that corresponds to the matrix.* The possible solution is found by using a method very similar to the Gauss-Jordan method. Recall that with the Gauss-Jordan method we pivot until we obtain a matrix in reduced row echelon form. We can find the value of the variable heading each column by reading down the column, turning at 1, and stopping at the end of the row.

$$
\begin{array}{cc}
x & y \\
\end{array}
$$

$$
\begin{bmatrix}
1 & 0 & 1 \\
0 & 1 & 2
\end{bmatrix}
$$

The first simplex matrix in the middle of this page is not in reduced row echelon form. Some of the columns (the s_1, s_2, and z columns) contain a single 1 and several 0's. These columns are the type that appear in reduced row echelon form, and we find the value of the variables heading them by reading down the column, turning at 1, and stopping at the end of the row. Some of the columns (the x_1 and x_2 columns) are not the type that appear in reduced row echelon form; the value of the variables heading these columns is zero.

$$
\begin{array}{c}
\begin{array}{ccccc} x_1 & x_2 & s_1 & s_2 & z \end{array} \\
\left[
\begin{array}{ccccc|c}
6 & 5 & 1 & 0 & 0 & 40 \\
200 & 100 & 0 & 1 & 0 & 1000 \\
-240 & -160 & 0 & 0 & 1 & 0
\end{array}
\right]
\begin{array}{l}
\rightarrow s_1 = 40 \\
\rightarrow s_2 = 1000 \\
\rightarrow z = 0
\end{array} \\
\begin{array}{cc} \uparrow & \uparrow \\ x_1 = 0 & x_2 = 0 \end{array}
\end{array}
$$

The possible solution that corresponds to the matrix is

$$(x_1, x_2, s_1, s_2) = (0, 0, 40, 1000) \quad \text{with } z = 0$$

Why do the numbers read from the matrix in this manner yield a possible solution? Clearly, $(x_1, x_2) = (0, 0)$ satisfies all four constraints, so it is a possible solution. If we substitute this solution into C_1 and C_2 and the objective function, we get the following:

$$
\begin{array}{ll}
C_1 \colon 6x_1 + 5x_2 + s_1 = 40 & C_2 \colon 200x_1 + 100x_2 + s_2 = 1000 \\
6 \cdot 0 + 5 \cdot 0 + s_1 = 40 & 200 \cdot 0 + 100 \cdot 0 + s_2 = 1000 \\
\qquad\qquad s_1 = 40 & \qquad\qquad s_2 = 1000
\end{array}
$$

$$z = 240x_1 + 160x_2 = 240 \cdot 0 + 160 \cdot 0 = 0$$

Thus, $(x_1, x_2, s_1, s_2) = (0, 0, 40, 1000)$ with $z = 0$ satisfies all the constraints and the objective function and is a possible solution.

The possible solution means "make no coffee tables, make no end tables, have 40 unused hours and 1000 unused dollars, and make no profit." When we solved this same problem with the geometric method in Section 2.4, one of the corner points was $(0, 0)$ (see Figure 3.2). We have just found that corner point with the simplex method rather than the geometric method (the geometric method does not find values of the slack variables). The simplex method continues to locate corner points until the optimal corner point is found.

FIGURE 3.2

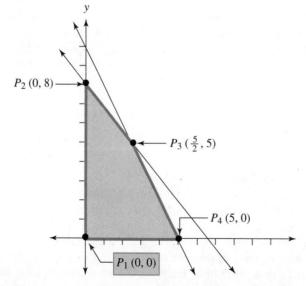

The possible solution is only one of many corner points; it is not the optimal corner point, and it does not maximize profit. In the next section, we will see how to build on this work to find the maximal solution.

Simplex Method Steps

1. *Model the problem.*
   a. List the independent variables.
   b. List the constraints and translate them into linear inequalities.
   c. Find the objective and translate it into a linear equation.
2. *Convert each constraint from an inequality to an equation.*
   a. Use one slack variable for each constraint.
   b. Determine what each slack variable measures.
3. *Rewrite the objective function with all variables on the left side.*
4. *Make a matrix out of the rewritten constraints and the rewritten objective function.* This is the first simple matrix.
5. *Determine the possible solution that corresponds to the matrix.*
   a. Columns containing only a 1 and 0's—turn at 1.
   b. Other columns—the value of the variable is zero.

The remaining steps, which build on this work to find the maximal solution, will be discussed in Section 3.2.

Comparison of the Gauss-Jordan and Simplex Methods

The Gauss-Jordan method solves systems of linear equations. The simplex method solves linear programming problems by optimizing linear objective functions subject to linear constraints.

The Gauss-Jordan method begins with a system of equations. The simplex method begins with a system of inequalities that is converted to a system of equations by the introduction of slack variables.

The Gauss-Jordan method does not produce a solution until you reach the final Gauss-Jordan matrix; none of the matrices that precede the final matrix have their own solution. The simplex method produces a series of possible solutions, one with each simplex matrix. Each possible solution would be a corner point of the region of possible solutions if we were using the geometric method.

We find the solution of a system of equations in the final Gauss-Jordan matrix by reading down each variable's column and turning at 1. Possible solutions of a linear programming problem are found in the same way *when a variable heads a column containing only a 1 and 0's*. The value of a variable that does not head such a column is zero.

The final Gauss-Jordan matrix is always in reduced row echelon form:

$$
\begin{array}{cc}
x & y \\
\end{array}
\begin{bmatrix}
1 & 0 & 33 \\
0 & 1 & 6
\end{bmatrix}
$$

The solution corresponding to this matrix is $(x, y) = (33, 6)$. A simplex matrix is not in reduced row echelon form; it could have the form

$$
\begin{array}{c}
\begin{array}{ccccc} x_1 & x_2 & s_1 & s_2 & z \end{array} \\
\left[
\begin{array}{c|cc|cc|c}
5 & 0 & 3 & 1 & 0 & 11 \\
-8 & 1 & 1 & 0 & 0 & 9 \\
2 & 0 & 7 & 0 & 1 & 15
\end{array}
\right]
\end{array}
$$

The solution corresponding to this matrix is $(x_1, x_2, s_1, s_2) = (0, 9, 0, 11)$ with $z = 15$.

Karmarkar's New Method

In 1984, Narendra Karmarkar, a mathematician at AT&T Bell Laboratories, invented a new method of linear programming that is an alternative to both the geometric method and the simplex method. The simplex method, as we saw in this section, starts at the corner point (0, 0) and methodically locates other corner points until the optimal corner point is found. A modern linear programming problem can involve thousands of corner points, and the simplex method involves locating a large proportion of these points. **Karmarkar's method,** on the other hand, makes use of quicker search routes through the interior of the region of possible solutions. This method has been found to be much quicker than the simplex method with some kinds of problems (see Figure 3.3).

Scientists at Bell Labs used Karmarkar's method to find the most economical way to build a telephone network that links U.S. cities, and to find the most economical way to route calls through the network once it was built. This involved a linear programming problem with 800,000 variables. The problem was not attempted with the simplex method because it would have taken weeks of computer time. It was solved with Karmarkar's method in less than one hour. Its application was quite profitable: AT&T estimates that its $15 billion system has an extra 9–10% of capacity because of the use of Karmarkar's method.

Airlines are currently attempting to use Karmarkar's method to quickly reroute airplanes and reschedule pilots and crews in the event of a major weather disruption. The simplex method is inappropriate here; it might necessitate a number of hours of computer time to find the answer. The Karmarkar method may yield the answer in a few minutes.

FIGURE 3.3

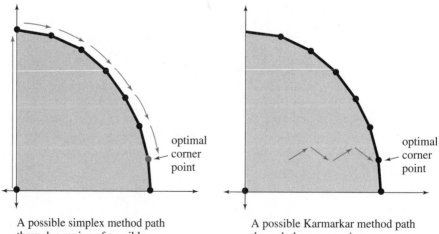

A possible simplex method path through a region of possible solutions with many corner points

A possible Karmarkar method path through the same region

Sometime in the not-so-distant future, you may be pleasantly surprised when your airline responds quickly and effectively to the closure of the airport at which you were to catch a connecting flight. When that happens, thank Narendra Karmarkar!

Narendra Karmarkar was born in 1956 in Gwalior, India, and grew up in Poona, near Bombay. He is a current example of the Hindu tradition of excellence in mathematics. Both his father and his uncle were mathematicians. He attended the California Institute of Technology and received his doctorate from the University of California at Berkeley.

3.1

EXERCISES

In Exercises 1–4, convert the linear inequality to a linear equation by introducing a slack variable.

1. $3x_1 + 2x_2 \le 5$

2. $512x_1 + 339x_2 \le 254$

3. $3.41x_1 + 9.20x_2 + 6.16x_3 \le 45.22$

4. $99.52x_1 + 21.33x_2 + 102.15x_3 \le 50.14$

In Exercises 5–10, (a) translate the information into a linear inequality, (b) convert the linear inequality into a linear equation by introducing a slack variable, and (c) determine what each variable (including the slack variable) measures.

5. A shopper has $42.15 in cash. She wants to purchase meat, at $6.99 a pound, cheese, at $3.15 a pound, and bread, at $1.98 a loaf. The shopper must pay for these items in cash.

6. A sharp dresser is at his favorite store, where everything is on sale. His budget will allow him to spend up to $425. Shirts are on sale for $19.99 each, slacks for $74.98 a pair, and sweaters for $32.98 each.

7. A plumber is starting an 8-hour day. Laying pipe takes her 5 minutes per foot, and installing elbows takes her 4 minutes per elbow.

8. An electrician is starting a $7\frac{1}{2}$-hour day. Installing conduit takes him 2.5 minutes per foot, connecting lines takes him 1 minute per connection, and installing circuit breakers takes him 4.5 minutes per circuit breaker.

9. A mattress warehouse has 50,000 cubic feet of space. Twin beds are 24 cubic feet, double beds are 36 cubic feet, queen-size beds are 56 cubic feet, and king-size beds are 72 cubic feet.

10. A nursery has 1000 square feet of outdoor display space. Trees need 16 square feet each, shrubs need 10 square feet each, and flowering plants need 1 square foot each.

In Exercises 11–16, find the possible solution that corresponds to the given matrix.

11.

x_1	x_2	s_1	s_2	z	
5	0	-3	1	0	12
3	0	19	0	1	22
-21	1	48	0	0	19

12.

x_1	x_2	s_1	s_2	z	
1	31	0	9	0	7.4
0	0	0	15	1	4.9
0	1	1	0	0	20

13.

x_1	x_2	s_1	s_2	s_3	z	
1.9	0.3	0	0	1	0	0.5
3.2	0.7	0	1	0	0	9.3
-5.5	0.8	1	0	0	0	7.8
-2.1	3.2	0	0	0	1	9.6

14.

x_1	x_2	s_1	s_2	s_3	z	
0	1	7	0	62	0	0.5
0	0	5	0	5	1	9.3
1	0	0	0	0	0	7.8
0	0	9	1	1	0	9.6

15.

x_1	x_2	s_1	s_2	s_3	s_4	z	
1	0	6	0	0	76	0	25
0	0	8	0	1	13	0	46
0	0	3	1	0	50	0	32
0	1	4	0	0	-9	0	73
0	0	4	0	0	12	1	63

16.

$$
\begin{array}{ccccccc}
x_1 & x_2 & s_1 & s_2 & s_3 & s_4 & z \\
\end{array}
$$

$$
\begin{bmatrix}
32 & 0 & 1 & 0 & 0 & 7 & 0 & 25 \\
7 & 0 & 0 & 0 & 1 & 3 & 0 & 46 \\
10 & 0 & 0 & 1 & 0 & 5 & 0 & 32 \\
65 & 1 & 0 & 0 & 0 & -9 & 0 & 73 \\
24 & 0 & 0 & 0 & 0 & 2 & 1 & 44 \\
\end{bmatrix}
$$

In Exercises 17–22, find the first simplex matrix and the possible solution that corresponds to it.

17. *Objective Function:*
$z = 2x_1 + 4x_2$
Constraints:
$3x_1 + 4x_2 \leq 40$
$4x_1 + 7x_2 \leq 50$
$x_1 \geq 0$
$x_2 \geq 0$

18. *Objective Function:*
$z = 2.4x_1 + 1.3x_2$
Constraints:
$6.4x_1 + x_2 \leq 360$
$x_1 + 9.5x_2 \leq 350$
$x_1 \geq 0$
$x_2 \geq 0$

19. *Objective Function:*
$z = 12.10x_1 + 43.86x_2$
Constraints:
$112x_1 - 3x_2 \leq 370$
$x_1 + x_2 \leq 70$
$47x_1 + 19x_2 \leq 512$
$x_1 \geq 0$
$x_2 \geq 0$

20. *Objective Function:*
$z = -3.52x_1 + 4.72x_2$
Constraints:
$2x_1 + 4x_2 \leq 7{,}170$
$32x_1 - 19x_2 \leq 1960$
$x_1 \leq 5$
$x_1 \geq 0$
$x_2 \geq 0$

21. *Objective Function:*
$z = 4x_1 + 7x_2 + 9x_3$
Constraints:
$5x_1 + 3x_2 + 9x_3 \leq 10$
$12x_1 + 34x_2 + 100x_3 \leq 10$
$52x_1 + 7x_2 + 12x_3 \leq 10$
$x_1 \geq 0$
$x_2 \geq 0$
$x_3 \geq 0$

22. *Objective Function:*
$z = 9.1x_1 + 3.5x_2 + 8.22x_3$
Constraints:
$x_1 + 16x_2 + 9.5x_3 \leq 1210$
$72x_1 + 3.01x_2 + 50x_3 \leq 1120$
$57x_1 + 87x_2 + 742x_3 \leq 309$
$x_1 \geq 0$
$x_2 \geq 0$
$x_3 \geq 0$

In Exercises 23–26, find the first simplex matrix and the possible solution that corresponds to it. Do not try to solve the problem.

23. Five friends, each of whom is an experienced baker, form a company that will make bread and cakes and sell them to local restaurants and specialty stores. Each loaf of bread requires 50 minutes of labor and ingredients costing $0.90 and can be sold for a $1.20 profit. Each cake requires 30 minutes of labor and ingredients costing $1.50 and can be sold for a $4.00 profit. The partners agree that no one will work more than 8 hours a day. Their financial resources do not allow them to spend more than $190 per day on ingredients. How many loaves of bread and how many cakes should they make each day in order to maximize their profit?

24. A craftswoman produces two products, floor lamps and table lamps. Production of one floor lamp requires 75 minutes of her labor and materials that cost $25. Production of one table lamp requires 50 minutes of labor and materials that cost $20. The craftswoman wants to work no more than 40 hours each week, and her financial resources allow her to pay no more than $900 for materials each week. If she can sell as many lamps as she can make and if her profit is $40 per floor lamp and $32 per table lamp, how many floor lamps and how many table lamps should she make each week to maximize her weekly profit?

25. Pete's Coffees sells two blends of coffee beans, Yusip Blend and Exotic Blend. Yusip Blend is one-half Costa Rican beans and one-half Ethiopian beans, and Exotic Blend is one-quarter Costa Rican beans and three-quarters Ethiopian beans. Profit on the Yusip Blend is $3.50 per pound, while profit on the Exotic Blend is $4 per pound. Each day, the shop can obtain 200 pounds of Costa Rican beans and 330 pounds of Ethiopian beans, and it uses those beans only in the two blends. If it can sell all that it makes, how many pounds of Yusip Blend and of Exotic Blend should Pete's Coffees prepare each day in order to maximize profit?

26. Pete's Coffees sells two blends of coffee beans, Morning Blend and South American Blend. Morning Blend is one-third Mexican beans and two-thirds Colombian beans, and South American Blend is two-thirds Mexican beans and one-third Colombian beans. Profit on the Morning Blend is

$3 per pound, while profit on the South American Blend is $2.50 per pound. Each day, the shop can obtain 100 pounds of Mexican beans and 80 pounds of Colombian beans, and it uses those beans only in the two blends. If it can sell all that it makes, how many pounds of Morning Blend and of South American Blend should Pete's Coffees prepare each day in order to maximize profit?

 Answer the following questions using complete sentences.

27. What was one of the earliest uses of Karmarkar's method of linear programming?

28. What motivated Karmarkar's invention?

3.2

THE SIMPLEX METHOD: COMPLETE PROBLEMS

In Section 3.1 we saw how to convert a linear programming problem to a matrix problem. In this section, we will see how to use the simplex method to solve a linear programming problem. We'll start by continuing where we left off with the craftsman problem. We have modeled the problem, introduced slack variables, and rewritten the objective function so all variables are on the left side.

Independent Variables	**Slack Variables**
x_1 = number of coffee tables	s_1 = unused hours
x_2 = number of end tables	s_2 = unused money
$(x_1, x_2, s_1, s_2 \geq 0)$	

Constraints

C_1: $6x_1 + 5x_2 + s_1 = 40$ The time constraint

C_2: $200x_1 + 100x_2 + s_2 + 1000$ The money constraint

Objective Function

$-240x_1 - 160x_2 + 0s_1 + 0s_2 + 1z = 0$ z measures profit

We have also made a matrix out of the rewritten constraints and the rewritten objective function (the first simplex matrix) and determined the possible solution that corresponds to that matrix. The first simplex matrix was as follows:

$$\begin{array}{ccccc} x_1 & x_2 & s_1 & s_2 & z \end{array}$$
$$\begin{bmatrix} 6 & 5 & 1 & 0 & 0 & 40 \\ 200 & 100 & 0 & 1 & 0 & 1000 \\ -240 & -160 & 0 & 0 & 1 & 0 \end{bmatrix} \begin{array}{l} \leftarrow C_1 \\ \leftarrow C_2 \\ \leftarrow \text{Objective function} \end{array}$$

The corresponding possible solution was

$$(x_1, x_2, s_1\ s_2) = (0, 0, 40, 1000) \quad \text{with } z = 0$$

This solution is a corner point of the region of possible solutions (see Figure 3.4 on page 132); it is point P_1 (0, 0) (the geometric method does not find values of the slack variables). However, it is not the optimal corner point; it is a possible solution but not the maximal solution. Our goal in the remaining steps of the simplex method is to find the maximal solution. We do this with a procedure identical to that used in the Gauss-Jordan method *except for how we tell where to pivot.* Recall that with the Gauss-Jordan method, pivoting is used to obtain a matrix of the form

FIGURE 3.4
Each simplex matrix has a corresponding possible solution. Each of these possible solutions is a corner point of the region of possible solutions. The possible solution $(x_1, x_2, s_1, s_2) = (0, 0, 40, 1000)$ is the corner point P_1 $(0, 0)$.

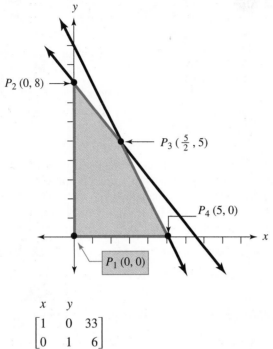

$$\begin{array}{cc} x & y \\ \begin{bmatrix} 1 & 0 & 33 \\ 0 & 1 & 6 \end{bmatrix} \end{array}$$

This is not the case with the simplex method.

Pivoting with the Simplex Method

Step 1 *Look at the last row, the objective function row. Select the most negative entry in that row. The column containing that entry is the **pivot column.** If the last row contains no negative entries, then no pivoting is necessary; the possible solution that corresponds to the matrix is the maximal solution.*

$$\begin{array}{ccccc} x_1 & x_2 & s_1 & s_2 & z \\ \begin{bmatrix} 6 & 5 & 1 & 0 & 0 & 40 \\ 200 & 100 & 0 & 1 & 0 & 1000 \\ -240 & -160 & 0 & 0 & 1 & 0 \end{bmatrix} \end{array}$$ ← Objective function

The only negative entries in the last row are -240 and -160; the most negative is -240. This entry is in the first column, so our pivot column is the first column. We will pivot on one of the entries of the pivot column.

$$\begin{array}{ccccc} x_1 & x_2 & s_1 & s_2 & z \\ \begin{bmatrix} 6 & 5 & 1 & 0 & 0 & 40 \\ 200 & 100 & 0 & 1 & 0 & 1000 \\ -240 & -160 & 0 & 0 & 1 & 0 \end{bmatrix} \end{array}$$
↑
Pivot column

Step 2 *Divide the last entry in each constraint row by the corresponding entry in the pivot column. The row that yields the smallest nonnegative such quotient is the **pivot row.***

$$\begin{bmatrix} 6 & 5 & 1 & 0 & 0 & 40 \\ 200 & 100 & 0 & 1 & 0 & 1000 \\ -240 & -160 & 0 & 0 & 1 & 0 \end{bmatrix} \begin{matrix} \leftarrow 40/6 \approx 6.67 \\ \leftarrow 1000/200 = 5 \\ \leftarrow \text{Not a constraint row} \end{matrix}$$

$$\uparrow$$
Pivot column

Our quotients are 6.67 and 5; the smallest nonnegative quotient is 5, which is in row two. Our pivot row is the second row.

Step 3 *Pivot on the entry in the pivot row and pivot column.*

$$\begin{bmatrix} \boxed{6} & 5 & 1 & 0 & 0 & 40 \\ \boxed{200} & \boxed{100} & \boxed{0} & \boxed{1} & \boxed{0} & \boxed{1000} \\ \boxed{-240} & -160 & 0 & 0 & 1 & 0 \end{bmatrix} \quad \boxed{\leftarrow Pivot\ row}$$

$$\boxed{\uparrow \atop Pivot\ column}$$

We will pivot on the entry "200." Now that we've determined *where* to pivot, we use the row operations as we would with the Gauss-Jordan method.

R2 ÷ 200
$$\begin{bmatrix} 6 & 5 & 1 & 0 & 0 & 40 \\ 1 & 0.5 & 0 & 0.005 & 0 & 5 \\ -240 & -160 & 0 & 0 & 1 & 0 \end{bmatrix}$$

R1 − 6R2 : R1
240R2 + R3 : R3
$$\begin{bmatrix} 0 & 2 & 1 & -0.03 & 0 & 10 \\ 1 & 0.5 & 0 & 0.005 & 0 & 5 \\ 0 & -40 & 0 & 1.2 & 1 & 1200 \end{bmatrix}$$

We have finished our first pivot. The above matrix, the result of the first pivot, is called the **second simplex matrix.**

When to Stop Pivoting

The simplex method requires us to explore a series of matrices, just as the geometric method requires us to explore a series of corner points. Each simplex matrix provides us with a corner point of the region of possible solutions; the final simplex matrix will provide us with the optimal corner point, the point that solves the problem.

The second simplex matrix

$$\begin{array}{ccccc} x_1 & x_2 & s_1 & s_2 & z \\ \begin{bmatrix} 0 & 2 & 1 & -0.03 & 0 & 10 \\ 1 & 0.5 & 0 & 0.005 & 0 & 5 \\ 0 & -40 & 0 & 1.2 & 1 & 1200 \end{bmatrix} \end{array}$$

provides us with a corner point. Because the x_1, s_1, and z columns contain only a 1 and 0's, their values are found as in the Gauss-Jordan method. The x_2 and s_2 columns are not of this type, so their values are zero. Thus, the corner point that corresponds to the second simplex matrix is

FIGURE 3.5

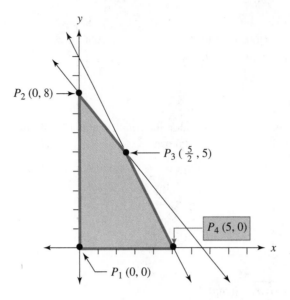

P_2 (0, 8)

P_3 ($\frac{5}{2}$, 5)

P_4 (5, 0)

P_1 (0, 0)

$(x_1, x_2, s_1, s_2) = (5, 0, 10, 0)$ with $z = 1{,}200$

This is corner point P_4 (see Figure 3.5).

Because x_1 is the number of coffee tables, x_2 the number of end tables, s_1 the number of unused hours, and s_2 the amount of unused money, this possible solution means "make five coffee tables and no end tables, have ten unused hours and no unused money, and make a profit of $1200."

Is this the optimal corner point, the point that solves the craftsman's problem? We answer this question by determining whether it is possible to pivot again. We must apply the steps in pivoting with the simplex method to the second simplex matrix.

Step 1 *Select the most negative entry in the last row, the objective function row. The column containing that entry is the pivot column. If the last row contains no negative entries, then no pivoting is necessary; the possible solution that corresponds to the matrix is the maximal solution.*

$$
\begin{array}{ccccc}
x_1 & x_2 & s_1 & s_2 & z \\
\end{array}
$$
$$
\begin{bmatrix}
0 & 2 & 1 & -0.03 & 0 & 10 \\
1 & 0.5 & 0 & 0.005 & 0 & 5 \\
0 & -40 & 0 & 1.2 & 1 & 1200
\end{bmatrix}
$$
$\leftarrow$ Objective function

Because there is a negative entry in the objective function row, we must pivot again. This negative entry is in the x_2 column, so our pivot column is the x_2 column.

$$
\begin{array}{ccccc}
x_1 & x_2 & s_1 & s_2 & z \\
\end{array}
$$
$$
\begin{bmatrix}
0 & 2 & 1 & -0.03 & 0 & 10 \\
1 & 0.5 & 0 & 0.005 & 0 & 5 \\
0 & -40 & 0 & 1.2 & 1 & 1200
\end{bmatrix}
$$

$\uparrow$
Pivot column

Step 2   *Divide the last entry in each constraint row by the corresponding entry in the pivot column. The row that yields the smallest nonnegative such quotient is the pivot row.*

$$\begin{bmatrix} 0 & 2 & 1 & -0.03 & 0 & 10 \\ 1 & 0.5 & 0 & 0.005 & 0 & 5 \\ 0 & -40 & 0 & 1.2 & 1 & 1200 \end{bmatrix}$$

$\leftarrow$ 10/2 = 5
$\leftarrow$ 5/0.5 = 10
$\leftarrow$ Not a constraint row

$\uparrow$
Pivot column

Our quotients are 5 and 10; the smallest nonnegative quotient is 5, which is in row one. Our pivot row is the first row.

Step 3   *Pivot on the entry in the pivot row and pivot column.*

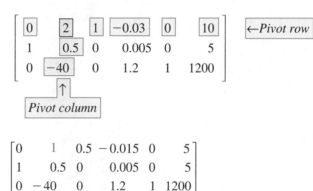

$$\begin{bmatrix} \boxed{0} & \boxed{2} & \boxed{1} & \boxed{-0.03} & \boxed{0} & \boxed{10} \\ 1 & \boxed{0.5} & 0 & 0.005 & 0 & 5 \\ 0 & \boxed{-40} & 0 & 1.2 & 1 & 1200 \end{bmatrix}$$

$\leftarrow$*Pivot row*

$\uparrow$
Pivot column

R1 ÷ 2

$$\begin{bmatrix} 0 & 1 & 0.5 & -0.015 & 0 & 5 \\ 1 & 0.5 & 0 & 0.005 & 0 & 5 \\ 0 & -40 & 0 & 1.2 & 1 & 1200 \end{bmatrix}$$

-0.5R1 + R2 : R2
40R1 + R3 : R3

$$\begin{array}{cccccc} x_1 & x_2 & s_1 & s_2 & z & \\ \begin{bmatrix} 0 & 1 & 0.5 & -0.015 & 0 & 5 \\ 1 & 0 & -0.25 & 0.0125 & 0 & 2.5 \\ 0 & 0 & 20 & 0.6 & 1 & 1400 \end{bmatrix} \end{array}$$

This matrix is our third *and last* simplex matrix, because it is not possible to pivot further—the bottom row contains no negative entries. The solution that corresponds to this matrix is the optimal corner point:

$$(x_1, x_2, s_1, s_2) = (2.5, 5, 0, 0) \quad \text{with } z = 1400$$

This solution means "each week make 2.5 coffee tables and 5 end tables, have no unused hours and no unused money, and make a profit of $1400" (see Figure 3.6 on page 136).

Why the Simplex Method Works

In the craftsman problem, our first simplex matrix was

$$\begin{array}{ccccc} x_1 & x_2 & s_1 & s_2 & z \\ \begin{bmatrix} 6 & 5 & 1 & 0 & 0 & 40 \\ 200 & 100 & 0 & 1 & 0 & 1000 \\ -240 & -160 & 0 & 0 & 1 & 0 \end{bmatrix} \end{array}$$

and its corresponding possible solution was

$$(x_1, x_2, s_1, s_2) = (0, 0, 40, 1000) \quad \text{with } z = 0$$

FIGURE 3.6

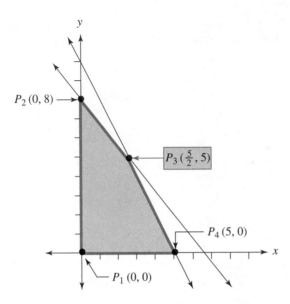

This means "make no tables and generate no profit." We selected the first column as our pivot column because it was the column with the most negative number in the bottom. Why do we select the column with the most negative number in the bottom?

The possible solution corresponding to the first simplex matrix involves no profit ($z = 0$). Certainly, the profit can be increased by making some tables. If the craftsman were to make only one type of table, which table would be best? Coffee tables generate more profit per table ($\$240$ versus $\$160$), so the craftsman would be better off making coffee tables. This choice of coffee tables over end tables, or of $\$240$ over $\$160$, was made in the pivoting process when we chose the first column as the pivot column. That choice was made because 240 is larger than 160.

After selecting the pivot column, we selected the second row as our pivot row because 5 is the smallest nonnegative quotient. Why do we select the row with the smallest nonnegative quotient?

$$\begin{bmatrix} 6 & 5 & 1 & 0 & 0 & 40 \\ 200 & 100 & 0 & 1 & 0 & 1000 \\ -240 & -160 & 0 & 0 & 1 & 0 \end{bmatrix}$$

$\leftarrow 40/6 = 6\frac{2}{3}$
$\leftarrow 1000/200 = 5$
$\leftarrow$ Not a constraint row

If the craftsman were to make only coffee tables, he should make the largest amount allowed by the constraints when 0 is substituted for x_2 (that is, when no end tables are made).

Time Constraint

$C_1: 6x_1 + 5x_2 \le 40$

$6x_1 + 5 \cdot 0 \le 40$

$6x_1 \le 40$

$x \le \dfrac{40}{6} = \dfrac{20}{3} = 6\dfrac{2}{3}$

Money Constraint

$C_2: 200x_1 + 100x_2 \le 1000$

$200x_1 + 100 \cdot 0 \le 1000$

$200x_1 \le 1000$

$x_1 \le \dfrac{1000}{200} = 5$

The time constraint will not be violated as long as $x_1 \leq 6\frac{2}{3}$; the money constraint will not be violated as long as $x_1 \leq 5$. The craftsman can make at most $x_1 = 5$ coffee tables and not violate either constraint. This choice of 5 over $6\frac{2}{3}$ was made in the pivoting process when we chose the second row as the pivot row. That choice was made because 5 is less than $6\frac{2}{3}$.

We have just used logic to determine that if the craftsman makes only one type of table, he should make coffee tables, and that he could make at most five coffee tables without violating his constraints. The simplex method leads to the same conclusion. The second simplex matrix was

$$
\begin{array}{ccccc}
x_1 & x_2 & s_1 & s_2 & z \\
\end{array}
$$
$$
\begin{bmatrix}
0 & 2 & 1 & -0.03 & 0 & 10 \\
1 & 0.5 & 0 & 0.005 & 0 & 5 \\
0 & -40 & 0 & 1.2 & 1 & 1200
\end{bmatrix}
$$

and its corresponding possible solution was

$$(x_1, x_2, s_1, s_2) = (5, 0, 10, 0) \quad \text{with } z = 1200$$

This does suggest that the craftsman make five coffee tables and no end tables. This matrix was not the final simplex matrix, so this possible solution was not the optimal solution. That is, the craftsman will increase his profit if he makes more than one type of table.

The last row of this matrix, the objective function row, is

$$0 \quad -40 \quad 0 \quad 1.2 \quad 1 \quad 1200$$

We pivoted again because of the presence of a negative number in this row. This pivoting resulted in an increased profit. Why does the presence of a negative number imply that if we profit again, the profit will be increased? The last row represents the equation

$$0x_1 + -40x_2 + 0s_1 + 1.2s_2 + 1z = 1200$$

which can be rewritten as

$$z = 1200 + 40x_2 - 1.2s_2$$

by solving for z. According to this last equation, the profit is $z = 1200$ if x_2 and s_2 are both zero. If x_2 is positive, then the profit could be larger than 1200. Thus, it is possible to achieve a larger profit by changing x_2 from zero to a positive number. Because x_2 measures the number of end tables, this means that the craftsman should make some end tables. Our last pivot did in fact result in his making some end tables.

The following list of steps includes those steps developed in Section 3.1, as well as those developed in this section. Notice that you can check your work at Step 7; if you have made an arithmetic error during your pivot, catch it here before you go any further.

Simplex Method Steps

1. *Model the problem.*
   a. List the independent variables.
   b. List the constraints and translate them into linear inequalities.
   c. Find the objective and translate it into a linear equation.
2. *Convert each constraint from an inequality to an equation.*
   a. Use one slack variable for each constraint.
   b. Determine what each slack variable measures.
3. *Rewrite the objective function with all variables on the left side.*
4. *Make a matrix out of the rewritten constraints and the rewritten objective function. This is the first simplex matrix.*
5. *Determine the possible solution that corresponds to the matrix.*
   a. Columns containing only a 1 and 0's—turn at 1.
   b. Other columns—the value of the variable is zero.
6. *Pivot to find a better possible solution.*
   a. The pivot column is the column with the most negative entry in the last row. In case of a tie, choose either.
   b. Divide the last entry of each constraint row by the entry in the pivot column. The pivot row is the row with the smallest nonnegative such quotient. In case of a tie, choose either.
   c. Pivot on the selected row and column. This gives a new simplex matrix.
7. *Determine the possible solution that corresponds to the matrix* (as in Step 5).
   ✔ Check your work by seeing if the solution substitutes into the objective function and constraints.
8. *Determine whether the current possible solution maximizes the objective function.*
   a. If the last row of the new simplex matrix has no negative entries, then the problem is solved, and the current solution is the maximal solution.
   b. If the last row of the new simplex matrix has one or more negative entries, then pivot again. Return to Step 6.
9. *Interpret the final solution.* Express the solution in words, as a solution to a real-world problem rather than a mathematical problem.

EXAMPLE 1 The Leather Factory has one sewing machine with which it makes coats and vests. Each coat requires 50 minutes on the sewing machine and uses 12 square feet of leather. Each vest requires 30 minutes on the sewing machine and uses 8 square feet of leather. The sewing machine is available 8 hours a day, and the Leather Factory can obtain 118 square feet of leather a day. The coats sell for \$175 each, and the vests sell for \$100 each. Find the daily production level that would yield the maximum income.

Solution Step 1 *Model the problem.*

Independent Variables

x_1 = number of coats made each day

x_2 = number of vests made each day

Constraints

C_1: sewing machine hours $\leq$ 8 hours $\qquad$ = 480 minutes

(coat sewing time) $\quad$ + $\quad$ (vest sewing time) $\quad$ $\leq$ 480 minutes

$$\begin{pmatrix} \text{time} \\ \text{per} \\ \text{coat} \end{pmatrix} \cdot \begin{pmatrix} \text{number} \\ \text{of} \\ \text{coats} \end{pmatrix} + \begin{pmatrix} \text{time} \\ \text{per} \\ \text{vest} \end{pmatrix} \cdot \begin{pmatrix} \text{number} \\ \text{of} \\ \text{vests} \end{pmatrix} \leq 480 \text{ minutes}$$

$$50 \qquad x_1 \qquad + \qquad 30 \qquad x_2 \qquad \leq 480$$

C_2: leather used $\leq$ 118 square feet

(coat leather) $\qquad\qquad$ + $\quad$ (vest leather) $\qquad$ $\leq$ 118 square feet

$$\begin{pmatrix} \text{leather} \\ \text{per} \\ \text{coat} \end{pmatrix} \cdot \begin{pmatrix} \text{number} \\ \text{of} \\ \text{coats} \end{pmatrix} + \begin{pmatrix} \text{leather} \\ \text{per} \\ \text{vest} \end{pmatrix} \cdot \begin{pmatrix} \text{number} \\ \text{of} \\ \text{vests} \end{pmatrix} \leq 118 \text{ square feet}$$

$$12 \qquad x_1 \qquad + \qquad 8 \qquad x_2 \qquad \leq 118$$

Objective Function

Maximize z = income = coat income + vest income.

$$z = \begin{pmatrix} \text{price} \\ \text{per} \\ \text{coat} \end{pmatrix} \cdot \begin{pmatrix} \text{number} \\ \text{of} \\ \text{coats} \end{pmatrix} + \begin{pmatrix} \text{price} \\ \text{per} \\ \text{vest} \end{pmatrix} \cdot \begin{pmatrix} \text{number} \\ \text{of} \\ \text{vests} \end{pmatrix}$$

$$= \quad 175 \qquad x_1 \qquad + \qquad 100 \qquad x_2$$

Step 2 *Convert each constraint from an inequality to an equation.*

C_1: $50x_1 + 30x_2 \leq 480$

$50x_1 + 30x_2 + s_1 = 480$

where s_1 = unused sewing machine time.

C_2: $12x_1 + 8x_2 \leq 118$

$12x_1 + 8x_2 + s_2 = 118$

where s_2 = unused leather.

Step 3 *Rewrite the objective function with all variables on the left side.*

$z = 175x_1 + 100x_2$

$-175x_1 - 100x_2 + 1z = 0$

Step 4 *Make a matrix out of the rewritten constraints and the rewritten objective function.* The first simple matrix is as follows:

x_1	x_2	s_1	s_2	z	
50	30	1	0	0	480
12	8	0	1	0	118
-175	-100	0	0	1	0

Step 5 *Determine the possible solution that corresponds to the first simplex matrix.* The solution is $(x_1, x_2, s_1, s_2) = (0, 0, 480, 118)$ with $z = 0$. This means "make no product and produce no income."

Step 6 *Pivot to find a better possible solution.*

$$
\begin{array}{ccccc}
x_1 & x_2 & s_1 & s_2 & z \\
\end{array}
$$
$$
\begin{bmatrix}
50 & 30 & 1 & 0 & 0 & 480 \\
12 & 8 & 0 & 1 & 0 & 118 \\
-175 & -100 & 0 & 0 & 1 & 0
\end{bmatrix}
$$
$$
\uparrow
$$
Pivot column

Our pivot column is the first column, because -175 is the most negative entry in the bottom row. To select the pivot row, divide the last entry in each constraint row by the entry in the pivot column and select the row with the smallest nonnegative such quotient.

$$
\begin{bmatrix}
50 & 30 & 1 & 0 & 0 & 480 \\
12 & 8 & 0 & 1 & 0 & 118 \\
-175 & -100 & 0 & 0 & 1 & 0
\end{bmatrix}
\qquad
\begin{array}{l}
\leftarrow 480/50 = 9.6 \\
\leftarrow 118/12 \approx 9.8333 \\
\leftarrow \text{Not a constraint row}
\end{array}
$$

Because 9.6 is the smallest nonnegative quotient, our pivot row is the first row.

$$
\begin{bmatrix}
\boxed{50} & \boxed{30}\;\boxed{1}\;\boxed{0}\;\boxed{0}\;\boxed{480} \\
\boxed{12} & 8 \quad 0 \quad 1 \quad 0 \quad 118 \\
\boxed{-175} & -100 \quad 0 \quad 0 \quad 1 \quad 0
\end{bmatrix}
\quad \boxed{\leftarrow Pivot\ row}
$$
$$
\uparrow
$$
$$
\boxed{Pivot\ column}
$$

We will pivot on the 50.

$$
R1 \div 50 \qquad
\begin{bmatrix}
1 & 0.6 & 0.02 & 0 & 0 & 9.6 \\
12 & 8 & 0 & 1 & 0 & 118 \\
-175 & -100 & 0 & 0 & 1 & 0
\end{bmatrix}
$$

$$
\begin{array}{l}
-12R1 + R2 : R2 \\
175R1 + R3 : R3
\end{array}
\qquad
\begin{array}{ccccc}
x_1 & x_2 & s_1 & s_2 & z \\
\end{array}
$$
$$
\begin{bmatrix}
1 & 0.6 & 0.02 & 0 & 0 & 9.6 \\
0 & 0.8 & -0.24 & 1 & 0 & 2.8 \\
0 & 5 & 3.5 & 0 & 1 & 1680
\end{bmatrix}
$$

Step 7 *Determine the possible solution that corresponds to the matrix.* The possible solution is $(x_1, x_2, s_1, s_2) = (9.6, 0, 0, 2.8)$ with $z = 1680$.

> ✔ Check your work by seeing if the solution substitutes into the objective function and constraints:
>
> $C_1\colon 50x_1 + 30x_2 + s_1 = 50 \cdot 9.6 + 30 \cdot 0 + 0 = 480$
>
> $C_2\colon 12x_1 + 8x_2 + s_2 = 12 \cdot 9.6 + 8 \cdot 0 + 2.8 = 118$
>
> Objective function: $z = 175x_1 + 100x_2 = 175 \cdot 9.6 + 100 \cdot 0$
> $= 1680$

If any of these substitutions failed, we would know to stop and find our error.

Step 8 *Determine whether the current possible solution maximizes the objective function.* The last row contains no negative entries, so we don't need to pivot again, and the problem is finished. Some linear programming problems involve only one pivot, some involve two, and some involve more. This problem involved only one pivot.

Step 9 *Interpret the final solution.* Recall that x_1 measures the number of coats made per day, x_2 measures the number of vests, s_1 measures the number of slack machine hours, and s_2 measures the amount of unused leather. Thus, the Leather Factory should make 9.6 coats per day and no vests. (If workers spend all day making coats, they will make 9.6 coats. They'll finish the tenth coat the next day.) This will result in no unused machine hours and 2.8 square feet of unused leather, and will generate a maximum income of $1680 per day. These profit considerations indicate that the Leather Factory should increase the cost of its leather vests if it wants to sell vests. •

3.2

EXERCISES

In Exercises 1–4, determine where to pivot.

1.

x_1	x_2	s_1	s_2	z	
5	0	3	1	0	12
3	0	19	0	10	22
−21	1	−48	0	0	19

2.

x_1	x_2	s_1	s_2	z	
1	31	0	9	0	7.4
0	20	0	15	1	4.9
0	−1	1	0	0	20

3.

x_1	x_2	s_1	s_2	s_3	z	
19	3	0	0	1	0	5
32	7	0	1	0	0	93
−55	8	1	0	0	0	8
−21	32	0	0	0	1	96

4.

x_1	x_2	s_1	s_2	s_3	z	
0	1	7	0	62	0	5
0	0	5	1	5	0	93
1	0	0	0	0	0	78
0	0	−9	1	−1	1	96

In Exercises 5–8, (a) determine where to pivot, (b) pivot, and (c) determine the solution that corresponds to the resulting matrix.

5.

x_1	x_2	s_1	s_2	z	
1	2	1	0	0	3
4	1	0	1	0	2
−6	−4	0	0	1	0

6.

x_1	x_2	s_1	s_2	z	
8	2	1	0	0	4
5	1	0	1	0	3
−2	4	0	0	1	0

7.

x_1	x_2	s_1	s_2	z	
1	1	5	0	0	3
3	0	1	1	0	12
−6	0	−2	0	1	6

8.

x_1	x_2	s_1	s_2	z	
0	2	1	5	0	8
1	3	0	9	0	6
0	−4	0	2	2	12

Use the simplex method to solve Exercises 9–16. (Some of these exercises were started in Section 3.1.)

9. Five friends, each of whom is an experienced baker, form a company that will make bread and cakes and sell them to local restaurants and specialty stores. Each loaf of bread requires 50 minutes of labor and ingredients costing $0.90 and can be sold for $1.20 profit. Each cake requires 30 minutes of labor and ingredients costing $1.50 and can be sold for $4.00 profit. The partners agree that no one will work more than 8 hours a day. Their financial resources do not allow them to spend more than $190 per day on ingredients. How many loaves of bread and how many cakes should they make each day in order to maximize their profit? What is the maximum profit? Will this leave any extra time or money? If so, how much?

10. A craftswoman produces two products, floor lamps and table lamps. Production of one floor lamp requires 75 minutes of her labor and materials that cost $25. Production of one table lamp requires 50 minutes of labor and materials that cost $20. The craftswoman wants to work no more than 40 hours each week, and her financial resources allow her to pay no more than $900 for materials each week. If she can sell as many lamps as she can make and if her profit is $40 per floor lamp and $32 per table lamp, how many floor lamps and

how many table lamps should she make each week to maximize her weekly profit? What is that maximum profit? Will this leave any unused time or money? If so, how much?

11. A furniture manufacturing firm makes sofas and chairs, each of which is available in several styles. Each sofa, regardless of style, requires 8 hours in the upholstery shop and 4 hours in the carpentry shop and can be sold for a profit of $450. Each chair requires 6 hours in the upholstery shop and 3.5 hours in the carpentry shop and can be sold for a profit of $375. There are nine people working in the upholstery shop and five in the carpentry shop, each of whom can work no more than 40 hours per week. How many sofas and how many chairs should the firm make each week in order to maximize its profit? What is that maximum profit? Would this leave any extra time in the upholstery shop or the carpentry shop? If so, how much?

12. City Electronics Distributors handles two lines of televisions, the Packard and the Bell. The company purchases up to $57,000 worth of television sets from the manufacturers each month, which it stores in its 9,000-cubic-foot warehouse. The Packards come in 36-cubic-foot packing crates, and the Bells come in 30-cubic-foot crates. The Packards cost City Electronics $200 each and can be sold to a retailer for a $200 profit, while the Bells cost $250 each and can be sold for a $260 profit. How many sets should City Electronics order each month in order to maximize profit? What is that maximum profit? Would this leave any unused storage space or money? If so, how much?

13. J & M Winery makes two jug wines, House White and Premium White, which it sells to restaurants. House White is a blend of 75% French colombard grapes and 25% sauvignon blanc grapes, and Premium White is 75% sauvignon blanc grapes and 25% French colombard grapes. J & M also makes a Sauvignon Blanc, which is 100% sauvignon blanc grapes. Profit on the House White is $1.00 per liter, profit on the Premium White is $1.50 per liter, and profit on the Sauvignon Blanc is $2.00 per liter. This season, J & M can obtain 30,000 pounds of French colombard grapes and 20,000 pounds of sauvignon blanc grapes. It takes 2 pounds of grapes to make 1 liter of wine. If it can sell all that it makes, how many liters of House White, of Premium White, and of Sauvignon Blanc should J & M prepare in order to maximize profit? What is that maximum profit? Would this leave any extra grapes? If so, what amount?

14. J & M Winery makes two jug wines, House Red and Premium Red, which it sells to restaurants. House Red is a blend of 20% cabernet sauvignon grapes and 80% gamay grapes. Premium Red is 60% cabernet sauvignon grapes and 40% gamay grapes. J & M also makes a Cabernet Sauvignon, which is 100% cabernet sauvignon grapes.

Profit on the House Red is $0.90 per liter, profit on the Premium Red is $1.60 per liter, and profit on the Cabernet Sauvignon is $2.50 per liter. This season, J & M can obtain 30,000 pounds of gamay grapes and 22,000 pounds of cabernet sauvignon grapes. It takes 2 pounds of grapes to make 1 liter of wine. If it can sell all that it makes, how many liters of House Red, of Premium Red, and of Cabernet Sauvignon should J & M prepare in order to maximize profit? What is that maximum profit? Would this leave any extra grapes? If so, what amount?

15. Pete's Coffees sells two blends of coffee beans, Smooth Sipper and Kona Blend. Smooth Sipper is composed of equal amounts of Kona, Colombian, and Arabian beans, and Kona Blend is one-half Kona beans and one-half Colombian beans. Profit on the Smooth Sipper is $3 per pound, while profit on the Kona Blend is $4 per pound. Each day, the shop can obtain 100 pounds of Kona beans, 200 pounds of Colombian beans, and 200 pounds of Arabian beans. It uses those beans only in the two blends. If it can sell all that it makes, how many pounds of Smooth Sipper and of Kona Blend should Pete's Coffees prepare each day in order to maximize profit? What is that maximum profit? Would this leave any extra beans? If so, what amount?

16. Pete's Coffees sells two blends of coffee beans, African Blend and Major Thompson's Blend. African Blend is one-half Tanzanian beans and one-half Ethiopian beans, and Major Thompson's Blend is one-quarter Tanzanian beans and three-quarters Colombian beans. Profit on the African Blend is $4.25 per pound, while profit on Major Thompson's Blend is $3.50 per pound. Each day, the shop can obtain 300 pounds of Tanzanian beans, 200 pounds of Ethiopian beans, and 450 pounds of Colombian beans. It uses those beans only in the two blends. If it can sell all that it makes, how many pounds of African Blend and of Major Thompson's Blend should Pete's Coffees prepare each day in order to maximize profit? What is that maximum profit? Would this leave any extra beans? If so, what amount?

 Answer the following questions using complete sentences.

17. Who invented the simplex method of linear programming?

18. Compare and contrast the geometric method of linear programming with the simplex method of linear programming. Give advantages and disadvantages of each.

19. Compare and contrast the simplex method of linear programming with Karmarkar's method of linear programming. Give advantages and disadvantages of each.

20. Explain how the maximal value of the objective function can be found without looking at the number in the lower-right corner of the final simplex matrix.

21. Explain why the entries in the last row have had their signs changed, but the entries in all of the other rows have not had their signs changed.

TECHNOLOGY AND THE SIMPLEX METHOD

The simplex method has two major advantages over the geometric method: It will work with problems that have more than two independent variables, and it is amenable to the use of a computer. Frequently, real-world linear programming problems have so many constraints and variables that the calculations involved in pivoting can be overwhelming if done by hand. This difficulty is eliminated by use of the Amortrix computer software introduced in Chapter 2 or a graphing calculator.

EXAMPLE 2   Our friend the craftsman now owns his own shop. He still makes coffee tables and end tables, but he makes each in three different styles: antique, art deco, and modern. The amount of labor and the cost of the materials required by each product, along with the profit they generate, are shown in Figure 3.7. The craftsman now has two employees, each of whom can work up to 40 hours a week. The craftsman himself frequently has to work more than 40 hours a week, but he will not allow himself to exceed 50 hours a week. His new bank loan allows him to spend up to $4000 a week on materials. How many coffee tables and end tables of each style should he make each week to maximize his profit?

FIGURE 3.7

Item	Style	Hours of Labor	Cost of Materials	Profit
Coffee Table	antique	6.00	$230	$495
Coffee Table	art deco	6.25	$220	$500
Coffee Table	modern	5.00	$190	$430
End Table	antique	5.25	$125	$245
End Table	art deco	5.75	$120	$250
End Table	modern	4.5	$105	$250

Solution

Independent Variables

x_1 = number of antique coffee tables

x_2 = number of art deco coffee tables

x_3 = number of modern coffee tables

x_4 = number of antique end tables

x_5 = number of art deco end tables

x_6 = number of modern end tables

Constraints

C_1: hours worked $\leq 2 \cdot 40 + 50 = 130$

$$\begin{pmatrix} \text{antique} \\ \text{coffee} \\ \text{table} \\ \text{hours} \end{pmatrix} + \cdots + \begin{pmatrix} \text{modern} \\ \text{end} \\ \text{table} \\ \text{hours} \end{pmatrix} \leq 130$$

$6x_1 + 6.25x_2 + 5x_3 + 5.25x_4 + 5.75x_5 + 4.5x_6 \leq 130$

$6x_1 + 6.25x_2 + 5x_3 + 5.25x_4 + 5.75x_5 + 4.5x_6 + s_1 = 130$

C_2: money spent ≤ 4000

$$\begin{pmatrix} \text{antique} \\ \text{coffee} \\ \text{table} \\ \text{money} \end{pmatrix} + \cdots + \begin{pmatrix} \text{modern} \\ \text{end} \\ \text{table} \\ \text{money} \end{pmatrix} \leq 4000$$

$230x_1 + 220x_2 + 190x_3 + 125x_4 + 120x_5 + 105x_6 \leq 4000$

$230x_1 + 220x_2 + 190x_3 + 125x_4 + 120x_5 + 105x_6 + s_2 = 4000$

where s_1 = unused hours and s_2 = unused money for materials.

Objective Function

Maximize z = profit

$$= \left(\begin{array}{c}\text{antique coffee}\\ \text{table profit}\end{array}\right) + \cdots + \left(\begin{array}{c}\text{modern end}\\ \text{table profit}\end{array}\right)$$

$$= 495x_1 + 500x_2 + 430x_3 + 245x_4 + 250x_5 + 250x_6$$

$$-495x_1 - 500x_2 - 430x_3 - 245x_4 - 250x_5 - 250x_6 + 0s_1 + 0s_2 + 1z = 0$$

The first simplex matrix:

$$\begin{array}{cccccccccc} x_1 & x_2 & x_3 & x_4 & x_5 & x_6 & s_1 & s_2 & z \\ \left[\begin{array}{ccccccccc} 6.00 & 6.25 & 5.00 & 5.25 & 5.75 & 4.50 & 1 & 0 & 0 & 130 \\ 230 & 220 & 190 & 125 & 120 & 105 & 0 & 1 & 0 & 4000 \\ -495 & -500 & -430 & -245 & -250 & -250 & 0 & 0 & 1 & 0 \end{array}\right] \end{array}$$

Its corresponding possible solution:

$$(x_1, x_2, x_3, x_4, x_5, x_6, s_1, s_2) = (0, 0, 0, 0, 0, 0, 130, 4000) \quad \text{with } z = 0$$

At this point, you enter the first simplex matrix into the computer or a graphing calculator.

$$\begin{array}{cccccccccc} x_1 & x_2 & x_3 & x_4 & x_5 & x_6 & s_1 & s_2 & z \\ \left[\begin{array}{ccccccccc} 6.00 & \boxed{6.25} & 5.00 & 5.25 & 5.75 & 4.50 & 1 & 0 & 0 & 130 \\ 230 & \boxed{220} & \boxed{190} & \boxed{125} & \boxed{120} & \boxed{105} & \boxed{0} & \boxed{1} & \boxed{0} & \boxed{4000} \\ -495 & \boxed{-500} & -430 & -245 & -250 & -250 & 0 & 0 & 1 & 0 \end{array}\right] \end{array} \begin{array}{l} \leftarrow 130/6.25 = 20.8 \\ \leftarrow 4000/220 \approx 18.2 \end{array}$$

After determining that the appropriate pivot is row 2, column 2, you instruct the computer or graphing calculator to divide row 2 by 220. After completing this pivot and performing all further pivots, check that your answer is reasonable by substituting it into the constraints and objective function. (See Exercise 26.)

In some linear programming problems, decimal answers make sense. In the example above, x_1 measures the number of antique coffee tables made each week. A decimal answer such as 4.5 would make sense here, because the fifth table would be halfway done at the end of the week and finished at the beginning of the next week.

In some linear programming problems, decimal answers do not make sense. If x_1 measured the number of loaves of bread a bakery made per day, then x_1 would have to be a whole number. A baker cannot make a portion of a loaf of bread. There are specific linear programming methods for problems in which the solution must be a whole number, but such methods are beyond the scope of this book. For our purposes, we will round off decimal answers to the nearest whole number when a decimal answer would not be reasonable.

EXERCISES

Use the Amortrix computer software or a graphing calculator to solve the linear programming problems in Exercises 22–25.

22. *Objective Function:*
Maximize $z = 25x_1 + 53x_2 + 18x_3 + 7x_4$
Constraints:
$3x_1 + 2x_2 + 5x_3 + 12x_4 \leq 28$
$4x_1 + 5x_2 + x_3 + 7x_4 \leq 32$
$x_1 + 7x_2 + 9x_3 + 10x_4 \leq 25$

23. *Objective Function:*
Maximize $z = 275x_1 + 856x_2 + 268x_3 + 85x_4$
Constraints:
$5.2x_1 + 9.8x_2 + 7.2x_3 + 3.7x_4 \leq 33.6$
$3.9x_1 + 5.3x_2 + 1.4x_3 + 2.5x_4 \leq 88.3$
$5.2x_1 + 7.7x_2 + 4.6x_3 + 4.6x_4 \leq 24.7$

24. *Objective Function:*
Maximize $z = 37x_1 + 19x_2 + 53x_3 + 49x_4$
Constraints:
$6.32x_1 + 7.44x_2 + 8.32x_3 + 1.46x_4 + 9.35x_5 \leq 63$
$8.36x_1 + 5.03x_2 + x_3 + 5.25x_5 \leq 32$
$1.14x_1 + 9.42x_2 + 9.39x_3 + 10.42x_4 + 9.32x_5 \leq 14.7$

25. *Objective Function:*
Maximize $z = 17x_1 + 26x_2 + 85x_3 + 63x_4 + 43x_5$
Constraints:
$72x_1 + 46x_2 + 73x_3 + 26x_4 + 54x_5 \leq 185$
$37x_1 + 84x_2 + 45x_3 + 83x_4 + 85x_5 \leq 237$

Use the Amortrix computer software or a graphing calculator to solve exercises 26–33. Interpret each solution; that is, explain the values of each slack variable and z in addition to answering the question.

26. Finish Example 2 from this section.

27. The Leather Factory has one sewing machine with which it makes coats and vests. Each coat requires 50 minutes on the sewing machine and uses 12 square feet of leather. Each vest requires 30 minutes on the sewing machine and uses 8 square feet of leather. The sewing machine is available 8 hours a day, and the Leather Factory can obtain 118 square feet of leather a day. The coats generate a profit of $175 each, and the vests generate a profit of $100 each. Find the daily production level that would yield maximum profit.

28. The five friends have been so successful with their baking business that they have opened their own shop—the Five Friends Bakery. The bakery sells two types of breads: Nine Grain and Sour Dough; two types of cakes: Chocolate and Poppy Seed; and two types of muffins: Blueberry and Apple Cinnamon. Production data are given in Figure 3.8. The five partners have hired four workers. No one works more than 40 hours a week. Their business success (together with their new bank loan) allows them to spend up to $2000 per week on ingredients. What weekly production schedule should they follow in order to maximize their profit?

29. Fiat Lux, Inc., manufactures floor lamps, table lamps, and desk lamps. Production data are given in Figure 3.9. There are seven employees in the wood shop, five in the metal shop, three in the electrical shop, and one in testing; each works no more than 40 hours per week. If Fiat Lux, Inc., can sell all the lamps it produces, how many should it produce each week?

FIGURE 3.8

	Labor	Cost of Ingredients	Profit
Nine Grain Bread	50 min. per loaf	$1.05 per loaf	$0.60 per loaf
Sour Dough Bread	50 min. per loaf	$0.95 per loaf	$0.70 per loaf
Chocolate Cake	35 min. per cake	$2.00 per cake	$2.50 per cake
Poppy Seed Cake	30 min. per cake	$1.55 per cake	$2.00 per cake
Blueberry Muffins	15 min. per dozen	$1.60 per dozen	$16.10 per dozen
Apple Cinnamon Muffins	15 min. per dozen	$1.30 per dozen	$14.40 per dozen

FIGURE 3.9

	Wood Shop	Metal Shop	Electrical Shop	Testing	Profit
Floor Lamps	20 min.	30 min.	15 min.	5 min.	$55
Table Lamps	25 min.	15 min.	12 min.	5 min.	$45
Desk Lamps	30 min.	10 min.	11 min.	4 min.	$40

FIGURE 3.11

	Size	Cost	Profit
Packard 18″	30 cubic ft.	$150	$175
Packard 24″	36 cubic ft.	$200	$200
Packard rear projection	65 cubic ft.	$450	$720
Bell 18″	28 cubic ft.	$150	$170
Bell 30″	38 cubic ft.	$225	$230

30. A furniture manufacturing firm makes sofas, love seats, easy chairs, and recliners. Each piece of furniture is constructed in the carpentry shop, then upholstered, and finally coated with a protective coating. Production data are given in Figure 3.10. Fifteen people work in the upholstery shop, nine work in the carpentry shop, and one person applies the protective coating, and each person can work no more than 40 hours per week. How many pieces of furniture should the firm manufacture each week in order to maximize its profit?

FIGURE 3.10

	Carpentry	Upholstery	Coating	Profit
Sofas	4 hours	8 hours	20 min.	$450
Love Seats	4 hours	7.5 hours	20 min.	$350
Easy Chairs	3 hours	6 hours	15 min.	$300
Recliners	6 hours	6.5 hours	15 min.	$475

31. City Electronics Distributors handles two lines of televisions, the Packard and the Bell. The Packard comes in three sizes: 18-inch, 24-inch, and rear projection; the Bell comes in two sizes: 18-inch and 30-inch. City Electronics purchases up to $57,000 worth of television sets from manufacturers each month and stores them in its 9000-cubic-foot warehouse. Storage and financial data are given in Figure 3.11. How many sets should City Electronics order each month in order to maximize profit?

32. J & M Winery makes two jug wines, House White and Premium White, and two higher-quality wines, Sauvignon Blanc and Chardonnay, which it sells to restaurants, supermarkets, and liquor stores. House White is a blend of 40% French colombard grapes, 40% chenin blanc grapes, and 20% sauvignon blanc grapes, while Premium White is 75% sauvignon blanc grapes and 25% French colombard grapes. J & M's Sauvignon Blanc is 100% sauvignon blanc grapes, and its Chardonnay is 90% chardonnay grapes and 10% chenin blanc grapes. Profit on the House White is $1.00 per liter, profit on the Premium White is $1.50 per liter, profit on the Sauvignon Blanc is $2.25 per liter, and profit on the Chardonnay is $3.00 per liter. This season, J & M can obtain 30,000 pounds of French colombard grapes, 25,000 pounds of chenin blanc grapes, and 20,000 pounds each of sauvignon blanc grapes and chardonnay grapes. It takes 2 pounds of grapes to make 1 liter of wine. If the company can sell all it makes, how many liters of its various products should J & M prepare in order to maximize profit?

33. J & M Winery makes two jug wines, House Red and Premium Red, and two higher-quality wines, Cabernet Sauvignon and Zinfandel, which it sells to restaurants, supermarkets, and liquor stores. House Red is a blend of 20% pinot noir grapes, 30% zinfandel grapes, and 50% gamay grapes. Premium Red is 60% cabernet sauvignon grapes and 20% each pinot noir and gamay grapes. J & M's Cabernet Sauvignon is 100% cabernet sauvignon grapes, and its Zinfandel is 85% zinfandel grapes and 15% gamay grapes. Profit on the House Red is $0.90 per liter, profit on the Premium Red is $1.60 per liter, profit on the Zinfandel is $2.25 per liter, and profit on the Cabernet Sauvignon is $3.00 per liter. This season, J & M can obtain 30,000 pounds each of pinot noir, zinfandel, and gamay grapes and 22,000 pounds of cabernet sauvignon grapes. It takes 2 pounds of grapes to make 1 liter of wine. If the company can sell all that it makes, how many liters of its various products should J & M prepare in order to maximize profit?

3.3

MIXED CONSTRAINTS AND MINIMIZATION

So far, our simplex method problems have been carefully selected; they have all been maximization problems, and the constraints have all been "≤ inequalities" (that is, inequalities of the form $a_1x_1 + a_2x_2 + \cdots \leq b$, where all constants and variables are nonnegative). In this section we will discuss minimization problems and problems with **mixed constraints** (that is, constraints that are not all "≤ inequalities").

Mixed Constraints

Suppose that a new customer wishes to order 8 tables a week on a regular basis from our old friend the craftsman. This customer could potentially become a long-term customer, so the craftsman wishes to fill her order, even if it means a decrease in profit. This requirement to make at least 8 tables a week is a new constraint on production. The new constraint is

C_3: (coffee tables) + (end tables) ≥ 8

If the total number of tables is greater than 8, then some "surplus" tables (or tables beyond the necessary 8) are made, and we could say that

(coffee tables) + (end tables) − (surplus tables) = 8

C_3 can be rewritten as

$x_1 + x_2 - s_3 = 8$

where s_3 measures surplus tables and $s_3 \geq 0$.

In Section 3.2 all of our inequalities were "≤ inequalities." The left side of a "≤ inequality" is the smaller side, so we add a nonnegative variable to that side in order to turn the inequality into an equation; such a variable is called a slack variable. In this section, some of our inequalities are "≥ inequalities." The left side of a "≥ inequality" is the larger side, so we *subtract* a nonnegative variable from that side in order to turn the inequality into an equation. Such a variable is called a **surplus variable.** C_3 is a "≥ inequality," and s_3 is a surplus variable.

The rewritten model is:

Independent Variables	Slack/Surplus Variables
x_1 = number of coffee tables	s_1 = unused hours
x_2 = number of end tables	s_2 = unused money
$(x_1, x_2, s_1, s_2, s_3 \geq 0)$	s_3 = surplus tables

Constraints

C_1: $6x_1 + 5x_2 + s_1 = 40$ The time constraint

C_2: $200x_1 + 100x_2 + s_2 = 1000$ The money constraint

C_3: $x_1 + x_2 - s_3 = 8$ The new customer constraint

Objective Function

$-240x_1 - 160x_2 + 0s_1 + 0s_2 + 0s_3 + 1z = 0$

The first simplex matrix is:

$$\begin{array}{cccccc}
x_1 & x_2 & s_1 & s_2 & s_3 & z \\
\end{array}$$

$$\left[\begin{array}{cccccc|c}
6 & 5 & 1 & 0 & 0 & 0 & 40 \\
200 & 100 & 0 & 1 & 0 & 0 & 1000 \\
1 & 1 & 0 & 0 & -1 & 0 & 8 \\
-240 & -160 & 0 & 0 & 0 & 1 & 0
\end{array}\right]
\begin{array}{l}
\leftarrow C_1 \\
\leftarrow C_2 \\
\leftarrow C_3 \\

\end{array}$$

We cannot determine the solution that corresponds to this matrix in the normal way, because the s_3 column doesn't have a "1" at which to turn. To remedy that, multiply row 3 by -1, and get:

$$\begin{array}{cccccc}
x_1 & x_2 & s_1 & s_2 & s_3 & z \\
\end{array}$$

$$\left[\begin{array}{cccccc|c}
6 & 5 & 1 & 0 & 0 & 0 & 40 \\
200 & 100 & 0 & 1 & 0 & 0 & 1000 \\
-1 & -1 & 0 & 0 & 1 & 0 & -8 \\
-240 & -160 & 0 & 0 & 0 & 1 & 0
\end{array}\right]$$

The solution that corresponds with this matrix is

$$(x_1, x_2, s_1, s_2 \ s_3) = (0, 0, 40, 1000, -8) \quad \text{with } z = 0.$$

This solution is *not a possible solution,* let alone the one that maximizes profit. There are two ways to tell that this solution is not possible.

- A variable is negative (in particular, $s_3 = -8$). Recall that x_1, x_2, s_1, s_2, and s_3 must all be nonnegative because they count things.
- The solution will not successfully substitute into each constraint. In particular, it will not successfully substitute into C_3:

$$C_3: x_1 + x_2 \geq 8 \qquad \text{The original constraint } C_3$$
$$0 + 0 \geq 8 \qquad \text{Since } x_1 = x_2 = 0$$

Substituting the solution into C_3 results in a false statement: $0 \geq 8$. And C_3 is the "new customer" constraint, the one that assures that the craftsman makes enough tables to fill the new customer's standing order. If the craftsman makes $x_1 = 0$ coffee tables and $x_2 = 0$ end tables, then he cannot fill this standing order.

Previously, when we've done this work, each solution from each simplex matrix was a possible solution. We cannot begin to solve this problem as we normally do until we get a possible solution.

Mixed Constraints: Pivoting

Mixed constraint problems have two phases:

- **Phase I:** Pivot until a possible solution is found
- **Phase II:** Pivot in the normal manner until the maximum is found

In phase II, there are specific rules that tell us where to pivot (the column with the most negative entry in the bottom row, the row with the smallest nonnegative quotient). In phase I, there are *no specific rules that tell us where to pivot*. Instead, we use one general guideline: *pivot to replace the column that causes problems and gives a solution that's not possible.* Frequently we have to make some arbitrary choices in following this guideline. (This is a very simplified version of the two-phase method. The complete two-phase method, covered in upper division classes, does have specific pivoting rules.)

Phase I: Pivot until a possible solution is found. Observe that, in the first simplex matrix, it is the s_3 column that causes problems and gives us a solution that's not possible. The thing to do is to change another column into the column:

$$\begin{bmatrix} 0 \\ 0 \\ 1 \\ 0 \end{bmatrix}$$

It would be easiest to change either the x_1 or the x_2 column, since they both have a 1 in the right spot. Let's change the x_1 column (this choice is arbitrary—we could just as well have chosen the x_2 column). Working from the first simplex matrix:

R1 − 6R3 : R1
R2 − 200R3 : R2
240R3 + R4 : R4

$$\begin{array}{cccccc} x_1 & x_2 & s_1 & s_2 & s_3 & z \\ \begin{bmatrix} 0 & -1 & 1 & 0 & 6 & 0 & -8 \\ 0 & -100 & 0 & 1 & 200 & 0 & -600 \\ 1 & 1 & 0 & 0 & -1 & 0 & 8 \\ 0 & 80 & 0 & 0 & -240 & 1 & 1920 \end{bmatrix} \end{array}$$

The solution that corresponds with this matrix is

$$(x_1, x_2, s_1, s_2, s_3) = (8, 0, -8, -600, 0) \quad \text{with } z = 1920.$$

This is not a possible solution. There are two ways to tell that this solution is not possible:

- Both s_1 and s_2 are negative.
- The solution will not successfully substitute into each constraint. For example, it will not successfully substitute into C_1:

$$C_1: 6x_1 + 5x_2 \le 40 \qquad \text{The original constraint } C_1$$
$$6 \cdot 8 + 5 \cdot 0 \le 40 \qquad \text{Since } x_1 = 8 \text{ and } x_2 = 0$$
$$48 + 0 \le 40$$

Substituting the solution into C_1 results in a false statement: $48 \le 40$; we must phase I pivot again. Both the s_1 column and the s_2 column cause problems and give us a solution that's not possible. We'll focus on the s_1 column (an arbitrary decision) and change another column into the column:

$$\begin{bmatrix} 1 \\ 0 \\ 0 \\ 0 \end{bmatrix}$$

Let's change the x_2 column.

$-1R1 : R1$

$$\begin{array}{c} x_1 \quad x_2 \quad s_1 \quad s_2 \quad s_3 \quad z \\ \begin{bmatrix} 0 & 1 & -1 & 0 & -6 & 0 & 8 \\ 0 & -100 & 0 & 1 & 200 & 0 & -600 \\ 1 & 1 & 0 & 0 & -1 & 0 & 8 \\ 0 & 80 & 0 & 0 & -240 & 1 & 1920 \end{bmatrix} \end{array}$$

$100R1 + R2 : R2$
$-1R1 + R3 : R3$
$-80R1 + R4 : R4$

$$\begin{array}{c} x_1 \quad x_2 \quad s_1 \quad s_2 \quad s_3 \quad z \\ \begin{bmatrix} 0 & 1 & -1 & 0 & -6 & 0 & 8 \\ 0 & 0 & -100 & 1 & -400 & 0 & 200 \\ 1 & 0 & 1 & 0 & 5 & 0 & 0 \\ 0 & 0 & 80 & 0 & 240 & 1 & 1280 \end{bmatrix} \end{array}$$

The solution that corresponds with this matrix is

$$(x_1, x_2, s_1, s_2, s_3) = (0, 8, 0, 200, 0) \quad \text{with } z = 1280.$$

This solution is possible because it has no negative values (and because it successfully substitutes into the constraints and objective function). And it is maximal because the last row has no negative entries. (If the last row had a negative entry, we would enter phase II and pivot in the normal manner until the maximum is found.)

Thus, each day the craftsman should make no coffee tables and 8 end tables and have no unused hours, 200 unused dollars, and no "surplus tables"—that is, no tables in excess of the new customer's regular order—for a maximum profit of $1280.

Minimization

Not all quantities should be maximized; any business that maximizes its costs would go bankrupt very quickly. So far, we've used only the simplex method to solve maximization problems. We can solve minimization problems by converting them to maximization problems.

Consider the following pair of geometric method problems: minimize $w = 2x + 3y$, and maximize $z = -w = -2x - 3y$, where the region of possible solutions is shown in Figure 3.12.

FIGURE 3.12

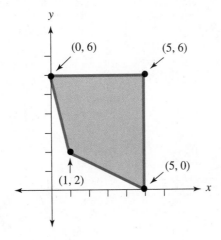

The region is bounded, so we substitute the corner points into the objective functions.

Corner Point	$w = 2x + 3y$	$z = -2x - 3y$
$(5, 0)$	$2 \cdot 5 + 3 \cdot 0 = 10$	$-2 \cdot 5 - 3 \cdot 0 = -10$
$(5, 6)$	$2 \cdot 5 + 3 \cdot 6 = 28$	$-2 \cdot 5 - 3 \cdot 6 = -28$
$(0, 6)$	$2 \cdot 0 + 3 \cdot 6 = 18$	$-2 \cdot 0 - 3 \cdot 6 = -18$
$(1, 2)$	$2 \cdot 1 + 3 \cdot 2 = 8$	$-2 \cdot 1 - 3 \cdot 2 = -8$

The minimum value of w is 8, which occurs at $(1, 2)$. The maximum value of z is -8, which also occurs at $(1, 2)$ ["greater than" means "to the right of on the number line," so -8 is greater than -28]. This always happens. The minimum value of a quantity w occurs at the same location as does the maximum value of a quantity $z = -w$.

EXAMPLE   J & M Winery produces two different Chardonnays, one under their economy label and one under their premium label. They have contracted to sell 30,000 cases of Chardonnay, and they are certain that they can sell more. The demand for their economy label will be at least twice that for the premium label, due to its lower selling price. They already have enough grapes for 8000 cases of the premium wine, and they can purchase more. Production costs for the premium label wine run $72 a case; for the economy label, $36 a case. How many cases of each Chardonnay should J & M produce in order to minimize production costs?

Solution

Independent Variables

x_1 = number of cases of premium label Chardonnay

x_2 = number of cases of economy label Chardonnay

Constraints

C_1: total cases of Chardonnay must be at least 30,000

$$x_1 + x_2 \geq 30{,}000 \rightarrow x_1 + x_2 - s_1 = 30{,}000$$

where s_1 = cases of Chardonnay in excess of their current orders.

C_2: the demand for the economy wine will be at least twice that for the premium wine

$$x_2 \geq 2x_1 \rightarrow -2x_1 + x_2 \geq 0 \rightarrow -2x_1 + x_2 - s_2 = 0$$

where s_2 = the amount by which the supply of economy Chardonnay will exceed twice the supply of premium Chardonnay.

C_3: already have enough grapes for 8000 cases of the premium wine

$$x_1 \geq 8000 \rightarrow x_1 - s_3 = 8000$$

where s_3 = amount of premium grapes J & M will have to purchase, in cases of finished wine.

Objective Function

Minimize w = cost

$$= \text{premium Chardonnay cost} + \text{economy Chardonnay cost}$$
$$= (\text{cost per case})(\text{number of premium cases}) +$$
$$(\text{cost per case})(\text{number of economy cases})$$
$$= 72x_1 + 36x_2$$

Minimizing w is equivalent to maximizing $z = -w$:

$$z = -72x_1 - 36x_2$$
$$72x_1 + 36x_2 + 1z = 0$$

The first simplex matrix:

$$
\begin{array}{ccccccc}
x_1 & x_2 & s_1 & s_2 & s_3 & z & \\
\end{array}
$$
$$
\begin{bmatrix}
1 & 1 & -1 & 0 & 0 & 0 & 30{,}000 \\
-2 & 1 & 0 & -1 & 0 & 0 & 0 \\
1 & 0 & 0 & 0 & -1 & 0 & 8{,}000 \\
72 & 36 & 0 & 0 & 0 & 1 & 0
\end{bmatrix}
$$

Its corresponding solution:

$$(x_1, x_2, s_1, s_2, s_3) = (0, 0, -30000, 0, -8000), \quad z = 0$$

The variables s_1 and s_3 have negative values, so this solution is not possible. We're in phase I, so we'll pivot to eliminate the negative value of s_3 by changing another column into

$$
\begin{bmatrix}
0 \\
0 \\
1 \\
0
\end{bmatrix}
$$

It would be easiest to change the x_1 column, since it has a 1 in the right spot.

R1 − R3 : R1
R2 + 2R3 : R2
−72R3 + R4 : R4

$$
\begin{array}{ccccccc}
x_1 & x_2 & s_1 & s_2 & s_3 & z \\
\begin{bmatrix}
0 & 1 & -1 & 0 & 1 & 0 & 22{,}000 \\
0 & 1 & 0 & -1 & -2 & 0 & 16{,}000 \\
1 & 0 & 0 & 0 & -1 & 0 & 8{,}000 \\
0 & 36 & 0 & 0 & 72 & 1 & -576{,}000
\end{bmatrix}
\end{array}
$$

This matrix's corresponding solution is

$$(x_1, x_2, s_1, s_2, s_3) = (8000, 0, -22000, -16000, 0), \quad z = 576{,}000$$

This solution is not possible, since it has negative values for s_1 and s_2. We are still in phase I, and we must pivot again to eliminate the negative value of s_1. We'll change another column into

$$\begin{bmatrix} 1 \\ 0 \\ 0 \\ 0 \end{bmatrix}$$

It would be easiest to change the x_2 column, since it has a 1 and a 0 in the right spots.

−R1 + R2 : R2
−36R1 + R4 : R4

$$
\begin{array}{ccccccc}
x_1 & x_2 & s_1 & s_2 & s_3 & z \\
\begin{bmatrix}
0 & 1 & -1 & 0 & 1 & 0 & 22{,}000 \\
0 & 0 & 1 & -1 & -3 & 0 & -6{,}000 \\
1 & 0 & 0 & 0 & -1 & 0 & 8{,}000 \\
0 & 0 & 36 & 0 & 36 & 1 & -1{,}368{,}000
\end{bmatrix}
\end{array}
$$

This matrix's corresponding solution is

$$(x_1, x_2, s_1, s_2, s_3) = (8000, 22000, 0, 6000, 0), \quad z = -1{,}368{,}000$$

This solution is possible, since it has no negative values.

✔ Check your work by seeing if the solution substitutes into the objective function and constraints.

Objective function:
$$z = -72x_1 - 36x_2$$
$$-1{,}368{,}000 = -72 \cdot 8000 - 36 \cdot 22{,}000 \quad ✔$$

C_1:　$x_1 + x_2 - s_1 = 30{,}000$
$$8000 + 22{,}000 - 0 = 30{,}000 \quad ✔$$

C_2:　$-2x_1 + x_2 - s_2 = 0$
$$-2 \cdot 8000 + 22{,}000 - 6000 = 0 \quad ✔$$

C_3:　$x_1 - s_3 = 8000$
$$8000 - 0 = 8000 \quad ✔$$

The last row has no negative entries (except for z, which must be negative); no further pivoting is needed and the solution is maximal for z (and minimal for w) at $z = -1,368,000$, $w = 1,368,000$. (If the last row had a negative entry, we would enter phase II and pivot in the normal manner until the maximum is found.)

Incorporate the values and the meaning of each of the variables:

$x_1 = 8000$	← Number of cases of premium label Chardonnay
$x_2 = 22,000$	← Number of cases of economy label Chardonnay
$s_1 = 0$	← Cases of Chardonnay in excess of current orders
$s_2 = 6000$	← The amount by which the supply of economy Chardonnay will exceed twice the supply of premium Chardonnay
$s_3 = 0$	← Premium grapes to purchase
$z = -1,368,000$	
$w = -z = 1,368,000$	← Minimum production costs

The solution to the problem is: In order to minimize production costs at $1,368,000, J & M Winery should produce 8000 cases of premium Chardonnay and 22,000 cases of economy Chardonnay. This will entail producing no cases in excess of their current orders. The supply of economy Chardonnay will exceed twice the supply of premium Chardonnay by 6000 cases. J & M will not have to purchase any premium grapes.   •

The following flowchart addresses maximization and minimization problems as well as mixed constraints problems.

Simplex Method Flowchart

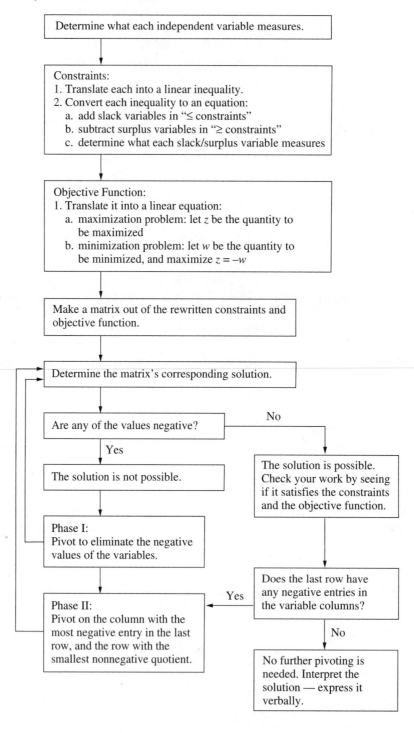

3.3

EXERCISES

1. Minimize $w = 2x_1 + 7x_2$ subject to the constraints:
$$x_1 + 4x_2 \le 40$$
$$4x_1 + x_2 \le 40$$
$$x_1 + x_2 \ge 5$$
$$x_1, x_2 \ge 0$$

2. Minimize $w = 2.70x_1 + 1.85x_2$ subject to the constraints:
$$6x_1 + x_2 \le 360$$
$$x_1 + 5x_2 \le 350$$
$$x_1 + 2x_2 \ge 20$$
$$x_1, x_2 \ge 0$$

3. Maximize $z = 5.10x_1 + 1.26x_2$ subject to the constraints:
$$7x_1 + 3x_2 \le 370$$
$$x_1 + x_2 \le 70$$
$$x_1 \ge 10$$
$$x_2 \ge 20$$

4. Maximize $z = 36x_1 + 42x_2$ subject to the constraints:
$$2x_1 + x_2 \le 770$$
$$3x_1 + x_2 \le 960$$
$$x_1 \ge 120$$
$$x_2 \ge 210$$

5. Maximize $z = 5.2x_1 + 1.3x_2$ subject to the constraints:
$$3x_1 - 2x_2 \ge 25$$
$$3x_1 + x_2 \le 145$$
$$x_1 \ge 20$$
$$0 \le x_2 \le 38$$

6. Minimize $w = 31x_1 + 79x_2$ subject to the constraints:
$$2x_1 - 5x_2 \le 24$$
$$2x_1 - x_2 \le 8$$
$$x_1 \ge 10$$
$$x_2 \ge 9$$
$$x_2 \le 12$$

7. Minimize $w = 2x_1 + 3x_2$ subject to the constraints:
$$15x_1 + x_2 \ge 25$$
$$x_1 + 6x_2 \ge 20$$
$$8x_1 + 7x_2 \ge 78$$
$$x_1, x_2 \ge 0$$

8. Minimize $w = 9x_1 + 7x_2$ subject to the constraints:
$$x_1 + x_2 \ge 100$$
$$51x_1 + 23x_2 \ge 3672$$
$$x_1, x_2 \ge 0$$

9. The Monterey Canning Co. produces canned whole Brussels sprouts and canned Brussels sprouts pieces. As an experiment, they are going to also produce canned Brussels sprouts salsa. Market analyses indicate that sales of sprouts salsa will be at least 25% as large as those of whole sprouts. Regular customers will order at least 400,000 pounds of canned whole sprouts and 90,000 pounds of canned sprouts pieces. A new customer has committed to purchasing at least 200,000 pounds of canned sprout products. It costs $1.25 to produce one pound of canned whole sprouts, $1.75 to produce one pound of canned sprouts pieces, and $3.00 to produce one pound of canned sprouts salsa. Find the amount of each product that the company should produce in order to minimize its production costs.

10. J & M Winery produces two different Cabernet Sauvignons, one under their economy label and one under their premium label. They have contracted to sell 20,000 cases of Cabernet, and they are certain that they can sell more. The demand for their economy label will be at least 50% more than for the premium label, due to its lower selling price. They already have enough grapes for 7000 cases of the premium wine, and they can purchase more. Production costs for the premium label wine run $72 a case; for the economy label, $36 a case. How many cases of each Cabernet should J & M produce in order to minimize production costs?

11. City Electronics Distributors handles two lines of televisions, the Packard and the Bell. They purchase up to $57,000 worth of television sets from the manufacturers each month, and store them in their 9000-cubic-foot warehouse. The Packards come in 36-cubic-foot packing crates, and the Bells come in 30-cubic-foot crates. The Packards cost City Electronics $200 each and can be sold to a retailer for a $200 profit, while the Bells cost $250 each and can be sold for $260 profit. City Electronics must stock enough sets to meet its regular customers' standing orders.
   a. If City Electronics has standing orders for 250 sets in addition to orders from other retailers, how many sets should City Electronics order each month in order to maximize profit?
   b. If City Electronics' standing orders increase to 260 sets, how many sets should City Electronics order each month in order to maximize profit?

12. Compucraft sells personal computers and printers made by Peach Computers. The computers come in 12-cubic-foot boxes, and the printers come in 8-cubic-foot boxes. Compucraft's owner estimates that at least 30 computers can be sold each month, and that the number of computers sold will be at least 50% more than the number of printers. The computers cost Compucraft $1000 each, and can be sold at a $1000 profit, while the printers cost $300 each and can be sold for a $350 profit. Compucraft has 1000 cubic feet of storage available for the Peach personal computers and printers and sufficient financing to spend $70,000 each month on computers and printers. How many computers and printers should Compucraft order from Peach each month in order to maximize profit?

 Answer the following using complete sentences.

13. Explain why the bottom row of the first simplex matrix of a minimization problem contains the coefficients of the objective function, but the bottom row of the first simplex matrix of a maximization problem contains the negatives of the coefficients.

14. Explain why a slack variable is subtracted but a surplus variable is added.

TECHNOLOGY AND TRANS-PORTATION PROBLEMS

Linear programming was invented in 1947 by George Dantzig as part of an air force research project in computing the most efficient and economical way to distribute men, weapons, and supplies to the various fronts of World War II. Problems involving the efficient distribution of goods from one set of locations to a second set of locations remain an important application of the simplex method.

EXAMPLE 2

Sunny Electronics has two warehouses in its western division: one in Los Angeles and one in Portland. Sunny has an order from The Stereo Guys for its new Boomman portable stereo. The Stereo Guys needs at least 180 stereos at its Los Angeles store and at least 120 at its San Francisco store. The Los Angeles warehouse has a stock of 275 stereos, while the Portland warehouse has a stock of 210 stereos. The cost of shipping one stereo is given in Figure 3.13. Find how many stereos Sunny should ship from each of its warehouses to The Stereo Guys' stores in order to minimize shipping costs.

FIGURE 3.13

	To Los Angeles Store	To San Francisco Store
From Los Angeles Warehouse	$2.10	$5.25
From Portland Warehouse	$6.50	$5.60

Solution

Independent Variables:

x_1 = number of Boommans sent from LA warehouse to LA store

x_2 = number of Boommans sent from LA warehouse to SF store

x_3 = number of Boommans sent from Portland warehouse to LA store

x_4 = number of Boommans sent from Portland warehouse to SF store

Constraints:

C_1: LA store needs at least 180

$x_1 + x_3 \geq 180 \rightarrow x_1 + x_3 - s_1 = 180$

where s_1 = surplus stereos sent to LA store.

C_2: SF store needs at least 120

$x_2 + x_4 \geq 120 \rightarrow x_2 + x_4 - s_2 = 120$

where s_2 = surplus stereos sent to SF store.

C_3: LA warehouse has 275 in stock

$x_1 + x_2 \leq 275 \rightarrow x_1 + x_2 + s_3 = 275$

where s_3 = stereos at LA warehouse not sent anywhere.

C_4: Portland warehouse has 210 in stock

$x_3 + x_4 \leq 210 \rightarrow x_3 + x_4 + s_4 = 210$

where s_4 = stereos at Portland warehouse not sent anywhere.

Objective Function:

Minimize w = shipping costs

$$= 2.10x_1 + 5.25x_2 + 6.50x_3 + 5.60x_4$$

Minimizing w is equivalent to maximizing $z = -w$.

$$z = -2.10x_1 - 5.25x_2 - 6.50x_3 - 5.60x_4$$
$$2.10x_1 + 5.25x_2 + 6.50x_3 + 5.60x_4 + 1z = 0$$

The first simplex matrix:

x_1	x_2	x_3	x_4	s_1	s_2	s_3	s_4	z	
1	0	1	0	-1	0	0	0	0	180
0	1	0	1	0	-1	0	0	0	120
1	1	0	0	0	0	1	0	0	275
0	0	1	1	0	0	0	1	0	210
2.10	5.25	6.50	5.60	0	0	0	0	1	0

Its corresponding solution:

$$(x_1, x_2, x_3, x_4, s_1, s_2, s_3, s_4) = (0, 0, 0, 0, -180, -120, 275, 210), \quad z = 0$$

We are now ready to enter the first simplex matrix into the computer or graphing calculator. •

EXERCISES

15. Complete Example 2.

16. Central State Edison Co. has two electric power plants that supply the power needs of two cities, Nutley and Orgone Heights. The Two Mile Island Plant can supply 35 million kilowatt-hours (kwh) of electricity, and the Three Corners plant can supply 45 million kwh. Nutley's minimum power demand is 40 million kwh, and Orgone Heights' minimum power demand is 32 million kwh. The costs of supplying 1 million kwh are given in Figure 3.14. Find how Central State Edison can deliver the needed power at a minimal cost.

FIGURE 3.14

Power Plant	Nutley	Orgone Heights
Two Mile Island	$8000	$6000
Three Corners	$7000	$9000

17. Two reservoirs supply the water needs of Crockett and Valona. The south reservoir can supply up to 20 million gallons per day, and the north reservoir can supply up to 40 million gallons per day. Crockett needs at least 30 million gallons per day, and Valona needs at least 10 million gallons per day. The costs of supplying 1 million gallons are given in Figure 3.15. Find how the water can be delivered at minimal cost.

FIGURE 3.15

Reservoir	Crockett	Valona
South	$7000	$9000
North	$11,000	$10,000

3.4

SHADOW VALUES

In Sections 3.1 and 3.2 we encountered a craftsman who produces coffee tables and end tables. We used the simplex method to determine that he could use his limited resources of time and money to generate a maximum profit of $1400 per week if he made 2.5 coffee tables and 5 end tables each week. We also found that this level of production would utilize all of his available time (40 hours a week) and money ($1000 for materials each week). Production data are given in Figure 3.16.

FIGURE 3.16

	Labor (per table)	Cost of Materials (per table)	Profit (per table)
Coffee Tables	6 hours	$200	$240
End Tables	5 hours	$100	$160

The craftsman followed our advice. He has now returned to us, asking if he can further increase his profits. He knows that to do this he must increase either the time or the money invested. He could increase the time by hiring an assistant, but he is concerned that any increase in profit would not be enough to pay an assistant's wages. He could increase the money for materials by obtaining a loan, but he is concerned that any increase in profit would not be enough to cover the extra expense. What should he do?

EXAMPLE 1 Determine the effect on the craftsman's weekly production and profit if he increases the time invested from 40 hours per week to 41 hours per week. Also determine the circumstances under which he should hire an assistant.

Solution Only the time constraint needs to be altered; the rest of the model that we devised in Section 3.2 remains intact.

Independent Variables

x_1 = number of coffee tables

x_2 = number of end tables

$(x_1, x_2, s_1, s_2 \geq 0)$

Slack Variables

s_1 = unused hours

s_2 = unused money

Constraints

C_1: $6x_1 + 5x_2 + s_1 = 41$ (up from 40)   The new time constraint

C_2: $200x_1 + 100x_2 + s_2 = 1000$ The money constraint

Objective Function

$-240x_1 - 160x_2 + 0s_1 + 0s_2 + 1z = 0$ z measures profit

The first simplex matrix:

$$\begin{array}{ccccc} x_1 & x_2 & s_1 & s_2 & z \end{array}$$
$$\begin{bmatrix} 6 & 5 & 1 & 0 & 0 & 41 \\ 200 & 100 & 0 & 1 & 0 & 1000 \\ -240 & -160 & 0 & 0 & 1 & 0 \end{bmatrix}$$

Its corresponding possible solution:

$$(x_1, x_2, s_1, s_2) = (0, 0, 41, 1000), \quad z = 0$$

Pivot on the 200:
$$\begin{bmatrix} 6 & 5 & 1 & 0 & 0 & 41 \\ 200 & 100 & 0 & 1 & 0 & 1000 \\ -240 & -160 & 0 & 0 & 1 & 0 \end{bmatrix}$$
$\uparrow$

     $\leftarrow 41/6 = 6.83\ldots$
     $\leftarrow 1000/200 = 5$
     $\leftarrow$ Not a constraint row

R2 ÷ 200 : R2
$$\begin{bmatrix} 6 & 5 & 1 & 0 & 0 & 41 \\ 1 & 0.5 & 0 & 0.005 & 0 & 5 \\ -240 & -160 & 0 & 0 & 1 & 0 \end{bmatrix}$$

R1 − 6R2 : R1
240R2 + R3 : R3
$$\begin{bmatrix} 0 & 2 & 1 & -0.03 & 0 & 11 \\ 1 & 0.5 & 0 & 0.005 & 0 & 5 \\ 0 & -40 & 0 & 1.2 & 1 & 1200 \end{bmatrix}$$

This is the second simplex matrix; its corresponding possible solution is

$$(x_1, x_2, s_1, s_2) = (5, 0, 11, 0), \quad z = 1200.$$

Pivot on the 2:
$$\begin{bmatrix} 0 & 2 & 1 & -0.03 & 0 & 11 \\ 1 & 0.5 & 0 & 0.005 & 0 & 5 \\ 0 & -40 & 0 & 1.2 & 1 & 1200 \end{bmatrix}$$
$\uparrow$

     $\leftarrow 11/2 = 5.5$
     $\leftarrow 5/0.5 = 10$
     $\leftarrow$ Not a constraint row

R1 ÷ 2 : R1
$$\begin{bmatrix} 0 & 1 & 0.5 & -0.015 & 0 & 5.5 \\ 1 & 0.5 & 0 & 0.005 & 0 & 5 \\ 0 & -40 & 0 & 1.2 & 1 & 1200 \end{bmatrix}$$

−0.5R1 + R2 : R2
40R1 + R3 : R3
$$\begin{array}{ccccc} x_1 & x_2 & s_1 & s_2 & z \end{array}$$
$$\begin{bmatrix} 0 & 1 & 0.5 & -0.015 & 0 & 5.5 \\ 1 & 0 & -0.25 & 0.0125 & 0 & 2.25 \\ 0 & 0 & 20 & 0.6 & 1 & 1420 \end{bmatrix}$$

This is the third and final simplex matrix; its corresponding possible solution is

$$(x_1, x_2, s_1, s_2) = (2.25, 5.5, 0, 0), \quad z = 1420$$

Compare this to the solution from Section 3.2, when the craftsman was working a total of 40 hours per week:

$$(x_1, x_2, s_1, s_2) = (2.5, 5, 0, 0), \quad z = 1400$$

If the craftsman increases the time invested from 40 hours per week to 41 hours per week, his coffee table production will go down 0.25 tables from 2.5 to 2.25, his end table production will go up 0.5 tables from 5 to 5.5, and his weekly profit will go up $20 from $1400 to $1420. He should hire an assistant if doing so would cost him less than $20 per hour, because the assistant would generate $20 in profit per hour.   •

The changes in production and profit found in Example 1 were found by altering the time constraint and recomputing all of the resulting simplex matrices. Fortunately, there is a shortcut that avoids this recomputation. Those changes can be found by simply looking in the final simplex matrix of the original craftsman problem from Section 3.2. They are displayed in that matrix's s_1 column.

$$\begin{array}{ccccc} x_1 & x_2 & s_1 & s_2 & z \\ \begin{bmatrix} 0 & 1 & 0.5 & -0.015 & 0 & 5 \\ 1 & 0 & -0.25 & 0.0125 & 0 & 2.5 \\ 0 & 0 & 20 & 0.6 & 1 & 1400 \end{bmatrix} \end{array}$$

It's easy to tell what each entry measures if you think of the s_1 column as an alteration to the last column. The 5 in the last column is the final value of x_2, and the 0.5 in the s_1 column alters that 5, so the production of end tables would be increased by 0.5 tables per day if labor increased by 1 hour. The 2.5 in the last column is the final value of x_1, and the -0.25 in the s_1 column alters that 2.5, so the production of coffee tables would be decreased by 0.25 tables per day if labor hours increased by 1. The 1400 in the last column is the final value of z, and the 20 in the s_1 column alters that 1400, so the profit would be increased by $20 if labor hours increased by 1. This $20 is called the **shadow value** of labor.

EXAMPLE 2   Determine the effect on the craftsman's weekly production and profit if he increases the money for materials from $1000 per week to $1001 per week. Also determine the circumstances under which he should spend more on materials.

Solution   We could make this determination by altering a constraint and recomputing the resulting simplex matrices, as we did in Example 1. Instead, we will take advantage of the shortcut discussed above, and find the changes in the final simplex matrix of the original craftsman problem from section 3.2, when the craftsman was working a total of 40 hours per week. They are displayed in that matrix's s_2 column, because s_2 measures slack money. (In Example 1, the changes resulting from increasing the time invested were displayed in the s_1 column, because s_1 measures slack time.)

$$\begin{array}{ccccc} x_1 & x_2 & s_1 & s_2 & z \\ \begin{bmatrix} 0 & 1 & 0.5 & -0.015 & 0 & 5 \\ 1 & 0 & -0.25 & 0.0125 & 0 & 2.5 \\ 0 & 0 & 20 & 0.6 & 1 & 1400 \end{bmatrix} \end{array}$$

The 5 in the last column is the final value of x_2, and the -0.015 in the s_2 column alters that 5, so the production of end tables would be decreased by 0.015 tables per day if the money for materials were increased by $1. The 2.5 in the last column is the final value of x_1, and the 0.0125 in the s_2 column alters that 2.5, so the production of coffee tables would be increased by 0.0125 tables per day if the money for materials were increased by $1. The 1400 in the last column is the final value of z, and the 0.6 in the s_2 column alters that 1400, so the profit would be increased by $0.60 to $1400.60 if the money for materials were increased by $1. If he spends more on materials, his profit will increase by $0.60 for each $1 invested. He should do so only if the extra $0.60 is sufficient to cover the extra expense of borrowing.   •

3.4

EXERCISES

1. The SitOnIt Furniture Co. makes chairs and sofas. The manufacture of one chair requires 6 hours of carpentry work, 1 hour of finishing work, and 2 hours of upholstery work. The manufacture of one sofa requires 3 hours of carpentry work, 1 hour of finishing work, and 6 hours of upholstery work. The factory's resources are such that at most 96 hours of carpentry work, 18 hours of finishing work, and 72 hours of upholstery work are available each day. Each chair sells for a profit of $160, and each sofa for a profit of $140.
   a. How many chairs and how many sofas should be produced each day to maximize the profit? What is the maximum profit?
   b. Determine the effect on the daily production and profit if one extra hour of carpentry becomes available.
   c. Determine the effect on the daily production and profit if one extra hour of finishing becomes available.
   d. Determine the effect on the daily production and profit if one extra hour of upholstery becomes available.
   e. If SitOnIt were to hire one more worker, who should it be? How much could it pay this worker and still profit?
   f. Should SitOnIt consider retraining any of its workers? Why?

2. Woody's Woodworks manufactures wooden puzzles and wooden cars. Puzzles require 2 hours of finishing work and 1 hour of carpentry work. Cars require 1 hour of finishing work and 1 hour of carpentry work. Woody's resources are such that at most 80 hours of carpentry work and 100 hours of finishing work are available each week. Puzzles can be sold for a $3 profit, and cars can be sold for a $2 profit.
   a. How many puzzles and how many cars should be produced each week to maximize the profit? What is the maximum profit?
   b. Determine the effect on the weekly production and profit if one extra hour of carpentry becomes available.

   c. Determine the effect on the weekly production and profit if one extra hour of finishing becomes available.

3. In Example 1 we found that the craftsman's profit would increase $20 if the time invested increased by 1 hour. In this exercise you will show that this pattern continues, that is, that 2 extra hours would generate $40, and 3 extra hours would generate $60.
   a. Change constraint C_1 so that the time invested increases from 40 hours to $40 + x$ hours.
   b. Find the resulting first simplex matrix.
   c. Compute the resulting final simplex matrix.
   d. Use part (c) to determine the effect on the craftsman's weekly production and profit if he increases the time invested from 40 hours per week to 42 hours per week.
   e. Use part (c) to determine the effect on the craftsman's weekly production and profit if he increases the time invested from 40 hours per week to 43 hours per week.
   f. Determine the effect on the craftsman's weekly production and profit if he increases the time invested from 40 hours per week to 48 hours per week and he increases the money invested from $1000 per week to $1200 per week.

4. In Example 2 we found that the craftsman's profit would increase $0.60 if the money invested increased by $1. In this exercise you will show that there is a limitation to this type of profit increase.
   a. What type of table production would be decreased by an increase in money invested?
   b. What is the size of this decrease?
   c. How many such decreases are possible?
   d. How large of an increase in the money invested would still generate this type of increase in profit?

5. In Example 1 we found that the craftsman's profit would increase $20 if the time invested increased by 1 hour. In this

exercise you will show that there is a limitation to this type of profit increase.

   a. What type of table production would be decreased by an increase in time for labor?

   b. What is the size of this decrease?

   c. How many such decreases are possible?

   d. How large an increase in the time invested would still generate this type of increase in profit?

6. Determine the effect on the craftsman's weekly production and profit if he increases the money invested by the amount found in Exercise 4 and he increases the time invested by the amount found in Exercise 5.

7. This exercise refers to the SitOnIt Furniture Company in Exercise 1.

   a. How large an increase in the carpentry hours would still generate the type of increase in profit found in 1(b)?

   b. How large an increase in the finishing hours would still generate the type of increase in profit found in 1(c)?

   c. How large an increase in the upholstery hours would still generate the type of increase in profit found in 1(d)?

   d. Determine the effect on the production and profit if the company increases all three categories of hours by the above amounts.

8. This exercise refers to Woody's Woodworks in Exercise 2.

   a. How large an increase in the carpentry hours would still generate this type of increase in profit?

   b. How large an increase in the finishing hours would still generate this type of increase in profit?

   c. Determine the effect on the production and profit if the company increases both categories of hours by the above amounts.

9. This exercise refers to the five friends of Exercise 9 in Section 3.2.

   a. Determine the effect on the daily production and profit if 1 extra hour of labor becomes available.

   b. How large an increase in the time invested would still generate this type of increase in profit?

   c. Determine the effect on the daily production and profit if 1 extra dollar becomes available for materials.

   d. How large an increase in the money invested would still generate this type of increase in profit?

10. This exercise refers to the craftswoman of Exercise 10 in Section 3.2.

   a. Determine the effect on the weekly production and profit if 1 extra hour of labor becomes available.

   b. How large an increase in the time invested would still generate this type of increase in profit?

   c. Determine the effect on the weekly production and profit if 1 extra dollar becomes available for materials.

   d. How large an increase in the money invested would still generate this type of increase in profit?

11. This exercise refers to the J & M Winery of Exercise 13 in Section 3.2.

   a. Determine the effect on the production and profit if 1 extra pound of French colombard grapes becomes available.

   b. How large an increase in availability of French colombard grapes would still generate this type of increase in profit?

   c. Determine the effect on the production and profit if 1 extra pound of sauvignon blanc grapes becomes available for materials.

   d. How large an increase in availability of sauvignon blanc grapes would still generate this type of increase in profit?

12. This exercise refers to the J & M Winery of Exercise 14 in Section 3.2.

   a. Determine the effect on the production and profit if 1 extra pound of gamay grapes becomes available.

   b. How large an increase in availability of gamay grapes would still generate this type of increase in profit?

   c. Determine the effect on the production and profit if 1 extra pound of cabernet sauvignon grapes becomes available for materials.

   d. How large an increase in availability of cabernet sauvignon grapes would still generate this type of increase in profit?

3.5

DUALITY

Any linear programming problem has a second linear programming problem associated with it. This second problem is called the **dual problem;** the original problem is called the **primal problem.** So far, we've looked only at primal problems. A primal problem fits the stated constraints and objective; the dual problem has its own constraints and objective, different from those of the primal problem.

The solution of a primal problem can be found in that problem's final simplex matrix, as we have always done. It can also be found in the final simplex matrix of the dual problem. This pairing of a primal problem with its dual problem has several important features:

- It provides us with an alternative method of solving a given problem: we can solve a primal problem by actually solving its dual. If a primal problem is overwhelming, its dual might be simple.
- It provides us with interesting economic insights.
- It is vital to understanding advanced topics in linear programming.

We'll use the Leather Factory problem from Example 1 in Section 3.2, summarized below, to introduce duality.

The Primal Problem

The Leather Factory produces two products: coats and vests. Production data are given in Figure 3.17. If the Leather Factory's sewing machine is available 8 hours a day and it can obtain 118 square feet of leather a day, how many coats and how many vests should it make each week to maximize its daily revenue?

FIGURE 3.17

	Required Sewing Machine Time	Required Leather	Sales Price (Per Item)
Coats	50 minutes	12 square feet	$175
Vests	30 minutes	8 square feet	$100

The primal problem fits the stated constraints of limited sewing machine time and limited leather availability, as well as the stated objective of maximizing weekly profit. The primal model was as follows.

Independent Variables

x_1 = number of coats made each day
x_2 = number of vests made each day

Constraints

C_1: $50x_1 + 30x_2 \leq 480$ The sewing machine time constraint
C_2: $12x_1 + 8x_2 \leq 118$ The leather constraint

Objective Function

Maximize $z = 175x_1 + 100x_2$ z measures daily revenue

We solved the primal problem in Section 3.2; we found that daily revenue is maximized at $1680 if the Leather Factory makes 9.6 coats and no vests each day, leaving no unused machine hours and 2.8 square feet of unused leather.

The Dual Problem

The dual problem has its own constraints and objective, different from those of the primal problem. The following example is the Leather Factory's dual problem.

EXAMPLE 1 PerSuede Me, a leather clothing manufacturer, has been unable to make enough of its product to keep up with demand. Thus, its owners are interested in buying some or all of the Leather Factory's sewing machine time and some or all of its leather supply. The Leather Factory's owner is interested in this proposal, but only if it increases his profit. What minimum rates should he charge for sewing machine time and for leather supplies? What is the minimum weekly revenue that the Leather Factory's owner should demand if he sells all of his resources?

Solution In order to distinguish the dual problem from the primal problem, we will use y's (rather than x's) for the independent variables, and r's (rather than s's) for the slack and surplus variables.

Independent Variables

y_1 = price for 1 minute of sewing machine time

y_2 = price for 1 square foot of leather

Constraints

What constraints does the Leather Factory's owner face in setting prices for its resources? These prices must reflect the value of the resources to their owner. In particular, PerSuede Me must offer at least $175 for a combination of resources that includes 50 minutes of sewing machine time and 12 square feet of leather, because the Leather Factory could use these resources to produce a coat that can be sold for $175. If PerSuede Me offered less than $175, the Leather Factory would have no reason to sell the resources.

Coat resource income ≥ 175

$$(50 \text{ minutes}) \cdot \left(\frac{y_2 \text{ dollars}}{\text{minute}}\right) + (12 \text{ square feet}) \cdot \left(\frac{y_2 \text{ dollars}}{\text{square foot}}\right) \geq \$175$$

$$50 \qquad y_1 \qquad + \qquad 12 \qquad y_2 \qquad \geq 175$$

$C_1\text{: } 50y_1 + 12y_2 \geq 175$

$$50y_1 + 12y_2 - r_1 = 175$$

where r_1 = extra revenue generated by selling the resources necessary to make one coat. PerSuede Me also must offer at least $100 for a combination of resources that includes 30 minutes of sewing machine time and 8 square feet of leather, because the Leather Factory could use these resources to produce a vest that can be sold for $100.

$$(30 \text{ minutes}) \cdot \left(\frac{y_1 \text{ dollars}}{\text{minute}}\right) + (8 \text{ square feet}) \cdot \left(\frac{y_2 \text{ dollars}}{\text{square foot}}\right) \geq \$100$$

$$30 \qquad y_1 \qquad + \qquad 8 \qquad y_2 \qquad \geq 175$$

$C_2\text{: } 30y_1 + 8y_2 \geq 100$

$$30y_1 + 8y_2 - r_2 = 100 \qquad \text{The vest constraint}$$

where r_2 = extra revenue generated by selling the resources necessary to make one vest.

Objective Function

Our objective is to find the minimum acceptable weekly revenue. If we let w = revenue, we get

w = (sewing machine revenue) + (leather revenue)

$$= (480 \text{ minutes}) \cdot \left(\frac{y_1 \text{ dollars}}{\text{minute}}\right) + (118 \text{ square feet}) \cdot \left(\frac{y_2 \text{ dollars}}{\text{square foot}}\right)$$

$$= \quad 480 \qquad\quad y_1 \qquad + \qquad 118 \qquad\quad y_2$$

We will minimize $w = 480y_1 + 118y_2$ by maximizing

$$z = -w = -480y_1 - 118y_2$$

The first simplex matrix is as follows:

$$
\begin{array}{ccccc}
y_1 & y_2 & r_1 & r_2 & z \\
\end{array}
$$
$$
\left[
\begin{array}{ccccc|c}
50 & 12 & -1 & 0 & 0 & 175 \\
30 & 8 & 0 & -1 & 0 & 100 \\
480 & 118 & 0 & 0 & 1 & 0 \\
\end{array}
\right]
$$

The corresponding solution is $(y_1, y_2, r_1, r_2) = (0, 0, -175, -100)$, $z = 0$. The variables r_1 and r_2 have negative values, so the current solution is not possible. First, we'll phase I pivot to eliminate the negative value of r_1. We'll change the y_1 column into

$$
\begin{bmatrix}
1 \\
0 \\
0 \\
\end{bmatrix}
$$

R1 ÷ 50 : R1
$$
\left[
\begin{array}{ccccc|c}
1 & 0.24 & -0.02 & 0 & 0 & 3.5 \\
30 & 8 & 0 & -1 & 0 & 100 \\
480 & 118 & 0 & 0 & 1 & 0 \\
\end{array}
\right]
$$

−30R1 + R2 : R2
−480R1 + R3 : R3
$$
\begin{array}{ccccc}
y_1 & y_2 & r_1 & r_2 & z \\
\end{array}
$$
$$
\left[
\begin{array}{ccccc|c}
1 & 0.24 & -0.02 & 0 & 0 & 3.5 \\
0 & 0.8 & 0.6 & -1 & 0 & -5 \\
0 & 2.8 & 9.6 & 0 & 1 & -1680 \\
\end{array}
\right]
$$

The corresponding solution is $(y_1, y_2, r_1, r_2) = (3.5, 0, 0, 5)$, $z = -1680$. The variables all have nonnegative values, so the current solution is possible. The bottom row has no negative entries (except for the -1680, which must be negative because we are minimizing), so the current solution is optimal. And w measures the minimum acceptable weekly revenue; $w = -z = -(-1680) = 1680$.

This means that the Leather Factory's owner should charge at least $3.50 per minute of sewing machine time, and he should not necessarily charge for leather, if he sells some of his resources. If he sells all of his resources, he should demand a weekly revenue of at least $1680. ●

Interpreting the Dual's Solution

In Example 1 we found that the Leather Factory's owner should not necessarily charge for leather if he sells some of his resources to PerSuede Me. How does this make sense? A business doesn't survive by giving away resources.

We can make sense of part of Example 1's solution by recalling the solution to the primal problem. In Section 3.2 we found that the Leather Factory's daily revenue is maximized at $1680 if it makes 9.6 coats and no vests each day, leaving no unused machine time and

2.8 square feet of unused leather. Thus, under its normal operating conditions, the Leather Factory has a surplus of leather. It could give away some of this surplus leather without affecting its revenue, as we found in Example 1. The primal solution shows that the Leather Factory does not have a surplus of machine time, so it could not give away time. This is in agreement with the solution of Example 1, where we found that the Leather Factory must charge at least $3.50 per minute of sewing machine time. Furthermore, the primal solution shows that the Leather Factory's daily revenue is $1680. This is in agreement with the solution of Example 1, where we found that the Leather Factory's owner should demand a weekly revenue of at least $1680 if he sells all of his resources. If PerSuede Me offered less than $1680, the Leather Factory would have no reason to sell the resources.

We can make further sense of the solution of Example 1 by finding the shadow values of a minute of machine time and a square foot of leather.

EXAMPLE 2 **a.** Find the shadow value of one minute of machine time, and justify this with the results of Example 1.

b. Find the shadow value of one square foot of leather, and justify this with the results of Example 1.

Solution **a.** We are asked to determine the effect on the Leather Factory's daily revenue of increasing the amount of available sewing machine time from 480 minutes to 481 minutes. We can find this effect in the final simplex matrix of the original Leather Factory problem from Section 3.2. It is displayed in that matrix's s_1 column, because s_1 measures unused sewing machine time.

$$
\begin{array}{ccccc}
x_1 & x_2 & s_1 & s_2 & z \\
\end{array}
$$
$$
\begin{bmatrix}
1 & 0.6 & 0.02 & 0 & 0 & 9.6 \\
0 & 0.8 & -0.24 & 1 & 0 & 2.8 \\
0 & 5 & 3.5 & 0 & 1 & 1680
\end{bmatrix}
$$

The 1680 in the last column is the final value of z, and the 3.5 in the s_1 column alters that 1680, so the revenue would be increased by $3.50 if machine time increased by 1 minute. This $3.50 is the shadow value of 1 minute of sewing machine time; it measures the increase in the Leather Factory's daily revenue that would result from increasing available sewing machine time by 1 minute.

In Example 1 we found that the Leather Factory's owner should charge at least $3.50 per minute of sewing machine time. It should make sense to you that these two results are the same. The Leather Factory's charge for 1 minute of machine time should be at least as large as the revenue it would otherwise make from that time.

b. We are asked to determine the effect on the Leather Factory's daily revenue of increasing the amount of available leather from 118 square feet to 119 square feet. We can find this effect in the final simplex matrix of the original Leather Factory problem from Section 3.2. It is displayed in that matrix's s_2 column, because s_2 measures unused leather.

$$
\begin{array}{ccccc}
x_1 & x_2 & s_1 & s_2 & z \\
\end{array}
$$
$$
\begin{bmatrix}
1 & 0.6 & 0.02 & 0 & 0 & 9.6 \\
0 & 0.8 & -0.24 & 1 & 0 & 2.8 \\
0 & 5 & 3.5 & 0 & 1 & 1680
\end{bmatrix}
$$

The 1680 in the last column is the final value of z, and the last 0 in the s_2 column alters that 1680, so the revenue would not change if the amount of available leather increased by 1 square foot. This $0 is the shadow value of 1 square foot of leather; it measures the increase in the Leather Factory's daily revenue that would result from increasing the amount of available leather by 1 square foot. The revenue would not increase because there was a surplus of leather; no additional leather is needed.

In Example 1 we found that the Leather Factory's owner should not necessarily charge for leather if he sells some of his resources to PerSuede Me. This is in agreement with the shadow value of leather. •

The Dual's Final Simplex Matrix Compared with the Primal's Final Simplex Matrix

All of the data present in the final simplex matrix for a primal problem (including shadow values and values of the independent variables) is also present in the final simplex matrix for the corresponding dual problem. This is illustrated for the Leather Factory problem in Figure 3.18.

FIGURE 3.18

From the Primal Problem		From the Dual Problem
Maximum daily revenue is $z =$	$1680	= minimum daily charge if Leather Factory sells all of its resources
Optimal daily coat production is $x_1 =$	9.6 coats	= shadow value of r_1
Optimal daily vest production is $x_2 =$	0 vests	= shadow value of r_2
Shadow value of $s_1 =$ (revenue generated by one more minute of sewing machine time)	$3.50	= y_1, the minimum charge for 1 minute of sewing machine time
Shadow value of $s_2 =$ (revenue generated by 1 more square foot of leather)	$0	= y_2, the minimum charge for 1 square foot of leather

$$
\begin{array}{ccccc}
x_1 & x_2 & s_1 & s_2 & z \\
\left[\begin{array}{ccccc}
1 & 0.6 & 0.02 & 0 & 0 \\
0 & 0.8 & -0.24 & 1 & 0 \\
0 & 5 & 3.5 & 0 & 1
\end{array}\right. & & & & \left.\begin{array}{c}
9.6 \\
2.8 \\
1680
\end{array}\right]
\end{array}
$$

$$
\begin{array}{ccccc}
y_1 & y_2 & r_1 & r_2 & z \\
\left[\begin{array}{ccccc}
1 & 0.24 & -0.02 & 0 & 0 \\
0 & 0.8 & 0.6 & -1 & 0 \\
0 & 2.8 & 9.6 & 0 & 1
\end{array}\right. & & & & \left.\begin{array}{c}
3.5 \\
-5 \\
-1680
\end{array}\right]
\end{array}
$$

The Method of Duals

In Example 1 we generated the dual problem by finding the minimum acceptable selling prices for the Leather Factory's resources. It is also possible to generate the dual problem by manipulating the primal problem's model, without being concerned about the actual meaning of the dual. This yields an alternative way of solving a primal problem, called the **method of duals.**

This conversion of a primal problem into its dual can be done only under the conditions described in Figure 3.19.

FIGURE 3.19

If the Primal Problem Is:	Then the Dual Problem Will Be:
A maximization problem with "≤ constraints"	A minimization problem with "≥ constraints"
A minimization problem with "≥ constraints"	A maximization problem with "≤ constraints"

EXAMPLE 3   Determine the dual problem of the Leather Factory problem by manipulating the primal problem's model.

Solution   **Step 1**   *Generate the primal model.* We have already done this. The model is given below.

Independent Variables

x_1 = number of coats made each day

x_2 = number of vests made each day

Constraints

C_1: $50x_1 + 30x_2 \leq 480$ The sewing machine time constraint

C_2: $12x_1 + 8x_2 \leq 118$ The leather constraint

Objective Function

Maximize $z = 175x_1 + 100x_2$ z measures daily revenue

Notice that the primal problem is a maximization problem with "≤ constraints." This tells us that the dual problem will be a minimization problem with "≥ constraints."

Step 2   *Make a matrix of the primal model.* Basically, we copy down the constraints and the objective function without the variables. The constraints go in the first rows, and the objective function goes in the last row.

> **WARNING:** Do not include slack variables, and do not rewrite the objective function with all variables on the left side.

$$\begin{bmatrix} 50 & 30 & 480 \\ 12 & 8 & 118 \\ 175 & 100 & 0 \end{bmatrix}$$

Step 3   *Alter the matrix from Step 2 by changing its rows into columns.*

$$\begin{bmatrix} 50 & 12 & 175 \\ 30 & 8 & 100 \\ 480 & 118 & 0 \end{bmatrix}$$

Step 4   *Convert the matrix from Step 3 into the dual model.* Convert the top rows into constraints and the bottom row into an objective function. This is essentially Step 2 in reverse. The top row becomes our first constraint:

C_1: $50y_1 + 12y_2 \geq 175$

The second row becomes our second constraint:

$C_2: 30y_1 + 8y_2 \geq 100$

The last row becomes our objective function:

minimize $w = 480y_1 + 118y_2$

This is identical to the model generated in Example 1.

In the following example, we use the method of duals to solve a linear programming problem. We also follow the lettering convention given in Figure 3.20. This convention helps decrease confusion arising from the interplay between a primal problem and its dual.

FIGURE 3.20
Lettering Convention for
the Method of Duals

	Maximization Problems	**Minimization Problems**
Independent Variables	x_1, x_2, x_3, etc.	y_1, y_2, y_3, etc.
Slack and Surplus Variables	s_1, s_2, s_3, etc.	r_1, r_2, r_3, etc.
Objective Function	z	w

EXAMPLE 4   A cattle rancher needs to provide his cattle with a daily dietary supplement. Each cow needs at least 6 units of protein and 5 units of carbohydrates. Bossy Buffer costs 32 cents per ounce and each ounce provides 4 units of protein and 8 units of carbohydrates. Science Cow costs 12 cents per ounce and each ounce provides 3 units of protein and 2 units of carbo-hydrates. Use the method of duals to determine the mixture of Bossy Buffer and Science Cow that will meet the daily requirements at minimum cost.

Solution   **Step 1**   *Generate the primal model.* The primal problem is a minimization problem, so we will use y's for the independent variables, r's for the slack and surplus variables, and w for the objective function. The dual problem will be a maximization problem, and we will use x's for the independent variables, s's for the slack and surplus variables, and z for the objective function.

Independent Variables

y_1 = ounces of Bossy Buffer to be added daily to the feed
y_2 = ounces of Science Cow to be added daily to the feed

Constraints

C_1: total units of protein ≥ 6
(protein from Bossy Buffer)   + (protein from Science Cow) ≥ 6
$\left(\dfrac{4 \text{ units of protein}}{\text{ounce}} \right) \cdot (y_1 \text{ ounces}) + \left(\dfrac{3 \text{ units of protein}}{\text{ounce}} \right) \cdot (y_2 \text{ ounces}) \geq 6$
$C_1: 4y_1 + 3y_2 \geq 6$

C_2: total units of carbohydrates ≥ 5

(carbos from Bossy Buffer)   + (carbos from Science Cow) ≥ 5

$$\left(\frac{8 \text{ units of carbo}}{\text{ounce}}\right) \cdot (y_1 \text{ ounces}) + \left(\frac{2 \text{ units of carbo}}{\text{ounce}}\right) \cdot (y_2 \text{ ounces}) \geq 5$$

C_2: $8y_1 + 2y_2 \geq 5$

Objective Function

Minimize $w = \text{cost} = \text{Bossy Buffer cost} + \text{Science Cow cost}$

$$w = \left(\frac{32 \text{ cents}}{\text{ounce}}\right) \cdot (y_1 \text{ ounces}) + \left(\frac{12 \text{ cents}}{\text{ounce}}\right) \cdot (y_2 \text{ ounces})$$

$w = 32y_1 + 12y_2$

Notice that the primal problem is a minimization problem with "$\geq$ constraints." This tells us that the dual problem will be a maximization problem with "$\leq$ constraints."

Step 2   *Make a matrix of the primal model.* We copy down the constraints and the objective function without the variables. The constraints go in the first rows, and the objective function goes in the last row.

> **WARNING:** Do not include slack variables, and do not rewrite the objective function with all variables on the left side.

$$\begin{bmatrix} 4 & 3 & 6 \\ 8 & 2 & 5 \\ 32 & 12 & 0 \end{bmatrix}$$

Step 3   *Alter the matrix from Step 2 by changing its rows into columns.*

$$\begin{bmatrix} 4 & 8 & 32 \\ 3 & 2 & 12 \\ 6 & 5 & 0 \end{bmatrix}$$

Step 4   *Convert the matrix from Step 3 into the dual model.* The top row becomes our first constraint:

$$C_1: 4x_1 + 8x_2 \leq 32 \to 4x_1 + 8x_2 + 1s_1 = 32$$

The second row becomes our second constraint:

$$C_2: 3x_1 + 2x_2 \leq 12 \to 3x_1 + 2x_2 + 1s_2 = 12$$

The last row becomes our objective function:

$$\text{Maximize } z = 6x_1 + 5x_2 \to -6x_1 - 5x_2 + z = 0$$

Step 5   *Solve the dual problem.* The first simplex matrix:

$$
\begin{array}{ccccc}
x_1 & x_2 & s_1 & s_2 & z \\
\end{array}
$$

$$
\begin{bmatrix}
4 & 8 & 1 & 0 & 0 & 32 \\
3 & 2 & 0 & 1 & 0 & 12 \\
-6 & -5 & 0 & 0 & 1 & 0
\end{bmatrix}
$$

Its corresponding solution is $(x_1, x_2, s_1, s_2) = (0, 0, 32, 12)$, $z = 0$. This solution is possible, because all of the variables have nonnegative values. We pivot on the 3.

$$
\begin{array}{ccccc}
x_1 & x_2 & s_1 & s_2 & z \\
\end{array}
$$

$$
\begin{bmatrix}
4 & 8 & 1 & 0 & 0 & 32 \\
3 & 2 & 0 & 1 & 0 & 12 \\
-6 & -5 & 0 & 0 & 1 & 0
\end{bmatrix}
\begin{array}{l}
\leftarrow 32/4 = 8 \\
\leftarrow 12/3 = 4 \\
\;
\end{array}
$$

$$\uparrow$$

R2 ÷ 3 : R2

$$
\begin{bmatrix}
4 & 8 & 1 & 0 & 0 & 32 \\
1 & 2/3 & 0 & 1/3 & 0 & 4 \\
-6 & -5 & 0 & 0 & 1 & 0
\end{bmatrix}
$$

R1 − 4R2 : R1
6R2 + R3 : R3

$$
\begin{bmatrix}
0 & 16/3 & 1 & -4/3 & 0 & 16 \\
1 & 2/3 & 0 & 1/3 & 0 & 4 \\
0 & -1 & 0 & 2 & 1 & 24
\end{bmatrix}
\begin{array}{l}
\leftarrow 16/(16/3) = 3 \\
\leftarrow 4/(2/3) = 6 \\
\;
\end{array}
$$

$$\uparrow$$

We pivot on the 16/3.

R1 · 3/16 : R1

$$
\begin{bmatrix}
0 & 1 & 3/16 & -1/4 & 0 & 3 \\
1 & 2/3 & 0 & 1/3 & 0 & 4 \\
0 & -1 & 0 & 2 & 1 & 24
\end{bmatrix}
$$

−2/3 · R1 + R2 : R2
R1 + R3 : R3

$$
\begin{array}{ccccc}
x_1 & x_2 & s_1 & s_2 & z \\
\end{array}
$$

$$
\begin{bmatrix}
0 & 1 & 3/16 & -1/4 & 0 & 3 \\
1 & 0 & -1/8 & 1/2 & 0 & 2 \\
0 & 0 & 3/16 & 7/4 & 1 & 27
\end{bmatrix}
$$

This is the final simplex matrix, because the last row has no negative entries. Its corresponding (dual) solution is $(x_1, x_2, s_1, s_2) = (2, 3, 0, 0)$, $z = 27$. Be careful—this is *not* the primal problem's solution.

Step 6 *Solve and interpret the primal problem.* The corresponding primal solution is $(y_1, y_2) = (3/16, 7/4)$, with $w = 27$. This means that the rancher should add 3/16 ounces of Bossy Buffer and $7/4 = 1\ 3/4$ ounces of Science Cow to the daily feed at a minimum cost of 27 cents per ounce.

The Method of Duals

Step 1 *Generate the primal model.*
Follow the lettering convention.

Step 2 *Make a matrix of the primal model.*
  a. In the first rows copy the coefficients of the constraints, without slack variables.
  b. In the last row copy the coefficients of the objective function in their original order.

WARNING: Do not include slack variables, and do not rewrite the objective function with all variables on the left side.

Step 3 *Alter the matrix from Step 2 by changing its rows into columns.*

Step 4 *Convert the matrix from Step 3 into the dual model.*
  a. Convert the top rows into constraints and the bottom row into an objective function.
  b. Follow the lettering convention.
  c. If the primal problem is a maximization problem with "≤ constraints," the dual problem must be a minimization problem with "≥ constraints." If the primal problem is a minimization problem with "≥ constraints," the dual problem must be a maximization problem with "≤ constraints."

Step 5 *Solve the dual problem.*

Step 6 *Solve and interpret the primal problem.*

3.5

EXERCISES

In Exercises 1–4, determine the dual problem of the given linear programming problem. Do not solve either problem.

1. Maximize $z = 5x_1 + 3x_2 + 4x_3$ subject to the constraints:

$$8x_1 + 4x_2 + 9x_3 \leq 12$$
$$7x_1 + 11x_2 + 913x_3 \leq 48$$

2. Maximize $z = 19x_1 + 42x_2$ subject to the constraints:

$$11x_1 + 16x_2 \leq 86$$
$$5x_1 + 10x_2 \leq 90$$
$$18x_1 + 73x_2 \leq 100$$

3. Minimize $w = 36y_1 + 16y_2$ subject to the constraints:

$$4y_1 + 7y_2 \geq 84$$
$$15y_1 + 21y_2 \geq 68$$
$$2y_1 + 17y_2 \geq 19$$

4. Minimize $w = 100y_1 + 112y_2 + 42y_3$ subject to the constraints:

$$8y_1 + 3y_2 + 412y_3 \geq 26$$
$$9y_1 + 2y_2 + 22y_3 \geq 45$$
$$3y_1 + y_2 + 3y_3 \geq 24$$

In Exercises 5–8, the final simplex matrix for the dual of a linear programming problem is given.

    a. Give the solution (independent variables and objective function) to the dual problem. Is the dual's objective to maximize z or to minimize w?

    b. Give the solution (independent variables and objective function) to the primal problem. Is the primal objective to maximize z or to minimize w?

5.
$$\begin{array}{ccccc} y_1 & y_2 & r_1 & r_2 & z \\ \end{array}$$
$$\begin{bmatrix} 0.87 & 0 & 1 & 0.72 & 0 & 5.7 \\ 0.32 & 1 & 0 & 0.98 & 0 & 9.7 \\ 0.55 & 0 & 0 & 0.33 & 1 & -1120 \end{bmatrix}$$

6.
$$\begin{array}{ccccc} y_1 & y_2 & r_1 & r_2 & z \\ \end{array}$$
$$\begin{bmatrix} 0 & 1.2 & 1 & 0 & 0 & 3.9 \\ 1 & 5.4 & 0 & 9.8 & 0 & 4.2 \\ 0 & 8.6 & 0 & 2.2 & 1 & -1002 \end{bmatrix}$$

7.
$$\begin{array}{ccccc} x_1 & x_2 & s_1 & s_2 & z \\ \end{array}$$
$$\begin{bmatrix} 1 & 0 & 0.32 & 0.87 & 0 & 19.2 \\ 0 & 1 & 0.73 & 1.22 & 0 & 18.7 \\ 0 & 0 & 0.91 & 0.58 & 1 & 12.8 \end{bmatrix}$$

8.
$$\begin{array}{ccccc} x_1 & x_2 & s_1 & s_2 & z \\ \end{array}$$
$$\begin{bmatrix} 1 & 0.6 & 1.3 & 0 & 0 & 55 \\ 0 & 1.2 & 0.7 & 1 & 0 & 19 \\ 0 & 2.3 & 5.1 & 0 & 1 & 123 \end{bmatrix}$$

9. Use the method of duals to minimize $w = 2y_1 + 3y_2$ subject to the constraints:
$$15y_1 + y_2 \geq 25$$
$$y_1 + 6y_2 \geq 20$$
$$8y_1 + 7y_2 \geq 78$$
$$y_1, y_2 \geq 0$$

10. Use the method of duals to minimize $w = 9y_1 - 7y_2$ subject to the constraints:
$$y_1 + y_2 \geq 100$$
$$51y_1 + 23y_2 \geq 3672$$
$$y_1, y_2 \geq 0$$

Use the method of duals to solve Exercises 11 and 12.

11. J & M Winery produces two different Chardonnays, one under their economy label and one under their premium label. They have contracted to sell 30,000 cases of Chardonnay, and they are certain that they can sell more. The demand for their economy label will be at least twice that for the premium label, due to its lower selling price.

They already have enough grapes for 8000 cases of the premium wine, and they can purchase more. Production costs for the premium label wine run $72 a case; for the economy label, $36 a case. How many cases of each Chardonnay should J & M produce in order to minimize production costs? (This is Example 1 from Section 3.3.)

12. J & M Winery produces two different Cabernet Sauvignons, one under their economy label and one under their premium label. They have contracted to sell 20,000 cases of Cabernet, and they are certain that they can sell more. The demand for their economy label will be at least 50% more than for the premium label, due to its lower selling price. They already have enough grapes for 7000 cases of the premium wine, and they can purchase more. Production costs for the premium label wine run $72 a case; for the economy label, $36 a case. How many cases of each Cabernet should J & M produce in order to minimize production costs? (This is Exercise 10 from Section 3.3.)

13. The University of Erewhon's cafeteria dinners always consist of chili macs and whipped potatoes. The cafeteria is required (by the College Cafeteria Association's guidelines) to serve dinners that provide at least 3500 calories, 120 grams of fat, and 750 mg of sodium. The nutritional content and preparation costs of the cafeteria's food is given in Figure 3.21. Use the method of duals to determine the serving sizes that minimize the cafeteria's preparation costs but still meet the College Cafeteria Association's guidelines.

FIGURE 3.21

	Calories per Ounce	Fat per Ounce	Sodium per Ounce	Cost per Ounce
Chili Macs	92	5.8 g	221 mg	15¢
Potatoes	13	0.4 g	72 mg	2¢

14. Solve Exercise 13 without using the method of duals. Compare the dimensions of the first simplex matrix from Exercise 13 with those of Exercise 14. Compare the number of pivots with the two exercises. Which method is quicker?

15. The Monterey Canning Co. produces canned whole Brussels sprouts and canned Brussels sprouts pieces. As an experiment, they are going to also produce canned Brussels sprouts salsa. Market analyses indicate that sales of sprouts salsa will be at least 25% as large as those of whole sprouts.

Regular customers will order at least 400,000 pounds of canned whole sprouts and 90,000 pounds of canned sprouts pieces. A new customer has committed to purchasing at least 200,000 pounds of canned sprout products. It costs $1.25 to produce one pound of canned whole sprouts, $1.75 to produce one pound of canned sprouts pieces, and $3.00 to produce one pound of canned sprouts salsa. Use the method of duals to find the amount of each product that the company should produce in order to minimize its production costs.

CHAPTER 3 REVIEW

Terms

dual problem	Karmarkar's method	phase I pivot	simplex method
first (and second) simplex matrix	linear programming	phase II pivot	slack variable
	method of duals	primal problem	surplus variable
geometric method	mixed constraints	shadow value	

Review Exercises

Use the simplex method to solve the linear programming problems in Exercises 1 and 2.

1. Mowson Audio Co. makes stereo speaker assemblies. It purchases speakers from a speaker manufacturing firm and installs them in its own cabinets. Mowson's model 110 speaker assembly, which sells for $200, has a tweeter and a midrange speaker. The model 330 assembly, which sells for $350, has two tweeters, a midrange speaker, and a woofer. Mowson currently has in stock 90 tweeters, 60 midrange speakers, and 44 woofers.
   a. How many speaker assemblies should Mowson make in order to maximize its income?
   b. What is that maximum income?
   c. Find and interpret the shadow value of a tweeter.
   d. Find and interpret the shadow value of a midrange speaker.
   e. Find and interpret the shadow value of a woofer.

2. Mowson Audio Co. has introduced a new speaker assembly. The new model 220 has a tweeter, a midrange speaker, and a woofer and sells for $280. If Mowson Audio has 140 tweeters, 90 midrange speakers, and 66 woofers in

stock, how many speaker assemblies should it make in order to maximize its income? Interpret the solution; that is, explain the values of each slack variable in addition to answering the question.

3. The Stereo Guys store sells two lines of personal stereos: the Sunny and the Iwa. The Sunny comes in a 12-cubic-foot box and can be sold for a $220 profit, and the Iwa comes in an 8-cubic-foot box and can be sold for a $200 profit. The Stereo Guys' marketing department estimates that at least 600 personal stereos can be sold each month and that, due to Sunny's reputation for quality, the demand for the Sunny unit is at least twice that for the Iwa. The Stereo Guys warehouse has 12,000 cubic feet of space available for personal stereos.
   a. How many Sunnys and Iwas should the company stock each month to maximize profit?
   b. What is that maximum profit?
   c. Find and interpret the shadow value of a cubic foot of warehouse space.

4. A chinchilla farmer needs to provide his chinchillas with a daily dietary supplement. Each chinchilla needs at lest 8 units of protein, 5 units of carbohydrates, and 2 units of vitamins.

Chinchilla Vanilla costs 32 cents per ounce, and each ounce provides 2 units of protein, 2 units of carbohydrates, and 1 unit of vitamins. Science Chinchilla costs 40 cents per ounce, and each ounce provides 5 units of protein, 3 units of carbohydrates, and 1 unit of vitamins.

a. Determine the mixture of Chinchilla Vanilla and Science Chinchilla that will meet the daily requirements at minimum cost, without using the method of duals. Interpret the solution; that is, explain the values of each slack variable and w in addition to answering the question.

b. Determine the mixture of Chinchilla Vanilla and Science Chinchilla that will meet the daily requirements at minimum cost, using the method of duals.

 Answer the following questions using complete sentences.

5. Who invented the simplex method of linear programming? Under what circumstances was it invented?

6. How does the simplex method differ from the geometric method of linear programming?

7. How does Karmarkar's method differ from the simplex method of linear programming?

MATRIX EQUATIONS 4

In Chapter 2 we used matrices when we solved systems of equations with the Gauss-Jordan method. In Chapter 3 we used matrices when we solved linear programming problems by converting a problem's linear inequalities into linear equations and applying the simplex method to the resulting system of equations.

The interrelationship between matrices and systems of equations is one of the most prominent topics in this book. In this chapter we will continue to investigate this interrelationship by solving systems of equations using a matrix method quite different from the Gauss-Jordan method. This new method, called the *method of matrix equations,* is in some instances

simpler and easier than the Gauss-Jordan method. We will also apply matrix equations to Leontief's *input-output analysis,* which is used by both corporations and governments to determine their most desirable levels of production.

In Chapters 7 and 8 we will investigate two more powerful applications of matrices, Markov chains and game theory. Markov chains are used in business, sociology, the physical sciences, and biology to analyze trends and to predict the outcome of those trends. Game theory provides a method of analyzing competitive situations where opposing sides must choose strategies that maximize their own winnings or minimize their opponents' winnings. Game theory is applicable to business and military problems, as well as to games of recreation.

Clearly, matrices are powerful mathematical tools that have a wide variety of important applications in business, biology, sociology, the physical sciences, and military science.

4.0

MATRIX ARITHMETIC

Like numbers, matrices can be added, subtracted, and multiplied; however, they cannot be divided. In this section we will investigate the arithmetic of matrices, including addition, subtraction, and multiplication. We will then use that arithmetic to solve matrix equations.

Addition and Subtraction of Matrices

To add two matrices, simply add corresponding elements.

EXAMPLE 1   Add

$$\begin{bmatrix} 5 & -2 \\ 3 & 7 \end{bmatrix} + \begin{bmatrix} 3 & 6 \\ 0 & -1 \end{bmatrix}$$

Solution
$$\begin{bmatrix} 5 & -2 \\ 3 & 7 \end{bmatrix} + \begin{bmatrix} 3 & 6 \\ 0 & -1 \end{bmatrix} = \begin{bmatrix} 5 + 3 & -2 + 6 \\ 3 + 0 & 7 + -1 \end{bmatrix}$$

$$= \begin{bmatrix} 8 & 4 \\ 3 & 6 \end{bmatrix}$$

EXAMPLE 2   Add

$$\begin{bmatrix} 5 & -2 \\ 3 & 7 \end{bmatrix} + \begin{bmatrix} 3 & 6 \\ 0 & -1 \\ 2 & 3 \end{bmatrix}$$

Solution
$$\begin{bmatrix} 5 & -2 \\ 3 & 7 \end{bmatrix} + \begin{bmatrix} 3 & 6 \\ 0 & -1 \\ 2 & 3 \end{bmatrix} = \begin{bmatrix} 5 + 3 & -2 + 6 \\ 3 + 0 & 7 + -1 \\ ? + 2 & ? + 3 \end{bmatrix}$$

These two matrices cannot be added. Matrices must be of the same dimensions if they are to be added.

Matrix subtraction is quite similar to matrix addition. To subtract two matrices, subtract corresponding elements. Matrices must be of the same dimensions if they are to be subtracted.

Scalar Multiplication

To multiply a matrix by a number, multiply each element of the matrix by the number. This form of multiplication is called **scalar multiplication,** and the multiplier is called a **scalar.**

EXAMPLE 3   Perform the scalar multiplication

$$3\begin{bmatrix} 5 & 6 \\ 3 & 7 \end{bmatrix}$$

Solution $3\begin{bmatrix} 5 & 6 \\ 3 & 7 \end{bmatrix} = \begin{bmatrix} 3 \cdot 5 & 3 \cdot 6 \\ 3 \cdot 3 & 3 \cdot 7 \end{bmatrix} = \begin{bmatrix} 15 & 18 \\ 9 & 21 \end{bmatrix}$

The 3 is a scalar because it is multiplied by the above matrix.   ●

Matrix Equations

Two matrices are **equal** if and only if their corresponding elements are equal.

EXAMPLE 4   Find the values of x, y, z, and w in the matrix equation

$$\begin{bmatrix} 3x & y + 3 \\ 8 & 0 \end{bmatrix} = \begin{bmatrix} 2x & 5 \\ z - 2 & w \end{bmatrix}$$

Solution   The corresponding elements of these two matrices must be equal. In particular, $3x$ and $2x$ are in corresponding positions, so $3x = 2x$. Thus we obtain the following four equations.

upper-left elements	upper-right elements
$3x = 2x$	$y + 3 = 5$
$3x - 2x = 0$	$y = 2$
$x = 0$	

lower-left elements	lower-right elements
$8 = z - 2$	$0 = w$
$z = 10$	$w = 0$

●

A **matrix equation** is an equation that has matrices in it. Matrix equations can be solved in a manner similar to that used to solve algebraic equations.

EXAMPLE 5   Solve the matrix equation given below, and check its solution.

$$2\begin{bmatrix} w & x \\ y & z \end{bmatrix} + \begin{bmatrix} 5 & 6 \\ 3 & 7 \end{bmatrix} = \begin{bmatrix} 7 & 10 \\ 3 & 13 \end{bmatrix}$$

Solution   This equation is very similar to the equation $2x + 3 = 7$; the only difference is that the equation is composed of matrices rather than numbers. These two equations are solved with the same sequence of steps. Although we only want the solution of the matrix equation, both equations are solved below to illustrate the similarity of the steps.

$2\begin{bmatrix} w & x \\ y & z \end{bmatrix} + \begin{bmatrix} 5 & 6 \\ 3 & 7 \end{bmatrix} = \begin{bmatrix} 7 & 10 \\ 3 & 13 \end{bmatrix}$	← *The equation* →	$2x + 3 = 7$
$2\begin{bmatrix} w & x \\ y & z \end{bmatrix} = \begin{bmatrix} 7 & 10 \\ 3 & 13 \end{bmatrix} - \begin{bmatrix} 5 & 6 \\ 3 & 7 \end{bmatrix}$	Step 1: Subtract the constant term from each side.	$2x = 7 - 3$
$2\begin{bmatrix} w & x \\ y & z \end{bmatrix} = \begin{bmatrix} 2 & 4 \\ 0 & 6 \end{bmatrix}$	Step 2: Compute the result of subtracting.	$2x = 4$
$\frac{1}{2} \cdot 2\begin{bmatrix} w & x \\ y & z \end{bmatrix} = \frac{1}{2} \cdot \begin{bmatrix} 2 & 4 \\ 0 & 6 \end{bmatrix}$	Step 3: Multiply each side by 1/2.	$\frac{1}{2} \cdot 2x = \frac{1}{2} \cdot 4$
$\begin{bmatrix} w & x \\ y & z \end{bmatrix} = \begin{bmatrix} 1 & 2 \\ 0 & 3 \end{bmatrix}$	Step 4: Compute the result of multiplying by 1/2 and give the solution.	$x = 2$

✔   The solution to Example 5 can easily be checked by substituting it back into the original equation:

$$2\begin{bmatrix} w & x \\ y & z \end{bmatrix} + \begin{bmatrix} 5 & 6 \\ 3 & 7 \end{bmatrix} = 2\begin{bmatrix} 1 & 2 \\ 0 & 3 \end{bmatrix} + \begin{bmatrix} 5 & 6 \\ 3 & 7 \end{bmatrix}$$

$$= \begin{bmatrix} 2 & 4 \\ 0 & 6 \end{bmatrix} + \begin{bmatrix} 5 & 6 \\ 3 & 7 \end{bmatrix}$$

$$= \begin{bmatrix} 7 & 10 \\ 3 & 13 \end{bmatrix}$$

Matrix Multiplication

With matrices, there are two different types of multiplication: scalar multiplication is the multiplication of a matrix by a number; **matrix multiplication** is the multiplication of two matrices. Matrix multiplication is a process that at first seems rather strange and obscure. Its usefulness will become apparent when we apply it to Leontief input-output matrices later in this chapter and to Markov chains in Chapter 8.

EXAMPLE 6   If

$$A = \begin{bmatrix} 3 & -2 \\ 4 & 1 \end{bmatrix} \quad \text{and} \quad B = \begin{bmatrix} 1 & 5 \\ 0 & -7 \end{bmatrix}$$

find the product AB.

Solution $AB = \begin{bmatrix} 3 & -2 \\ 4 & 1 \end{bmatrix}\begin{bmatrix} 1 & 5 \\ 0 & -7 \end{bmatrix} = ?$

We start with row one of A and column one of B. We multiply corresponding elements together, add the products, and put the sum in row one, column one of the product matrix.

$$AB = \begin{bmatrix} \boxed{3} & \boxed{-2} \\ 4 & 1 \end{bmatrix} \begin{bmatrix} \boxed{1} & 5 \\ \boxed{0} & -7 \end{bmatrix} = \begin{bmatrix} \boxed{3 \cdot 1} + \boxed{-2 \cdot 0} & ? \\ ? & ? \end{bmatrix} = \begin{bmatrix} \boxed{3} & ? \\ ? & ? \end{bmatrix}$$

We continue in this manner until each row of A has been multiplied by each column of B. Thus, we now multiply row one of A and column two of B and put the result in row one, column two of the product matrix.

$$AB = \begin{bmatrix} \boxed{3} & \boxed{-2} \\ 4 & 1 \end{bmatrix} \begin{bmatrix} 1 & \boxed{5} \\ 0 & \boxed{-7} \end{bmatrix} = \begin{bmatrix} 3 & \boxed{3 \cdot 5} + \boxed{-2 \cdot -7} \\ ? & ? \end{bmatrix} = \begin{bmatrix} 3 & \boxed{29} \\ ? & ? \end{bmatrix}$$

We're finished with row one; what remains is to multiply row two of A first by column one of B and then by column two. Multiplying row two of A by column one of B gives

$$AB = \begin{bmatrix} 3 & -2 \\ \boxed{4} & \boxed{1} \end{bmatrix} \begin{bmatrix} \boxed{1} & 5 \\ \boxed{0} & -7 \end{bmatrix} = \begin{bmatrix} 3 & 29 \\ \boxed{4 \cdot 1} + \boxed{1 \cdot 0} & ? \end{bmatrix} = \begin{bmatrix} 3 & 29 \\ \boxed{4} & ? \end{bmatrix}$$

Finally, we multiply row two of A by column two of B and put the result in row two, column two of the product matrix:

$$AB = \begin{bmatrix} 3 & -2 \\ \boxed{4} & \boxed{1} \end{bmatrix} \begin{bmatrix} 1 & \boxed{5} \\ 0 & \boxed{-7} \end{bmatrix} = \begin{bmatrix} 3 & 29 \\ 4 & \boxed{4 \cdot 5} + \boxed{1 \cdot -7} \end{bmatrix} = \begin{bmatrix} 3 & 29 \\ 4 & \boxed{13} \end{bmatrix}$$

The product of A and B is

$$AB = \begin{bmatrix} 3 & 29 \\ 4 & 13 \end{bmatrix}$$

A matrix is really just a table. Consider the table in Figure 4.1, which gives prices of recorded music.

FIGURE 4.1

	Sale Price	Regular Price
CD	$12	$16
LP	$7	$9
Tape	$8	$11

The table could be rewritten as the matrix

$$\begin{bmatrix} 12 & 16 \\ 7 & 9 \\ 8 & 11 \end{bmatrix}$$

Matrix multiplication is a natural way to manipulate data in tables. If we were to purchase three CDs, two LPs, and three tapes, we would calculate the sale price as

$$3 \cdot \$12 + 2 \cdot \$7 + 3 \cdot \$8 = \$74$$

and the regular price as

$$3 \cdot \$16 + 2 \cdot \$9 + 3 \cdot \$11 = \$99$$

This calculation is matrix multiplication; it's the same as

$$[3 \ 2 \ 3] \begin{bmatrix} 12 & 16 \\ 7 & 9 \\ 8 & 11 \end{bmatrix} = [3 \cdot 12 + 2 \cdot 7 + 3 \cdot 8 \quad 3 \cdot 16 + 2 \cdot 9 + 3 \cdot 11]$$

$$= [74 \quad 99]$$

It is not always possible to multiply two matrices together. In the matrix multiplication problem above, if the first matrix had one extra element, we wouldn't be able to multiply.

$$[3 \quad 2 \quad 3 \quad 77] \begin{bmatrix} 12 & 16 \\ 7 & 9 \\ 8 & 11 \end{bmatrix}$$

$$= [3 \cdot 12 + 2 \cdot 7 + 3 \cdot 8 + 77 \cdot ? \quad 3 \cdot 16 + 2 \cdot 9 + 3 \cdot 11 + 77 \cdot ?]$$

It's easy to tell when two matrices cannot be multiplied: If you run out of elements to pair together when you're multiplying the first row by the first column, then the two matrices cannot be multiplied. In the above problem, we paired 3 and 12, 2 and 7, and 3 and 8, but we had nothing to pair with 77. This tells us that the two matrices cannot be multiplied together.

Some people prefer to determine whether it's possible to multiply two matrices before they actually start multiplying. In the above problem, we could not multiply a 1×4 matrix by a 3×2.

$$[3 \quad 2 \quad 3 \quad 77] \begin{bmatrix} 12 & 16 \\ 7 & 9 \\ 8 & 11 \end{bmatrix}$$

$$\begin{array}{ccc} 1 \times 4 & & 3 \times 2 \\ \uparrow & & \uparrow \end{array}$$

Different, so we can't multiply

Previously, when 77 wasn't there, we *were* able to multiply.

$$[3 \quad 2 \quad 3] \begin{bmatrix} 12 & 16 \\ 7 & 9 \\ 8 & 11 \end{bmatrix}$$

$$\begin{array}{ccc} 1 \times 3 & & 3 \times 2 \\ \uparrow & & \uparrow \end{array}$$

The same, so we can multiply

EXAMPLE 7 $A = \begin{bmatrix} 3 & -2 \\ 4 & 1 \end{bmatrix}$, $B = [1 \ \ 0]$, and $C = \begin{bmatrix} 1 \\ 0 \end{bmatrix}$.

a. Find AB, if it exists.

b. Find AC, if it exists.

Solution   **a.** $AB = \begin{bmatrix} 3 & -2 \\ 4 & 1 \end{bmatrix} \begin{bmatrix} 1 & 0 \end{bmatrix} = \begin{bmatrix} 3 \cdot 1 + -2 \cdot ? & 3 \cdot 0 + -2 \cdot ? \\ 4 \cdot 1 + 1 \cdot ? & 4 \cdot 0 + 1 \cdot ? \end{bmatrix}$

The product AB does not exist, because we ran out of elements to pair together. Alternatively, A is a 2×2 matrix, and B is a 1×2 matrix.

$$\begin{bmatrix} 3 & -2 \\ 4 & 1 \end{bmatrix} \begin{bmatrix} 1 & 0 \end{bmatrix}$$

$2 \times 2 \quad 1 \times 2$

↑ ↑

Different, so we can't multiply

b. $AC = \begin{bmatrix} 3 & -2 \\ 4 & 1 \end{bmatrix} \begin{bmatrix} 1 \\ 0 \end{bmatrix} = \begin{bmatrix} 3 \cdot 1 + -2 \cdot 0 \\ 4 \cdot 1 + 1 \cdot 0 \end{bmatrix} = \begin{bmatrix} 3 \\ 4 \end{bmatrix}$

The product of A and C is

$$\begin{bmatrix} 3 \\ 4 \end{bmatrix}$$

If we had checked the dimensions in advance, we would have found that A and C can be multiplied:

$$\begin{bmatrix} 3 & -2 \\ 4 & 1 \end{bmatrix} \begin{bmatrix} 1 \\ 0 \end{bmatrix}$$

$2 \times 2 \quad 2 \times 1$

↑ ↑

The same, so we can multiply

This dimension check also gives the dimensions of the product matrix. The matching twos "cancel" and leave a 2×1 matrix.   ●

Matrix Multiplication

If matrix A is an $m \times n$ matrix and matrix B is an $n \times p$ matrix, then the product AB exists and is an $m \times p$ matrix.

(*Think: $m \times n \cdot n \times p$ yields $m \times p$.*)

To find the entry in the ith row and the jth column of the product matrix AB, multiply each element of A's ith row by the corresponding element of B's jth column and add the results.

Properties of Matrix Multiplication

When you multiply numbers together, you use certain properties so automatically that you don't even realize you are using them. In particular, multiplication of real numbers is **commutative;** that is, $ab = ba$ (the order in which you multiply doesn't matter). Multiplication of real numbers is also **associative;** that is, $a(bc) = (ab)c$ (if you multiply three numbers together, it doesn't matter which two you multiply first).

Matrix multiplication is certainly different from real number multiplication. It's important to know whether matrix multiplication is commutative and associative, because people tend to use those properties without being aware of doing so.

EXAMPLE 8   Determine whether matrix multiplication is commutative by finding the product BA for the matrices B and A given in Example 6 and comparing BA with AB.

Solution
$$BA = \begin{bmatrix} 1 & 5 \\ 0 & -7 \end{bmatrix} \begin{bmatrix} 3 & -2 \\ 4 & 1 \end{bmatrix} = \begin{bmatrix} 1 \cdot 3 + 5 \cdot 4 & 1 \cdot -2 + 5 \cdot 1 \\ 0 \cdot 3 + -7 \cdot 4 & 0 \cdot -2 + -7 \cdot 1 \end{bmatrix}$$

$$= \begin{bmatrix} 23 & 3 \\ -28 & -7 \end{bmatrix}$$

The product of B and A is

$$BA = \begin{bmatrix} 23 & 3 \\ -28 & -7 \end{bmatrix}$$

In Example 6, we found that

$$AB = \begin{bmatrix} 3 & 29 \\ 4 & 13 \end{bmatrix}$$

Clearly, AB is not the same as BA; that is, $AB \neq BA$.   ●

Matrix multiplication is not commutative; in general, $AB \neq BA$. This means that you have to be careful to multiply in the right order—it's easy to be careless and find BA when you're asked to find AB.

EXAMPLE 9   Check the associative property for matrix multiplication by finding $A(BC)$ and $(AB)C$ for matrices A, B, and C given below.

$$A = \begin{bmatrix} 2 & -3 \end{bmatrix} \quad B = \begin{bmatrix} 4 & 0 & -3 \\ 2 & -1 & 5 \end{bmatrix} \quad C = \begin{bmatrix} -3 & 5 \\ 2 & 0 \\ 1 & -1 \end{bmatrix}$$

Solution   To find $A(BC)$, we first find BC and then multiply it by A. Notice that B's dimensions are 2×3 and C's dimensions are 3×2. This tells us that B and C can be multiplied and that the product will be a 2×2 matrix.

$$BC = \begin{bmatrix} 4 & 0 & -3 \\ 2 & -1 & 5 \end{bmatrix} \begin{bmatrix} -3 & 5 \\ 2 & 0 \\ 1 & -1 \end{bmatrix}$$

$$= \begin{bmatrix} 4 \cdot -3 + & 0 \cdot 2 + -3 \cdot 1 & 4 \cdot 5 + & 0 \cdot 0 + -3 \cdot -1 \\ 2 \cdot -3 + & -1 \cdot 2 + & 5 \cdot 1 & 2 \cdot 5 + -1 \cdot 0 + & 5 \cdot -1 \end{bmatrix}$$

$$= \begin{bmatrix} -15 & 23 \\ -3 & 5 \end{bmatrix}$$

To find $A(BC)$, we place A on the *left* of BC and multiply. Notice that A's dimensions are 1×2 and BC's dimensions are 2×2, so A and BC can be multiplied, and the product will be a 1×2 matrix.

$$A(BC) = [2 \quad -3] \begin{bmatrix} -15 & 23 \\ -3 & 5 \end{bmatrix}$$

$$= [2 \cdot -15 + -3 \cdot -3 \quad 2 \cdot 23 + -3 \cdot 5]$$

$$= [-21 \quad 31]$$

Now that we've found $A(BC)$, we need to find $(AB)C$ and then see if they're the same. To find $(AB)C$, we first find AB and then multiply it by C. Notice that A's dimensions are 1×2 and B's dimensions are 2×3, so A and B can be multiplied.

$$AB = [2 \quad -3] \begin{bmatrix} 4 & 0 & -3 \\ 2 & -1 & 5 \end{bmatrix}$$

$$= [2 \cdot 4 + -3 \cdot 2 \quad 2 \cdot 0 + -3 \cdot -1 \quad 2 \cdot -3 + -3 \cdot 5]$$

$$= [2 \quad 3 \quad -21]$$

To find $(AB)C$, we place AB on the *left* of C and multiply:

$$(AB)C = [2 \quad 3 \quad -21] \begin{bmatrix} -3 & 5 \\ 2 & 0 \\ 1 & -1 \end{bmatrix}$$

$$= [2 \cdot -3 + 3 \cdot 2 + -21 \cdot 1 \quad 2 \cdot 5 + 3 \cdot 0 + -21 \cdot -1]$$

$$= [-21 \quad 31]$$

Therefore, $A(BC) = (AB)C$. ●

In Example 9, $A(BC) = (AB)C$, even though the work involved in finding $A(BC)$ was different from the work involved in finding $(AB)C$. This always happens; $A(BC) = (AB)C$, provided the dimensions of A, B, and C are such that they can be multiplied together. Matrix multiplication is associative.

Identity Matrices

EXAMPLE 10   If

$$I = \begin{bmatrix} 1 & 0 \\ 0 & 1 \end{bmatrix} \quad \text{and} \quad A = \begin{bmatrix} 2 & 3 \\ 4 & 5 \end{bmatrix}$$

find IA.

Solution

$$IA = \begin{bmatrix} 1 & 0 \\ 0 & 1 \end{bmatrix} \begin{bmatrix} 2 & 3 \\ 4 & 5 \end{bmatrix}$$

$$= \begin{bmatrix} 1 \cdot 2 + 0 \cdot 4 & 1 \cdot 3 + 0 \cdot 5 \\ 0 \cdot 2 + 1 \cdot 4 & 0 \cdot 3 + 1 \cdot 5 \end{bmatrix}$$

$$= \begin{bmatrix} 2 & 3 \\ 4 & 5 \end{bmatrix}$$

In Example 10, notice that $IA = A$. If you were to multiply in the opposite order, you would find that $AI = A$. The matrix I is similar to the number 1, because $I \cdot A = A \cdot I = A$, just as $1 \cdot a = a \cdot 1 = a$.

The matrix $\begin{bmatrix} 1 & 0 \\ 0 & 1 \end{bmatrix}$ is called an identity matrix, because multiplying this matrix by any other 2×2 matrix A yields a product *identical* to A. The matrix $\begin{bmatrix} 1 & 0 & 0 \\ 0 & 1 & 0 \\ 0 & 0 & 1 \end{bmatrix}$ is also called an identity matrix, because multiplying this 3×3 matrix by any other 3×3 matrix A yields a product identical to A. An **identity matrix** is a square matrix I that has ones for each entry in the diagonal (the diagonal that starts in the upper-left corner) and zeros for all other entries. The product of any matrix A and an identity matrix is always A, as long as the dimensions of the two matrices are such that they can be multiplied.

Properties of Matrix Multiplication

1. *There is no commutative property:* In general, $AB \neq BA$. You must be careful about the order in which you multiply.
2. *Associative property:* $A(BC) = (AB)C$, provided the dimensions of A, B, and C are such that they can be multiplied together.
3. *Identity property:* An *identity matrix* is a square matrix I that has ones for each entry in the diagonal (the diagonal that starts in the upper-left corner) and zeros for all other entries. If I and A have the same dimensions, then $IA = AI = A$.

Historical Note

Arthur Cayley & James Joseph Sylvester

1821–1895   1814–1897

Arthur Cayley

James Sylvester

The theory of matrices was a product of the unique partnership of Arthur Cayley and James Sylvester. They met in their twenties and were friends, colleagues, and coauthors for the rest of their lives.

Cayley's mathematical ability was recognized at an early age, and he was encouraged to study the subject. He graduated from Cambridge

University at the top of his class. After graduation, Cayley was awarded a three-year fellowship that allowed him to do as he pleased. During this time he made several trips to Europe, where he spent his time taking walking tours, mountaineering, painting, reading novels, and studying architecture, as well as reading and writing mathematics. He

wrote 25 papers in mathematics, papers that were well received by the mathematical community.

When Cayley's fellowship expired, he found that no position as a mathematician was open to him unless he entered the clergy, so he left mathematics and prepared for a legal career. When he was admitted to the bar, he met James Joseph Sylvester.

Sylvester's mathematical ability had also been recognized at an early age. He studied mathematics at the University of London at the age of 14, under Augustus De Morgan (see Chapter 5 for more information on this famous mathematician). He entered Cambridge University at the age of 17 and won several prizes. However, Cambridge would not award him his degrees because he was Jewish. He completed his bachelor's and master's degrees at Trinity College in Dublin. Many years

4.0

EXERCISES

In Exercises 1–4, the following information applies:

$$M = \begin{bmatrix} 3 & 7 \\ 2 & 1 \end{bmatrix} \qquad N = \begin{bmatrix} 5 & -2 \\ 0 & 3 \end{bmatrix} \qquad P = \begin{bmatrix} 6 & 0 \\ 2 & 4 \end{bmatrix}$$

1. a. Find $M + N$.   **b.** Find $N + M$.
   c. Do you think matrix addition is commutative? Why?

2. a. Find $M - N$.   **b.** Find $N - M$.
   c. Do you think matrix subtraction is commutative? Why?

3. a. Find $M + (N + P)$.   **b.** Find $(M + N) + P$.
   c. Do you think matrix addition is associative? Why?

4. a. Find $M - (N - P)$.   **b.** Find $(M - N) - P$.
   c. Do you think matrix subtraction is associative? Why?

later, Cambridge changed its discriminatory policy and gave Sylvester his degrees.

Sylvester taught science at University College in London for two years, found that he didn't like teaching science, and quit. He went to the United States, where he got a job teaching mathematics at the University of Virginia. After three months, he quit when the administration refused to discipline a student who had insulted him. After several unsuccessful attempts to obtain a teaching position, he returned to England and worked for an insurance firm as an actuary, retaining his interest in mathematics only through tutoring. Florence Nightingale was one of his private pupils. When Sylvester became thoroughly bored with insurance work, he studied for a legal career and met Cayley.

Cayley and Sylvester revived and intensified each other's interest in mathematics, and each started to write mathematics again. During his 14 years spent practicing law, Cayley wrote almost 300 papers. Each frequently expressed gratitude to the other for assistance and

inspiration. In one of his papers, Sylvester wrote that "the theorem above enunciated was in part suggested in the course of a conversation with Mr. Cayley (to whom I am indebted for my restoration to the enjoyment of mathematical life)." In another, he said, "Mr. Cayley habitually discourses pearls and rubies."

Cayley joyfully departed from the legal profession when Cambridge offered him a professorship in mathematics, even though his income suffered as a result. He was finally able to spend his life studying, teaching, and writing mathematics. He became quite famous as a mathematician, writing almost 1000 papers in algebra and geometry, often in collaboration with Sylvester. Many of these papers are pioneering works of scholarship. Cayley also played an important role in changing Cambridge's policy that had prohibited the admission of women as students.

Sylvester was repeatedly honored for his pioneering work in algebra. He left the law but was unable to obtain a professorship in

mathematics at a prominent institution until late in his life. At the age of 62, he accepted a position at the newly founded Johns Hopkins University in Baltimore as its first professor of mathematics. While there, he founded the *American Journal of Mathematics,* introduced graduate work in mathematics into American universities, and generally stimulated the development of mathematics in America. He also arranged for Cayley to spend a semester at Johns Hopkins as guest lecturer. At the age of 70, Sylvester returned to England to become Savilian Professor of Geometry at Oxford University.

Cayley and Sylvester were responsible for the theory of matrices, including the operation of matrix multiplication. Sixty-seven years after the invention of matrix theory, Heisenberg recognized it as the perfect tool for his revolutionary work in quantum mechanics. The work of Cayley and Sylvester in algebra became quite important for modern physics, particularly in the theory of relativity. Cayley also wrote on non-Euclidean geometry.

In Exercises 5–12, find the indicated sums, differences, or products (if they exist) of the following matrices.

$$A = \begin{bmatrix} 3 & 1 \\ 4 & 6 \\ 8 & 2 \end{bmatrix} \quad B = \begin{bmatrix} 7 & 0 \\ -1 & 2 \end{bmatrix} \quad C = \begin{bmatrix} 6 \\ 3 \\ 5 \end{bmatrix}$$

$$D = \begin{bmatrix} 9 & 3 & 2 \\ -4 & 8 & -7 \end{bmatrix} \quad F = \begin{bmatrix} 5 & 2 \end{bmatrix} \quad G = \begin{bmatrix} 1 & 0 \\ -2 & 8 \end{bmatrix}$$

$$H = \begin{bmatrix} 8 \\ 0 \end{bmatrix} \quad J = \begin{bmatrix} 8 & 2 & -3 \\ -4 & 9 & 0 \end{bmatrix}$$

5. a. $B + G$ **b.** $2G + 3B$ **c.** BG **d.** GB

6. a. $2A - D$ **b.** AD **c.** DA

7. a. FG **b..** GF **c.** $F + 4G$

8. a. GH **b.** HG **c.** $3G - H$

9. a. $A + B$ **b.** AB **c.** BA

10. a. $3D - J$ **b.** $2J - D$ **c.** DJ

11. a. CD **b.** DC **c.** $C - 3D$

12. a. FH **b.** HF **c.** $5F + H$

13. a. Find $F(B + G)$.   **b.** Find $FB + FG$.
   c. Do you think matrix multiplication is distributive? Why?

14. a. Find $3(D + J)$.   **b.** Find $3D + 3J$.

 c. Do you think scalar multiplication is distributive? Why?

15. Use matrix multiplication and Figure 4.2 to find the sale price and the regular price of purchasing 5 CDs, 0 LPs, and 3 tapes.

FIGURE 4.2

	Sale Price	Regular Price
CD	$12	$16
LP	$7	$9
Tape	$8	$11

16. Use matrix multiplication and Figure 4.2 to find the sale price and the regular price of purchasing 4 CDs, 2 LPs, and 1 tape.

17. Use matrix multiplication and Figure 4.3 to find the price of purchasing 2 slices of pizza and 1 cola at Blondie's and at SliceMan's.

FIGURE 4.3

	Blondie's Price	SliceMan's Price
Pizza	$1.25	$1.30
Cola	$ .95	$1.10

18. Use matrix multiplication and Figure 4.3 to find the price of purchasing 6 slices of pizza and 2 colas at Blondie's and at SliceMan's.

19. a. Use the information in Figure 4.2 and *matrix addition* to find the sale price and the regular price of 2 CDs, 2 LPs, and 2 tapes.

 b. Use the information in Figure 4.2 and *scalar multiplication* to find the sale price and the regular price of 2 CDs, 2 LPs, and 2 tapes.

 c. Given a matrix A, how could $2A$ be described using matrix addition?

20. a. Use the information in Figure 4.3 and *matrix addition* to find Blondie's price and SliceMan's price for 3 pizzas and 3 colas.

 b. Use the information in Figure 4.3 and *scalar multiplication* to find Blondie's price and SliceMan's price for 3 pizzas and 3 colas.

 c. Given a matrix A, how could $3A$ be described using matrix addition?

21. José, Eloise, and Sylvie are each enrolled in the same four courses: English, math, history, and business. English and math are 4-unit courses, while history and business are 3-unit courses. The students' grades are given in Figure 4.4.

FIGURE 4.4

	English	Math	History	Business
José	A	B	B	C
Eloise	C	A	A	C
Sylvie	B	A	B	A

Counting A's as 4 grade points, B's as 3 grade points, and C's as 2 grade points, use matrix multiplication and scalar multiplication to compute José's and Sylvie's grade point average (GPA).

HINT: Start by computing José's total grade points without using matrices. Then set up two matrices, one that is a matrix version of Figure 4.4 with grade points rather than letter grades, and one that gives the 4 courses' units. Set up these two matrices so that their product mirrors your hand calculation of José's total grade points. Finally, incorporate an appropriate scalar multiplication so that the result is GPA, not total grade points.

22. Dave, Johnny, and Larry are each enrolled in the same four courses: communications, broadcasting, marketing, and contracts. Communications and contracts are 4-unit courses, while broadcasting and marketing are 3-unit courses. The students' grades are given in Figure 4.5.

FIGURE 4.5

	Communications	Broadcasting	Marketing	Contracts
Dave	A	B	C	A
Jay	A	B	C	B
Johnny	A	A	A	A
Larry	B	A	B	C

Counting A's as 4 grade points, B's as 3 grade points, and C's as 2 grade points, use matrix multiplication and scalar multiplication to compute Dave's, Jay's, Johnny's, and Larry's grade point average (GPA). (See Exercise 21.)

23. Verify the associative property by finding $A(BC)$ and $(AB)C$.

$$A = \begin{bmatrix} -4 & 5 \\ 2 & 3 \end{bmatrix} \quad B = \begin{bmatrix} 3 & 0 & -1 \\ 1 & 4 & -2 \end{bmatrix}$$

$$C = \begin{bmatrix} 0 \\ 5 \\ -1 \end{bmatrix}$$

24. Verify the associative property by finding $A(BC)$ and $(AB)C$.

$$A = \begin{bmatrix} 2 & 0 \\ -1 & 1 \\ 3 & 2 \end{bmatrix} \quad B = \begin{bmatrix} 5 & 3 & -2 \\ 2 & 0 & 1 \end{bmatrix}$$

$$C = \begin{bmatrix} 1 \\ 2 \\ 0 \end{bmatrix}$$

In Exercises 25–30, find the product (if it exists).

25. $\begin{bmatrix} 1 & 0 \\ 0 & 1 \end{bmatrix} \cdot \begin{bmatrix} 3 & -2 \\ 4 & 0 \end{bmatrix}$

26. $\begin{bmatrix} 1 & 0 & 0 \\ 0 & 1 & 0 \\ 0 & 0 & 1 \end{bmatrix} \cdot \begin{bmatrix} -4 & 5 & 2 \\ 8 & -1 & 9 \\ -2 & 27 & 4 \end{bmatrix}$

27. $\begin{bmatrix} 1 & 0 & 0 \\ 0 & 1 & 0 \\ 0 & 0 & 1 \end{bmatrix} \cdot \begin{bmatrix} 4 & 1 & -1 \\ 5 & 12 & 3 \end{bmatrix}$

28. $\begin{bmatrix} 27 & 19 \\ 42 & 25 \end{bmatrix} \cdot \begin{bmatrix} 1 & 0 \\ 0 & 1 \end{bmatrix}$

29. $\begin{bmatrix} 19 & 7 & 34 \\ 74 & 0 & -11 \\ 13 & -2 & 44 \end{bmatrix} \cdot \begin{bmatrix} 1 & 0 & 0 \\ 0 & 1 & 0 \\ 0 & 0 & 1 \end{bmatrix}$

30. $\begin{bmatrix} 1 & 0 \\ 0 & 1 \end{bmatrix} \cdot \begin{bmatrix} 6 & 18 & -3 \\ 0 & 1 & 5 \\ -14 & 5 & -2 \end{bmatrix}$

31. Find the values of x, y, z, and w in the matrix equation

$$\begin{bmatrix} 8x & 2y + 3 \\ 8 & 10 \end{bmatrix} = \begin{bmatrix} 2x + 1 & 5 \\ z - 4 & w \end{bmatrix}$$

Check your solution; answers are not given in the back.

32. Find the values of x and y in the matrix equation

$$[5x + 8 \quad 2y - 7] = [23x - 2 \quad 8 - 5y]$$

Check your solution; answers are not given in the back.

In Exercises 33–36, solve the given matrix equation. Check your solution; answers are not given in the back.

33. $7\begin{bmatrix} w & x \\ y & z \end{bmatrix} - \begin{bmatrix} 6 & 3 \\ 2 & -4 \end{bmatrix} = \begin{bmatrix} 5 & 7 \\ 2 & 14 \end{bmatrix}$

34. $3\begin{bmatrix} w & x \\ y & z \end{bmatrix} + \begin{bmatrix} 6 & -9 \\ 2 & 0 \end{bmatrix} = \begin{bmatrix} 7 & -1 \\ 1 & 5 \end{bmatrix}$

35. $4\begin{bmatrix} w & x \\ y & z \end{bmatrix} + \begin{bmatrix} 5 & 6 \\ 3 & 7 \end{bmatrix} = -5\begin{bmatrix} w & x \\ y & z \end{bmatrix} - \begin{bmatrix} 2 & 9 \\ 0 & 1 \end{bmatrix}$

36. $7\begin{bmatrix} w & x \\ y & z \end{bmatrix} - \begin{bmatrix} 9 & 0 \\ 2 & -3 \end{bmatrix} = 2\begin{bmatrix} w & x \\ y & z \end{bmatrix} + \begin{bmatrix} -4 & 2 \\ -1 & 1 \end{bmatrix}$

Answer the following questions using complete sentences.

37. Compare and contrast Mr. Cayley and Mr. Sylvester. Discuss why each turned his back on mathematics, why each returned to mathematics, their relative success in schooling, and their relative success in employment.

38. Describe the two ways of determining whether two matrices can be multiplied.

39. Why is an identity matrix so named?

40. In some of Exercises 1, 2, 3, 4, 13, and 14 you are able to definitely determine whether the given operation has the given property, and in some of those exercises you are not able to make a definite determination. In which exercises can you not make a definite determination? Why?

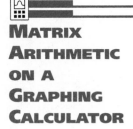

MATRIX ARITHMETIC ON A GRAPHING CALCULATOR

Some graphing calculators (including the TI-82, TI-83, TI-85, and TI-86) can work with matrices. In particular, they can perform all forms of matrix arithmetic. Consider the matrices

$$A = \begin{bmatrix} 1 & 2 \\ 3 & 4 \end{bmatrix} \quad \text{and} \quad B = \begin{bmatrix} 1 & 5 \\ 0 & -7 \end{bmatrix}$$

We will discuss how to find their sum and their product.

The TI-82 and TI-83 require that you name these matrices "[A]" and "[B]" with brackets; the TI-85 and TI-86 require that you name them "A" and "B" without brackets.

TI-85/86 users: When your calculator is in "ALPHA" mode, some buttons will generate a letter of the alphabet. If your calculator has a flashing "A" on the screen, then it is in "ALPHA" mode. One way to put your calculator in "ALPHA" mode is to press the $\boxed{\text{ALPHA}}$ button. However, some other commands will automatically put your calculator in "ALPHA" mode. Don't confuse the $\boxed{\text{2nd}}$ and $\boxed{\text{ALPHA}}$ buttons. The $\boxed{\text{2nd}}$ button refers to the mathematical labels above and to the *left* of the buttons.

Entering a Matrix

To enter matrix A, select "EDIT" from the "MATRX" menu and then matrix A from the "MATRX EDIT" menu:

TI-82/83:	• Press $\boxed{\text{MATRX}}$
	• Use the $\boxed{\rightarrow}$ button to highlight "EDIT"
	• Highlight option 1 "[A]"[*]
	• Press $\boxed{\text{ENTER}}$
TI-85/86:	• Press $\boxed{\text{2nd}}$ $\boxed{\text{MATRX}}$
	• Press $\boxed{\text{EDIT}}$ (i.e., $\boxed{\text{F2}}$)
	• Press $\boxed{\text{A}}$ $\boxed{\text{ENTER}}$[**]

Then enter the dimensions of our 2×2 matrix A by typing

2 $\boxed{\text{ENTER}}$ 2 $\boxed{\text{ENTER}}$

To enter the elements, type

1 $\boxed{\text{ENTER}}$ 2 $\boxed{\text{ENTER}}$ 3 $\boxed{\text{ENTER}}$ 4 $\boxed{\text{ENTER}}$

---

[*]Option 1 is automatically highlighted. If we were selecting some other option, we would use the $\boxed{\uparrow}$ and $\boxed{\downarrow}$ buttons to highlight it.

[**]The $\boxed{\text{LOG}}$ button becomes the $\boxed{\text{A}}$ button when the calculator is in "ALPHA" mode. Usually this requires preceding the $\boxed{\text{A}}$ button with the $\boxed{\text{ALPHA}}$ button. However, the TI-85/86 was automatically placed in "ALPHA" mode when you pressed $\boxed{\text{EDIT}}$, as shown by the flashing "A" on the screen.

In a similar manner, enter matrix *B*. When you're done entering, type 2nd QUIT .

Viewing a Matrix

To view matrix *A,* type:

TI-82/83: • Press MATRX
  • Highlight option 1: "[A]"
  • Press ENTER and "[A]" will appear on the screen
  • Press ENTER and the matrix itself will appear on the screen, as shown in Figure 4.6 below

TI-85/86: Type ALPHA A ENTER and matrix *A* will appear on the screen, as shown in Figure 4.6 below.

FIGURE 4.6

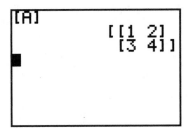

Adding Two Matrices

To calculate *A* + *B,* make the screen read "A + B" or "[A] + [B]" by typing:

TI-82/83:   MATRX 1 + MATRX 2 ENTER

TI-85/86:   ALPHA A + ALPHA B ENTER

Multiplying Two Matrices

To calculate *AB,* make the screen read "A * B" or "[A] * [B]" by typing:

TI-82/83:   MATRX 1 × MATRX 2 ENTER

TI-85/86:   ALPHA A × ALPHA B ENTER

EXERCISES

Use a graphing calculator to complete the following exercises.

41. Exercise 5   **42.** Exercise 6

43. Exercise 7   **44.** Exercise 8

45. Exercise 9   **46.** Exercise 10

47. Exercise 11   **48.** Exercise 12

49. National Express, a nationwide air freight service, has flights connecting Atlanta, Chicago, Denver, San Francisco, and Los Angeles, as shown in Figure 4.7. This arrangement of interconnecting flights can be represented by the **incidence matrix** N, also shown in Figure 4.7. That matrix

FIGURE 4.7

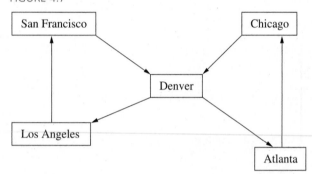

$$N = D \begin{array}{c} \quad \\ A \\ C \\ D \\ L \\ S \end{array} \begin{array}{ccccc} A & C & D & L & S \\ \left[\begin{array}{ccccc} 0 & 1 & 0 & 0 & 0 \\ 0 & 0 & 1 & 0 & 0 \\ 1 & 0 & 0 & 1 & 0 \\ 0 & 0 & 0 & 0 & 1 \\ 0 & 0 & 1 & 0 & 0 \end{array}\right] \end{array}$$

has a row and a column for each of the five listed cities, where the row indicates a flight's origin and the column indicates its destination. For example, the 1 in row 1, column 2 indicates a direct flight from Atlanta to Chicago, and the 0 in row 2, column 1 indicates the lack of a direct flight from Chicago to Atlanta.

a. Create an incidence matrix C for connecting flights that require one connection. The only such flight out of Atlanta is to Denver (with a connection at Chicago), so the only 1 in C's Atlanta row would be in the Denver column. Be certain that in creating matrix C you list the cities in the order used in matrix N.

b. Compute matrix N^2. What do you observe?

c. Find $N + N^2$.

d. Interpret the entries of $N + N^2$.

e. Find $N + N^2 + N^3$, $N + N^2 + N^3 + N^4$, etc. until you find a matrix with no zero entries. Interpret this matrix.

f. If National Express executives were to augment their flight schedule, what flight(s) would you advise them to add? Why?

50. National Express now services New York City and Washington, D.C., in addition to the cities listed in Exercise 49. Their new arrangement of interconnecting flights is illustrated in Figure 4.8.

a. Find N.

b. Find $N + N^2$, $N + N^2 + N^3$, $N + N^2 + N^3 + N^4$, etc. until you find a matrix with no zero entries. Interpret this matrix.

c. If National Express executives were to augment their flight schedule, what flight(s) would you advise them to add? Why?

FIGURE 4.8

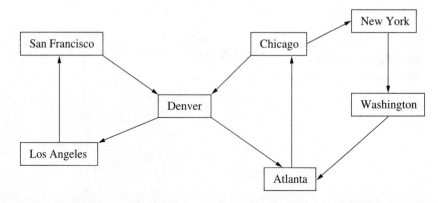

4.1

**INVERSE
MATRICES**

In Section 4.0 we discussed matrix arithmetic. In particular, we discussed the addition, subtraction, and multiplication of matrices. However, we did not discuss how to divide one matrix by another, because *there is no such thing as matrix division.*

In Section 4.0 we also found that matrix equations can be solved in a manner similar to that used to solve algebraic equations. The matrix equation

$$\begin{bmatrix} 4 & 2 \\ 5 & 3 \end{bmatrix} \begin{bmatrix} w & x \\ y & z \end{bmatrix} = \begin{bmatrix} 2 & 4 \\ 6 & 8 \end{bmatrix}$$

is similar to the algebraic equation $2x = 8$. In solving this algebraic equation we can either multiply each side by 1/2 or divide each side by 2. In solving the matrix equation

$$\begin{bmatrix} 4 & 2 \\ 5 & 3 \end{bmatrix} \begin{bmatrix} w & x \\ y & z \end{bmatrix} = \begin{bmatrix} 2 & 4 \\ 6 & 8 \end{bmatrix}$$

we can do something analogous to multiplying each side by 1/2, but we cannot do something analogous to dividing each side by 2, because there is no such thing as matrix division.

The number 1/2 is called the **multiplicative inverse** of 2, because $\frac{1}{2} \cdot 2 = 1$. Similarly, a matrix B is called the **inverse** of a matrix A if $BA = I$ and $AB = I$.

EXAMPLE 1   Verify that $A = \begin{bmatrix} 4 & 2 \\ 5 & 3 \end{bmatrix}$ and $B = \begin{bmatrix} 3/2 & -1 \\ -5/2 & 2 \end{bmatrix}$ are inverses.

Solution   We must show that $BA = I$ and that $AB = I$.

$$BA = \begin{bmatrix} 3/2 & -1 \\ -5/2 & 2 \end{bmatrix} \begin{bmatrix} 4 & 2 \\ 5 & 3 \end{bmatrix}$$

$$= \begin{bmatrix} (3/2) \cdot 4 + -1 \cdot 5 & (3/2) \cdot 2 + -1 \cdot 3 \\ (-5/2) \cdot 4 + 2 \cdot 5 & (-5/2) \cdot 2 + 2 \cdot 3 \end{bmatrix}$$

$$= \begin{bmatrix} 6 - 5 & 3 - 3 \\ -10 + 10 & -5 + 6 \end{bmatrix}$$

$$= \begin{bmatrix} 1 & 0 \\ 0 & 1 \end{bmatrix} = I$$

$$AB = \begin{bmatrix} 4 & 2 \\ 5 & 3 \end{bmatrix} \begin{bmatrix} 3/2 & -1 \\ -5/2 & 2 \end{bmatrix}$$

$$= \begin{bmatrix} 4 \cdot (3/2) + 2 \cdot (-5/2) & 4 \cdot -1 + 2 \cdot 2 \\ 5 \cdot (3/2) + 3 \cdot (-5/2) & 5 \cdot -1 + 3 \cdot 2 \end{bmatrix}$$

$$= \begin{bmatrix} 6 - 5 & -4 + 4 \\ 15/2 - 15/2 & -5 + 6 \end{bmatrix}$$

$$= \begin{bmatrix} 1 & 0 \\ 0 & 1 \end{bmatrix} = I$$

When we find the inverse of a matrix A, we usually label it as A^{-1} (read "A inverse"). In the above example, we found that the inverse of $A = \begin{bmatrix} 4 & 2 \\ 5 & 3 \end{bmatrix}$ is $\begin{bmatrix} 3/2 & -1 \\ -5/2 & 2 \end{bmatrix}$. Thus $A^{-1} = \begin{bmatrix} 3/2 & -1 \\ -5/2 & 2 \end{bmatrix}$.

The following list of properties of matrix multiplication includes those from Section 4.0 as well as the property of inverse covered in this section.

Properties of Matrix Multiplication

1. *There is no commutative property.* In general, $AB \neq BA$. You must be careful about the order in which you multiply.
2. *Associative property:* $A(BC) = (AB)C$, provided the dimensions of A, B, and C are such that they can be multiplied together.
3. *Identity property:* An *identity matrix* is a square matrix I that has 1's for each entry in the diagonal (the diagonal that starts in the upper-left corner) and 0's for all other entries. If I and A have the same dimensions, then $IA = AI = A$.
4. *Distributive property:* $A(B + C) = AB + AC$, provided the dimensions of A, B, and C are such that they can be combined in this manner.
5. *Inverse property:* The *inverse* of a matrix A is a matrix A^{-1}, where $A^{-1}A = I$ and $AA^{-1} = I$.

Using Inverses to Solve Matrix Equations

EXAMPLE 2   Solve the equation

$$\begin{bmatrix} 4 & 2 \\ 5 & 3 \end{bmatrix} \begin{bmatrix} w & x \\ y & z \end{bmatrix} = \begin{bmatrix} 2 & 4 \\ 6 & 8 \end{bmatrix}$$

Solution   For convenience, we will give each of these matrices a name:

$$A = \begin{bmatrix} 4 & 2 \\ 5 & 3 \end{bmatrix} \qquad X = \begin{bmatrix} w & x \\ y & z \end{bmatrix} \qquad B = \begin{bmatrix} 2 & 4 \\ 6 & 8 \end{bmatrix}$$

The equation to be solved is then $AX = B$. This equation is very similar to the equation $2x = 8$, and they can both be solved with the same sequence of steps. Although we only want the solution of the matrix equation, both equations are solved below to illustrate the similarity of the steps.

$AX = B$	$\leftarrow$ *The equation* $\rightarrow$	$2x = 8$
$A^{-1}(AX) = A^{-1}B$	Step 1. Multiply each side by an inverse (A's inverse in the matrix equation, 2's inverse in the algebraic equation).	$\frac{1}{2}(2x) = \frac{1}{2} \cdot 8$
$(A^{-1}A)X = A^{-1}B$	Step 2. Apply the associative property.	$(\frac{1}{2} \cdot 2)x = \frac{1}{2} \cdot 8$
$IX = A^{-1}B$	Step 3. Apply the inverse property.	$1x = 4$
$X = A^{-1}B$	Step 4. Apply the identity property.	$x = 4$

Recall that $A = \begin{bmatrix} 4 & 2 \\ 5 & 3 \end{bmatrix}$, $X = \begin{bmatrix} w & x \\ y & z \end{bmatrix}$, and $B = \begin{bmatrix} 2 & 4 \\ 6 & 8 \end{bmatrix}$. Also recall that in Example 1

above, we verified that the inverse of $A = \begin{bmatrix} 4 & 2 \\ 5 & 3 \end{bmatrix}$ is $A^{-1} = \begin{bmatrix} 3/2 & -1 \\ -5/2 & 2 \end{bmatrix}$.

Thus, $X = A^{-1}B$ becomes

$$\begin{bmatrix} w & x \\ y & z \end{bmatrix} = \begin{bmatrix} 3/2 & -1 \\ -5/2 & 2 \end{bmatrix} \begin{bmatrix} 2 & 4 \\ 6 & 8 \end{bmatrix}$$

$$= \begin{bmatrix} (3/2)\cdot 2 + -1\cdot 6 & (3/2)\cdot 4 + -1\cdot 8 \\ (-5/2)\cdot 2 + 2\cdot 6 & (-5/2)\cdot 4 + 2\cdot 8 \end{bmatrix}$$

$$= \begin{bmatrix} 3 - 6 & 6 - 8 \\ -5 + 12 & -10 + 16 \end{bmatrix}$$

$$= \begin{bmatrix} -3 & -2 \\ 7 & 6 \end{bmatrix}$$

•

Notice that in Step 1 above, it would be incorrect to write

$$A^{-1}(AX) = BA^{-1}$$

Matrix multiplication is not commutative, so we must be careful to multiply in the same order. Notice also that if we had multiplied on the right by A^{-1}, we would have

$$(AX)A^{-1} = BA^{-1}$$

and it would be impossible to cancel the A and the A^{-1}.

✔ The solution to Example 2 can be easily checked.

$$\begin{bmatrix} 4 & 2 \\ 5 & 3 \end{bmatrix} \begin{bmatrix} w & x \\ y & z \end{bmatrix} = \begin{bmatrix} 4 & 2 \\ 5 & 3 \end{bmatrix} \begin{bmatrix} -3 & -2 \\ 7 & 6 \end{bmatrix}$$

$$= \begin{bmatrix} 4\cdot -3 + 2\cdot 7 & 4\cdot -2 + 2\cdot 6 \\ 5\cdot -3 + 3\cdot 7 & 5\cdot -2 + 3\cdot 6 \end{bmatrix}$$

$$= \begin{bmatrix} -12 + 14 & -8 + 12 \\ -15 + 21 & -10 + 18 \end{bmatrix}$$

$$= \begin{bmatrix} 2 & 4 \\ 6 & 8 \end{bmatrix}$$

Finding the Inverse of a 2 × 2 Matrix

Most square matrices have inverses. A general method for finding the inverse of a square matrix will be discussed in Section 4.2; that method uses a Gauss-Jordan–like procedure. However, there is a shortcut to that general method that will work only with 2 × 2 matrices. The shortcut method refers to the **primary diagonal,** which is the diagonal that starts in the upper-left corner, and the **secondary diagonal,** which is the diagonal that starts in the upper-right corner, as illustrated below.

primary diagonal $\longrightarrow \begin{bmatrix} 4 & 2 \\ 5 & 3 \end{bmatrix}$ secondary diagonal

EXAMPLE 3 Find the inverse of

$$A = \begin{bmatrix} 4 & 2 \\ 5 & 3 \end{bmatrix}$$

Solution **Step 1** *Interchange the two elements in the primary diagonal.*

$$\begin{bmatrix} 3 & 2 \\ 5 & 4 \end{bmatrix}$$

Step 2 *Change the signs of the two elements in the secondary diagonal.*

$$\begin{bmatrix} 3 & -2 \\ -5 & 4 \end{bmatrix}$$

Step 3 *Find the determinant of the matrix in step 2.* The **determinant** of a 2 × 2 matrix is

det = (primary diagonal's product) − (secondary diagonal's product)

The determinant of the matrix in Step 2 is

det = (primary diagonal's product) − (secondary diagonal's product)

$$= \qquad (3 \cdot 4) \qquad - \qquad (-2 \cdot -5)$$

$$= 12 - 10 = 2$$

Step 4 *Compute the inverse of the original matrix.* The inverse is

$$\frac{1}{\det} \cdot (\text{the matrix from Step 2})$$

$$= \frac{1}{2} \cdot \begin{bmatrix} 3 & -2 \\ -5 & 4 \end{bmatrix}$$

$$= \begin{bmatrix} 3/2 & -1 \\ -5/2 & 2 \end{bmatrix}$$

In Example 1, we verified that this is in fact the inverse of the original matrix. ●

> ### The Inverse of a 2 × 2 Matrix
>
> To find the inverse of
> $$A = \begin{bmatrix} a & b \\ c & d \end{bmatrix}$$
>
> **Step 1** *Interchange the two elements in the primary diagonal.* This gives
> $$\begin{bmatrix} d & b \\ c & a \end{bmatrix}$$
>
> **Step 2** *Change the signs of the two elements in the secondary diagonal.* This gives
> $$\begin{bmatrix} d & -b \\ -c & a \end{bmatrix}$$
>
> **Step 3** *Find the determinant of the matrix in Step 2.* The *determinant* of a 2 × 2 matrix is
> $$\det = \text{(primary diagonal's product)} - \text{(secondary diagonal's product)}$$
> $$= \qquad da \qquad - \qquad bc$$
>
> **Step 4** *Compute the inverse of the original matrix.* The inverse is
> $$\frac{1}{\det} \cdot \text{(the matrix from Step 2)}$$
> $$= \frac{1}{da - bc} \begin{bmatrix} d & -b \\ -c & a \end{bmatrix}$$

Using a Matrix Equation to Solve a System of Equations

A matrix equation is really a system of equations. For example, consider the matrix equation
$$\begin{bmatrix} 2 & 3 \\ 4 & -1 \end{bmatrix} \begin{bmatrix} x \\ y \end{bmatrix} = \begin{bmatrix} 11 \\ 1 \end{bmatrix}$$

If we multiply the first two matrices together, we get
$$\begin{bmatrix} 2x + 3y \\ 4x - y \end{bmatrix} = \begin{bmatrix} 11 \\ 1 \end{bmatrix}$$

These two matrices are equal if and only if their corresponding elements are equal, so we have the system
$$2x + 3y = 11$$
$$4x - y = 1$$

EXAMPLE 4 Solve the system
$$5x - 2y = -1$$
$$4x + y = 7$$

by solving a matrix equation.

Solution The given system is equivalent to the matrix equation

$$\begin{bmatrix} 5 & -2 \\ 4 & 1 \end{bmatrix} \begin{bmatrix} x \\ y \end{bmatrix} = \begin{bmatrix} -1 \\ 7 \end{bmatrix}$$

For convenience, we will give each of these matrices a name:

$$A = \begin{bmatrix} 5 & -2 \\ 4 & 1 \end{bmatrix} \qquad X = \begin{bmatrix} x \\ y \end{bmatrix} \qquad B = \begin{bmatrix} -1 \\ 7 \end{bmatrix}$$

The equation to be solved is then $AX = B$.

$A^{-1}(AX) = A^{-1}B$	Multiplying on the left by A^{-1}
$(A^{-1}A)X = A^{-1}B$	Associative property
$IX = A^{-1}B$	Inverse property
$X = A^{-1}B$	Identity property

In order to find the matrix X, we will first find the inverse of A and then multiply it by B (being careful to multiply in the correct order).

$$\begin{bmatrix} 1 & -2 \\ 4 & 5 \end{bmatrix} \qquad \text{Interchanging the two elements in } A\text{'s primary diagonal}$$

$$\begin{bmatrix} 1 & 2 \\ -4 & 5 \end{bmatrix} \qquad \text{Changing the signs of the two elements in } A\text{'s secondary diagonal}$$

The determinant of this matrix is

$$\begin{aligned} \det &= (\text{primary diagonal's product}) - (\text{secondary diagonal's product}) \\ &= \qquad\quad (1 \cdot 5) \qquad\quad - \qquad\quad (2 \cdot -4) \\ &= 5 - (-8) = 13 \end{aligned}$$

A's inverse is

$$A^{-1} = \frac{1}{13} \cdot \begin{bmatrix} 1 & 2 \\ -4 & 5 \end{bmatrix}$$

$$= \begin{bmatrix} 1/13 & 2/13 \\ -4/13 & 5/13 \end{bmatrix}$$

The solution to the matrix equation is

$$\begin{aligned} X &= \qquad A^{-1} \qquad\quad B \\ &= \begin{bmatrix} 1/13 & 2/13 \\ -4/13 & 5/13 \end{bmatrix} \begin{bmatrix} -1 \\ 7 \end{bmatrix} \\ &= \begin{bmatrix} 1/13 \cdot -1 + 2/13 \cdot 7 \\ -4/13 \cdot -1 + 5/13 \cdot 7 \end{bmatrix} \end{aligned}$$

$$= \begin{bmatrix} -1/13 + 14/13 \\ 4/13 + 35/13 \end{bmatrix}$$

$$= \begin{bmatrix} 1 \\ 3 \end{bmatrix}$$

Recalling that $X = \begin{bmatrix} x \\ y \end{bmatrix}$, we have

$$\begin{bmatrix} x \\ y \end{bmatrix} = \begin{bmatrix} 1 \\ 3 \end{bmatrix}$$

or, written in a more normal style,

$$(x, y) = (1, 3)$$

✔ The solution to Example 4 can be easily checked by substituting $(1, 3)$ for (x, y) in the original system.

$$5x - 2y = 5 \cdot 1 - 2 \cdot 3 = 5 - 6 = -1 \quad ✔$$
$$4x + y = 4 \cdot 1 + 3 = 7 \quad ✔$$

•

Why Solve Systems with Matrix Equations?

We've already encountered a number of different methods for solving systems of equations. In particular, we've encountered the elimination method, the Gauss-Jordan method, and now the matrix equation method. Why are there so many methods?

Usually, the existence of a number of different methods for solving one type of problem is the result of two factors:

- That type of problem comes up a lot
- Each method has a distinct advantage

This is the case here. Systems of equations *do* come up a lot, and each of the methods of solving a system of equations *does* have a distinct advantage.

The advantage of the elimination method is that it's easy, if the system is small. It's usually the easiest way to solve a system of two equations in two unknowns.

The advantage of the Gauss-Jordan method is that it's easier for a larger system. It tends to be easier than the elimination method if the system is larger than two equations in two unknowns. Furthermore, the Gauss-Jordan method is amenable to computer or graphing calculator usage, whereas the elimination method is not.

The matrix equation method has two advantages. It too is amenable to computer and graphing calculator usage, at least as much as the Gauss-Jordan method is. Its other advantage is illustrated in the following example.

EXAMPLE 5   Solve the system

$$5x - 2y = 2$$
$$4x + y = 4$$

by solving a matrix equation.

Solution   Notice that this system is the same as that in Example 4, except the numbers on the right side of the equations are different. The given system is equivalent to the matrix equation

$$\begin{bmatrix} 5 & -2 \\ 4 & 1 \end{bmatrix} \begin{bmatrix} x \\ y \end{bmatrix} = \begin{bmatrix} 2 \\ 4 \end{bmatrix}$$

We can name the first two matrices A and X, as we did in Example 4, and we can name the third matrix C. The equation to be solved is then $AX = C$. Its solution is $X = A^{-1}C$. In Example 4, we found that

$$A^{-1} = \begin{bmatrix} 1/13 & 2/13 \\ -4/13 & 5/13 \end{bmatrix}$$

Thus, $X = \quad A^{-1} \quad C$

$$= \begin{bmatrix} 1/13 & 2/13 \\ -4/13 & 5/13 \end{bmatrix} \begin{bmatrix} 2 \\ 4 \end{bmatrix}$$

$$= \begin{bmatrix} 1/13 \cdot 2 + 2/13 \cdot 4 \\ -4/13 \cdot 2 + 5/13 \cdot 4 \end{bmatrix}$$

$$= \begin{bmatrix} 2/13 + 8/13 \\ -8/13 + 20/13 \end{bmatrix}$$

$$= \begin{bmatrix} 10/13 \\ 12/13 \end{bmatrix}$$

Recalling that $X = \begin{bmatrix} x \\ y \end{bmatrix}$, we have

$$\begin{bmatrix} x \\ y \end{bmatrix} = \begin{bmatrix} 10/13 \\ 12/13 \end{bmatrix}$$

or, written in a more normal style,

$$(x, y) = (10/13, 12/13)$$

The only work that we had to do in Example 5 was multiplying A^{-1} by C, and that work was quite quick and easy. Most of the work had been done in the previous example when we found A^{-1}. The advantage of the matrix equation method of solving a system of equations is that once the inverse is found, it can be used to solve any new system formed by changing only the constant terms.

Cryptography

According to Webster's dictionary, **cryptography** is "the art of writing or deciphering messages in code." A very simplistic way to encode a message is assign the letter A to the number 1, the letter B to the number 2, and so on, as shown in Figure 4.9.

FIGURE 4.9

A	B	C	D	E	F	G	H	I	J	K	L	M	N	O	P	Q	R	S	T	U	V	W	X	Y	Z	(blank)
1	2	3	4	5	6	7	8	9	10	11	12	13	14	15	16	17	18	19	20	21	22	23	24	25	26	27

The message "Hi Al" would translate into the sequence   8   9   27   1   12.

This message could be decoded quite easily. If, however, an encoding matrix is applied to the above sequence, the message can be difficult to decode. For example, the above sequence is contained in the matrix $M = \begin{bmatrix} 8 & 27 & 12 \\ 9 & 1 & 27 \end{bmatrix}$ (notice the extra blank inserted in the matrix to fill it out). If this matrix is multiplied by an **encoding matrix,** such as $E = \begin{bmatrix} 5 & 7 \\ 3 & -2 \end{bmatrix}$, the result is

$$EM = \begin{bmatrix} 5 & 7 \\ 3 & -2 \end{bmatrix} \begin{bmatrix} 8 & 27 & 12 \\ 9 & 1 & 27 \end{bmatrix} = \begin{bmatrix} 103 & 142 & 249 \\ 6 & 79 & -18 \end{bmatrix}$$

and the message becomes the sequence   103   6   142   79   249 -18.

Notice that the number 27 has a consistent meaning in the matrix M; in each of its two occurrences, it translates into a blank space. This is not the case in the matrix EM; no number occurs twice, even though there are two blanks in the message. This is one reason that the application of an encoding matrix to a message makes the message hard to decode.

To decode this message, put it back into matrix form and multiply it by the **decoding matrix** E^{-1}.

4.1

EXERCISES

In Exercises 1–4, verify that the given matrices are inverses of each other by computing AB *and* BA.

1. $A = \begin{bmatrix} 7 & 1 \\ -3 & 2 \end{bmatrix}$ and $B = \begin{bmatrix} 2/17 & -1/17 \\ 3/17 & 7/17 \end{bmatrix}$

2. $A = \begin{bmatrix} 5 & -6 \\ 2 & -3 \end{bmatrix}$ and $B = \begin{bmatrix} 1 & -2 \\ 2/3 & -5/3 \end{bmatrix}$

3. $A = \begin{bmatrix} 9 & 2 & -1 \\ 3 & -2 & 4 \\ 1 & 0 & -1 \end{bmatrix}$ and $B = \begin{bmatrix} 1/15 & 1/15 & 1/5 \\ 7/30 & -4/15 & -13/10 \\ 1/15 & 1/15 & -4/5 \end{bmatrix}$

4. $A = \begin{bmatrix} 4 & 1 & 2 \\ -2 & 2 & -1 \\ 2 & 5 & 4 \end{bmatrix}$ and $B = \begin{bmatrix} 13/30 & 1/5 & -1/6 \\ 1/5 & 2/5 & 0 \\ -7/15 & -3/5 & 1/3 \end{bmatrix}$

In Exercises 5–8, solve the given equation. Use the inverses given in Exercises 1–4. Check your answers by substituting them back in. (Answers are not given at the back of the book.)

5. $\begin{bmatrix} 7 & 1 \\ -3 & 2 \end{bmatrix} \begin{bmatrix} w & x \\ y & z \end{bmatrix} = \begin{bmatrix} 2 & 4 \\ 6 & 8 \end{bmatrix}$

6. $\begin{bmatrix} 5 & -6 \\ 2 & -3 \end{bmatrix} \begin{bmatrix} w & x \\ y & z \end{bmatrix} = \begin{bmatrix} 2 & 4 \\ 6 & 8 \end{bmatrix}$

7. $\begin{bmatrix} 9 & 2 & -1 \\ 3 & -2 & 4 \\ 1 & 0 & -1 \end{bmatrix} \begin{bmatrix} x_{11} & x_{12} & x_{13} \\ x_{21} & x_{22} & x_{23} \\ x_{31} & x_{32} & x_{33} \end{bmatrix} = \begin{bmatrix} 60 & 30 & -15 \\ -90 & -45 & 30 \\ 0 & 15 & -90 \end{bmatrix}$

8. $\begin{bmatrix} 4 & 1 & 2 \\ -2 & 2 & -1 \\ 2 & 5 & 4 \end{bmatrix} \begin{bmatrix} x_{11} & x_{12} & x_{13} \\ x_{21} & x_{22} & x_{23} \\ x_{31} & x_{32} & x_{33} \end{bmatrix} = \begin{bmatrix} 60 & 30 & -15 \\ -90 & -45 & 30 \\ 0 & 15 & -90 \end{bmatrix}$

In Exercises 9–12, solve the given equation. Check your answers by substituting them back in. (Answers are not given at the back of the book.)

9. $\begin{bmatrix} 3 & 6 \\ 5 & 1 \end{bmatrix} \begin{bmatrix} w & x \\ y & z \end{bmatrix} = \begin{bmatrix} 9 & 6 \\ 2 & 3 \end{bmatrix}$

10. $\begin{bmatrix} 2 & 1 \\ 0 & 3 \end{bmatrix} \begin{bmatrix} w & x \\ y & z \end{bmatrix} = \begin{bmatrix} 4 & 5 \\ 1 & 9 \end{bmatrix}$

11. $\begin{bmatrix} 4 & -1 \\ 2 & -2 \end{bmatrix} \begin{bmatrix} x \\ y \end{bmatrix} = \begin{bmatrix} 3 \\ -1 \end{bmatrix}$

12. $\begin{bmatrix} 5 & -2 \\ 0 & 3 \end{bmatrix} \begin{bmatrix} x \\ y \end{bmatrix} = \begin{bmatrix} 4 \\ 3 \end{bmatrix}$

13. a. Convert the matrix equation in Exercise 11 into a system of equations.
  b. Check that the solution to Exercise 11 is in fact the solution to the system found in part (a).

14 a. Convert the matrix equation in Exercise 12 into a system of equations.
  b. Check that the solution to Exercise 12 is in fact the solution to the system found in part (a).

In Exercises 15–22, solve the given system of equations by solving a matrix equation.

15. $5x - 7y = 27$
   $3x + 2y = 10$

16. $3x + 7y = 41$
   $4x - 9y = -37$

17. $5x - 7y = 1$
   $3x + 2y = -18$
   HINT: Use the result of Exercise 15.

18. $3x + 7y = 58$
   $4x - 9y = -51$
   HINT: Use the result of Exercise 16.

19. $5x - 7y = 53$
   $3x + 2y = 38$
   HINT: Use the result of Exercise 15.

20. $3x + 7y = 69$
   $4x - 9y = -73$
   HINT: Use the result of Exercise 16.

21. $9x + 2y - z = 20$
   $3x - 2y + 4z = -23$
   $x - z = 6$
   HINT: Use the result of Exercise 3.

22. $4x + y + 2z = 24$
   $-2x + 2y - z = -7$
   $2x + 5y + 4z = 42$
   HINT: Use the result of Exercise 4.

Use the matrix equation method to solve Exercises 23–26.

23. The Wet Pleasure Kayak Co. makes two types of kayaks: the river model and the oceangoing model. Each river model requires 40 minutes in the cutting department and 20 minutes in the assembly department, and each oceangoing model requires 50 minutes in the cutting department and 30 minutes in the assembly department. The cutting department has a maximum of 460 hours available each week, and the assembly department has 240 hours available each week. How many kayaks should be produced each week for the plant to operate at full capacity?

24. The Eternal Slumber Casket Co. makes two types of caskets: the luxury model and the pine box. Each luxury model requires 120 minutes in the carpentry department and 180 minutes in the finishing department, and each pine box requires 30 minutes in the carpentry department and 30 minutes in the finishing department. The carpentry department has a maximum of 22 hours available each week, and the finishing department has 32 hours available each week. How many caskets should be produced each week for the plant to operate at full capacity?

25. A hospital patient is on a strict diet that is to include exactly 740 calories and 28 grams of fat. The hospital dietician plans to accomplish this by serving the patient only beef liver and Velveeta surprise. Each ounce of beef liver has 40 calories and 1.4 grams of fat, and each cup of Velveeta surprise has 340 calories and 14 grams of fat. What will the patient be served each day?

26. Dogs at a kennel are to be fed exactly 15 IUs of vitamin E and 60 mg of vitamin C each day. Each Pup-O-Vit tablet contains 5 IUs of vitamin E and 20 mg of vitamin C. Each

Dogamins tablet contains 10 IUs of vitamin E and 40 mg of vitamin C. How many tablets should a dog be fed each day?

The messages in Exercises 27 and 28 were encoded with the encoding matrix E.

27. Decode the message: 72   37   156   44   170   71   141   66   185   −13   158   −23   226   55   314   21   136   −30   68   −15   259   −12

28. Decode the message: 93   0   133   24   135   50   198   63   284   3   177   −24   259   −12

29. In this exercise, you will prove that the method of finding inverses discussed in this section is valid.

   a. Find the inverse of $A = \begin{bmatrix} a & b \\ c & d \end{bmatrix}$.

   b. Compute AA^{-1}.
   c. Compute $A^{-1}A$.

30. In the boxed instructions for finding the inverse of a 2 × 2 matrix, Step 3 is to find the determinant of the matrix in Step 2. Actually, it doesn't matter if you find the determinant of the matrix in Step 2, or the matrix in Step 1, or the original matrix A. Why does this not matter?

 Answer the following using complete sentences.

31. In verifying that two matrices A and B are inverses, why is it necessary to compute both AB and BA?

32. How is the inverse of a matrix similar to the multiplicative inverse of a number?

33. Some square matrices do not have inverses. Do any numbers not have multiplicative inverses? What do you think would happen in computing the inverse of a 2 × 2 matrix if that matrix does not actually have an inverse?

INVERSES ON A GRAPHING CALCULATOR

To find the inverse of a matrix previously entered as matrix A, make the screen read "A^{-1}" or "$[A]^{-1}$."

EXAMPLE 6   Use a graphing calculator to find the inverse of

$$A = \begin{bmatrix} 4 & 2 \\ 5 & 3 \end{bmatrix}$$

Solution   First, enter matrix A into the calculator, as discussed in Section 4.0. Then type:

TI-82/83:	MATRX 1 x⁻¹ ENTER
TI-85/86:	ALPHA A 2nd x⁻¹ ENTER

The results are shown in Figure 4.10:

FIGURE 4.10

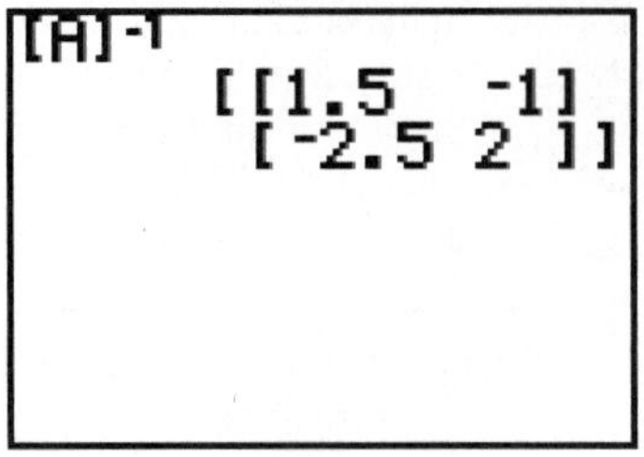

```
[A]⁻¹
    [[1.5   -1]
     [-2.5  2 ]]
```

EXERCISES

Use a graphing calculator and matrix inverses to solve the following systems. Check your answers by substituting them back into the equations. (Answers are not given in the back of the book.)

34. $5x - 7y = 8$
$2x + 3y = 4$

35. $9x + 1y = 12$
$6x - 2y = 13$

36. $8x + 7y - 2z = 11$
$9x - 3y + 12z = 4$
$15x + 17y - 22z = 19$

37. $4.1x + 6.2y - 8.9z = 1.5$
$3.7x - 2.7y + 5.8z = 8.5$
$3.6x + 7.4y - 4.7z = 30$

38. $4.6x + 7.8y - z + 2.4w = 19$
$3.1x - 2.5y - 7.8z + 4.8w = 9.2$
$43x - 22y + 91z = 43w - 7$
$w + x + y + z = 1$

39. a. Create a 5 × 5 encoding matrix. Find the determinant of this matrix.
   HINT: Use the "det" command on your graphing calculator.
   If the determinant is 0, create a different 5 × 5 encoding matrix.

   b. Write a message for a classmate, and use your encoding matrix to encode the message. Give your classmate the encoding matrix and the encoded message. Give your instructor the decoded message and the name of your classmate.

   c. Receive an encoding matrix and an encoded message from a classmate. Decode the message. Give your instructor the decoded message and the name of the classmate.

   d. Why must the encoding matrix's determinant not be 0?

4.2

THE GAUSS–JORDAN METHOD AND INVERSES

In Section 4.1 we discussed a method of finding the inverse of a 2 × 2 matrix. That method involved switching the elements of the primary diagonal, changing the signs of the secondary diagonal, and dividing the result by the determinant. It is important to realize that this method applies only to 2 × 2 matrices; it does not work with larger matrices. In this section, we discuss a general method for finding the inverse of a square matrix of any dimensions.

To find the inverse of $A = \begin{bmatrix} 1 & 2 \\ 3 & 4 \end{bmatrix}$, we must find a matrix $A^{-1} = \begin{bmatrix} a & b \\ c & d \end{bmatrix}$ such that

$$A \cdot A^{-1} = I$$

$$\begin{bmatrix} 1 & 2 \\ 3 & 4 \end{bmatrix} \cdot \begin{bmatrix} a & b \\ c & d \end{bmatrix} = \begin{bmatrix} 1 & 0 \\ 0 & 1 \end{bmatrix}$$

Multiplying the left side yields:

$$\begin{bmatrix} 1a + 2c & 1b + 2d \\ 3a + 4c & 3b + 4d \end{bmatrix} = \begin{bmatrix} 1 & 0 \\ 0 & 1 \end{bmatrix}$$

Two matrices are equal if and only if corresponding elements are equal, so:

$$1a + 2c = 1 \qquad 1b + 2d = 0$$
$$3a + 4c = 0 \qquad 3b + 4d = 1$$

This system of four equations in four unknowns naturally breaks down into two separate problems, one involving the variables a and c, the other involving b and d. We can solve each of these two separate-but-similar problems with the Gauss-Jordan method.

	Problem 1: $\begin{aligned} 1a + 2c &= 1 \\ 3a + 4c &= 0 \end{aligned}$	Problem 2: $\begin{aligned} 1b + 2d &= 0 \\ 3b + 4d &= 1 \end{aligned}$
The System		
The System in Matrix Form	$\begin{array}{cc} a & c \end{array}$ $\begin{bmatrix} 1 & 2 & 1 \\ 3 & 4 & 0 \end{bmatrix}$	$\begin{array}{cc} b & d \end{array}$ $\begin{bmatrix} 1 & 2 & 0 \\ 3 & 4 & 1 \end{bmatrix}$
$-3 \cdot R1 + R2 : R2$	$\begin{bmatrix} 1 & 2 & 1 \\ 0 & -2 & -3 \end{bmatrix}$	$\begin{bmatrix} 1 & 2 & 0 \\ 0 & -2 & 1 \end{bmatrix}$
$-1/2 \cdot R2 : R2$	$\begin{bmatrix} 1 & 2 & 1 \\ 0 & 1 & 3/2 \end{bmatrix}$	$\begin{bmatrix} 1 & 2 & 0 \\ 0 & 1 & -1/2 \end{bmatrix}$
$-2 \cdot R2 + R1 : R1$	$\begin{array}{cc} a & c \end{array}$ $\begin{bmatrix} 1 & 0 & -2 \\ 0 & 1 & 3/2 \end{bmatrix}$	$\begin{array}{cc} b & d \end{array}$ $\begin{bmatrix} 1 & 0 & 1 \\ 0 & 1 & -1/2 \end{bmatrix}$
The Solution	$a = -2, c = 3/2$	$b = 1, d = -1/2$

This means that $A^{-1} = \begin{bmatrix} a & b \\ c & d \end{bmatrix} = \begin{bmatrix} -2 & 1 \\ 3/2 & -1/2 \end{bmatrix}$

Notice that the exact same row operations were used in Problems 1 and 2, and that the results of those row operations differed only in the last column. This always happens. In practice, the Gauss-Jordan work for the two problems is combined into a single Gauss-Jordan problem by inserting the last column from problem 2's matrix into problem 1's matrix, as shown below.

The System	$1a + 2c = 1$ $3a + 4c = 0$	$1a + 2c = 0$ $3a + 4c = 1$
The System in Matrix Form	$\begin{bmatrix} 1 & 2 & 1 & 0 \\ 3 & 4 & 0 & 1 \end{bmatrix}$	*Notice the new last column*
$-3 \cdot R1 + R2 : R2$	$\begin{bmatrix} 1 & 2 & 1 & 0 \\ 0 & -2 & -3 & 1 \end{bmatrix}$	
$-1/2 \cdot R2 : R2$	$\begin{bmatrix} 1 & 2 & 1 & 0 \\ 0 & 1 & 3/2 & -1/2 \end{bmatrix}$	
$-2 \cdot R2 + R1 : R1$	$\begin{bmatrix} 1 & 0 & -2 & 1 \\ 0 & 1 & 3/2 & -1/2 \end{bmatrix}$	
The Solution	$A^{-1} = \begin{bmatrix} a & b \\ c & d \end{bmatrix} = \begin{bmatrix} -2 & 1 \\ 3/2 & -1/2 \end{bmatrix}$	

By combining the two problems into one, as shown above, we eliminate redundant calculations. Even though we have one more column than we're used to, this is still the Gauss-Jordan method, and we are still using the row operations to achieve reduced row echelon form. We start with a matrix whose left half is A and whose right half is I; we finish with a matrix whose left half is I and whose right half is A^{-1}.

EXAMPLE 1 Find the inverse of

$$B = \begin{bmatrix} 5 & 4 & 8 \\ 6 & 7 & 2 \\ 1 & 3 & 2 \end{bmatrix}$$

Solution Start with a matrix whose left half is B and whose right half is I.

$$\begin{bmatrix} 5 & 4 & 8 & 1 & 0 & 0 \\ 6 & 7 & 2 & 0 & 1 & 0 \\ 1 & 3 & 2 & 0 & 0 & 1 \end{bmatrix}$$

Use the row operations to achieve reduced row echelon form.

$1/5 \cdot R1 : R1$
$$\begin{bmatrix} 1 & 4/5 & 8/5 & 1/5 & 0 & 0 \\ 6 & 7 & 2 & 0 & 1 & 0 \\ 1 & 3 & 2 & 0 & 0 & 1 \end{bmatrix}$$

$-6 \cdot R1 + R2 : R2$
$-1 \cdot R1 + R3 : R3$
$$\begin{bmatrix} 1 & 4/5 & 8/5 & 1/5 & 0 & 0 \\ 0 & 11/5 & -38/5 & -6/5 & 1 & 0 \\ 0 & 11/5 & 2/5 & -1/5 & 0 & 1 \end{bmatrix}$$

R2 − R3 : R3
$$\begin{bmatrix} 1 & 4/5 & 8/5 & 1/5 & 0 & 0 \\ 0 & 11/5 & -38/5 & -6/5 & 1 & 0 \\ 0 & 0 & -8 & -1 & 1 & -1 \end{bmatrix}$$

5/11 · R2 : R2
$$\begin{bmatrix} 1 & 4/5 & 8/5 & 1/5 & 0 & 0 \\ 0 & 1 & -38/11 & -6/11 & 5/11 & 0 \\ 0 & 0 & -8 & -1 & 1 & -1 \end{bmatrix}$$

−4/5 · R2 + R1 : R1
$$\begin{bmatrix} 1 & 0 & 48/11 & 7/11 & -4/11 & 0 \\ 0 & 1 & -38/11 & -6/11 & 5/11 & 0 \\ 0 & 0 & -8 & -1 & 1 & -1 \end{bmatrix}$$

−1/8 · R3 : R3
$$\begin{bmatrix} 1 & 0 & 48/11 & 7/11 & -4/11 & 0 \\ 0 & 1 & -38/11 & -6/11 & 5/11 & 0 \\ 0 & 0 & 1 & 1/8 & -1/8 & 1/8 \end{bmatrix}$$

−48/11 · R3 + R1 : R1
38/11 · R3 + R2 : R2
$$\begin{bmatrix} 1 & 0 & 0 & 1/11 & 2/11 & -6/11 \\ 0 & 1 & 0 & -5/44 & 1/44 & 19/44 \\ 0 & 0 & 1 & 1/8 & -1/8 & 1/8 \end{bmatrix}$$

The left half of this final simplex matrix is I, and the right half is B^{-1}. That is,

$$B^{-1} = \begin{bmatrix} 1/11 & 2/11 & -6/11 \\ -5/44 & 1/44 & 19/44 \\ 1/8 & -1/8 & 1/8 \end{bmatrix}$$

4.2

EXERCISES

In Exercises 1–4, find the inverse of the given matrix.

1. $\begin{bmatrix} 5 & 7 & -2 \\ 3 & -1 & 1 \\ 8 & 6 & 9 \end{bmatrix}$

2. $\begin{bmatrix} -8 & 2 & 7 \\ 19 & 4 & 0 \\ -3 & 5 & 12 \end{bmatrix}$

3. $\begin{bmatrix} 12 & 3 & -2 & 8 \\ 6 & -1 & 4 & 9 \\ 3 & 5 & -20 & 7 \\ 18 & -11 & 12 & -3 \end{bmatrix}$

4. $\begin{bmatrix} 2 & 0 & 1 & 9 \\ 5 & 2 & -3 & 7 \\ 6 & 8 & 2 & 3 \\ 5 & 6 & 8 & -1 \end{bmatrix}$

In Exercises 5–8, use matrix inverses and the results of Exercises 1–4 to solve the given system.

5. a. $5x + 7y - 2z = 3$
$3x - y + z = 2$
$8x + 6y + 9z = 14$

b. $5x + 7y - 2z = 12$
$3x - y + z = 11$
$8x + 6y + 9z = 19$

6. a. $-8x + 2y + 7z = 12$
   $19x + 4y = 3$
   $-3x + 5y + 12z = 7$
 b. $-8x + 2y + 7z = 82$
   $19x + 4y = -23$
   $-3x + 5y + 12z = -48$

7. a. $12x + 3y - 2z + 8w = 5$
   $6x - y + 4z + 9w = 17$
   $3x + 5y - 20z + 7w = 99$
   $18x - 11y + 12z - 3w = 52$
 b. $12x + 3y - 2z + 8w = 87$
   $6x - y + 4z + 9w = 42$
   $3x + 5y - 20z + 7w = -66$
   $18x - 11y + 12z - 3w = -76$

8. a. $2x + z + 9w = 12$
   $5x + 2y - 3z + 7w = 59$
   $6x + 8y + 2z + 3w = -19$
   $5x + 6y + 8z - w = 42$
 b. $2x + z + 9w = 9$
   $5x + 2y - 3z + 7w = 4$
   $6x + 8y + 2z + 3w = -6$
   $5x + 6y + 8z - w = -11$

In Exercises 9–10, use matrix inverses to solve the given system.

9. $3x - y - z = 12$
   $2x + 7y - 8z = 8$
   $5x + y - 3z = 90$

10. $0.3x - 1.2y - 8.3z = 5.4$
   $9.1x - 2.2y - 8.3z = 12$
   $5x - 0.3y + 1.1z = 8.4$

4.3

LEONTIEF INPUT-OUTPUT MODELS

Robinson Crusoe— A Fractured Fairy Tale

When Robinson Crusoe was shipwrecked, he had only the clothes on his back, a package of corn seed, and a copy of Nobel prize winner Wassily Leontief's book *Input-Output Economics*. At first, Robinson's diet consisted primarily of seaweed, which he disliked but was readily available. He supplemented this with the occasional fish he was able to catch and the few ears of corn he was able to grow. In his first year on the island, he trapped a number of fish in a lagoon by using rocks to separate the lagoon from the ocean. This created a fish farm. By this time, Robinson's cornfield was reasonably productive, and he fed corn to his captive fish. Furthermore, he used fish as fertilizer in his cornfield. Robinson found that it took 5 ears of corn to feed one fish for one year, and that one fish was sufficient fertilizer for 10 ears of corn. He also found that one ear of corn, when used as seed, would yield 100 ears of corn.

EXAMPLE 1   In one year, Robinson produced 15,000 ears of corn and 2900 fish.

a. How much of this was consumed in the production process?
b. How much remained for Robinson Crusoe and his friend Friday to eat?

Solution   **a.** The production data given above is summarized in Figure 4.11.

FIGURE 4.11

Production of 1 Ear of Corn Requires:	Production of 1 Fish Requires:
1/100 of an ear of corn	5 ears of corn
1/10 of a fish	0 fish

Robinson Crusoe and his wrecked ship.

In producing 15,000 ears of corn, $\frac{1}{100} \cdot 15{,}000 = 150$ ears were consumed. In producing 2900 fish, $5 \cdot 2900 = 14{,}500$ ears of corn were consumed. Thus, a total of $150 + 14{,}500 = 14{,}650$ ears were consumed. Similarly, the production process consumed $\frac{1}{10} \cdot 15{,}000 = 1500$ fish.

b. Fifteen thousand ears of corn were grown, and 14,650 were consumed in production, leaving $15{,}000 - 14{,}650 = 350$ ears of corn to eat; 2900 fish were produced, and 1500 were consumed in production, leaving 1400 fish to eat. •

Robinson had been reading his copy of Leontief's *Input-Output Economics* and realized that the calculations done in Example 1(a) are essentially the same as those that Leontief did by multiplying two matrices:

$$\begin{bmatrix} 0.01 & 5 \\ 0.1 & 0 \end{bmatrix} \begin{bmatrix} 15{,}000 \\ 2{,}900 \end{bmatrix} = \begin{bmatrix} 0.01 \cdot 15{,}000 + 5 \cdot 2{,}900 \\ 0.1 \ \cdot 15{,}000 + 0 \cdot 2{,}900 \end{bmatrix} = \begin{bmatrix} 14{,}650 \\ 1{,}500 \end{bmatrix}$$

Leontief called the first of the two multiplied matrices a **technological matrix,** because it gives the technological requirements of production; he called the second matrix a **production matrix,** because it gives levels of production. The production matrix, which we will call *P*, must be a column matrix, with an entry for each production category.

The technological matrix, which we will call T, must be a square matrix. T must have a row and a column for each production category. A typical technological matrix looks like this (possibly with more or less categories):

Input requirements of:

	Category 1	Category 2	Category 3

$$\text{from} \begin{cases} \text{Category 1} \\ \text{Category 2} \\ \text{Category 3} \end{cases} \begin{bmatrix} \cdots & \cdots & \cdots \\ \cdots & \cdots & \cdots \\ \cdots & \cdots & \cdots \end{bmatrix}$$

The technological matrix is merely a matrix version of the chart given in Example 1. Its rows and columns must be in the same order as P's rows. (Here, we consistently list corn first and fish second.) Labeling T and P as shown below makes this ordering straightforward.

$$T = \begin{bmatrix} 0.01 & 5 \\ 0.1 & 0 \end{bmatrix} \begin{matrix} C \\ F \end{matrix} \qquad P = \begin{bmatrix} 15,000 \\ 2,900 \end{bmatrix} \begin{matrix} C \\ F \end{matrix}$$

As discussed above, the product TP gives the amount consumed in the production process.

$$TP = \begin{bmatrix} 0.01 & 5 \\ 0.1 & 0 \end{bmatrix} \begin{bmatrix} 15,000 \\ 2,900 \end{bmatrix} = \begin{bmatrix} 14,650 \\ 1,500 \end{bmatrix} \begin{matrix} C \\ F \end{matrix}$$

EXAMPLE 2   Describe the calculation performed in Example 1(b) as a matrix calculation.

Solution   In Example 1(b) we subtracted the amount consumed in the production process from the original amount produced. Thus,

$$\begin{bmatrix} 15,000 \\ 2,900 \end{bmatrix} - \begin{bmatrix} 14,650 \\ 1,500 \end{bmatrix} = \begin{bmatrix} 350 \\ 1,400 \end{bmatrix} \begin{matrix} C \\ F \end{matrix}$$
$$\quad\uparrow \qquad\qquad \uparrow \qquad\qquad \uparrow$$
$$\quad P \quad\; - \quad\; TP \;\; = \quad\; D$$

The result of this subtraction is called the **demand matrix** D because it measures the part of production that can be used to satisfy the consumers' demand. (Here, the consumers are Robinson Crusoe and Friday.) Notice that the demand matrix is labeled in the order used in labeling the matrices T and P. •

Robinson Crusoe and Friday had 350 ears of corn and 1400 fish with which they had to feed themselves for one year. The fish production was almost sufficient to meet their needs, as it gave each man almost two fish a day. However, the corn production was insufficient; it meant that Robinson and Friday had to split one ear a day. They agreed that they should each have two fish and two ears of corn a day. They needed to know the production levels that would meet this demand.

In Example 2, we were given the production matrix P and the technological matrix T. We calculated the demand matrix D:

$$D = P - TP$$

We're now in a situation where we are given the technological matrix T and the demand matrix D, and we need to calculate the production matrix P.

EXAMPLE 3   Determine a formula for the production level necessary to meet a given demand.

Solution   We need to solve the equation $P - TP = D$ for P.

$$P - TP = D$$
$$IP - TP = D \qquad \text{Since } IP = P \text{ by the identity property}$$
$$(I - T)P = D \qquad \text{Factoring, using the distributive property}$$
$$(I - T)^{-1}(I - T)P = (I - T)^{-1}D \qquad \text{Multiplying by } (I - T)^{-1}$$
$$P = (I - T)^{-1}D \qquad \text{Inverse property} \qquad \bullet$$

EXAMPLE 4   Use the production level formula derived in Example 3 to find the level that would satisfy Robinson's and Friday's demands.

Solution   Each wanted two fish and two ears of corn a day, or $2 \cdot 2 \cdot 365 = 1460$ ears of corn per year and 1460 fish per year. Thus, the demand matrix is:

$$D = \begin{bmatrix} 1460 \\ 1460 \end{bmatrix}$$

The necessary production matrix is:

$$P = (I - T)^{-1}D$$

The technological matrix is:

$$T = \begin{bmatrix} 0.01 & 5 \\ 0.1 & 0 \end{bmatrix}$$

Thus:

$$I - T = \begin{bmatrix} 1 & 0 \\ 0 & 1 \end{bmatrix} - \begin{bmatrix} 0.01 & 5 \\ 0.1 & 0 \end{bmatrix} = \begin{bmatrix} 0.99 & -5 \\ -0.1 & 1 \end{bmatrix}$$

$$(I - T)^{-1} = \frac{1}{(0.99)(1) - (0.1)(5)} \begin{bmatrix} 1 & 5 \\ 0.1 & 0.99 \end{bmatrix}$$

$$= \frac{1}{0.49} \begin{bmatrix} 1 & 5 \\ 0.1 & 0.99 \end{bmatrix}$$

$$P = (I - T)^{-1}D = \frac{1}{0.49} \begin{bmatrix} 1 & 5 \\ 0.1 & 0.99 \end{bmatrix} \begin{bmatrix} 1460 \\ 1460 \end{bmatrix}$$

$$= \frac{1}{0.49} \begin{bmatrix} 1 \cdot 1460 + 5 \cdot 1460 \\ 0.1 \cdot 1460 + 0.99 \cdot 1460 \end{bmatrix}$$

$$= \frac{1}{0.49} \begin{bmatrix} 8760 \\ 1591.4 \end{bmatrix}$$

$$= \begin{bmatrix} 17,877.55 \ldots \\ 3,247.75 \ldots \end{bmatrix}$$

$$\approx \begin{bmatrix} 17,878 \\ 3,248 \end{bmatrix} \begin{matrix} C \\ F \end{matrix}$$

This means that if Robinson Crusoe and Friday produced 17,878 ears of corn and 3248 fish in one year, the production process would result in 1460 ears of corn and 1460 fish for them to eat.
●

Graphing calculators can generate identity matrices. To generate a 2×2 identity matrix, type:

TI-82:	MATRX (scroll to "MATH") 5	2 ENTER
TI-83:	MATRX (scroll to "MATH") 5	2) ENTER
TI-85/86:	2nd MATRX F4 F3	2 ENTER
	The above makes "identity" appear on the screen	*The above makes the identity matrix a 2 × 2 matrix*

Thus, to calculate $(I - T)^{-1}D$:

• enter the 2×2 matrix T and store it (as [A] on a TI-82/83, as T on a TI-85/86)
• enter the 2×1 matrix D and store it (as [D] on a TI-82/83, as D on a TI-85/86)
• generate a 2×2 identity matrix as described above and then type:

TI-82/83:	− MATRX 1 ENTER x⁻¹ ENTER × MATRX 4 ENTER
TI-85/86:	− ALPHA T ENTER x⁻¹ ENTER × ALPHA D ENTER

✔ The result of Example 4 can be checked in the manner shown in Example 1. In producing 17,878 ears of corn, $\frac{1}{100} \cdot 17{,}878 \approx 179$ ears were consumed. In producing 3248 fish, $5 \cdot 3248 = 16{,}240$ ears of corn were consumed. Thus, a total of $179 + 16{,}240 = 16{,}419$ ears were consumed, and $17{,}878 - 16{,}419 = 1469$ ears remained to be eaten. (This number does not exactly match 1460 because of the accumulated round-off error.) Similarly, the production process consumed $\frac{1}{10} \cdot 17{,}878 \approx 1788$ fish, leaving $3248 - 1788 = 1460$ fish to be eaten.

Historical Note

Wassily Leontief

1906–

Wassily Leontief was born in Russia in 1906. He received his master's in economics from the University of Leningrad and his Ph.D. in economics from the University of Berlin. Wassily left Russia for Berlin to study at the university. Later his father, also an economist, escaped from Russia with his wife when he was sent to Berlin as a representative of the Soviet Ministry of Finance.

Wassily emigrated to the United States when he was invited to join the staff of the National Bureau of Economic Research. Two months after starting at the bureau, he gave a lecture at Harvard University; as a result, he was offered a position there. At the age of 42 he became director of Harvard's Economic Research Project. Later, he became professor of economics at New York University and director of NYU's Institute for Economic Analysis.

Leontief is best known for the development of **input-output analysis,** as described in his 1966 book *Input-Output Economics.* In that book, he analyzed the 1958 U.S. economy by dividing the economy into 81 categories; thus, his technological matrix was an 81×81 matrix.

Input-output analysis was an integral part of the total economic planning of the Soviet economy. It is now employed by most developed nations and many corporations. In 1973, Leontief was awarded the Nobel Prize in economics. In his Nobel memorial lecture, he applied input-output analysis to the global environment. The United Nations sponsored this research to help member states reduce poverty and unemployment in ways that preserve and improve the global environment.

Wassily Leontief and his 81×81 technological matrix

INPUT-OUTPUT ECONOMICS

WASSILY LEONTIEF

NEW YORK OXFORD UNIVERSITY PRESS 1966

4.3

EXERCISES

1. Once the fish farm and cornfield were producing sufficient quantities, Robinson sent for his wife to join him. Naturally, this necessitated an increase in production. His wife, a vegetarian, wanted 3 ears of corn a day and no fish. Determine the corresponding production levels for the island's 3 inhabitants.

2. Once Crusoe's wife arrived, Friday decided to send for his wife also. Ms. Friday wanted the same food allotment that the men received. Determine the corresponding production levels for the island's 4 inhabitants (see Exercise 1).

3. Robinson decided that he was overfeeding his fish. He cut the fish feed from 5 ears of corn per year to 4 ears per year. Find the production level that would satisfy Robinson's and Friday's personal demands.

4. Friday realized that Crusoe had cut the fish feed too much and increased it to 4 1/2 ears per year. Find the production level that would satisfy Robinson's and Friday's personal demands.

5. a. Given $T = \begin{bmatrix} 0.1 & 0.2 \\ 0.1 & 0.3 \end{bmatrix}$ and $D = \begin{bmatrix} 2000 \\ 3000 \end{bmatrix}$, find P.

   b. If the production categories in part (a) are coal and steel, in that order, explain the entries of the matrices T, D, and P.

6. a. Given $T = \begin{bmatrix} 0.3 & 0.1 \\ 0.2 & 0.4 \end{bmatrix}$ and $D = \begin{bmatrix} 30{,}000 \\ 40{,}000 \end{bmatrix}$, find P.

   b. If the production categories in part (a) are water and electricity, in that order, explain the entries of the matrices T, D, and P.

7. a. Given $T = \begin{bmatrix} 0.5 & 0 & 0.1 \\ 0.3 & 0.2 & 0 \\ 0.2 & 0.2 & 0.1 \end{bmatrix}$ and $D = \begin{bmatrix} 5000 \\ 5000 \\ 6000 \end{bmatrix}$, find P.

   b. If the production categories in part (a) are automobiles, steel and plastic, in that order, explain the entries of the matrices T, D, and P.

8. a. Given $T = \begin{bmatrix} 0 & 0 & 0.2 \\ 0.3 & 0.1 & 0.4 \\ 0.1 & 0.1 & 0.2 \end{bmatrix}$ and $D = \begin{bmatrix} 50{,}000 \\ 60{,}000 \\ 70{,}000 \end{bmatrix}$, find P.

   b. If the production categories in part (a) are chemicals, medicine, and solvents, in that order, explain the entries of the matrices T, D, and P.

9. An economy is based on two sectors: agriculture and energy. Production of a dollar's worth of agriculture requires inputs of $0.10 from agriculture and $0.25 from energy. Production of a dollar's worth of energy requires input of $0.35 from agriculture and $0.10 from energy.

   a. Find the output for each sector that is needed to satisfy a demand of $25 billion for agriculture and $25 million for energy.

   b. Check your answer as shown in Example 4.
   (Answers are not given in the back of the book.)

10. An economy is based on two sectors: manufacturing and energy. Production of a dollar's worth of manufacturing requires inputs of $0.31 from manufacturing and $0.35 from energy. Production of a dollar's worth of energy requires inputs of $0.28 from manufacturing and $0.20 from energy.

    a. Find the output for each sector that is needed to satisfy a demand of $44 million for manufacturing and $62 million for energy.

    b. Check your answer as shown in Example 4.

11. An economy is based on three sectors: agriculture, manufacturing, and energy. Production of a dollar's worth of agriculture requires inputs of $0.30 from agriculture, $0.25 from manufacturing, and $0.40 from energy. Production of a dollar's worth of manufacturing requires inputs of $0.23 from agriculture, $0.48 from manufacturing, and $0.50 from energy. Production of a dollar's worth of energy requires inputs of $0.30 from agriculture, $0.35 from manufacturing, and $0.05 from energy. Find the output for each sector that is needed to satisfy a demand of $25 million for agriculture, $35 million for manufacturing, and $15 million for energy.

12. An economy is based on three sectors: agriculture, manufacturing, and energy. Production of a dollar's worth of agriculture requires inputs of $0.12 from agriculture, $0.31 from manufacturing, and $0.09 from energy. Production of a dollar's worth of manufacturing requires inputs of $0.01 from agriculture, $0.31 from manufacturing, and $0.35 from energy. Production of a dollar's worth of energy requires inputs of $0.08 from agriculture, $0.49 from manufacturing, and $0.10 from energy. Find the output for each sector that is needed to satisfy a demand of $14 billion for agriculture, $22 billion for manufacturing, and $39 billion for energy.

CHAPTER 4 REVIEW

Terms

associative	element	input-output analysis	production matrix
column	encoding and decoding	inverse of a matrix	row
commutative	matrices	matrix equation	scalar
cryptography	entry	matrix multiplication	scalar multiplication
demand matrix	equal matrices	multiplicative inverse	technological matrix
determinant	identity matrix	primary and secondary	
dimensions	incidence matrix	diagonals	

Review Exercises

In Exercises 1–10, find the indicated sums, differences, or products (if they exist) of the following matrices.

$$A = \begin{bmatrix} 5 & 7 \\ 6 & -1 \end{bmatrix} \qquad B = \begin{bmatrix} 3 & -5 \\ 8 & 2 \end{bmatrix} \qquad C = \begin{bmatrix} 5 \\ 4 \end{bmatrix} \qquad D = \begin{bmatrix} 4 & 1 \end{bmatrix}$$

1. AB

2. BA

3. CD

4. DC

5. AC

6. BD

7. $5A + C$

8. $5A - B$

9. Determinant of A

10. $A^{-1}B$

In Exercises 11–13, answer the given question and give an equation illustrating that answer. For example:

> *Question: Is matrix addition commutative?*
>
> *Answer: Yes, $A + B = B + A$, as long as A and B are of the same dimensions.*

11. Is matrix addition associative?

12. Is matrix subtraction:
  a. commutative?
  b. associative?

13. Is matrix multiplication:
  a. commutative?
  b. associative?
  c. distributive?

14. Al and Tipper are each enrolled in the same two courses: political science and speech. Political science is a 4-unit course, and speech is a 3-unit course. Al and Tipper's grades are given in Figure 4.12.

FIGURE 4.12

	Political Science	**Speech**
Al	A	B
Tipper	C	A

Counting A's as 4 grade points and B's as 3 grade points, use matrix multiplication and scalar multiplication to compute Al's and Tipper's grade point average (GPA).

In Exercises 15–16, solve the given matrix equation. Check your solution; answers are not given in the back of the book.

15. $3\begin{bmatrix} w & x \\ y & z \end{bmatrix} - \begin{bmatrix} 5 & 7 \\ 2 & 1 \end{bmatrix} = \begin{bmatrix} 3 & -2 \\ 0 & 2 \end{bmatrix}$

16. $\begin{bmatrix} 3 & 2 \\ 0 & -1 \end{bmatrix}\begin{bmatrix} w & x \\ y & z \end{bmatrix} - \begin{bmatrix} 5 & -2 \\ 5 & 7 \end{bmatrix} = \begin{bmatrix} 6 & 3 \\ -1 & 4 \end{bmatrix}$

17. Pacific Rim Air Express, an international air freight service, has flights connecting Hong Kong, Los Angeles, Portland, San Francisco, and Tokyo, as shown in Figure 4.13.

FIGURE 4.13

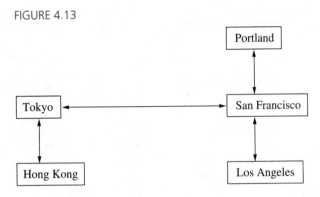

a. Find the incidence matrix N.
b. Find and interpret N^2.
c. Find and interpret $N + N^2$.
d. Find $N + N^2 + N^3$, $N + N^2 + N^3 + N^4$, etc. until you find a matrix with no zero entries. Interpret this matrix.
f. If Pacific Rim Air Express executives were to augment their flight schedule, what flight(s) would you advise them to add? Why?

18. $A = \begin{bmatrix} 3 & 7 & 6 \\ -1 & 0 & 2 \\ 5 & 3 & -7 \end{bmatrix}$. Find A^{-1}.

19. Solve the equation
$$\begin{bmatrix} 3 & 7 & 6 \\ -1 & 0 & 2 \\ 5 & 3 & -7 \end{bmatrix} \begin{bmatrix} x_{11} & x_{12} & x_{13} \\ x_{21} & x_{22} & x_{23} \\ x_{31} & x_{32} & x_{33} \end{bmatrix} = \begin{bmatrix} 40 & -20 & 13 \\ 8 & 70 & -60 \\ 10 & 51 & 9 \end{bmatrix}$$
Check your answers by substituting them back in. (Answers are not given at the back of the book.)

20. Solve the system
$5x + 2y - z = 12$
$3x + 4y + 7z = 8$
$9x - y - z = 10$

by converting the system to a matrix equation and solving that matrix equation. Check your answers by substituting them back in. (Answers are not given at the back of the book.)

21. Decode the following message
   88   128   −22   119   101   160   144   72   66   162
   126   62   141   99   84   147   169   168   70   124   −46
using the following encoding matrix
$$\begin{bmatrix} 4 & 3 & 1 \\ 2 & -1 & 5 \\ 8 & 0 & -2 \end{bmatrix}$$

22. Explain the derivation of the matrix equation $P - TP = D$ used in input-output analysis.

23. Solve the matrix equation $P - TP = D$ for P, and explain why this result is important.

24. An economy is based on two sectors: agriculture and machinery. Production of a dollar's worth of agriculture requires inputs of $0.17 from agriculture and $0.34 from machinery. Production of a dollar's worth of machinery requires an input of $0.22 from machinery.
   a. Find the output for each sector that is needed to satisfy a demand of $25 billion for agriculture and $25 million for machinery.
   b. Check your answer. (Answers are not given in the back of the book.)

Answer the following questions using complete sentences.

25. What two mathematicians invented the theory of matrices? Briefly describe their success in mathematics.

26. Do all matrices have inverses? Why or why not?

27. Who invented input-output analysis?

SETS AND COUNTING 5

Recently, 1000 college seniors were asked whether they favored increasing the state's gasoline tax to generate funds to improve highways and whether they favored increasing the state's alcohol tax to generate funds to improve the public education system. The responses were tallied, and the following results were printed in the campus newspaper: 750 favored an increase in the gasoline tax, 600 favored an increase in the alcohol tax, and 450 favored increases in both taxes. How many of these 1000 students favored an increase in at least one of the taxes? How many favored increasing only the gasoline tax, increasing only the alcohol tax, or increasing neither tax? The mathematical tool designed to answer questions like these is the

set. Although you might be able to answer the given questions without any formal knowledge of sets, the mental reasoning involved in obtaining your answers uses some of the basic principles of sets. (Incidentally, the answers are 900, 300, 150, and 100, respectively.)

The branch of mathematics that deals with sets is called **set theory.** Set theory can be helpful in solving both mathematical and nonmathematical problems. It is an important tool in analyzing the results of consumer surveys, marketing analyses, and political polls. Standardized admissions tests such as the Graduate Record Examination (G.R.E.) ask questions that can be answered with set theory. In this text, we use sets extensively in Chapter 6 on probability.

5.1

SETS AND SET OPERATIONS

A **set** is a collection of objects or things. The objects or things in the set are called **elements** (or **members**) of the set. In our example above, we could talk about the *set* of students who favor increasing only the gasoline tax or the *set* of students who do not favor increasing either tax. In geography, we can talk about the *set* of all state capitals or the *set* of all states west of the Mississippi. It is easy to determine whether something is in these sets; for example, Des Moines is an element of the set of state capitals, whereas Dallas is not. Such sets are called **well-defined** because there is a way of determining for sure whether a particular item is an element of the set.

EXAMPLE 1   Which of the following sets are well-defined?

a. the set of all movies directed by Alfred Hitchcock
b. the set of all great rock-and-roll bands
c. the set of all possible two-person committees selected from a group of five people

Solution   a. This set is well-defined; either a movie was directed by Hitchcock, or it was not.
b. This set is *not* well-defined; membership is a matter of opinion. Some people would say that the Ramones (one of the pioneer punk bands of the late 1970s) are a member, while others might say they are not.
c. This set is well-defined; either the two people are from the group of five, or they are not. •

Notation

By tradition, a set is denoted by a capital letter, frequently one that will serve as a reminder of the contents of the set. **Roster notation** (also called **listing notation**) is a method of describing a set by listing each element of the set inside the symbols { and }, which are called **set braces.** In a listing of the elements of a set, each distinct element is listed only once, and the order of the elements doesn't matter.

The symbol $\in$ stands for the phrase *is an element of,* and $\notin$ stands for *is not an element of.* The **cardinal number** of a set A is the number of elements in the set and is denoted by $n(A)$. Thus, if H is the set of all letters in the name "Heartbreakers," then $H = \{$h, e, a, r, t, b, k, s$\}$. Notice that k is an element of the set H, z is not an element of H, and H has eight elements. In symbols, $k \in H$, $z \notin H$, and $n(H) = 8$.

Two sets are **equal** if they contain exactly the same elements. The order in which the elements are listed does not matter. If B is the set of all letters in the name "Herb Baskets," then $B = \{$h, e, r, b, a, s, k, t$\}$. This set contains exactly the same elements as the set H of letters in the name "Heartbreakers." Therefore, $B = H$.

Often it is not appropriate, or not possible, to describe a set in roster notation. For extremely large sets, such as the set V of all registered voters in Detroit, or for sets that contain an infinite number of elements, such as the set G of all negative real numbers, the roster method would be either too cumbersome or impossible to use. Although V could be expressed via the roster method (since each county compiles a list of all registered voters in its jurisdiction), it would take hundreds or even thousands of pages to list everyone who is registered to vote in Detroit! In the case of the set G of all negative real numbers, no list, no matter how long, is capable of listing all members of the set; there is an infinite number of negative numbers.

In such cases, it is often necessary, or at least more convenient, to use **set-builder notation,** which lists the rules that determine whether an object is an element of the set rather than the actual elements. A set-builder description of set G above is

Tom Petty and the "Heartbreakers," or Tom Petty and the "Herb Baskets"? The set *H* of all letters in the name "Heartbreakers" is the same as the set *B* of all letters in the name "Herb Baskets." Consequently, the sets are equal; *B* = *H*.

$$G = \{x \mid x < 0 \quad \text{and} \quad x \in \Re\}$$

which is read as "the set of all *x* such that *x* is less than zero and *x* is a real number." A set-builder description of set *V* above is

$$V = \{\text{persons} \mid \text{the person is a registered voter in Detroit}\}$$

which is read as "the set of all persons such that the person is a registered voter in Detroit." In set-builder notation, the vertical line stands for the phrase "such that." Whatever is on the left side of the line is the general type of thing in the set, and the rules about set membership are listed on the right.

EXAMPLE 2   Describe each of the following in words.

a. $\{x \mid x > 0 \text{ and } x \in \Re\}$
b. $\{\text{persons} \mid \text{the person is a living former U.S. president}\}$
c. $\{\text{women} \mid \text{the woman is a former U.S. president}\}$

Solution   **a.** the set of all *x* such that *x* is a positive real number
b. the set of all people such that the person is a living former U.S. president
c. the set of all women such that the woman is a former U.S. president   ●

The set listed in part (c) of Example 2 has no elements; there are no women who are former U.S. presidents. If we let *W* equal "the set of all women such that the woman is a former U.S. president," then $n(W) = 0$. A set that has no elements is called an **empty set** and is denoted by $\varnothing$ or by { }. Notice that since the empty set has no elements, $n(\varnothing) = 0$. In contrast, the set {0} is not empty; it has one element, the number zero, so $n(\{0\}) = 1$.

Universal Set and Subsets

When we work with sets, we must define a universal set. For any given problem, the **universal set,** denoted by *U,* is the set of all possible elements of any set used in the problem. For example, when we spell words, *U* is the set of all letters in the alphabet. When every element of one set is also a member of another set, we say that the first set is a *subset* of the second; for instance, {p, i, n} is a subset of {p, i, n, e}. In general, we say that *A* is a **subset** of *B,* denoted by $A \subseteq B$, if for every $x \in A$ it follows that $x \in B$. Alternatively, $A \subseteq B$ if *A* contains no elements that are not in *B*. If *A* contains an element that is not in *B*, then *A* is not a subset of *B* (symbolized as $A \nsubseteq B$).

EXAMPLE 3 Let *B* = {countries | the country has a permanent seat on the U.N. Security Council} Determine if *A* is a subset of *B*.

a. *A* = {Russia, United States}
b. *A* = {China, Japan}
c. *A* = {United States, France, China, United Kingdom, Russia}
d. *A* = { }

Solution We use the roster method to list the elements of set *B*.

$$B = \{\text{China, France, Russia, United Kingdom, United States}\}$$

a. Since every element of *A* is also an element of *B*, *A* is a subset of *B*; $A \subseteq B$.
b. Since *A* contains an element (Japan) that is not in *B*, *A* is not a subset of *B*; $A \nsubseteq B$.
c. Since every element of *A* is also an element of *B* (note that *A* = *B*), *A* is a subset of *B* (and *B* is a subset of *A*); $A \subseteq B$ (and $B \subseteq A$). In general, every set is a subset of itself; $A \subseteq A$ for any set *A*.
d. Does *A* contain an element that is not in *B?* No! Therefore, *A* (an empty set) is a subset of *B*; $A \subseteq B$. In general, the empty set is a subset of all sets; $\varnothing \subseteq A$ for any set *A*. ●

FIGURE 5.1
A is a subset of *B*
$A \subseteq B$

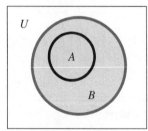

We can express the relationship $A \subseteq B$ visually by drawing a Venn diagram, as shown in Figure 5.1. A **Venn diagram** consists of a rectangle, representing the universal set, and various circles within the rectangle, each representing a set.

If two sets are equal, they contain exactly the same elements. It then follows that each is a subset of the other. For example, if *A* = *B*, then every element of *A* is an element of *B* (and vice versa). In this case, *A* is called an **improper subset** of *B*. (Likewise, *B* is an improper subset of *A*.) Every set is an improper subset of itself; for example, $A \subseteq A$. On the other hand, if *A* is a subset of *B* and *B* contains an element not in *A* (that is, $A \neq B$), then *A* is called a **proper subset** of *B*. To indicate a proper subset, the symbol $\subset$ is used. While it is acceptable to write $\{1, 2\} \subseteq \{1, 2, 3\}$, the relationship of a proper subset is stressed when it is written $\{1, 2\} \subset \{1, 2, 3\}$. Notice the similarities between the subject symbols, $\subset$ and $\subseteq$, and the inequality symbols, $<$ and $\leq$, used in algebra; it is acceptable to write $1 \leq 3$, but writing $1 < 3$ is more informative.

Intersection of Sets

Sometimes an element of one set is also an element of another set; that is, the sets may overlap. This overlap is called the **intersection** of the sets. If an element is in two sets *at the same time,* it is in the intersection of the sets.

Historical Note

John Venn

1834–1923

John Venn is considered by many to be one of the originators of modern symbolic logic. Venn received his degree in mathematics from the University at Cambridge at the age of 23. He was then elected a fellow of the college and held this fellowship until his death, some 66 years later. Two years after receiving his degree, Venn accepted a teaching position at Cambridge: college lecturer in moral sciences.

During the latter half of the 19th century, the study of logic experienced a rebirth in England. Mathematicians were attempting to symbolize and quantify the central concepts of logical thought. Consequently, Venn chose to focus on the study of logic during his tenure at Cambridge. In addition, he investigated the field of probability and published *The Logic of Chance*, his first major work, in 1866.

Venn was well read in the works of his predecessors, including the noted logicians Augustus De Morgan, George Boole, and Charles Dodgson (a.k.a. Lewis Carroll). Boole's pioneering work on the marriage of logic and algebra proved to be a strong influence on Venn; in fact, Venn used the type of diagram that now bears his name in an 1876 paper in which he examined Boole's system of symbolic logic.

Venn was not the first scholar to use the diagrams that now bear his name. Gottfried Leibniz, Leonhard Euler, and others utilized similar diagrams years before Venn did. Examining each author's diagrams, Venn was critical of their lack of uniformity. He developed a consistent, systematic explanation of the general use of geometrical figures in the analysis of logical arguments. Today, these geometrical figures are known by his name and are used extensively in elementary set theory and logic.

Venn's writings were held in high esteem. His textbooks, *Symbolic Logic* (1881) and *The Principles of Empirical Logic* (1889), were used during the late 19th and early 20th centuries. In addition to

his works on logic and probability, Venn also conducted much research into historical records, especially those of his college and those of his family.

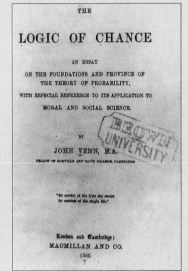

Set theory and the cardinal numbers of sets are used extensively in the study of probability. Although he was a professor of logic, Venn investigated the foundations and applications of theoretical probability. Venn's first major work, *The Logic of Chance*, exhibited the diversity of his academic interests.

FIGURE 5.2
The intersection $A \cap B$

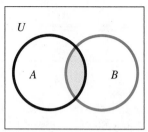

Intersection of Sets

The *intersection* of set A and set B, denoted by $A \cap B$, is

$$A \cap B = \{x \mid x \in A \quad \text{and} \quad x \in B\}$$

The intersection of two sets consists of those elements that are common to both sets.

For example, given the sets $A = \{$Chuckie, Elvira, Freddie, Jason$\}$ and $B = \{$Ash, Freddie, Jamie, Jason$\}$, their intersection is $A \cap B = \{$Freddie, Jason$\}$.

Venn diagrams are useful in depicting the relationship between sets. The Venn diagram in Figure 5.2 illustrates the intersection of two sets; the shaded region represents $A \cap B$.

Mutually Exclusive Sets

Sometimes a pair of sets has no overlap. Consider an ordinary deck of playing cards. Let $D = \{$cards $\mid$ the card is a diamond$\}$ and $S = \{$cards $\mid$ the card is a spade$\}$. Certainly, *no* cards are both diamonds and spades *at the same time;* that is, $S \cap D = \emptyset$.

Two sets A and B are **mutually exclusive** (or **disjoint**) if they have no elements in common, that is, if $A \cap B = \emptyset$. The Venn diagram in Figure 5.3 illustrates mutually exclusive sets.

FIGURE 5.3
Mutually exclusive sets
$(A \cap B = \emptyset)$

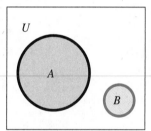

Union of Sets

What does it mean when we ask, "How many of the 500 college students in a transportation survey own an automobile or a motorcycle?" Does it mean "How many students own either an automobile or a motorcycle *or both?*" or does it mean "How many students own either an automobile or a motorcycle, *but not both?*" The former is called the *inclusive or,* because it includes the possibility of owning both; the latter is called the *exclusive or.* In logic and in mathematics, the word *or* refers to the *inclusive or,* unless you are told otherwise.

The meaning of the word *or* is important to the concept of union. The **union** of two sets is a new set formed by joining those two sets together, just as the union of the states is the joining together of fifty states to form one nation.

Union of Sets

The *union* of set A and set B, denoted by $A \cup B$, is

$$A \cup B = \{x \mid x \in A \quad \text{or} \quad x \in B\}$$

The union of A and B consists of all elements that are in either A or B or both, that is, all elements that are in at least one of the sets.

FIGURE 5.4
The union $A \cup B$

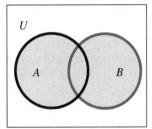

For example, given the sets $A = \{$David, Jay$\}$ and $B = \{$Rosie, Geraldo, Oprah, Roseanne$\}$, their union is $A \cup B = \{$Rosie, David, Geraldo, Jay, Oprah, Roseanne$\}$, and their intersection is $A \cap B = \varnothing$. The Venn diagram in Figure 5.4 illustrates the union of two sets; the shaded region represents $A \cup B$.

Because $A \cup B$ consists of all elements that are in A or B (or both), to find $n(A \cup B)$, we add $n(A)$ plus $n(B)$. However, doing so results in an answer that may be too big; that is, if A and B have elements in common, these elements will be counted twice (once as a part of A and once as a part of B). Therefore, to find the cardinal number of $A \cup B$, we add the cardinal number of A to the cardinal number of B and then *subtract* the cardinal number of $A \cap B$ (so that the overlap is not counted twice).

Cardinal Number Formula for the Union of Sets

For any two sets A and B, the number of elements in their union is $n(A \cup B)$, where

$$n(A \cup B) = n(A) + n(B) - n(A \cap B)$$

As long as any three of the four quantities in the general formula are known, the missing quantity can be found by algebraic manipulation.

EXAMPLE 4 Given $n(U) = 169$, $n(A) = 81$, and $n(B) = 66$, find the following.

a. If $n(A \cap B) = 47$, find $n(A \cup B)$ and draw a Venn diagram depicting the composition of the universal set.

b. If $n(A \cup B) = 147$, find $n(A \cap B)$ and draw a Venn diagram depicting the composition of the universal set.

Solution

FIGURE 5.5

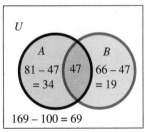

a. We must use the cardinal number formula for the union of sets. Substituting the three given quantities, we have

$$n(A \cup B) = n(A) + n(B) - n(A \cap B)$$
$$= 81 + 66 - 47$$
$$= 100$$

The Venn diagram in Figure 5.5 illustrates the composition of U.

b. We must use the cardinal number formula for the union of sets. Substituting the three given quantities, we have

$$n(A \cup B) = n(A) + n(B) - n(A \cap B)$$
$$147 = 81 + 66 - n(A \cap B)$$
$$147 = 147 - n(A \cap B)$$
$$n(A \cap B) = 147 - 147$$
$$n(A \cap B) = 0$$

FIGURE 5.6

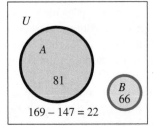

Therefore, A and B have no elements in common; they are mutually exclusive. The Venn diagram in Figure 5.6 illustrates the composition of U. ●

EXAMPLE 5 A recent transportation survey of 500 college students (the universal set U) yielded the following information: 291 own an automobile (A), 179 own a motorcycle (M), and 85 own both an automobile and a motorcycle ($A \cap M$). What percent of these students own an automobile or a motorcycle?

Solution Recall that "automobile or motorcycle" means "automobile or motorcycle or both" (the inclusive *or*) and that *or* implies union. Hence, we must find $n(A \cup M)$, the cardinal number of the union of sets A and M. We are given that $n(A) = 291$, $n(M) = 179$, and $n(A \cap M) = 85$. Substituting the given values into the cardinal number formula for the union of sets, we have

$$n(A \cup M) = n(A) + n(M) - n(A \cap M)$$
$$= 291 + 179 - 85$$
$$= 385$$

Therefore, 385 of the 500 students surveyed own an automobile or a motorcycle. Expressed as a percent, $385/500 = 0.77$; therefore, 77% of the students own an automobile or a motorcycle (or both). •

Complement of a Set

In certain situations, it might be important to know how many things are *not* in a given set. For instance, when playing cards, you might want to know how many cards are not ranked lower than a five, or when taking a survey, you might want to know how many people did not vote for a specific proposition. The set of all elements in the universal set that are *not* in a specific set is called the **complement** of the set.

Complement of a Set

The *complement* of set A, denoted by A' (read "A prime" or "the complement of A"), is

$$A' = \{x \mid x \in U \quad \text{and} \quad x \notin A\}$$

The complement of a set consists of all elements that are in the universal set but not in the given set.

FIGURE 5.7
The complement A'

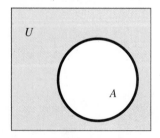

For example, given that $U = \{1, 2, 3, 4, 5, 6, 7, 8, 9\}$ and $A = \{1, 3, 5, 7, 9\}$, the complement of A is $A' = \{2, 4, 6, 8\}$. What is the complement of A'? Just as $-(-x) = x$ in algebra, $(A')' = A$ in set theory. The Venn diagram in Figure 5.7 illustrates the complement of set A; the shaded region represents A'.

Suppose A is a set of elements drawn from a universal set U. If x is an element of the universal set ($x \in U$), then exactly one of the following must be true: (1) x is an element of A ($x \in A$), or (2) x is not an element of A ($x \notin A$). Since no element of the universal set can be in both A and A' at the same time, it follows that A and A' are mutually exclusive sets whose union equals the entire universal set. Therefore, the sum of the cardinal numbers of A and A' equals the cardinal number of U.

It is often quicker to count the elements that are *not* in a set rather than counting those that are. Consequently, to find the cardinal number of a set, we can subtract the cardinal number of its complement from the cardinal number of the universal set; that is, $n(A) = n(U) - n(A')$.

Cardinal Number Formula for the Complement of a Set

For any set A and its complement A',

$$n(A) + n(A') = n(U)$$

where U is the universal set.
Alternatively,

$$n(A) = n(U) - n(A') \quad \text{and} \quad n(A') = n(U) - n(A).$$

EXAMPLE 6 How many letters in the alphabet precede the letter w?

Solution Rather than counting all the letters that precede w, we will take a shortcut by counting all the letters that do *not* precede w. Let $L = \{\text{letters} \mid \text{the letter precedes w}\}$. Therefore, $L' = \{\text{letters} \mid \text{the letter does not precede w}\}$. Now $L' = \{\text{w, x, y, z}\}$ and $n(L') = 4$; therefore, we have

$$n(L) = n(U) - n(L') \qquad \text{Cardinal number formula for the complement}$$
$$= 26 - 4$$
$$= 22$$

There are 22 letters preceding the letter w.

5.1

EXERCISES

1. State whether the given set is well-defined.
   a. the set of all black automobiles
   b. the set of all inexpensive automobiles
   c. the set of all prime numbers
   d. the set of all large numbers

2. Suppose $A = \{2, 5, 7, 9, 13, 25, 26\}$.
   a. Find $n(A)$.
   b. True or false: $7 \in A$
   c. True or false: $9 \notin A$
   d. True or false: $20 \notin A$

In Exercises 3–6, list all subsets of the given set. Identify which subsets are proper and which are improper.

3. $B = \{\text{Lennon, McCartney}\}$

4. $N = \{0\}$

5. $S = \{\text{yes, no, undecided}\}$

6. $M = \{\text{classical, country, jazz, rock}\}$

In Exercises 7–12, the universal set is $U = \{Monday, Tuesday, Wednesday, Thursday, Friday, Saturday, Sunday\}$. If $A = \{Monday, Tuesday, Wednesday, Thursday, Friday\}$ and $B = \{Friday, Saturday, Sunday\}$, find the indicated set.

7. $A \cap B$ **8.** $A \cup B$ **9.** B'
10. A' **11.** $A' \cup B$ **12.** $A \cap B'$

In Exercises 13–22, use a Venn diagram like the one in Figure 5.8 to shade in the region corresponding to the indicated set.

FIGURE 5.8

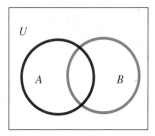

13. $A \cap B$ **14.** $A \cup B$
15. A' **16.** B'

17. $A \cup B'$ **18.** $A' \cup B$

19. $A' \cap B$ **20.** $A \cap B'$

21. $A' \cup B'$ **22.** $A' \cap B'$

23. Suppose $n(U) = 150$, $n(A) = 37$, and $n(B) = 84$.
   a. If $n(A \cup B) = 100$, find $n(A \cap B)$ and draw a Venn diagram illustrating the composition of U.
   b. If $n(A \cup B) = 121$, find $n(A \cap B)$ and draw a Venn diagram illustrating the composition of U.

24. Suppose $n(U) = w$, $n(A) = x$, $n(B) = y$, and $n(A \cup B) = z$.
   a. Why must x be less than or equal to z?
   b. If $A \neq U$ and $B \neq U$, fill in the blank with the most appropriate symbol: $<$, $>$, $\leq$, or $\geq$.
   w _____ z, w _____ y, y _____ z, x _____ w
   c. Find $n(A \cap B)$ and draw a Venn diagram illustrating the composition of U.

25. In a recent transportation survey, 500 high school seniors were asked to check the appropriate box or boxes on the following form:

☐ I own an automobile.
☐ I own a motorcycle.

The results were tabulated as follows: 91 checked the automobile box, 123 checked the motorcycle box, and 29 checked both boxes.
   a. Draw a Venn diagram illustrating the results of the survey.
   b. What percent of these students own an automobile or a motorcycle?

26. In a recent market research survey, 500 married couples were asked to check the appropriate box or boxes on the following form:

☐ We own a VCR.
☐ We own a microwave oven.

The results were tabulated as follows: 248 checked the VCR box, 314 checked the microwave oven box, and 166 checked both boxes.
   a. Draw a Venn diagram illustrating the results of the survey.
   b. What percent of these couples own a VCR or a microwave oven?

27. In a recent socioeconomic survey, 750 married women were asked to check the appropriate box or boxes on the following form:

☐ I have a career.
☐ I have a child.

The results were tabulated as follows: 529 checked the child box, 213 checked the career box, and 143 were blank (no boxes were checked).
   a. Draw a Venn diagram illustrating the results of the survey.
   b. What percent of these women had both a child and a career?

28. In a recent health survey, 750 single men in their twenties were asked to check the appropriate box or boxes on the following form:

☐ I am a member of a private gym.
☐ I am a vegetarian.

The results were tabulated as follows: 374 checked the gym box, 92 checked the vegetarian box, and 332 were blank (no boxes were checked).
   a. Draw a Venn diagram illustrating the results of the survey.
   b. What percent of these men were both members of a private gym and vegetarians?

For Exercises 29–32, let

$U = \{x \mid x$ is the name of one of the states in the United States$\}$.

$A = \{x \mid x \in U$ and x begins with the letter A$\}$

$I = \{x \mid x \in U$ and x begins with the letter I$\}$

$M = \{x \mid x \in U$ and x begins with the letter M$\}$

$N = \{x \mid x \in U$ and x begins with the letter N$\}$

$O = \{x \mid x \in U$ and x begins with the letter O$\}$

29. Find $n(M')$.   **30.** Find $n(A \cup N)$.

31. Find $n(I' \cap O')$.   **32.** Find $n(M \cap I)$.

Exercises 33–36, let

$U = \{x \mid x$ is the name of one of the months in a year$\}$

$J = \{x \mid x \in U$ and x begins with the letter J$\}$

$Y = \{x \mid x \in U$ and x ends with the letter Y$\}$

$V = \{x \mid x \in U$ and x begins with a vowel$\}$

$R = \{x \mid x \in U$ and x ends with the letter R$\}$

33. Find $n(R')$.   **34.** Find $n(J \cap V)$.

35. Find $n(J \cup Y)$.   **36.** Find $n(V \cap R)$.

In Exercises 37–46, determine how many cards, in an ordinary deck of 52, fit the description. (If you are unfamiliar with playing cards, see pages 269–270 for the description of a standard deck.)

37. spades or aces   **38.** clubs or twos

39. face cards or black   **40.** face cards or diamonds

41. face cards and black   **42.** face cards and diamonds

43. aces or eights   **44.** three or sixes

45. aces and eights   **46.** threes and sixes

47. Suppose $A = \{1, 2, 3\}$ and $B = \{1, 2, 3, 4, 5, 6\}$.
   a. Find $A \cap B$.
   b. Find $A \cup B$.
   c. In general, if $E \cap F = E$, what must be true concerning sets E and F?
   d. In general, if $E \cup F = F$, what must be true concerning sets E and F?

48. Fill in the blank, and give an example to support your answer.
   a. If $A \subset B$, then $A \cap B =$ _____.
   b. If $A \subset B$, then $A \cup B =$ _____.

49. a. List all subsets of $A = \{a\}$. How many subsets does A have?
   b. List all subsets of $A = \{a, b\}$. How many subsets does A have?
   c. List all subsets of $A = \{a, b, c\}$. How many subsets does A have?
   d. List all subsets of $A = \{a, b, c, d\}$. How many subsets does A have?
   e. Is there a relationship between the cardinal number of set A and the number of subsets of set A?
   f. How many subsets does $A = \{a, b, c, d, e, f\}$ have?
   HINT: Use your answer to part (e).

50. Prove the cardinal number formula for the complement of a set.
   HINT: Apply the cardinal number formula for the union of sets to A and A'.

 Answer the following questions using complete sentences.

51. If $A \cap B = \varnothing$, what is the relationship between sets A and B?

52. If $A \cup B = \varnothing$, what is the relationship between sets A and B?

53. Explain the difference between $\{0\}$ and $\varnothing$.

54. Explain the difference between 0 and $\{0\}$.

55. Is it possible to have $A \cap A = \varnothing$?

56. What is the difference between proper and improper subsets?

57. A set can be described by two methods: the roster method and set-builder notation. When is it advantageous to use the roster method? When is it advantageous to use set-builder notation?

58. John Venn was a professor in what academic field? Where did he teach?

59. What was one of John Venn's main contributions to the field of logic? What new benefits did it offer?

Many graduate schools require applicants to take the Graduate Record Examination (G.R.E.). This exam is intended to measure verbal, quantitative, and analytical skills developed throughout a person's life. There are many classes and study guides available to help people prepare for the exam. The remaining questions are typical of those found in the study guides and on the exam itself.

Exercises 60–64 refer to the following: Two collectors, John and Juneko, are each selecting a group of three posters from a group of seven movie posters: J, K, L, M, N, O, and P. No poster can be in both groups. The selections made by John and Juneko are subject to the following restrictions:

- If K is in John's group, M must be in Juneko's group.
- If N is in John's group, P must be in Juneko's group.
- J and P cannot be in the same group.
- M and O cannot be in the same group.

60. Which of the following pairs of groups selected by John and Juneko conform to the restrictions?

	John	Juneko
a.	*J, K, L*	*M, N, O*
b.	*J, K, P*	*L, M, N*
c.	*K, N, P*	*J, M, O*
d.	*L, M, N*	*K, O, P*
e.	*M, O, P*	*J, K, N*

61. If N is in John's group, which of the following could not be in Juneko's group?
   a. *J*   **b.** *K*   **c.** *L*   **d.** *M*   **e.** *P*

62. If K and N are in John's group, Juneko's group must consist of which of the following?
   a. *J, M,* and *O*   **b.** *J, O,* and *P*
   c. *L, M,* and *P*   **d.** *L, O,* and *P*
   e. *M, O,* and *P*

63. If J is in Juneko's group, which of the following is true?
   a. *K* cannot be in John's group.
   b. *N* cannot be in John's group.
   c. *O* cannot be in Juneko's group.
   d. *P* must be in John's group.
   e. *P* must be in Juneko's group.

64. If K is in John's group, which of the following is true?
   a. *J* must be in John's group.
   b. *O* must be in John's group.
   c. *L* must be in Juneko's group.
   d. *N* cannot be in John's group.
   e. *O* cannot be in Juneko's group.

5.2

APPLICATIONS OF VENN DIAGRAMS

As we have seen, Venn diagrams are very useful tools for visualizing the relationships be-tween sets. They can be used to establish general formulas involving set operations and to determine the cardinal numbers of sets. Venn diagrams are particularly useful in survey analysis.

Surveys

Surveys are often used to divide people or objects into categories. Because the categories sometimes overlap, people can fall into more than one category. Venn diagrams and the for-mulas for cardinal numbers can help researchers organize the data.

EXAMPLE 1 Has the advent of the VCR affected attendance at movie theaters? To study this question, Professor Redrum's film class conducted a survey of people's movie-watching habits. He had his students ask hundreds of people between the ages of sixteen and forty-five to check the appropriate box or boxes on the following form:

> ☐ I watched a movie in a theater during the past month.
> ☐ I watched a movie on a videocassette during the past month.

After the professor had collected the forms and tabulated the results, he told the class that 388 people checked the theater box, 495 checked the videocassette box, 281 checked both boxes, and 98 of the forms were blank. Giving the class only this information, Professor Redrum posed the following three questions.

a. What percent of the people surveyed watched a movie in a theater or on a videocassette during the past month?
b. What percent of the people surveyed watched a movie in a theater only?
c. What percent of the people surveyed watched a movie on a videocassette only?

Solution **a.** In order to calculate the desired percentages, we must determine $n(U)$, the total number of people surveyed. This can be accomplished by drawing a Venn diagram. Because the survey divides people into two categories (those who watched a movie in a theater and those who watched a movie on a videocassette), we need to define two sets. Let

$T = \{\text{people} \mid \text{the person watched a movie in a theater}\}$

$C = \{\text{people} \mid \text{the person watched a movie on a videocassette}\}$

Now translate the given survey information into the symbols for the sets and attach their given cardinal numbers: $n(T) = 388$, $n(C) = 495$, and $n(T \cap C) = 281$.

Our first goal is to find $n(U)$. To do so, we will fill in the cardinal numbers of all regions of a Venn diagram consisting of two overlapping circles (because we are dealing with two sets). The intersection of T and C consists of 281 people, so we draw two over-lapping circles and fill in 281 as the number of elements in common (see Figure 5.9).

Because we were given $n(T) = 388$ and know that $n(T \cap C) = 281$, the difference $388 - 281 = 107$ tells us that 107 people watched a movie in a theater but did not watch a movie on a videocassette. We fill in 107 as the number of people who watched a movie only in a theater (see Figure 5.10).

FIGURE 5.9

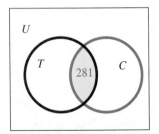

FIGURE 5.10

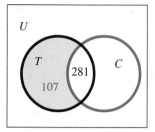

FIGURE 5.11

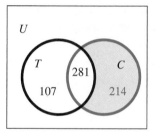

FIGURE 5.12

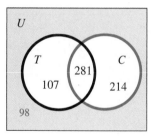

Because $n(C) = 495$, the difference $495 - 281 = 214$ tells us that 214 people watched a movie on a videocassette but not in a theater. We fill in 214 as the number of people who watched a movie only on a videocassette (see Figure 5.11).

The only region remaining to be filled in is the region outside both circles. This region represents people who didn't watch a movie in a theater or on a videocassette and is symbolized by $(T \cup C)'$. Because 98 people didn't check either box on the form, $n[(T \cup C)'] = 98$ (see Figure 5.12).

After we have filled in the Venn diagram with all the cardinal numbers, we readily see that $n(U) = 98 + 107 + 281 + 214 = 700$. Therefore, 700 people were in the survey.

To determine what *percent* of the people surveyed watched a movie in a theater *or* on a videocassette during the past month, simply divide $n(T \cup C)$ by $n(U)$:

$$\frac{n(T \cup C)}{n(U)} = \frac{107 + 281 + 214}{700}$$

$$= \frac{602}{700}$$

$$= 0.86$$

Therefore, 86% of the people surveyed watched a movie in a theater or on a videocassette during the past month.

b. To find what *percent* of the people surveyed watched a movie in a theater only, divide 107 (the number of people who watched a movie in a theater only) by $n(U)$:

$$\frac{107}{700} = 0.152857142 \ldots$$

Approximately 15% of the people surveyed watched a movie in a theater only.

c. Because 214 people watched a movie on videocassette only, $214/700 = 0.305714285 \ldots$, or approximately 31%, of the people surveyed watched a movie on videocassette only. •

When you solve a cardinal number problem (a problem that asks, "How many?" or "What percent?") involving a universal set that is divided into various categories (for instance, a survey), use the following general steps.

Solving a Cardinal Number Problem

A cardinal number problem is a problem in which you are asked, "How many?" or "What percent?"

1. Define a set for each category in the universal set. If a category and its negation are both mentioned, define one set A and utilize its complement A'.
2. Draw a Venn diagram with as many overlapping circles as the number of sets you have defined.
3. Write down all the given cardinal numbers corresponding to the various given sets.
4. Starting with the innermost overlap, fill in each region of the Venn diagram with its cardinal number.

FIGURE 5.13

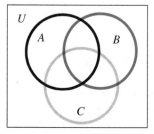

When we are working with three sets, we must account for all possible intersections of the sets. Hence, in such cases we will use the Venn diagram shown in Figure 5.13.

EXAMPLE 2   A consumer survey was conducted to examine patterns in ownership of microwave ovens, answering machines, and VCRs. The following data were obtained: 213 people had microwaves, 294 had answering machines, 337 had VCRs, 109 had all three, 64 had none, 198 had answering machines and VCRs, 382 had answering machines or microwaves, and 61 had microwaves and VCRs but no answering machines.

a. What percent of the people surveyed owned a microwave but no VCR or answering machine?

b. What percent of the people surveyed owned a VCR but no microwave or answering machine?

Solution   **a.** In order to calculate the desired percentages, we must determine $n(U)$, the total number of people surveyed. This can be accomplished by drawing a Venn diagram. Because the survey divides people into three categories (those who own a microwave, those who own an answering machine, and those who own a VCR), we need to define three sets. Let

M = {people | the person owns a microwave oven}
A = {people | the person owns an answering machine}
V = {people | the person owns a VCR}

Now translate the given survey information into the symbols for the sets and attach their given cardinal numbers:

213 people had microwaves ⟶	$n(M) = 213$
294 had answering machines ⟶	$n(A) = 294$
337 had VCRs ⟶	$n(V) = 337$
109 had all three ⟶ (M and A and V)	$n(M \cap A \cap V) = 109$
64 had none ⟶ (not M and not A and not V)	$n(M' \cap A' \cap V') = 64$
198 had answering machines and VCRs ⟶ (A and V)	$n(A \cap V) = 198$
382 had answering machines or microwaves ⟶ (A or M)	$n(A \cup M) = 382$
61 had microwaves and VCRs but no answering machines ⟶ (M and V and not A)	$n(M \cap V \cap A') = 61$

FIGURE 5.14

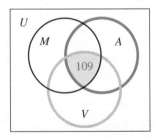

Our first goal is to find $n(U)$. To do so, we will fill in the cardinal numbers of all regions of a Venn diagram like that in Figure 5.13. We start by using information concerning membership in all three sets. Because the intersection of all three sets consists of 109 people, we fill in 109 in the region common to M and A and V (see Figure 5.14).

FIGURE 5.15

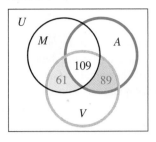

FIGURE 5.16

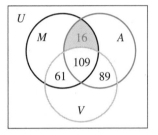

FIGURE 5.17

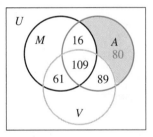

FIGURE 5.18

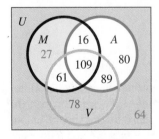

Next, we utilize any information concerning membership in two of the three sets. Because $n(A \cap V) = 198$, a total of 198 people are common to both A and V; some are in M, and some are not in M. Of these 198 people, 109 are in M (see Figure 5.14). Therefore, the difference $198 - 109 = 89$ gives the number not in M. Eighty-nine people are in A and V and *not* in M; that is, $n(A \cap V \cap M') = 89$. Concerning membership in the two sets M and V, we are given $n(M \cap V \cap A') = 61$. Therefore, we know that 61 people are in M and V and not in A (see Figure 5.15).

We are given $n(A \cup M) = 382$. From this number, we can calculate $n(A \cap M)$ by using the cardinal number formula for the union of sets:

$$n(A \cup M) = n(A) + n(M) - n(A \cap M)$$

$$382 = 294 + 213 - n(A \cap M)$$

$$n(A \cap M) = 125$$

Therefore, a total of 125 people are in A and M; some are in V, and some are not in V. Of these 125 people, 109 are in V (see Figure 5.14). Therefore, the difference $125 - 109 = 16$ gives the number not in V. Sixteen people are in A and M and *not* in V; that is, $n(A \cap M \cap V') = 16$ (see Figure 5.16).

Knowing that a total of 294 people are in A (given $n(A) = 294$), we are now able to fill in the last region of A. The missing region (people in A only) has $294 - 109 - 89 - 16 = 80$ members; $n(A \cap M' \cap V') = 80$ (see Figure 5.17).

In similar manner, we subtract the known pieces of M from $n(M) = 213$, which is given, and obtain $213 - 61 - 109 - 16 = 27$; therefore, 27 people are in M only. Likewise, to find the last region of V, we use $n(V) = 337$ (given) and obtain $337 - 89 - 109 - 61 = 78$; therefore, 78 people are in V only. Finally, the 64 people who own none of the items are placed "outside" the three circles (see Figure 5.18).

By adding up the cardinal numbers of all the regions in Figure 5.18, we find that the total number of people in the survey is 524; that is, $n(U) = 524$.

Now, to determine what *percent* of the people surveyed owned only a microwave (no VCR and no answering machine), we simply divide $n(M \cap V' \cap A')$ by $n(U)$:

$$\frac{n(M \cap V' \cap A')}{n(U)} = \frac{27}{524}$$

$$= 0.051526717 \ldots$$

Approximately 5% of the people surveyed owned a microwave and did not own a VCR or an answering machine.

b. To determine what *percent* of the people surveyed owned only a VCR (no microwave and no answering machine), we divide $n(V \cap M' \cap A')$ by $n(U)$:

$$\frac{n(V \cap M' \cap A')}{n(U)} = \frac{78}{524}$$

$$= 0.148854961 \ldots$$

Approximately 15% of the people surveyed owned a VCR and did not own a microwave or an answering machine. ●

De Morgan's Laws

One of the basic properties of algebra is the distributive property:

$$a(b + c) = ab + ac$$

Given $a(b + c)$, the operation outside the parentheses can be distributed over the operation inside the parentheses. It makes no difference whether you add b and c first and then multiply the sum by a or first multiply each pair, a and b, a and c, and then add their products; the same result is obtained. Is there a similar property for the complement, union, and intersection of sets?

EXAMPLE 3 Suppose $U = \{1, 2, 3, 4, 5\}$, $A = \{1, 2, 3\}$, and $B = \{2, 3, 4\}$.

a. For the given sets, does $(A \cup B)' = A' \cup B'$?
b. For the given sets, does $(A \cup B)' = A' \cap B'$?

Solution a. To find $(A \cup B)'$, we must first find $A \cup B$:

$$A \cup B = \{1, 2, 3\} \cup \{2, 3, 4\}$$
$$= \{1, 2, 3, 4\}$$

The complement of $A \cup B$ (relative to the given universal set U) is

$$(A \cup B)' = \{5\}$$

To find $A' \cup B'$, we must first find A' and B':

$$A' = \{4, 5\} \quad \text{and} \quad B' = \{1, 5\}$$

The union of A' and B' is

$$A' \cup B' = \{4, 5\} \cup \{1, 5\}$$
$$= \{1, 4, 5\}$$

Now, $\{5\} \neq \{1, 4, 5\}$; therefore, $(A \cup B)' \neq A' \cup B'$.

b. We find $(A \cup B)'$ as in part (a): $(A \cup B)' = \{5\}$. Now,

$$A' \cap B' = \{4, 5\} \cap \{1, 5\}$$
$$= \{5\}$$

For the given sets, $(A \cup B)' = A' \cap B'$. •

FIGURE 5.19

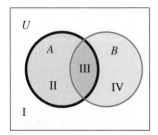

Part (a) of Example 3 shows that the operation of complementation *cannot* be explicitly distributed over the operation of union; that is, $(A \cup B)' \neq A' \cup B'$. However, part (b) of the example implies that there *may* be some relationship between the complement, union, and intersection of sets. The fact that $(A \cup B)' = A' \cap B'$ *for the given sets A and B* does not mean that it is true *for all sets A and B*. We will use a general Venn diagram to examine the validity of the statement $(A \cup B)' = A' \cap B'$.

When we draw two overlapping circles within a universal set, four regions are formed. Every element of the universal set U is in exactly one of the following regions, as shown in Figure 5.19:

I	in neither A nor B
II	in A and not in B
III	in both A and B
IV	in B and not in A

The set $A \cup B$ consists of all elements in regions II, III, and IV. Therefore, the complement $(A \cup B)'$ consists of all elements in region I. Now A' consists of all elements in regions I and IV, and B' consists of the elements in regions I and II. Therefore, the elements common to both A' and B' are those in region I; that is, the set $A' \cap B'$ consists of all elements in region I. Since $(A \cup B)'$ and $A' \cap B'$ contain exactly the same elements (those in region I), the sets are equal; that is, $(A \cup B)' = A' \cap B'$ is true for all sets A and B.

The relationship $(A \cup B)' = A' \cap B'$ is known as one of **De Morgan's laws.** Simply stated, "the complement of a union is the intersection of the complements." In a similar manner, it can be shown that $(A \cap B)' = A' \cup B'$ (see Exercise 29).

De Morgan's Laws

For any sets A and B,

$$(A \cup B)' = A' \cap B'$$

That is, the complement of a union is the intersection of the complements. Also,

$$(A \cap B)' = A' \cup B'$$

That is, the complement of an intersection is the union of the complements.

EXAMPLE 4 Suppose $U = \{0, 1, 2, 3, 4, 5, 6, 7, 8, 9\}$, $A = \{2, 3, 7\, 8\}$, and $B = \{0, 4, 5, 7, 8, 9\}$. Use De Morgan's law to find $(A' \cup B)'$.

Solution The complement of a union is equal to the intersection of the complements; therefore, we have

$$
\begin{aligned}
(A' \cup B)' &= (A')' \cap B' &&\text{De Morgan's law} \\
&= A \cap B' &&(A')' = A \\
&= \{2, 3, 7, 8\} \cap \{1, 2, 3, 6\} \\
&= \{2, 3\}
\end{aligned}
$$

Notice that this problem could be done without using De Morgan's law, but solving it would then involve finding first A', then $A' \cup B$, and finally $(A' \cup B)'$. This method would involve more work. (Try it!)

Historical Note

Augustus De Morgan

1806–1871

Being born blind in one eye did not stop Augustus De Morgan from becoming a well-read philosopher, historian, logician, and mathematician. De Morgan was born in Madras, India, when his father was working for the East India Company. After moving to England, De Morgan was educated at Cambridge, and at the age of 22 he became the first professor of mathematics at the newly opened University of London (later renamed University College).

De Morgan viewed all of mathematics as an abstract study of symbols and of systems of operations applied to these symbols. While studying the ramifications of symbolic logic, De Morgan formulated the general properties of complementation that now bear his name. Not limited to symbolic logic,

De Morgan's many works include books and papers on the foundation of algebra, differential calculus, and probability. He was known to be a jovial person who was fond of puzzles, and his witty and amusing book *A Budget of Paradoxes* still entertains readers today. Besides his accomplishments in the academic arena, De Morgan was an expert flutist, spoke five languages, and thoroughly enjoyed big-city life.

Knowing of his interest in probability, an actuary (someone who studies life expectancies and determines payments of premiums for insurance companies) once asked

5.2

EXERCISES

1. A survey of 200 people yielded the following information: 94 owned a VCR, 127 owned a microwave oven, and 78 owned both. How many people owned the following?
   a. a VCR or a microwave oven
   b. a VCR but not a microwave oven
   c. a microwave oven but not a VCR
   d. neither a VCR nor a microwave oven

2. A survey of 300 workers yielded the following information: 231 belonged to a union, and 195 were Democrats. If 172 of the union members were Democrats, how many workers were in the following situations?
   a. belonged to a union or were Democrats
   b. belonged to a union but were not Democrats

   c. were Democrats but did not belong to a union
   d. neither belonged to a union nor were Democrats

3. The records of 1492 high school graduates were examined, and the following information was obtained: 1072 took biology, and 679 took geometry. If 271 of those who took geometry did not take biology, how many graduates took the following?
   a. both classes
   b. at least one of the classes
   c. biology but not geometry
   d. neither class

4. A department store surveyed 428 shoppers, and the following information was obtained: 214 made a purchase,

De Morgan a question concerning the probability that a certain group of people would be alive at a certain time. In his response, De Morgan employed a formula containing the number π. In amazement, the actuary responded, "That must surely be a delusion! What can a circle have to do with the number of people alive at a certain time?" De Morgan replied that π has numerous applications and occurrences in many diverse areas of mathematics. Because it was first defined and used in geometry, people are conditioned to accept the mysterious number only in reference to a circle. However, in the history of mathematics, if probability had been systematically studied before geometry and circles, our present-day interpretation of the number π would be entirely different.

In addition to his accomplishments in logic and higher-level mathematics, De Morgan introduced a convention with which we are all familiar: In a paper written in 1845, he suggested the use of a slanted line to represent a fraction, such as 1/2 or 3/4.

De Morgan was a staunch defender of academic freedom and religious tolerance. While a student at Cambridge, his application for a fellowship was refused because he would not take and sign a theological oath. Later in life, he resigned his professorship as a protest against religious bias, because University College gave preferential treatment to members of the Church of England when textbooks were selected, and it did not have an open policy on religious philosophy. Augustus De Morgan

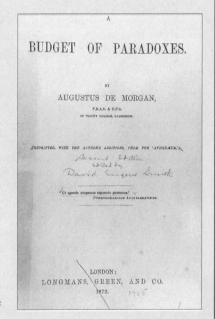

was a man unafraid to take a stand and make personal sacrifices when it came to principles he believed in.

---

and 299 were satisfied with the service they received. If 52 of those who made a purchase were not satisfied with the service, how many shoppers did the following?

a. made a purchase and were satisfied with the service

b. made a purchase or were satisfied with the service

c. were satisfied with the service but did not make a purchase

d. were not satisfied and did not make a purchase

5. In a survey, 674 adults were asked what television programs they had recently watched. The following information was obtained: 226 watched neither the Big Game nor the New Movie, and 289 watched the New Movie. If 183 of those who watched the New Movie did not watch the Big Game, how many of the surveyed adults watched the following?

a. both programs

b. at least one program

c. the Big Game

d. the Big Game but not the New Movie

6. A survey asked 816 college freshmen if they had been to a movie or eaten in a restaurant during the past week. The following information was obtained: 387 had been to neither a movie nor a restaurant, and 266 had been to a movie. If 92 of those who had been to a movie had not been to a restaurant, how many of the surveyed freshmen had been to the following?

a. both a movie and a restaurant

b. a movie or a restaurant

c. a restaurant

d. a restaurant but not a movie

7. A recent survey of w shoppers [that is, $n(U) = w$] yielded the following information: x shopped at Sears, y shopped at J.C. Penney's, and z shopped at both. How many people shopped at the following?

a. Sears or J.C. Penney's

b. only Sears

c. only J.C. Penney's

d. neither Sears nor J.C. Penney's

8. A recent transportation survey of *w* urban commuters [that is, $n(U) = w$] yielded the following information: *x* rode neither trains nor buses, *y* rode trains, and *z* rode only trains. How many people rode the following?

a. trains and buses

b. only buses

c. buses

d. trains or buses

9. A survey was conducted to examine patterns in ownership of microwave ovens, answering machines, and VCRs. The following data were obtained: 313 people had microwaves, 232 had answering machines, 269 had VCRs, 69 had all three, 64 had none, 98 had answering machines and VCRs, 57 had answering machines but no microwaves or VCRs, and 104 had microwaves and VCRs but no answering machines.

a. What percent of the people surveyed owned a microwave but no VCR or answering machine?

b. What percent of the people surveyed owned a VCR but no microwave or answering machine?

10. In a recent survey of monetary donations made by college graduates, the following information was obtained: 95 had donated to a political campaign, 76 had donated to assist medical research, 133 had donated to help preserve the environment, 25 had donated to all three, 22 had donated to none of the three, 38 had donated to a political campaign and to medical research, 46 had donated to medical research and to preserve the environment, and 54 had donated to a political campaign and to preserve the environment.

a. What percent of the college graduates donated to none of the three listed causes?

b. What percent of the college graduates donated to exactly one of the three listed causes?

11. Recently, U2, Sting, and Lollapalooza each toured the United States. A large group of college students was surveyed, and the following information was obtained: 1533 saw U2, 1127 saw Sting, 581 saw Lollapalooza, 219 saw all three, 1853 saw none, 766 saw only U2, 654 saw U2 and Sting, and 92 saw Sting and Lollapalooza but not U2.

a. What percent of the college students saw all three acts?

b. What percent of the college students saw at least two of the acts?

12. Dr. Hawk works in an allergy clinic, and his patients have the following allergies: 68 are allergic to dairy products, 93 are allergic to pollen, 91 are allergic to animal fur, 31 are allergic to all three, 29 are allergic only to pollen, 12 are allergic only to dairy products, 40 are allergic to dairy products and pollen.

a. What percent of Dr. Hawk's patients are allergic to animal fur?

b. What percent of Dr. Hawk's patients are allergic only to animal fur?

13. When the members of the Eye and I Photo Club discussed what type of film they had used during the past month, the following information was obtained: 77 used black and white, 24 used only black and white, 65 used color, 18 used only color, 101 used black and white or color, 27 used infrared, 9 used all three types, and 8 didn't use any film during the past month.

a. What percent of the members used only infrared film?

b. What percent of the members used at least two of the types of film?

14. After leaving the polls, many people are asked how they voted. (This is called an *exit poll*.) Concerning Propositions A, B, and C, the following information was obtained: 294 voted yes on A, 90 voted yes only on A, 346 voted yes on B, 166 voted yes only on B, 517 voted yes on A or B, 339 voted yes on C, no one voted yes on all three, and 72 voted no on all three.

a. What percent of the voters in the exit poll voted no on A?

b. What percent of the voters voted yes on more than one proposition?

15. In a recent survey, consumers were asked where they did their gift shopping. The following results were obtained: 621 shopped at Macy's, 513 shopped at Emporium, 367 shopped at Nordstrom, 723 shopped at Emporium or Nordstrom, 749 shopped at Macy's or Nordstrom, 776 shopped at Macy's or Emporium, 157 shopped at all three, 96 shopped at neither Macy's nor Emporium nor Nordstrom.

a. What percent of the consumers shopped at more than one store?

b. What percent of the consumers shopped exclusively at Nordstrom?

16. A company that specializes in language tutoring lists the following information concerning its English-speaking employees: 23 speak German, 25 speak French, 31 speak Spanish, 43 speak Spanish or French, 38 speak French or German, 46 speak German or Spanish, 8 speak Spanish, French, and German, and 7 office workers and secretaries speak English only.

a. What percent of the employees speak at least one language other than English?

b. What percent of the employees speak at least two languages other than English?

In Exercises 17 and 18, use a Venn diagram like the one in Figure 5.20.

17. A survey of 136 pet owners yielded the following information: 49 own fish; 55 own a bird; 50 own a cat; 68 own a dog; 2 own all four; 11 own only fish; 14 own only a bird; 10 own

FIGURE 5.20

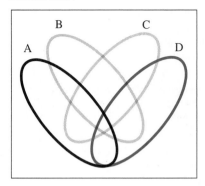

fish and a bird; 21 own fish and a cat; 26 own a bird and a dog; 27 own a cat and a dog; 3 own fish, a bird, a cat, and no dog; 1 owns fish, a bird, a dog, and no cat; 9 own fish, a cat, a dog, and no bird; and 10 own a bird, a cat, a dog, and no fish. How many of the surveyed pet owners have no fish, no birds, no cats, and no dogs? (They own other types of pets.)

18. An exit poll of 300 voters yielded the following information regarding voting patterns on Propositions A, B, C, and D: 119 voted yes on A; 163 voted yes on B; 129 voted yes on C; 142 voted yes on D; 37 voted yes on all four; 15 voted yes on A only; 50 voted yes on B only; 59 voted yes on A and B; 70 voted yes on A and C; 82 voted yes on B and D; 93 voted yes on C and D; 10 voted yes on A, B, and C and no on D; 2 voted yes on A, B, and D and no on C; 16 voted yes on A, C, and D and no on B; and 30 voted yes on B, C, and D and no on A. How many of the surveyed voters voted no on all four propositions?

In Exercises 19–22, given the sets U = {0, 1, 2, 3, 4, 5, 6, 7, 8, 9}, A = {0, 2, 4, 5, 9}, *and* B = {1, 2, 7, 8, 9}, *use De Morgan's laws to find the indicated sets.*

19. $(A' \cup B)'$ 20. $(A' \cap B)'$
21. $(A \cap B')'$ 22. $(A \cup B')'$

In Exercises 23–28, use a Venn diagram like the one in Figure 5.21 to shade in the region corresponding to the indicated set.

FIGURE 5.21

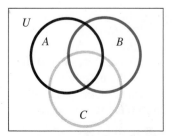

23. $A \cap B \cap C$ 24. $A \cup B \cup C$
25. $(A \cup B)' \cap C$ 26. $A \cap (B \cup C)'$
27. $B \cap (A \cup C')$ 28. $(A' \cup B) \cap C'$
29. Using Venn diagrams, prove De Morgan's law $(A \cap B)' = A' \cup B'$.
30. Using Venn diagrams, prove $A \cup (B \cap C) = (A \cup B) \cap (A \cup C)$.

 Answer the following questions using complete sentences.

31. What notation did De Morgan introduce in regard to fractions?
32. Why did De Morgan resign his professorship at University College?

Many graduate schools require applicants to take the Graduate Record Examination (G.R.E.). This exam is intended to measure verbal, quantitative, and analytical skills developed throughout a person's life. There are many classes and study guides available to help people prepare for the exam. The remaining questions are typical of those found in the study guides and on the exam itself.

Exercises 33–39 refer to the following: A nonprofit organization's board of directors, composed of four women (Angela, Betty, Carmine, and Delores) and three men (Ed, Frank, and Grant), holds frequent meetings. A meeting can be held at Betty's house, at Delores's house, or at Frank's house.

- Delores cannot attend any meetings at Betty's house.
- Carmine cannot attend any meetings on Tuesday or on Friday.
- Angela cannot attend any meetings at Delores's house.
- Ed can attend only those meetings that Grant also attends.
- Frank can attend only those meetings that both Angela and Carmine attend.

33. If all members of the board are to attend a particular meeting, under which of the following circumstances can it be held?
    a. Monday at Betty's
    b. Tuesday at Frank's
    c. Wednesday at Delores's
    d. Thursday at Frank's
    e. Friday at Betty's

34. Which of the following can be the group that attends a meeting on Wednesday at Betty's?
    a. Angela, Betty, Carmine, Ed, and Frank
    b. Angela, Betty, Ed, Frank, and Grant
    c. Angela, Betty, Carmine, Delores, and Ed
    d. Angela, Betty, Delores, Frank, and Grant
    e. Angela, Betty, Carmine, Frank, and Grant

35. If Carmine and Angela attend a meeting but Grant is unable to attend, which of the following could be true?

a. The meeting is held on Tuesday.
b. The meeting is held on Friday.
c. The meeting is held at Delores's.
d. The meeting is held at Frank's.
e. The meeting is attended by six of the board members.

36. If the meeting is held on Tuesday at Betty's, which of the following pairs can be among the board members who attend?
    a. Angela and Frank
    b. Ed and Betty
    c. Carmine and Ed
    d. Frank and Delores
    e. Carmine and Angela

37. If Frank attends a meeting on Thursday that is not held at his house, which of the following must be true?

a. The group can include, at most, two women.
b. The meeting is at Betty's house.
c. Ed is not at the meeting.
d. Grant is not at the meeting.
e. Delores is at the meeting.

38. If Grant is unable to attend a meeting on Tuesday at Delores's, what is the largest possible number of board members who can attend?
    a. 1   b. 2   c. 3   d. 4   e. 5

39. If a meeting is held on Friday, which of the following board members *cannot* attend?
    a. Grant   b. Delores   c. Ed   d. Betty   e. Frank

5.3

INTRODUCTION TO COMBINATORICS

If you went on a shopping spree and bought two pairs of jeans, three shirts, and two pairs of shoes, how many new outfits (consisting of a new pair of jeans, a new shirt, and a new pair of shoes) would you have? A compact disc buyers' club sends you a brochure saying that you can pick any five CDs from a group of 50 of today's hottest sounds for only $1.99. How many different combinations can you choose? Six local bands have volunteered to perform at a benefit concert, and there is some concern over the order in which the bands will perform. How many different lineups are possible? The answers to questions like these can be obtained by listing all the possibilities or by using three shortcut counting methods: the *fundamental principle of counting, combinations,* and *permutations.* Collectively, these methods are known as **combinatorics.** (Incidentally, the answers to the questions above are 12 outfits, 2,118,760 CD combinations, and 720 lineups.) In this section, we consider the first shortcut method.

Would you believe that there are more than 17,000,000 license plates that consist of three letters followed by three digits? Use the fundamental principle of counting to verify this.

The Fundamental Principle of Counting

Daily life requires that we make many decisions. For example, we must decide what food items to order from a menu, what items of clothing to put on in the morning, and what options to order when purchasing a new car. Often, we are asked to make a series of decisions: "Do you want soup or salad? What type of dressing? What type of vegetable?

What entrée? What beverage? What dessert?" These individual components of a complete meal lead to the question, "Given all the choices of soups, salads, dressings, vegetables, entrées, beverages, and desserts, what is the total number of possible dinner combinations?"

When making a series of decisions, how can you determine the total number of possible selections? One way is to list all the choices for each category and then match them up in all possible ways. To ensure that the choices are matched up in all possible ways, you can construct a **tree diagram.** A tree diagram consists of clusters of line segments, or **branches,** constructed as follows: A cluster of branches is drawn for each decision to be made such that the number of branches in each cluster equals the number of choices for the decision. For instance, if you must make two decisions and there are three choices for decision 1 and two choices for decision 2, the tree diagram would be similar to the one shown in Figure 5.22.

FIGURE 5.22

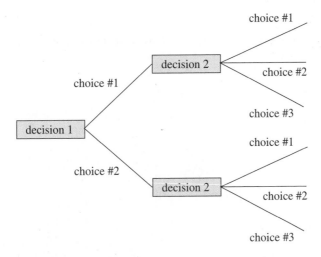

Although this method can be applied to all problems, it is very time-consuming and impractical when you are dealing with a series of many decisions, each of which contains numerous choices. Instead of actually listing all possibilities via a tree diagram, using a shortcut method might be desirable. The following example gives a clue to finding such a shortcut.

EXAMPLE 1 If you buy two pairs of jeans, three shirts, and two pairs of shoes, how many new outfits (consisting of a new pair of jeans, a new shirt, and a new pair of shoes) would you have?

Solution Because there are three categories, selecting an outfit requires a series of three decisions: you must select one pair of jeans, one shirt, and one pair of shoes. We will make our three decisions in the following order: jeans, shirt, and shoes. (The order in which the decisions are made does not affect the overall outfit.)

Our first decision (jeans) has two choices (jeans 1 or jeans 2); our tree starts with two branches, as in Figure 5.23.

FIGURE 5.23

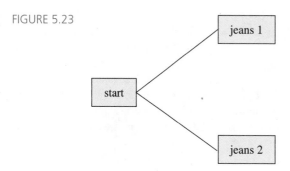

Our second decision is to select a shirt, for which there are three choices. At each pair of jeans on the tree, we draw a cluster of three branches, one for each shirt, as in Figure 5.24.

FIGURE 5.24

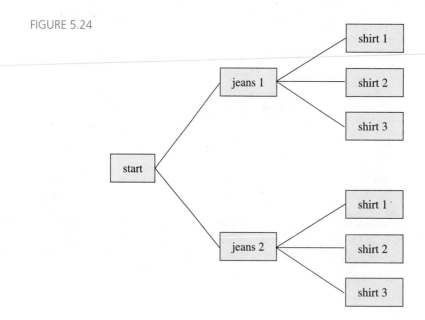

Our third decision is to select a pair of shoes, for which there are two choices. At each shirt on the tree, we draw a cluster of two branches, one for each pair of shoes, as in Figure 5.25.

We have now listed all possible ways of putting together a new outfit; twelve outfits can be formed from two pairs of jeans, three shirts, and two pairs of shoes.

FIGURE 5.25

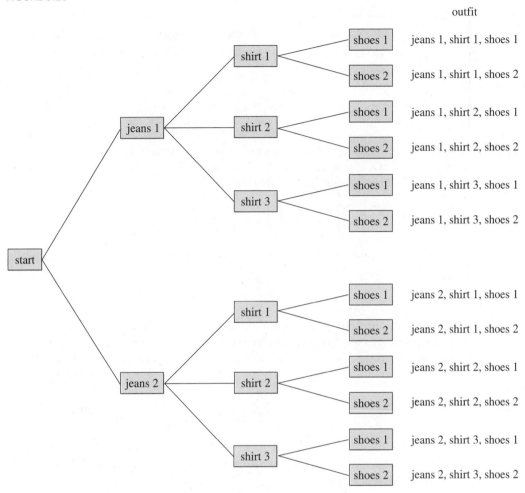

Referring to Example 1, note that each time a decision had to be made, the number of branches on the tree diagram was *multiplied* by the number of choices for the decision. Therefore, the total number of outfits could have been obtained by *multiplying* the number of choices for each decision:

$$2 \cdot 3 \cdot 2 = 12$$

jeans ———↑   ↑   ↑   �‾—outfits
shirts ————————┘   │
shoes ————————————┘

The generalization of this process of multiplication is called the **fundamental principle of counting.**

The Fundamental Principle of Counting

The total number of possible outcomes of a series of decisions (making selections from various categories) is found by multiplying the number of choices for each decision (or category) as follows:

1. Draw a box for each decision.
2. Enter the number of choices for each decision in the appropriate box and multiply.

EXAMPLE 2 A serial number consists of two consonants followed by three nonzero digits followed by a vowel (A, E, I, O, U): for example, "ST423E" and "DD666E." Determine how many serial numbers are possible given the following conditions.

   a. Letters and digits cannot be repeated in the same serial number.
   b. Letters and digits can be repeated in the same serial number.

Solution **a.** Because the serial number has six symbols, we must make six decisions. Consequently, we must draw six boxes:

There are 21 different choices for the first consonant. Because the letters cannot be repeated, there are only 20 choices for the second consonant. Similarly, there are 9 different choices for the first nonzero digit, 8 choices for the second, and 7 choices for the third. There are five different vowels, so the total number of possible serial numbers is

$$\boxed{21} \times \boxed{20} \times \boxed{9} \times \boxed{8} \times \boxed{7} \times \boxed{5} = 1{,}058{,}400$$

consonants   nonzero digits   vowel

There are 1,058,400 possible serial numbers when the letters and digits cannot be repeated within a serial number.

   b. Because letters and digits can be repeated, the number of choices does not decrease by one each time as in part (a). Therefore, the total number of possibilities is

$$\boxed{21} \times \boxed{21} \times \boxed{9} \times \boxed{9} \times \boxed{9} \times \boxed{5} = 1{,}607{,}445$$

consonants   nonzero digits   vowel

There are 1,067,445 possible serial numbers when the letters and digits cannot be repeated within a serial number. ●

Factorials

EXAMPLE 3 Three students rent a three-bedroom house near campus. One of the bedrooms is very desirable (it has its own bath), one has a balcony, and one is undesirable (it is very small). In how many ways can the housemates choose the bedrooms?

Solution   Three decisions must be made: who gets the room with the bath, who gets the room with the balcony, and who gets the small room. Using the fundamental principle of counting, we draw three boxes and enter the number of choices for each decision. There are three choices for who gets the room with the bath. Once that decision has been made, there are two choices for who gets the room with the balcony, and finally, there is only one choice for the small room.

$$\boxed{3} \times \boxed{2} \times \boxed{1} = 6$$

There are six different ways in which the three housemates can choose the three bedrooms. ●

Combinatorics often involve products of the type $3 \cdot 2 \cdot 1 = 6$, as seen in Example 3. This type of product is called a **factorial**, and the product $3 \cdot 2 \cdot 1$ is written as 3!. In this manner, $4! = 4 \cdot 3 \cdot 2 \cdot 1 \ (= 24)$, and $5! = 5 \cdot 4 \cdot 3 \cdot 2 \cdot 1 \ (= 120)$.

Factorials

If n is a positive integer, then *n factorial,* denoted by **n!**, is the product of all positive integers less than or equal to n.

$$n! = n \cdot (n - 1) \cdot (n - 2) \cdots 2 \cdot 1$$

As a special case, we define $0! = 1$.

Calculators and Factorials

Many calculators have a factorial button. Factorial buttons on scientific calculators are usually labeled "x!" or "n!". To calculate 6! on a scientific calculator, type "6" and then press the factorial button.

To calculate 6! on a Texas Instruments graphing calculator, use the "!" command to make the screen read "6!" and press $\boxed{\text{ENTER}}$. The "!" command is listed under the "MATH" menu.

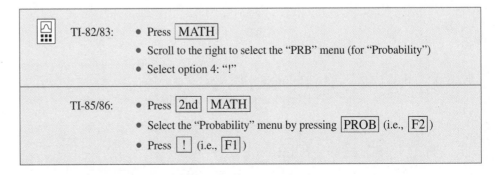

TI-82/83:   • Press $\boxed{\text{MATH}}$
            • Scroll to the right to select the "PRB" menu (for "Probability")
            • Select option 4: "!"

TI-85/86:   • Press $\boxed{\text{2nd}}$ $\boxed{\text{MATH}}$
            • Select the "Probability" menu by pressing $\boxed{\text{PROB}}$ (i.e., $\boxed{\text{F2}}$)
            • Press $\boxed{!}$ (i.e., $\boxed{\text{F1}}$)

In the future, we will refer to these keystrokes as $\boxed{\text{x!}}$.

EXAMPLE 4   Find the following values.

   a. 6!   **b.** $\dfrac{8!}{5!}$ **c.** $\dfrac{8!}{3! \cdot 5!}$

Solution   **a.** $6! = 6 \cdot 5 \cdot 4 \cdot 3 \cdot 2 \cdot 1$
                     $= 720$

Therefore, $6! = 720$.

   b. $\dfrac{8!}{5!} = \dfrac{8 \cdot 7 \cdot 6 \cdot 5 \cdot 4 \cdot 3 \cdot 2 \cdot 1}{5 \cdot 4 \cdot 3 \cdot 2 \cdot 1}$

$= \dfrac{8 \cdot 7 \cdot 6 \cdot \cancel{5} \cdot \cancel{4} \cdot \cancel{3} \cdot \cancel{2} \cdot \cancel{1}}{\cancel{5} \cdot \cancel{4} \cdot \cancel{3} \cdot \cancel{2} \cdot \cancel{1}}$

$= 8 \cdot 7 \cdot 6$

$= 336$

Therefore, $\frac{8!}{5!} = 336$.
Using a calculator, we obtain the same result.

8 x! ÷ 5 x! =

(With a graphing calculator press ENTER *instead of* = *.)*

   c. $\dfrac{8!}{3! \cdot 5!} = \dfrac{8 \cdot 7 \cdot 6 \cdot 5 \cdot 4 \cdot 3 \cdot 2 \cdot 1}{(3 \cdot 2 \cdot 1)(5 \cdot 4 \cdot 3 \cdot 2 \cdot 1)}$

$= \dfrac{8 \cdot 7 \cdot 6 \cdot \cancel{5} \cdot \cancel{4} \cdot \cancel{3} \cdot \cancel{2} \cdot \cancel{1}}{(3 \cdot 2 \cdot 1)(\cancel{5} \cdot \cancel{4} \cdot \cancel{3} \cdot \cancel{2} \cdot \cancel{1})}$

$= \dfrac{8 \cdot 7 \cdot 6}{3 \cdot 2 \cdot 1}$

$= 56$

Therefore, $\frac{8!}{3! \cdot 5!} = 56$.
Using a calculator, we obtain the same result.

8 x! ÷ (3 x! × 5 x!) =

(With a graphing calculator press ENTER *instead of* = *.)*

5.3

EXERCISES

1. A nickel, a dime, and a quarter are tossed.
   a. Use the fundamental principle of counting to determine how many different outcomes are possible.
   b. Construct a tree diagram to list all possible outcomes.

2. A die is rolled and a coin is tossed.
   a. Use the fundamental principle of counting to determine how many different outcomes are possible.
   b. Construct a tree diagram to list all possible outcomes.

3. Jamie has decided to buy either a Mega or a Better Byte personal computer. She also wants to purchase either Big Word, Word World, or Great Word word-processing software, and either Big Number or Number World spreadsheet software.
   a. Use the fundamental principle of counting to determine how many different packages of a computer and software Jamie has to choose from.
   b. Construct a tree diagram to list all possible packages of a computer and software.

4. Sammy's Sandwich Shop offers a soup, sandwich, and beverage combination at a special price. There are three sandwiches (turkey, tuna, and tofu), two soups (minestrone and split pea), and three beverages (coffee, milk and mineral water) to choose from.
   a. Use the fundamental principle of counting to determine how many different meal combinations are possible.
   b. Construct a tree diagram to list all possible soup, sandwich, and beverage combinations.

5. If you buy three pairs of jeans, four sweaters, and two pairs of boots, how many new outfits (consisting of a new pair of jeans, a new sweater, and a new pair of boots) will you have?

6. A certain model of automobile is available in six exterior colors, three interior colors, and three interior styles. In addition, the transmission can be either manual or automatic, and the engine can have either four or six cylinders. How many different versions of the automobile can be ordered?

7. In order to fulfill certain requirements for a degree, a student must take one course each from the following groups: health, civics, critical thinking, and elective. If there are four health, three civics, six critical thinking, and ten elective courses, how many different options for fulfilling the requirements does a student have?

8. To fulfill a requirement for a literature class, a student must read one short story by each of the following authors: Stephen King, Clive Barker, Edgar Allan Poe, and H. P. Lovecraft. If there are twelve King, six Barker, eight Poe, and eight Lovecraft stories to choose from, how many different combinations of reading assignments can a student choose from to fulfill the reading requirement?

9. A sporting goods store has fourteen lines of snow skis, seven types of bindings, nine types of boots, and three types of poles. Assuming that all items are compatible with each other, how many different complete ski equipment packages are available?

10. An audio equipment store has ten different amplifiers, four tuners, six turntables, eight tape decks, six compact disc players, and thirteen speakers. Assuming that all components are compatible with each other, how many different complete stereo systems are available?

11. A cafeteria offers a complete dinner that includes one serving each of appetizer, soup, entrée, and dessert for $6.99. If the menu has three appetizers, four soups, six entrées, and three desserts, how many different meals are possible?

12. A sandwich shop offers a "U-Chooz" special consisting of your choice of bread, meat, cheese, and special sauce (one each). If there are six different breads, eight meats, five cheeses, and four special sauces, how many different sandwiches are possible?

13. How many different social security numbers are possible? (A social security number consists of nine digits that can be repeated.)

14. In order to use an automated teller machine (ATM), a customer must enter his or her four-digit Personal Identification Number (PIN). How many different PINs are possible?

15. Every book published has an international standard book number (ISBN). The number is a code used to identify the specific book and is of the form X-XXX-XXXXX-X, where X is one of digits 0, 1, 2, . . . , 9. How many different ISBNs are possible?

16. How many different zip codes are possible using: (a) the old style (five digits) and (b) the new style (nine digits)? Why do you think the U.S. Postal Service introduced the new system?

17. Telephone area codes are three-digit numbers of the form XXX, where the first and third digits are neither 0 nor 1 and the second digit is either a 0 or 1. How many three-digit numbers of this type are possible?

18. The serial number on a dollar bill consists of a letter followed by eight digits and then a letter. How many different serial numbers are possible given the following conditions?
  a. Letters and digits cannot be repeated.
  b. Letters and digits can be repeated.
  c. The letters are nonrepeated consonants and the digits can be repeated.

19. Each student at State University has a student I.D. number consisting of four digits (the first digit is nonzero, and digits may be repeated) followed by three of the letters A, B, C, D, and E (letters may not be repeated). How many different student numbers are possible?

20. Each student at State College has a student I.D. number consisting of five digits (the first digit is nonzero, and digits may be repeated) followed by two of the letters A, B, C, D, and E (letters may not be repeated). How many different student numbers are possible?

In Exercises 21–36, find the indicated value.

21. $4!$ **22.** $5!$ **23.** $10!$ **24.** $8!$

25. $20!$ **26.** $25!$ **27.** $6! \cdot 4!$ **28.** $8! \cdot 6!$

29. a. $\dfrac{6!}{4!}$ **b.** $\dfrac{6!}{2!}$

30. a. $\dfrac{8!}{6!}$ **b.** $\dfrac{8!}{2!}$

31. $\dfrac{8!}{5! \cdot 3!}$ **32.** $\dfrac{9!}{5! \cdot 4!}$

33. $\dfrac{8!}{4! \cdot 4!}$ **34.** $\dfrac{6!}{3! \cdot 3!}$

35. $\dfrac{82!}{80! \cdot 2!}$ **36.** $\dfrac{77!}{74! \cdot 3!}$

37. Find the value of $\dfrac{n!}{(n-r)!}$ when $n = 16$ and $r = 14$.

38. Find the value of $\dfrac{n!}{(n-r)!}$ when $n = 19$ and $r = 16$.

39. Find the value of $\dfrac{n!}{(n-r)!}$ when $n = 5$ and $r = 5$.

40. Find the value of $\dfrac{n!}{(n-r)!}$ when $n = r$.

41. Find the value of $\dfrac{n!}{(n-r)!r!}$ when $n = 7$ and $r = 3$.

42. Find the value of $\dfrac{n!}{(n-r)!r!}$ when $n = 7$ and $r = 4$.

43. Find the value of $\dfrac{n!}{(n-r)!r!}$ when $n = 5$ and $r = 5$.

44. Find the value of $\dfrac{n!}{(n-r)!r!}$ when $n = r$.

Answer the following questions using complete sentences.

45. What is the fundamental principle of counting? When is it used?

46. What is a factorial? Who invented the symbol $n!$?

Many graduate schools require applicants to take the Graduate Record Examination (G.R.E.). This exam is intended to measure verbal, quantitative, and analytical skills developed throughout a person's life. There are many classes and study guides available to help people prepare for the exam. The remaining questions are typical of those found in the study guides and on the exam itself.

Exercises 47–51 refer to the following: In an executive parking lot, there are six parking spaces in a row, labeled 1 through 6. Exactly five cars of five different colors—black, gray, green, white, and yellow—are to be parked in the spaces. The cars can park in any of the spaces as long as the following conditions are met:

- The green car must be parked in space 3.
- The black car must be parked in a space next to the space in which the yellow car is parked.
- The gray car cannot be parked in a space next to the space in which the white car is parked.

47. If the yellow car is parked in space 1, how many acceptable parking arrangements are there for the five cars?
  a. 1   **b.** 2   **c.** 3   **d.** 4   **e.** 5.

48. Which of the following must be true of any acceptable parking arrangement?
  a. One of the cars is parked in space 2.
  b. One of the cars is parked in space 6.
  c. There is an empty space next to the space in which the gray car is parked.
  d. There is an empty space next to the space in which the yellow car is parked.
  e. Either the black car or the yellow car is parked in a space next to space 3.

49. If the gray car is parked in space 2, none of the cars can be parked in which space?
  a. 1   **b.** 3   **c.** 4   **d.** 5   **e.** 6

50. The white car could be parked in any of the spaces except which of the following?
  a. 1   **b.** 2   **c.** 4   **d.** 5   **e.** 6

51. If the yellow car is parked in space 2, which of the following must be true?
  a. None of the cars is parked in space 5.
  b. The gray car is parked in space 6.
  c. The black car is parked in a space next to the space in which the white car is parked.
  d. The white car is parked in a space next to the space in which the green car is parked.
  e. The gray car is parked in a space next to the space in which the black car is parked.

5.4

PERMUTATIONS AND COMBINATIONS

The fundamental principle of counting allows us to determine the total number of possible outcomes when a series of decisions (making selections from various categories) must be made. In Section 5.3, the examples and exercises involved selecting *one item each* from various categories; if you buy two pairs of jeans, three shirts, and two pairs of shoes, you will have twelve $(2 \cdot 3 \cdot 2 = 12)$ new outfits (consisting of a new pair of jeans, a new shirt, and a new pair of shoes). In this section, we examine the situation when *more than one* item is selected from a category. If more than one item is selected, the selections can be made either *with* or *without* replacement.

With versus Without Replacement

Selecting items *with replacement* means that the same item *can* be selected more than once; after a specific item has been chosen, it is put back into the pool of future choices. Selecting items *without replacement* means that the same item *cannot* be selected more than once; after a specific item has been chosen, it is not replaced.

Suppose you must select a four-digit personal identification number (PIN) for a bank account. In this case, the digits are selected with replacement; each time a specific digit is selected, the digit is put back into the pool of choices for the next selection. (Your PIN can be 3666; the same digit can be selected more than once.) When items are selected with replacement, we use the fundamental principle of counting to determine the total number of possible outcomes; there are $10 \cdot 10 \cdot 10 \cdot 10 = 10,000$ possible four-digit PINs.

In many situations, items cannot be selected more than once. For instance, when selecting a committee of three people from a group of 20, you cannot select the same person more than once. Once you have selected a specific person (say, Lauren), you do not put her back into the pool of choices. When selecting items without replacement, depending on whether the order of selection is important, *permutations* or *combinations* are used to determine the total number of possible outcomes.

Permutations

When more than one item is selected (without replacement) from a single category, and the order of selection *is* important, the various possible outcomes are called **permutations.** For example, when the rankings (first, second, and third place) in a talent contest are announced, the order of selection is important; Monte in first, Liat in second, and Ginny in third place is different from Ginny in first, Monte in second, and Liat in third. "Monte, Liat, Ginny" and "Ginny, Monte, Liat" are different permutations of the contestants. Naturally, these selections are made without replacement; we cannot select Monte for first place and reselect him for second place.

EXAMPLE 1 Six local bands have volunteered to perform at a benefit concert, but there is only enough time for four bands to play. There is also some concern over the order in which the chosen bands will perform. How many different lineups are possible?

Solution   We must select four of the six bands and put them in a specific order. The bands are selected without replacement; a band cannot be selected to play and then be reselected to play again. Because we must make four decisions, we draw four boxes and put the number of choices for each decision in each appropriate box. There are six choices for the opening band. Naturally, the opening band could not be the followup act, so there are only five choices for the next group. Similarly, there are four candidates for the third group and three choices for the closing band. The total number of different lineups possible is found by multiplying the number of choices for each decision:

$$\boxed{6} \times \boxed{5} \times \boxed{4} \times \boxed{3} = 360$$

Opening band   Closing band

With four out of six bands playing in the performance, 360 lineups are possible. Because the order of selecting the bands is important, the various possible outcomes, or lineups, are called permutations; there are 360 permutations of six items when the items are selected four at a time. •

The computation in Example 1 is similar to a factorial, but the factors do not go all the way down to 1; the product $6 \cdot 5 \cdot 4 \cdot 3$ is a "truncated" (cut-off) factorial. We can change this truncated factorial into a complete factorial in the following manner:

$$6 \cdot 5 \cdot 4 \cdot 3 = \frac{6 \cdot 5 \cdot 4 \cdot 3 \cdot (2 \cdot 1)}{(2 \cdot 1)} \qquad \text{Multiplying by } \tfrac{2}{2} \text{ and } \tfrac{1}{1}$$

$$= \frac{6!}{2!}$$

Notice that this last expression can be written as $\frac{6!}{2!} = \frac{6!}{(6-4)!}$. (Recall that we were selecting four out of six bands.) This result is generalized in the following box.

Permutation Formula

The number of *permutations*, or arrangements, of r items selected without replacement from a pool of n items ($r \leq n$), denoted by $_nP_r$, is

$$_nP_r = \frac{n!}{(n-r)!}$$

Permutations are used whenever more than one item is selected (without replacement) from a category and the order of selection is important.

Using the notation from the box and referring to Example 1, note that 360 possible lineups of four bands selected from a pool of six can be denoted by $_6P_4 = \frac{6!}{(6-4)!} = 360$. Other notations can be used to represent the number of permutations of a group of items. In particular, the notations $_nP_r$, $P(n, r)$, P^n_r, and $P_{n,r}$ all represent the number of possible permutations (or arrangements) of r items selected (without replacement) from a pool of n items.

EXAMPLE 2   Three door prizes (first, second, and third) are to be awarded at a ten-year high school reunion. Each of the 112 attendees puts his or her name in a hat. The first name drawn wins a two-night stay at the Chat 'n' Rest Motel, the second name wins dinner for two at Juju's Kitsch-Inn, and the third wins a pair of engraved mugs. How many different ways can the prizes be awarded?

Solution   We must select 3 out of 112 people (without replacement), and the order in which they are selected *is* important. (Winning dinner is different from winning the mugs.) Hence, we must find the number of permutations of 3 items selected from a pool of 112:

$$_{112}P_3 = \frac{112!}{(112 - 3)!}$$

$$= \frac{112!}{109!}$$

$$= \frac{112 \cdot 111 \cdot 110 \cdot 109 \cdot 108 \cdots \cdot 2 \cdot 1}{109 \cdot 108 \cdots \cdot 2 \cdot 1}$$

$$= 112 \cdot 111 \cdot 110$$

$$= 1,367,520$$

There are 1,367,520 different ways in which the three prizes can be awarded to the 112 people.   ●

In Example 2, if you try to use a calculator to find $\frac{112!}{109!}$ directly, you may not obtain an answer. Entering 112 and pressing $\boxed{x!}$ may result in a calculator error. (Try it.) Because factorials get very large quickly, some calculators are not able to find any factorial over 69!. ($69! = 1.711224524 \times 10^{98}$.)

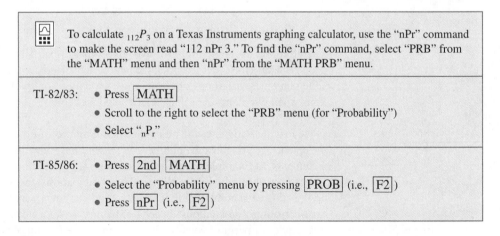

To calculate $_{112}P_3$ on a Texas Instruments graphing calculator, use the "nPr" command to make the screen read "112 nPr 3." To find the "nPr" command, select "PRB" from the "MATH" menu and then "nPr" from the "MATH PRB" menu.

TI-82/83: • Press $\boxed{\text{MATH}}$
   • Scroll to the right to select the "PRB" menu (for "Probability")
   • Select "$_nP_r$"

TI-85/86: • Press $\boxed{\text{2nd}}$ $\boxed{\text{MATH}}$
   • Select the "Probability" menu by pressing $\boxed{\text{PROB}}$ (i.e., $\boxed{\text{F2}}$)
   • Press $\boxed{\text{nPr}}$ (i.e., $\boxed{\text{F2}}$)

EXAMPLE 3   A bowling league has ten teams. How many different ways can the teams rank in the standings at the end of a tournament? (Ties are not allowed.)

Solution   Because order is important, we find the number of permutations of ten items selected from a pool of ten items:

$$_{10}P_{10} = \frac{10!}{(10 - 10)!}$$

$$= \frac{10!}{0!} \qquad \text{Recall that } 0! = 1$$

$$= \frac{10!}{1}$$

$$= 3,628,800$$

In a league containing ten teams, there are 3,628,800 different standings possible at the end of a tournament. •

Combinations

When items are selected from a group, the order of selection may or may not be important. If the order is important (as in Examples 1, 2, and 3), permutations are used to determine the total number of selections possible. What if the order of selection is *not* important? When more than one item is selected (without replacement) from a single category and the order of selection is not important, the various possible outcomes are called **combinations.**

EXAMPLE 4   Two adults are needed to chaperone a day care center's field trip. Marcus, Vivian, Frank, and Keiko are the four managers of the center. How many different groups of chaperones are possible?

Solution   In selecting the chaperones, the order of selection is *not* important; "Marcus and Vivian" is the same as "Vivian and Marcus." Hence, the permutation formula cannot be used. Because we do not yet have a shortcut for finding the total number of possibilities when the order of selection is not important, we must list all the possibilities:

| Marcus and Vivian | Marcus and Frank | Marcus and Keiko |
| Vivian and Frank | Vivian and Keiko | Frank and Keiko |

Therefore, six different groups of two chaperones are possible from the group of four managers. Because the order in which the people are selected is not important, the various possible outcomes, or groups of chaperones, are called *combinations;* there are six combinations when two items are selected from a pool of four. •

Just as $_nP_r$ denotes the number of *permutations* of r elements selected from a pool of n elements, $_nC_r$ denotes the number of *combinations* or r elements selected from a pool of n elements. In Example 4, we found that there are six combinations of two people selected from a pool of four by listing all six of the combinations; that is, $_4C_2 = 6$. If we had a larger pool, listing each combination in order to find out how many there are would be extremely time-consuming and tedious! Instead of listing, we take a different approach. We first find the number of permutations (with the permutation formula) and then alter that number to account for the distinction between permutations and combinations.

To find the number of combinations of two people selected from a pool of four, we first find the number of permutations:

$$_4P_2 = \frac{4!}{(4 - 2)!} = \frac{4!}{2!} = 12$$

This figure of 12 must be altered to account for the distinction between permutations and combinations.

In Example 4, we listed combinations; one such combination was "Marcus and Vivian." If we had listed permutations, we would have had to list both "Marcus and Vivian" and "Vivian and Marcus," because the *order* of selection matters with permutations. In fact, each combination of two chaperones listed in Example 4 generates two permutations; each pair of chaperones can be given in two different orders. Thus, there are twice as many permutations of two people selected from a pool of four as there are combinations. Alternatively, there are half as many combinations of two people selected from a pool of four as there are permutations. We used the permutation formula to find that $_4P_2 = 12$; thus,

$$_4C_2 = \frac{1}{2} \cdot {_4P_2} = \frac{1}{2}(12) = 6$$

This answer certainly fits with Example 4; we listed exactly six combinations.

What if three of the four managers were needed to chaperone the day care center's field trip? Rather than finding the number of combinations by listing each possibility, we first find the number of permutations and then alter that number to account for the distinction between permutations and combinations.

The number of permutations of three people selected from a pool of four is

$$_4P_3 = \frac{4!}{(4-3)!} = \frac{4!}{1!} = 24$$

We know that some of these permutations represent the same combination. For example, the combination "Marcus and Vivian and Keiko" generates $3! = 6$ different permutations (using initials, they are: MVK, MKV, KMV, KVM, VMK, VKM). Because each combination of three people generates six different permutations, there are one-sixth as many combinations as permutations. Thus,

$$_4C_3 = \frac{1}{6} \cdot {_4P_3} = \frac{1}{6}(24) = 4$$

This means that if three of the four managers were needed to chaperone the day care center's field trip, there would be $_4C_3 = 4$ possible combinations.

We just saw that when two items are selected from a pool of n items, each combination of two generates $2! = 2$ permutations, so

$$_nC_2 = \frac{1}{2!} \cdot {_nP_2}$$

We also saw that when three items are selected from a pool of n items, each combination of three generates $3! = 6$ permutations, so

$$_nC_3 = \frac{1}{3!} \cdot nP_3$$

More generally, when r items are selected from a pool of n items, each combination of r items generates $r!$ permutations, so

$$_nC_r = \frac{1}{r!} \cdot {_nP_r}$$

$$= \frac{1}{r!} \cdot \frac{n!}{(n-r)!} \qquad \text{Using the permutation formula}$$

$$= \frac{n!}{r! \cdot (n-r)!} \qquad \text{Multiplying the fractions together}$$

Combination Formula

The number of distinct *combinations* of r items selected without replacement from a pool of n items ($r \leq n$), denoted by $_nC_r$, is

$$_nC_r = \frac{n!}{(n-r)!r!}$$

Combinations are used whenever one or more items are selected (without replacement) from a category and the order of selection is not important.

EXAMPLE 5 A compact disc buyer's club sends you a brochure that offers any five CDs chosen from a group of 50 of today's hottest releases. How many different selections can you make?

Solution Because the order of selection is *not* important, we find the number of combinations when 5 items are selected from a pool of 50:

$$_{50}C_5 = \frac{50!}{(50-5)!5!}$$

$$= \frac{50!}{45!\,5!}$$

$$= \frac{50 \cdot 49 \cdot 48 \cdot 47 \cdot 46}{5 \cdot 4 \cdot 3 \cdot 2 \cdot 1}$$

$$= 2,118,760$$

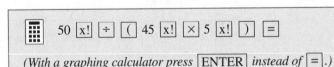

(With a graphing calculator press ENTER *instead of* = *.)*

In choosing 5 out of 50 compact discs, 2,118,760 different combinations of selections are possible.

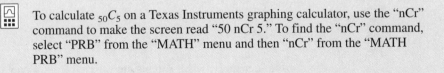

To calculate $_{50}C_5$ on a Texas Instruments graphing calculator, use the "nCr" command to make the screen read "50 nCr 5." To find the "nCr" command, select "PRB" from the "MATH" menu and then "nCr" from the "MATH PRB" menu.

EXAMPLE 6   A group consisting of twelve women and nine men must select a five-person committee. How many different committees are possible if it must consist of the following?

     a. three women and two men

     b. any mixture of men and women

Solution   a. Our problem involves two categories: women and men. The fundamental principle of counting tells us to draw two boxes (one for each category), enter the number of choices for each, and multiply.

$$\boxed{\begin{array}{c}\text{the number of ways} \\ \text{we can select three} \\ \text{out of twelve women}\end{array}} \times \boxed{\begin{array}{c}\text{the number of ways} \\ \text{we can select two} \\ \text{out of nine men}\end{array}} = \;?$$

Because the order of selecting the members of a committee is not important, we will use combinations:

$$(_{12}C_3) \cdot (_9C_2) = \frac{12!}{(12-3)! \cdot 3!} \cdot \frac{9!}{(9-2)! \cdot 2!}$$

$$= \frac{12!}{9! \cdot 3!} \cdot \frac{9!}{7! \cdot 2!}$$

$$= \frac{12 \cdot 11 \cdot 10}{3 \cdot 2 \cdot 1} \cdot \frac{9 \cdot 8}{2 \cdot 1}$$

$$= 220 \cdot 36$$

$$= 7920$$

There are 7920 different committees consisting of three women and two men.

     b. Because the gender of the committee members doesn't matter, our problem involves only one category: people. We must choose 5 out of the 21 people, and the order of selection is not important:

$$_{21}C_5 = \frac{21!}{(21-5)! \cdot 5!}$$

$$= \frac{21!}{16! \cdot 5!}$$

$$= \frac{21 \cdot 20 \cdot 19 \cdot 18 \cdot 17}{5 \cdot 4 \cdot 3 \cdot 2 \cdot 1}$$

$$= 20{,}349$$

There are 20,349 different committees consisting of five people.   ●

EXAMPLE 7   Find the value of $_5C_r$ for the following values of r.

     a. $r = 0$ b. $r = 1$ c. $r = 2$ d. $r = 3$ e. $r = 4$ f. $r = 5$

Solution **a.** $_5C_0 = \dfrac{5!}{(5-0)! \cdot 0!} = \dfrac{5!}{5! \cdot 0!} = 1$

b. $_5C_1 = \dfrac{5!}{(5-1)! \cdot 1!} = \dfrac{5!}{4! \cdot 1!} = 5$

c. $_5C_2 = \dfrac{5!}{(5-2)! \cdot 2!} = \dfrac{5!}{3! \cdot 2!} = 10$

d. $_5C_3 = \dfrac{5!}{(5-3)! \cdot 3!} = \dfrac{5!}{2! \cdot 3!} = 10$

e. $_5C_4 = \dfrac{5!}{(5-4)! \cdot 4!} = \dfrac{5!}{1! \cdot 4!} = 5$

f. $_5C_5 = \dfrac{5!}{(5-5)! \cdot 5!} = \dfrac{5!}{0! \cdot 5!} = 1$

The combinations generated in Example 7 exhibit a curious pattern. Notice that the values of $_5C_r$ are symmetric: $_5C_0 = {_5C_5}$, $_5C_1 = {_5C_4}$, and $_5C_2 = {_5C_3}$. Now examine the diagram in Figure 5.26. Each number in this "triangle" of numbers is the sum of two numbers in the row immediately above it. For example, $2 = 1 + 1$ and $10 = 4 + 6$, as shown by the inserted arrows. It is no coincidence that the values of $_5C_r$ found in Example 7 also appear as a row of numbers in this "magic" triangle. In fact, by labeling the first row of the triangle the "0th" (or initial) row, the 5th row contains all the values of $_5C_r$ for $r = 0, 1, 2, 3, 4,$ and 5. In general, the nth row of the triangle contains all the values of $_nC_r$ for $r = 0, 1, 2, \ldots, n$.

FIGURE 5.26

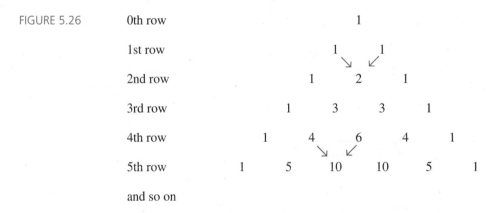

0th row

1st row

2nd row

3rd row

4th row

5th row

and so on

Historically, this triangular pattern of numbers is referred to as *Pascal's triangle,* in honor of the French mathematician, scientist, and philosopher Blaise Pascal (1623–1662). Pascal is a cofounder of probability theory (see the introduction to Chapter 6). Although the triangle has Pascal's name attached to it, this "magic" arrangement of numbers was known to other cultures hundreds of years before Pascal's time.

The most important part of any problem involving combinatorics is deciding which counting technique (or techniques) to use. The flowchart in Figure 5.27 (page 257) and the list of general steps on page 258 can help you decide which method or methods to use in a specific problem.

FIGURE 5.27   **"Which Counting Technique?" Flowchart**

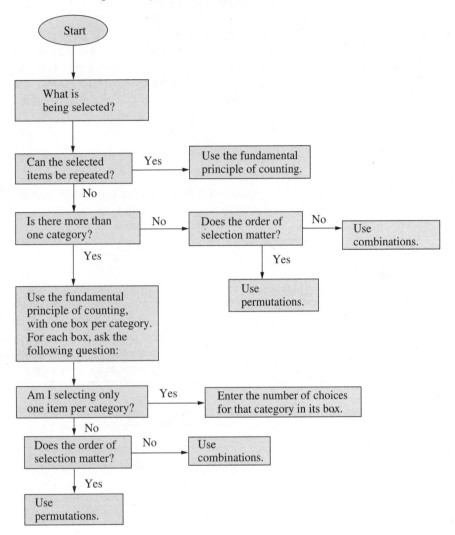

Which Counting Technique?

1. What is being selected?
2. If the selected items can be repeated, use the *fundamental principle of counting* and multiply the number of choices for each category.
3. If there is only one category, use:
   - *Combinations* if the order of selection does not matter—that is, r items can be selected from a pool of n items in

   $$_nC_r = \frac{n!}{(n-r)! \cdot r!}$$

   ways.
   - *Permutations* if the order of selection does matter—that is, r items can be selected from a pool of n items in

   $$_nP_r = \frac{n!}{(n-r)!}$$

   ways.
4. If there is more than one category, use the *fundamental principle of counting* with one box per category.
   a. If you are selecting one item per category, the number in the box for that category is the number of choices for that category.
   b. If you are selecting more than one item per category, the number in the box for that category is found by using step 3 above.

EXAMPLE 8   A standard deck of playing cards contains 52 cards.

  a. How many different five-card hands containing four kings are possible?
  b. How many different five-card hands containing four of a kind are possible?

Solution   a. We use the flowchart in Figure 5.27 and answer the given questions.
  Q: What is being selected?
  A: Playing cards.
  Q: Can the selected items be repeated?
  A: No.
  Q: Is there more than one category?
  A: Yes: Because we must have five cards, we need four kings and one non-king. Therefore, we need two boxes:

$$\boxed{\text{kings}} \times \boxed{\text{non-kings}}$$

  Q: Am I selecting only one item per category?
  A: *Kings:* no. Does the order of selection matter? No: Use combinations. Because there are $n = 4$ kings in the deck and we want to select $r = 4$, we must compute $_4C_4$. *Non-kings:* yes. Enter the number of choices for that category: There are 48 non-kings.

Historical Note

Chu Shih-chieh

CIRCA 1280–1303

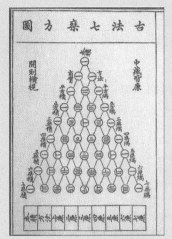

The "Pascal" triangle as depicted in 1303 at the front of Chu Shih-chieh's *Ssu-yüan yü-chien*. It is entitled "The Old Method Chart of the Seven Multiplying Squares" and tabulates the binomial coefficients up to the eighth power.

Chu Shih-chieh was the last and most acclaimed mathematician of the Sung Dynasty in China. Little is known of his personal life; the actual dates of his birth and death are unknown. His work appears to have flourished during the close of the 13th century. It is believed that Chu Shih-chieh spent many years as a wandering scholar, earning a living by teaching mathematics to those who wanted to learn.

Two of Chu Shih-chieh's works have survived the centuries. The first, *Suan-hsüeh ch'i-meng* (*Introduction to Mathematical Studies*), was written in 1299 and contains elementary mathematics. This work was very influential in Japan and Korea, although it was lost in China until the 19th century. Written in 1303, Chu's second work, *Ssu-yüan yü-chien* (*Precious Mirror of the Four Elements*), contains more advanced mathematics. The topics of *Precious Mirror* include the solving of simultaneous equations and the solving of equations up to the 14th degree.

Of the many diagrams in *Precious Mirror,* one has special interest: the arithmetic triangle. Chu Shih-chieh's triangle contains the first eight rows of what is known in the West as Pascal's triangle. However, Chu does not claim credit for the triangle; he refers to it as "a diagram of the *old* method for finding eighth and lower powers." "Pascal's" triangle was known to the Chinese well over 300 years before Pascal was born!

$$\boxed{\text{kings}} \times \boxed{\text{non-kings}} = \boxed{{}_4C_4} \times \boxed{48}$$

$$= \frac{4!}{(4-4)! \cdot 4!} \cdot 48$$

$$= \frac{4!}{0! \cdot 4!} \cdot 48$$

$$= 1 \cdot 48$$

$$= 48$$

There are 48 different five-card hands containing four kings.

b. *Four of a kind* means four cards of the same denomination (four kings, four queens, four sixes, etc.). Generalizing part (a) above, we conclude that the number of ways of getting four queens or four sixes is the same as the number of ways of getting four kings; there are 48 different five-card hands containing four queens, 48 hands containing four sixes, and so on. Because there are 13 denominations (two through ace), there are $13 \cdot 48 = 624$ possible five-card hands containing four of a kind.

5.4

EXERCISES

In Exercises 1–12, find the indicated value:

1. a. $_7P_3$ **b.** $_7C_3$

2. a. $_8P_4$ **b.** $_8C_4$

3. a. $_5P_5$ **b.** $_5C_5$

4. a. $_9P_0$ **b.** $_9C_0$

5. a. $_{14}P_1$ **b.** $_{14}C_1$

6. a. $_{13}C_3$ **b.** $_{13}C_{10}$

7. a. $_{100}P_3$ **b.** $_{100}C_3$

8. a. $_{80}P_4$ **b.** $_{80}C_4$

9. a. $_xP_{x-1}$ **b.** $_xC_{x-1}$

10. a. $_xP_1$ **b.** $_xC_1$

11. a. $_xP_2$ **b.** $_xC_2$

12. a. $_xP_{x-2}$ **b.** $_xC_{x-2}$

13. a. Find $_3P_2$.
  b. List all of the permutations of $\{a, b, c\}$ when the elements are taken two at a time.

14. a. Find $_3C_2$.
  b. List all of the combinations of $\{a, b, c\}$ when the elements are taken two at a time.

15. a. Find $_4C_2$.
  b. List all of the combinations of $\{a, b, c, d\}$ when the elements are taken two at a time.

16. a. Find $_4P_2$.
  b. List all of the permutations of $\{a, b, c, d\}$ when the elements are taken two at a time.

17. An art class consists of 12 students. All of them must present their portfolios and explain their work to the instructor and their classmates at the end of the semester.
  a. If their names are drawn from a hat to determine who goes first, second, and so on, how many presentation orders are possible?
  b. If their names are put in alphabetical order to determine who goes first, second, and so on, how many presentation orders are possible?

18. An English class consists of 24 students, and 3 are to be chosen to give speeches in a school competition. How many different ways can the teacher choose the team given the following conditions?
  a. The order of the speakers is important.
  b. The order of the speakers is not important.

19. In how many ways can the letters in the word "STOP" be arranged? (See the photograph below.)

Exercise 19: Right Letters, Wrong Order.
A woman in a passing vehicle got a chuckle out of the freshly painted 'SOTP' sign at a street corner in Anniston, Ala. A city paint crew quickly corrected the misspelling.

20. A committee of four is to be selected from a group of 15 people. How many different committees are possible given the following conditions?

    a. There is no distinction between the responsibilities of the members.

    b. One person is the chair, and the rest are general members.

    c. One person is the chair, one person is the secretary, one person is responsible for refreshments, and one person cleans up after meetings.

21. A softball league has 14 teams. If every team must play every other team once in the first round of league play, how many games must be scheduled?

22. In a group of 19 people, each person shakes hands once with each other person in the group. How many handshakes will occur?

23. A softball league has 14 teams. How many different end-of-the-season rankings of first, second, and third place are possible (disregarding ties)?

24. Two hundred people buy raffle tickets. Three winning tickets will be drawn at random.

    a. If first prize is $100, second prize is $50, and third prize is $20, in how many different ways can the prizes be awarded?

    b. If each prize is $50, in how many different ways can the prizes be awarded?

25. A group of eight women and six men must select a four-person committee. How many committees are possible if it must consist of the following?

    a. two women and two men

    b. any mixture of men and women

    c. a majority of women

26. A group of ten seniors, eight juniors, five sophomores, and five freshmen must select a committee of four. How many committees are possible if the committee must contain members as in the following list?

    a. one person from each class

    b. any mixture of the classes

    c. exactly two seniors

Exercises 27–31 refer to a deck of 52 playing cards (jokers not allowed). If you are unfamiliar with playing cards, see pages 269–270 for the description of a standard deck.

27. How many five-card poker hands are possible?

28. a. How many five-card poker hands consisting of all hearts are possible?

    b. How many five-card poker hands consisting of all cards of the same suit are possible?

29. a. How many five-card poker hands containing exactly three aces are possible?

b. How many five-card poker hands containing three of a kind are possible?

30. a. How many five-card poker hands consisting of three kings and two queens are possible?

    b. How many five-card poker hands consisting of three of a kind and a pair (a *full house*) are possible?

31. How many five-card poker hands containing exactly one pair are possible?

32. A 6/49 lottery requires choosing six of the numbers 1 through 49. How many different lottery tickets can you choose? (Order is not important, and the numbers do not repeat.)

33. A 6/53 lottery requires choosing six of the numbers 1 through 53. How many different lottery tickets can you choose? (Order is not important, and the numbers do not repeat.)

34. A 7/39 lottery requires choosing seven of the numbers 1 through 39. How many different lottery tickets can you choose? (Order is not important, and the numbers do not repeat.)

35. A 5/36 lottery requires choosing five of the numbers 1 through 36. How many different lottery tickets can you choose? (Order is not important, and the numbers do not repeat.)

36. Which lottery would be easier to win, a 6/49 or a 7/39? Why? HINT: See Exercises 32 and 34.

37. Which lottery would be easier to win, a 6/53 or a 5/36? Why? HINT: See Exercises 33 and 35.

38. For any given values of n and r, which is larger, $_nP_r$ or $_nC_r$? Why?

39. Suppose you want to know how many ways r items can be selected from a group of n items. What determines whether you should calculate $_nP_r$ or $_nC_r$?

40. a. Add adjacent entries of the fifth row of Pascal's triangle to obtain the sixth row.

    b. Find $_6C_r$ for $r = 0, 1, 2, 3, 4, 5$, and 6.

41. Use Pascal's triangle to answer the following.

    a. In which row would you find the value of $_4C_2$?

    b. In which row would you find the value of $_nC_r$?

    c. Is $_4C_2$ the second number in the fourth row?

    d. Is $_4C_2$ the third number in the fourth row?

    e. What is the location of $_nC_r$? Why?

42. Given the set $S = \{a, b, c, d\}$,

    a. how many one-element subsets does S have?

    b. how many two-element subsets does S have?

    c. how many three-element subsets does S have?

    d. how many four-element subsets does S have?

    e. how many zero-element subsets does S have?

    f. how many subsets does S have?

    g. If $n(S) = k$, how many subsets will S have?

Many graduate schools require applicants to take the Graduate Record Examination (G.R.E.). This exam is intended to measure verbal, quantitative, and analytical skills developed throughout a person's life. There are many classes and study guides available to help people prepare for the exam. The remaining questions are typical of those found in the study guides and on the exam itself.

Exercises 43–47 refer to the following: A baseball league has six teams—A, B, C, D, E, and F. All games are played at 7:30 P.M. on Fridays. Each team must play each other team exactly once, and the following conditions must be met:

- Team A plays team D first and team F second.
- Team B plays team E first and team C third.
- Team C plays team F first.

43. What is the total number of games that each team must play during the season?
  a. 3   **b.** 4   **c.** 5   **d.** 6   **e.** 7

44. On the first Friday, which of the following pairs of teams play each other?

  a. A and B; C and F; D and E
  b. A and B; C and E; D and F
  c. A and C; B and E; D and F
  d. A and D; B and C; E and F
  e. A and D; B and E; C and F

45. Which of the following teams must team B play second?
  a. A   **b.** C   **c.** D   **d.** E   **e.** F

46. The last set of games could be between which teams?
  a. A and B; C and F; D and E
  b. A and C; B and F; D and E
  c. A and D; B and C; E and F
  d. A and E; B and C; D and F
  e. A and F; B and E; C and D

47. If team D wins five games, which of the following must be true?
  a. Team A loses five games.
  b. Team A wins four games.
  c. Team A wins its first game.
  d. Team B wins five games.
  e. Team B loses at least one game.

CHAPTER 5 REVIEW

Terms

branch	element or member	intersection	roster or listing	tree diagram
cardinal number	empty set	mutually exclusive	notation	union
combination	equal sets	(or disjoint) sets	set	universal set
combinatorics	factorial	permutation	set-builder notation	Venn diagram
complement	improper subset	proper subset	subset	well-defined set

Review Exercises

1. What role did the following people play in the development of set theory?

- John Venn
- Augustus De Morgan

2. Given the sets
$U = \{0, 1, 2, 3, 4, 5, 6, 7, 8, 9\}$
$A = \{0, 2, 4, 6, 8\}$
$B = \{1, 3, 5, 7, 9\}$

find the following using the roster method.
  a. A' **b.** B' **c.** $A \cup B$ **d.** $A \cap B$

3. Given the sets $A = \{$Maria, Nobuko, Leroy, Mickey, Kelly$\}$ and $B = \{$Rachel, Leroy, Deanna, Mickey$\}$, find:
  a. $A \cup B$ **b.** $A \cap B$

4. List all subsets of $C = \{$Dallas, Chicago, Tampa$\}$. Identify which subsets are proper and which are improper.

5. Given $n(U) = 61$, $n(A) = 32$, $n(B) = 26$, and $n(A \cup B) = 40$,
  a. find $n(A \cap B)$.
  b. draw a Venn diagram illustrating the composition of U.

6. A survey of 2,000 college seniors yielded the following information: 1324 favored capital punishment, 937 favored stricter gun control, and 591 favored both.

a. How many favored capital punishment or stricter gun control?

b. How many favored capital punishment but not stricter gun control?

c. How many favored stricter gun control but not capital punishment?

d. How many favored neither capital punishment nor stricter gun control?

7. An exit poll yielded the following information concerning people's voting patterns on Propositions A, B, and C: 305 voted yes on A, 95 voted yes only on A, 393 voted yes on B, 192 voted yes only on B, 510 voted yes on A or B, 163 voted yes on C, 87 voted yes on all three, and 213 voted no on all three. What percent of the voters voted yes on more than one proposition?

8. Given the sets $U = \{a, b, c, d, e, f, g, h, i\}$, $A = \{b, d, f, g\}$, and $B = \{a, c, d, g, i\}$, use De Morgan's laws to find the following.

a. $(A' \cup B)'$ **b.** $(A \cap B')'$

9. Sid and Nancy are planning their anniversary celebration, which will include viewing an art exhibit, having dinner, and going dancing. They will go to either the Museum of Modern Art or the New Photo Gallery, dine either at Stars, at Johnny's, or at the Chelsea, and go dancing either at Le Club or at Lizards.

a. How many different ways can Sid and Nancy celebrate their anniversary?

b. Construct a tree diagram to list all possible ways in which Sid and Nancy can celebrate their anniversary.

10. A certain model of pickup truck is available in five exterior colors, three interior colors, and three interior styles. In addition, the transmission can be either manual or automatic, and the truck can have either two-wheel or four-wheel drive. How many different versions of the pickup truck can be ordered?

11. Each student at State University has a student I.D. number consisting of five digits (the first digit is nonzero, and digits can be repeated) followed by two of the letters *A, B, C,* and *D* (letters cannot be repeated). How many different student numbers are possible?

12. Find the value of each of the following.

a. $(17 - 7)!$ **b.** $(17 - 17)!$ **c.** $\dfrac{82!}{79!}$ **d.** $\dfrac{27!}{20!7!}$

13. In how many ways can you select three out of eleven items under the following conditions?

a. Order of selection is not important.

b. Order of selection is important.

14. Find the value of each of the following.

a. $_{15}P_4$ **b.** $_{15}C_4$ **c.** $_{15}P_{11}$

15. A group of ten women and twelve men must select a three-person committee. How many committees are possible if it must consist of the following?

a. one woman and two men

b. any mixture of men and women

c. a majority of men

16. A volleyball league has ten teams. If every team must play every other team once in the first round of league play, how many games must be scheduled?

17. A volleyball league has ten teams. How many different end-of-the-season rankings of first, second, and third place are possible (disregarding ties)?

18. Using a standard deck of 52 cards (no jokers), how many seven-card poker hands are possible?

19. Using a standard deck of 52 cards and two jokers, how many seven-card poker hands are possible?

20. A 6/42 lottery requires choosing six of the numbers 1 through 42. How many different lottery tickets can you choose?

21. What is the major difference between permutations and combinations?

22. Use Pascal's triangle to answer the following.

a. In which entry in which row would you find the value of $_7C_3$?

b. In which entry in which row would you find the value of $_7C_4$?

c. How is the value of $_7C_3$ related to the value of $_7C_4$?

d. What is the location of $_nC_r$? Why?

23. Given the set $S = \{a, b, c\}$,

a. how many one-element subsets does S have?

b. how many two-element subsets does S have?

c. how many three-element subsets does S have?

d. how many zero-element subsets does S have?

e. how many subsets does S have?

f. how is the answer to part (e) related to $n(S)$?

PROBABILITY 6

Probability theory began with a roll of the dice. Antoine Gombauld, the Chevalier de Méré, a French nobleman and successful gambler, made money over many years betting that he could roll at least one 6 in four rolls of a single die. He lost money, however, betting that he could roll at least one pair of 6's in 24 rolls of a pair of dice. He asked his friend, the French mathematician Blaise Pascal, to explain why. In 1645 a correspondence commenced between Pascal and another mathematician, Pierre de Fermat; in the ensuing communications the two answered Gombauld's query and laid the foundation of probability theory.

The letters between Pascal and Fermat, however, were not the first work in the field of probability. In 1545 the Italian mathematician Gerolamo Cardano wrote the first theoretical study of probabilities and gambling, *The Book on Games of Chance,* which was not widely recognized at the time. The work was published a century later, after the dialogue between Pascal and Fermat became well known.

Because of its initial association with gambling, probability theory was not viewed as a serious branch of mathematics. The development of an important business application helped change this negative view. In 1662, the Englishman John Graunt published his *Natural and Political Observations on the Bills of Mortality,* which used probabilities to analyze death records. English firms were selling the first life insurance policies, and they used mortality tables to make their fees appropriate to the risks involved.

The work of Gregor Mendel, an Austrian monk, was especially important in the history of probability theory. In 1865 he published the results of his experiments on pea plants. These results allowed for randomness in genetics and used probabilities to analyze the effect of that randomness. Although it was ignored during his lifetime, Mendel's work is considered the foundation of genetics.

Probability theory now plays an important role in business and genetics. Pollsters use it to measure the likelihood of a candidate winning an election. Market analysts use it to predict the success of a new product. Quality control analysts use it to measure the probability of a newly manufactured item being defective. Actuaries use it to set insurance rates that are commensurate with the risks involved. And sports writers use it to analyze the abilities of teams and their players.

This *Bills of Mortality* (records of death) was published shortly after John Graunt's analysis

6.1

INTRODUCTION TO PROBABILITY

Basic Terms of Probability

Much of the terminology and many of the computations of probability theory have their basis in set theory, because set theory contains the mathematical way of describing collections of objects and the size of those collections.

Basic Probability Terms

experiment: a process by which an observation, or **outcome,** is obtained

sample space: the set S of all possible outcomes of an experiment

event: any subset E of the sample space S

If a single die is rolled, the *experiment* is the rolling of the die. (*Die* is the singular of *dice.*) The possible *outcomes* are 1, 2, 3, 4, 5, and 6. The *sample space* (set of all possible outcomes) is $S = \{1, 2, 3, 4, 5, 6\}$. (The term *sample space* really means the same thing as *universal set;* the only distinction between the two ideas is that *sample space* is used only in probability theory, whereas *universal set* is used in any situation in which sets are used.) There are several possible *events* (subsets of the sample space), including the following:

$E_1 = \{3\}$	"a three comes up"
$E_2 = \{2, 4, 6\}$	"an even number comes up"
$E_3 = \{1, 2, 3, 4, 5, 6\}$	"a number between 1 and 6 inclusive comes up"

An early certain event

(THE FAR SIDE Copyright 1990
Universal Press Syndicate)

Early shell games

Notice that an event is not the same as an outcome. An event is a subset of the sample space; an outcome is an element of the sample space. "Rolling an odd number" is an event, not an outcome. It is the set $\{1, 3, 5\}$ that is composed of three separate outcomes. Some events are distinguished from outcomes only in that set brackets are used with events and not with outcomes. For example, $\{5\}$ is an event, and 5 is an outcome; either refers to "rolling a five."

The event E_3 ("a number between 1 and 6 inclusive comes up") is called a **certain event,** since $E_3 = S$. That is, E_3 is a sure thing. "Getting 17" is an **impossible event.** No outcome in the sample space $S = \{1, 2, 3, 4, 5, 6\}$ would result in 17, so this event is the null set.

Finding Probabilities and Odds

The probability of an event is a measure of the likelihood that the event will occur. If a single die is rolled, the outcomes are equally likely; a 3 is just as likely to come up as any other number. There are six possible outcomes, so a 3 should come up about one out of every six rolls. That is, the probability of event E_1 ("a 3 comes up") is 1/6. The 1 in the numerator is the number of elements in $E_1 = \{3\}$. The 6 in the denominator is the number of elements in $S = \{1, 2, 3, 4, 5, 6\}$.

If an experiment's outcomes are equally likely, then the probability of an event E is the number of outcomes in the event divided by the number of outcomes in the sample space, or $n(E)/n(S)$. Probability can be thought of as "success over total."

> **Probability of an Event with Equally Likely Outcomes**
>
> The **probability of an event E,** denoted by $p(E)$, is
>
> $$p(E) = \frac{n(E)}{n(S)}$$
>
> if the experiment's outcomes are equally likely.
> (*Think: success over total.*)

Many people use the words *probability* and *odds* interchangeably. However, the words have different meanings. The **odds** in favor of an event are the number of ways the event can occur compared to the number of ways the event can *fail* to occur, or "success compared to *failure*" (if the experiment's outcomes are equally likely). The odds of event E_1 ("a 3 comes up") are 1 to 5 (or 1:5), since a 3 can come up in one way and can fail to come up in five ways. Similarly, the odds of event E_3 ("a number between 1 and 6 inclusive comes up") are 6 to 0 (or 6:0), since a number between 1 and 6 can come up in six ways and can fail to come up in zero ways.

> **Odds of an Event with Equally Likely Outcomes**
>
> The *odds* of an event E with equally likely outcomes, denoted by $o(E)$, are given by
>
> $$o(E) = n(E){:}n(E')$$
>
> (*Think: success compared with failure*)

In addition to having the meaning given above, the word *odds* can also refer to "house odds," which has to do with how much you will be paid if you win a bet at a casino. The odds of an event are sometimes called the **true odds,** to distinguish them from the house odds.

EXAMPLE 1   A die is rolled. Find the following:

   a. the probability of rolling a 5
   b. the odds of rolling a 5
   c. the probability of rolling a number below 5
   d. the odds of rolling a number below 5

Solution   **a.** The sample space is $S = \{1, 2, 3, 4, 5, 6\}$. $E_1 = \{5\}$ ("rolling a 5"). The probability of E_1 is

$$p(E_1) = \frac{n(E_1)}{n(S)} = \frac{1}{6}$$

This means that one out of every six possible outcomes is a success (that is, a 5).

b. $E_1' = \{1, 2, 3, 4, 6\}$. The odds of E_1 are

$$o(E_1) = n(E_1){:}n(E_1') = 1{:}5$$

This means that there is one possible success for every five possible failures.

c. $E_2 = \{1, 2, 3, 4\}$ ("rolling a number below 5"). The probability of E_2 is

$$p(E_2) = \frac{n(E_2)}{n(S)} = \frac{4}{6} = \frac{2}{3}$$

This means that two out of every three possible outcomes are a success.

d. $E_2' = \{5, 6\}$. The odds of E_2 are

$$o(E_2) = n(E_2{:}n(E_2') = 4{:}2 = 2{:}1$$

This means that there are two possible successes for every one possible failure. Notice that odds are reduced in the same manner that a fraction is reduced. •

EXAMPLE 2 A coin is flipped. Find the following:

a. the sample space
b. the probability of event E_1, "getting heads"
c. the odds of event E_1, "getting heads"
d. the probability of event E_2, "getting heads or tails"
e. the odds of event E_2, "getting heads or tails"

Solution a. *Finding the sample space S:* The experiment is flipping a coin. The only possible outcomes are heads and tails. These outcomes are equally likely. The sample space S is the set of all possible outcomes, so $S = \{h, t\}$.

b. *Finding the probability of heads:*

$$E_1 = \{h\} \text{ ("getting heads")}$$

$$p(E_1) = \frac{n(E_1)}{n(S)} = \frac{1}{2}$$

This means that one out of every two possible outcomes is a success.

c. *Finding the odds of heads:*

$$E_1' = \{t\}$$

$$o(E_1) = n(E_1){:}n(E_1') = 1{:}1$$

This means that for every one possible success there is one possible failure.

d. *Finding the probability of heads or tails:*

$$E_2 = \{h, t\}$$

$$p(E_2) = \frac{n(E_2)}{n(S)} = \frac{2}{2} = \frac{1}{1}$$

This means that every outcome is a success. Notice that E_2 is a certain event.

e. *Finding the odds of heads or tails:*

$$E_2' = \emptyset$$

$$o(E_2) = n(E_2){:}n(E_2') = 2{:}0 = 1{:}0$$

This means that there are no possible failures. •

Relative Frequency versus Probability

So far, we have discussed probabilities only in a theoretical way. When we found that the probability of heads was $\frac{1}{2}$, we never actually tossed a coin. It doesn't always make sense to calculate probabilities theoretically; sometimes they must be found empirically, the way a batting average is calculated. For example, in 8389 times at bat, Babe Ruth had 2875 hits. His batting average was $\frac{2875}{8389} \approx 0.343$. In other words, his probability of getting a hit was 0.343.

Sometimes a probability can be found either theoretically or empirically. We've already found that the theoretical probability of heads is $\frac{1}{2}$. We could also flip a coin a number of times and calculate (number of heads)/(total number of flips); this can be called the **relative frequency** of heads, to distinguish it from the theoretical probability of heads.

Usually, the relative frequency of an outcome is not equal to its probability, but if the number of trials is large, the two tend to be close. If you tossed a coin a couple of times, anything could happen, and the fact that the probability of heads is $\frac{1}{2}$ would have no impact on the results. However, if you tossed a coin 100 times, you would probably find that the relative frequency of heads was close to $\frac{1}{2}$. If your friend tossed a coin 1000 times, she would probably find the relative frequency of heads to be even closer to $\frac{1}{2}$ than in your experiment. This relationship between probabilities and relative frequencies is called the **law of large numbers.**

Law of Large Numbers

If an experiment is repeated a large number of times, the relative frequency of an outcome will tend to be close to the probability of that outcome.

Probabilities represent the tendency of reality, or the best guess of what will happen. At a casino, probabilities are much more useful to the casino (the "house") than to an individual gambler, because the house performs the experiment for a much larger number of trials (in other words, plays the game more often). In fact, the house plays the game so many times that the relative frequencies will be almost exactly the same as the probabilities, with the result that the house isn't gambling at all—it knows what is going to happen! Similarly, a gambler with a "system" has to play the game for a long time for the system to be of use.

EXAMPLE 3 If a pair of coins is flipped, find the probability of getting exactly one heads.

Solution

The experiment is the flipping of a pair of coins. One possible outcome is that one coin is heads and the other is tails. A second and *different* outcome is that one coin is tails and the other is heads. These two outcomes seem the same. However, if one coin was painted, it would be easy to tell them apart. Outcomes of the experiment can be described by using ordered pairs in which the first component refers to the first coin and the second component refers to the second coin. The two different ways of getting one coin heads and one coin tails are (h, t) and (t, h).

The sample space is $S = \{(h, h), (h, t), (t, h), (t, t)\}$. The outcomes are equally likely. The event of getting exactly one heads is $E = \{(h, t), (t, h)\}$. Therefore,

$$p(E) = \frac{n(E)}{n(S)} = \frac{2}{4} = \frac{1}{2}$$

This means that if we were to toss a pair of coins a number of times, we should expect to get heads on one coin and tails on the other about half of the time. Realize that this is only a prediction, and we may in fact never get heads on one coin and tails on the other. ●

A deck of cards. Hearts are in the front, followed by clubs, diamonds, and spades.

Playing Cards   A modern deck of cards contains 52 cards, thirteen in each of four suits: hearts, diamonds, clubs, and spades (♥, ♦, ♣, ♠). Hearts and diamonds are red, and clubs and spades are black. Each suit consists of cards labeled 2 through 10, followed by jack, queen, king, and ace. **Face cards** are the jack, queen, and king, and **picture cards** are the jack, queen, king, and ace.

6.1

EXERCISES

In Exercises 1–14, use this information: A jar on your desk contains twelve black, eight red, ten yellow, and five green jellybeans. You pick a jellybean without looking.

1. What is the experiment?

2. What is the sample space?

In Exercises 3–14, find and interpret the following:

3. the probability that it is black

4. the probability that it is green

5. the probability that it is red or yellow

6. the probability that it is red or black

7. the probability that it is not yellow

8. the probability that it is not red

9. the probability that it is white

10. the probability that it is not white

11. the odds in favor of picking a black jellybean

12. the odds in favor of picking a green jellybean

13. the odds in favor of picking a red or yellow jellybean

14. the odds in favor of picking a red or black jellybean

15. **a.** How many hearts are there in a deck of cards?
    b. What fraction of a deck is hearts?

16. **a.** How many red cards are there in a deck of cards?
    b. What fraction of a deck is red?

17. **a.** How many face cards are there in a deck of cards?
    b. What fraction of a deck is face cards?

18. **a.** How many black cards are there in a deck of cards?
    b. What fraction of a deck is black?

19. **a.** How many kings are there in a deck of cards?
    b. What fraction of a deck is kings?

20. **a.** How many queens are there in a deck of cards?
    b. What fraction of a deck is queens?

In Exercises 21–34, one card is drawn from a well-shuffled deck of 52 cards (no jokers).

21. What is the experiment?

22. What is the sample space?

In Exercises 23–34, find and interpret (a) the probability and (b) the odds of drawing the following cards.

23. a black card

24. a heart

25. a queen

26. a two of clubs

27. a queen of spades

28. a club

29. a card below a 5 (count an ace as high)

30. a card below a 9 (count an ace as high)

31. a card above a 4 (count an ace as high)

32. a card above an 8 (count an ace as high)

33. a face card

34. a picture card

In Exercises 35–40, E is an event with equally likely outcomes.

35. If $p(E) = 1/5$, find $o(E)$.

36. If $p(E) = 8/9$, find $o(E)$.

37. If $p(E) = a/b$, find $o(E)$.

38. If $o(E) = 3:2$, find $p(E)$.

39. If $o(E) = 4:7$, find $p(E)$.

40. If $o(E) = a:b$, find $p(E)$.

In Exercises 41 and 42, find and interpret (a) the probability and (b) the odds of drawing the following cards. In finding the odds, use the formula developed in Exercise 40.

41. a red card

42. a jack

43. A family has two children. Using b to stand for boy and g for girl in ordered pairs, and assuming that boys and girls are equally likely, give:
 a. the sample space
 b. the event E that the family has exactly one daughter
 c. the event F that the family has at least one daughter
 d. the event G that the family has two daughters
 e. $p(E)$ **f.** $p(F)$ **g.** $p(G)$
 h. $o(E)$ **i.** $o(F)$ **j.** $o(G)$

44. Two coins are tossed. Using ordered pairs, give:
 a. the sample space
 b. the event E that exactly one is heads
 c. the event F that at least one is heads
 d. the event G that two are heads

 e. $p(E)$ **f.** $p(F)$ **g.** $p(G)$
 h. $o(E)$ **i.** $o(F)$ **j.** $o(G)$

45. A family has three children. Using b to stand for boy and g for girl, and assuming that boys and girls are equally likely, and using ordered triples such as (b, b, g), give:
 a. the sample space
 b. the event E that the family has exactly two daughters
 c. the event F that the family has at least two daughters
 d. the event G that the family has three daughters
 e. $p(E)$ **f.** $p(F)$ **g.** $p(G)$
 h. $o(E)$ **i.** $o(F)$ **j.** $o(G)$

46. Three coins are tossed. Using ordered triples, give:
 a. the sample space
 b. the event E that exactly two are heads
 c. the event F that at least two are heads
 d. the event G that all three are heads
 e. $p(E)$ **f.** $p(F)$ **g.** $p(G)$
 h. $o(E)$ **i.** $o(F)$ **j.** $o(G)$

47. A couple plans on having two children. Assume that boys and girls are equally likely.
 a. Find the probability of having two girls.
 b. Find the probability of having one girl and one boy.
 c. Find the probability of having two boys.
 d. Which is more likely: having two children of the same sex or two of different sexes? Why?

48. Two coins are tossed.
 a. Find the probability that both are heads.
 b. Find the probability that one is heads and one is tails.
 c. Find the probability that both are tails.
 d. Which is more likely: that the two coins match or that they don't match? Why?

49. A couple plans to have three children. Which is more likely: having three children of the same sex or of different sexes? Why? (Assume that boys and girls are equally likely.)

50. Three coins are tossed. Which is more likely: that the three coins match or that they don't match? Why?

In Exercises 51 and 52, use the following information: Lily Orteaga owns a video rental store called LilyO's Videos. Lily knows both the age and the address of all of her renters because she requires that renters have a copy of their driver's license on file. During the last three months, Lily accumulated data on the age of every renter and the part of town they live in. Those data are given in Figure 6.1 on page 272.

51. a. A renter is most likely to live in what region?
 b. What is the probability of part (a)?

52. a. A renter is most likely to be in what age bracket?
 b. What is the probability of part (a)?

FIGURE 6.1

Age	Parkview District	Central City	South of Town
Under 20	1287	1582	2038
20–40	3758	3677	3114
Over 40	2895	2156	2066

53. Roll a single die four times and record the number of times a 6 comes up. Repeat this ten times. If you had made the Chevalier de Méré's favorite bet (at $10 per game), would you have won or lost money? How much?

54. Roll a pair of dice 24 times and record the number of times a pair of 6's comes up. Repeat this five times. If you had made the Chevalier de Méré's bet (at $10 per game), would you have won or lost money? How much?

55. **a.** If you were to flip a pair of coins 30 times, approximately how many times do you think a pair of heads would come up? A pair of tails? One heads and one tails?

    b. Flip a pair of coins 30 times and record the number of times a pair of heads comes up, the number of times a pair of tails comes up, and the number of times one heads and one tails comes up. Do the results agree with your guess?

56. Flip a coin 20 times and find the relative frequency of heads. Compare this to the theoretical probability of heads.

57. Combine your classmates' results from Exercise 56 and find the relative frequency of heads. Compare this to the relative frequency found in Exercise 56, and compare it to the theoretical probability of heads.

 Answer the following questions using complete sentences.

58. Who started probability theory? How?

59. Why was probability theory not considered a serious branch of mathematics?

60. What did Gregor Mendel do with probabilities?

61. Who was Antoine Gombauld, and what was his role in probability theory?

62. Who was Gerolamo Cardano, and what was his role in probability theory?

63. Consider a "weighted die"—one that has a small weight in its interior. Such a weight would cause the face closest to the weight to come up less frequently and cause the face farthest from the weight to come up more frequently. Would the probabilities computed in Example 1 still be correct? Why or why not? Would the definition $p(E) = n(E)/n(S)$ still be appropriate? Why or why not?

64. Some dice have spots that are small, painted indentations; other dice have spots that are indentations filled with a different colored material. Which of these two types of dice is not fair? Why? What would be the most likely outcome of rolling this type of die? Why?
    HINT: 1 and 6 are on opposite faces, as are 2 and 5, and 3 and 4.

65. Explain how you would find the theoretical probability of rolling an even number on a single die. Explain how you would find the relative frequency of rolling an even number on a single die.

66. Give five examples of events whose probabilities must be found empirically rather than theoretically.

67. Does the theoretical probability of an event remain unchanged from experiment to experiment? Why? Does the relative frequency of an event remain unchanged from experiment to experiment? Why?

68. Write a paper in which you compare and contrast theoretical probability and relative frequency.

69. Write a paper in which you compare and contrast probability and odds.

6.2

PROBABILITY DISTRIBUTIONS

A spinner for a board game is designed so that the arrow can point to either 1, 2, 3, or 4, as shown in Figure 6.2. The sample space is $S = \{1, 2, 3, 4\}$. If event E_1 is "spinning a 4," it is *not* correct to reason that if $p(E_1) = n(E_1)/n(S) = 1/4$. It is intuitively clear that, since the 4 takes up half of the circle, $p(\text{spinning a 4}) = 1/2$, not 1/4. The formula $p(E) = n(E)/n(S)$ is appropriate only if the experiment's outcomes are equally likely; the outcomes of spinning the spinner in Figure 6.2 are not equally likely.

FIGURE 6.2

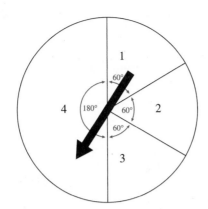

When outcomes are not equally likely, we can make a chart that lists each outcome and its probability. Such a chart is called a **probability distribution.** The probability of an event E is the sum of the probabilities of the outcomes that make up event E.

EXAMPLE 1 **a.** Create a probability distribution for spinning the spinner in Figure 6.2.
        b. Find the probability of spinning an even number.

Solution **a.** Since region 1 takes up 60° of the circle's 360°, the probability of spinning a 1 is 60/360 = 1/6. Similar calculations yield the following probability distribution.

Outcome	Probability
Spin a 1	1/6
Spin a 2	1/6
Spin a 3	1/6
Spin a 4	1/2

  b. The probability of spinning an even number is the sum of the probabilities of the appropriate outcomes:

$$p(\text{even number}) = p(\text{spin a } 2) + p(\text{spin a } 4) = 1/6 + 1/2 = 2/3$$ ●

Probability of an Event

If an experiment's outcomes are *equally likely,* then the **probability** of an event E is

$$p(E) = \frac{n(E)}{n(S)}$$

If an experiment's outcomes are *not equally likely,* then the **probability** of an event E is the sum of the probabilities of the outcomes that make up event E.

Historical Note

Gregor Johann Mendel

1822–1884

Johann Mendel was born to an Austrian peasant family. His interest in botany began on the family farm, where he helped his father graft fruit trees. He studied philosophy, physics, and mathematics at the University Philosophical Institute in Olmütz. However, he was unsuccessful in finding a job, so he quit school and returned to the farm. Depressed by the prospects of a bleak future, he became ill and stayed at home for a year.

Mendel later returned to Olmütz. After two years of study, he found the pressures of school and work to be too much, and his health again broke down. On the advice of his father and a professor, he entered the priesthood, even though he did not feel called to serve the church. His name was changed from Johann to Gregor.

Relieved of his financial difficulties, he was able to continue his studies. However, his nervous disposition interfered with his pastoral duties, and he was assigned to substitute teaching. He enjoyed this work and was popular with the staff and students. However, he failed the examination for certification as a teacher; ironically, his lowest grades were in biology. The Augustinians then sent him to the University of Vienna, where he became particularly interested in his plant physiology professor's unorthodox belief that new plant varieties can be caused by naturally arising variations. He was

also fascinated by his classes in physics, where he was exposed to the physicists' experimental and mathematical approach to their subject.

After further breakdowns and failures, Mendel returned to the monastery and was assigned the low-stress job of keeping the abbey garden. There he combined the experimental and mathematical approach of a physicist with his background in biology and performed a series of experiments designed to determine whether his professor was correct in his beliefs regarding the role of naturally arising variants in plants.

Mendel studied the transmission of specific traits of the pea plant—such as flower color and stem length—from parent plant to offspring. He pollinated the plants by hand and separated them until he had isolated each trait. For example, in his studies of flower color, he pollinated the plants until he

Mendel's Use of Probabilities

In his studies of flower color, Mendel pollinated peas until he produced pure-red plants (that is, plants that would produce only red-flowered offspring) and pure-white plants. He then cross-fertilized this first generation of pure reds and pure whites and obtained a second generation that had only red flowers. (The accepted theory of the day incorrectly predicted that he would have only pink flowers.) Finally, he cross-fertilized this second generation and obtained a third generation, approximately three-fourths of which had red flowers and one-fourth of which had white.

Mendel explained these results by postulating that there is a "determiner" responsible for flower color. These determiners are now called **genes.** Each plant has two flower color genes, one from each parent. If a plant inherits a red gene from each parent, the plant produced pure-red plants (plants that would produce only red-flowered offspring) and pure-white plants.

At the time, the accepted theory of heredity was that of blending. In this view, the characteristics of both parents blend together to form an individual. Mendel reasoned that if the blending theory was correct the union of a pure-red pea plant and a pure-white pea plant would result in a pink-flowered offspring. However, his experiments showed that such a union consistently resulted in red-flowered offspring.

Mendel crossbred a large number of peas that had different characteristics. In many cases an offspring would have a characteristic of one of its parents, undiluted by that of the other parent. Mendel concluded that the question of which parent's characteristics would be passed on was a matter of chance, and he successfully used probability theory to estimate the frequency with which characteristics would be passed on. In doing so, Mendel founded modern genetics. Mendel attempted similar experiments with bees, but these experiments were unsuccessful because he was unable to control the mating behavior of the queen bee.

Mendel was ignored when he published his paper "Experimentation in Plant Hybridization." Sixteen years after his death, his work was rediscovered by three European botanists who had reached similar conclusions in plant breeding, and the importance of his work was finally recognized.

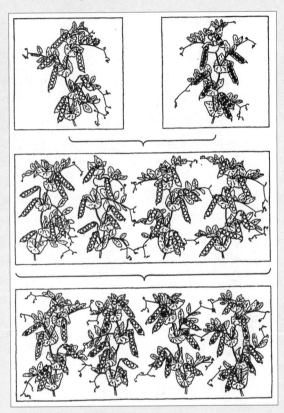

A 19th-century drawing illustrating Mendel's pea plants, showing the original cross, the first generation, and the second generation.

has red flowers. The plant also has red flowers if it inherits a red gene from one parent and a white gene from the other; that is, the red-flowered gene is **dominant** and the white-flowered gene is **recessive.** Which of its two genes a parent passes on to its offspring is strictly a matter of chance; Mendel successfully used probabilities to analyze this random aspect of heredity.

If we use R to stand for the red gene and w to stand for the white gene (with the capital letter indicating dominance), then the results of cross-fertilization can be described pictorially with a **Punnett square.** Figure 6.3 on page 276 shows the Punnett square for the crossing of the first generation of pure reds and pure whites. The four possible outcomes are the same: one red gene and one white gene, resulting in a red-flowered plant.

FIGURE 6.3

	R	R
w	(R, w)	(R, w)
w	(R, w)	(R, w)

← First parent's genes

← Offspring

← Offspring

↑
Second parent's genes

When the offspring of this experiment were cross-fertilized, Mendel found that approximately three-fourths of the offspring had red flowers and one-fourth had white. Mendel showed how to use probabilities to predict this result. Figure 6.4 shows the Punnett square for this second cross-fertilization. The sample space S consists of four outcomes:

$$S = \{(R, R), (R, w), (w, R), (w, w)\}$$

These four outcomes are equally likely, because each parent passes on one of its two color genes, and the color gene that is passed on is selected in a random manner.

FIGURE 6.4

	R	w
R	(R, R)	(R, w)
w	(w, R)	(w, w)

← First parent's genes

← Possible offspring

← Possible offspring

↑
Second parent's genes

Only one of the four possible outcomes, (w, w), results in a white-flowered plant, so event E_1 that the plant has white flowers is $E_1 = \{(w, w)\}$; therefore,

$$p(E_1) = \frac{n(E_1)}{n(S)} = \frac{1}{4}$$

This means that we should expect the actual relative frequency of white-flowered plants to be close to $\frac{1}{4}$.

Each of the other three outcomes, (R, R), (R, w), and (w, R), results in a red-flowered plant, because red dominates white. The event E_2 that the plant has red flowers is $E_2 = \{(R, R), (R, w), (w, R)\}$; therefore,

$$p(E_2) = \frac{n(E_2)}{n(S)} = \frac{3}{4}$$

Thus, we should expect the actual relative frequency of red-flowered plants to be close to $\frac{3}{4}$.

Ronald Fisher, a noted British statistician, used statistics to deduce that Mendel fudged his data. Mendel's relative frequencies were unusually close to the theoretical probabilities, and Fisher found that there was only about a 0.00007 chance of such close agreement. Others have suggested that perhaps Mendel did not willfully change his results but rather continued collecting data until the numbers were in close agreement with his expectations.[*]

---

[*]R. A. Fisher, "Has Mendel's Work Been Rediscovered?" *Annals of Science* 1, 1936, pp. 115–137.

Outcomes (R, w) and (w, R) are genetically identical; it doesn't matter which gene is inherited from which parent. For this reason, geneticists do not use the ordered-pair notation and instead refer to each of these two outcomes as "Rw." This convention results in a sample space S that consists of three outcomes that are not equally likely:

$$S = \{RR, Rw, ww\}$$

These outcomes are not equally likely because Rw can happen in either of two ways, whereas RR can happen in only way. The probability distribution for Mendel's second generation of pea plants is given below.

Outcome	Probability
RR	1/4
Rw	2/4 = 1/2
ww	1/4

The probability of an event is found by adding the probabilities of the outcomes that make up that event. Thus, the probability of event E_2 that the plant has red flowers is

$$p(E_2) = p(RR) + p(Rw)$$
$$= \frac{1}{4} + \frac{2}{4} = \frac{3}{4}$$

Probabilities in Genetics

Cystic fibrosis is an inherited disease characterized by abnormally functioning exocrine glands that secrete a thick mucus, clogging the pancreatic ducts and lung passages. Most patients with cystic fibrosis die of chronic lung disease; until recently, most died in early childhood. This early death makes it extremely unlikely that an afflicted person would ever parent a child. Only after the advent of Mendelian genetics did it become clear how a child could inherit the disease from two healthy parents.

In 1989 a team of Canadian and American doctors announced the discovery of the gene responsible for most cases of cystic fibrosis. As a result of that discovery, a new therapy for cystic fibrosis is being developed. Researchers splice a therapeutic gene into a cold virus and administer it through an affected person's nose. When the virus infects the lungs, the gene becomes active. It is hoped that this will result in normally functioning cells, without the damaging mucus. In April of 1993, a 23-year-old man with advanced cystic fibrosis became the first patient to receive this therapy.

In September 1996, a British team announced that eight volunteers with cystic fibrosis received this therapy; six were temporarily cured of the disease's debilitating symptoms. The team is now analyzing the results of a trial involving multiple doses, which could have a long-term effect.

Cystic fibrosis occurs in about 1 out of every 2000 births in the Caucasian population and only in about 1 in 250,000 births in the non-Caucasian population. It is one of the most common inherited diseases in North America. One in 25 Americans carries a single gene for cystic fibrosis. Children who inherit two such genes develop the disease; that is, cystic fibrosis is recessive.

EXAMPLE 2 Each of two prospective parents carries one cystic fibrosis gene.

a. Find the probability that their child would have cystic fibrosis.
b. Find the probability that their child would be healthy (i.e., free of symptoms).
c. Find the probability that their child would be free of symptoms but could pass the cystic fibrosis gene on to his or her own child.

Solution Dominant genes are indicated by capital letters and recessive genes by lowercase letters; we will denote the recessive cystic fibrosis gene with a c and the disease-free gene with a C. Each parent is Cc and thus does not have the disease. Figure 6.5 shows the Punnett square and the resulting probability distribution for the child.

FIGURE 6.5

	C	c
C	CC	Cc
c	Cc	cc

Outcome	Probability
CC	1/4
Cc	2/4 5 1/2
cc	1/4

a. Cystic fibrosis is recessive, so only the cc child will have the disease; the probability of such an event is 1/4.
b. The Cc and the CC children would be healthy; thus,

$$p(\text{healthy}) = p(\text{Cc}) + p(\text{CC}) = 2/4 + 1/4 = 3/4$$

c. The Cc child would never suffer from any symptoms, but could pass the cystic fibrosis gene on to his or her own child; such a person is called a **carrier.** (Both of the parents were carriers.) The probability of such an event is 1/2. ●

Sickle-cell anemia is an inherited disease characterized by a tendency of the red blood cells to become distorted and deprived of oxygen. Although it varies in severity, the disease can be fatal in early childhood. More often, patients have a shortened life span and chronic organ damage. Although there is no cure, newborns are now routinely screened for sickle-cell disease; daily penicillin can ward off life-threatening infections. Approximately 1 in 500 black babies is born with sickle-cell anemia, but only 1 in 160,000 nonblack babies has the disease. This disease is **codominant:** a person with two sickle-cell genes will have the disease, while a person with one sickle-cell gene will have a mild, nonfatal anemia called *sickle-cell trait.* Approximately 8–10% of the black population has sickle-cell trait.

Huntington's disease, caused by a dominant gene, is characterized by nerve degeneration causing spasmodic movements and progressive mental deterioration. The symptoms do not usually appear until well after reproductive age has been reached; the disease usually hits people in their forties. Death typically follows 12 to 15 years after the onset of the symptoms. There is no effective treatment available, but physicians can now assess with certainty whether someone will develop the disease, and they can estimate when the disease will strike. Many of those at risk choose not to undergo the test, especially if they have

Woody Guthrie's most famous song is "This Land Is Your Land." This folksinger, guitarist, and composer was a friend of Leadbelly, Pete Seeger, and Ramblin' Jack Elliott and exerted a strong influence on Bob Dylan. Guthrie died at the age of 55 of Huntington's disease.

already had children. Folk singer Arlo Guthrie is in this situation; his father, Woody Guthrie, died of Huntington's disease.

Genetic Screening

At this time, there are no conclusive tests that will tell a parent if he or she is a cystic fibrosis carrier, nor are there conclusive tests that will tell if a fetus has the disease. A new test resulted from the 1989 discovery of the location of most cystic fibrosis genes, but that test will detect only 85% to 95% of the cystic fibrosis genes, depending on the individual's ethnic background. The extent to which this test will be used has created quite a controversy.

Individuals who have relatives with cystic fibrosis are routinely informed about the availability of the new test. The controversial question is whether a massive genetic screening program should be instituted to identify cystic fibrosis carriers in the general population, regardless of family history. This is an important question, considering that four in five babies with cystic fibrosis are born to couples with no previous family history of the condition.

Opponents of routine screening cite a number of important concerns. The existing test is rather inaccurate; 5% to 15% of the cystic fibrosis carriers would be missed. It is not known how health insurers would use this genetic information—insurance firms could raise rates or refuse to carry people if a screening test indicated a presence of cystic fibrosis. Also, some experts question the adequacy of quality assurance for the diagnostic facilities and for the tests themselves.

Supporters of routine testing say that the individual should be allowed to decide whether to be screened. Failing to inform people denies them the opportunity to make a personal choice about their reproductive future. An individual found to be a carrier could

Historical Note

Nancy Wexler

In 1993 scientists working together at six major research centers located the gene that causes Huntington's disease. This discovery will enable people to learn whether they carry the Huntington's gene, and it will allow pregnant women to determine whether their child carries the gene. The discovery could eventually lead to a treatment.

The collaboration of research centers was organized largely by Nancy Wexler, a Columbia University professor of neuropsychology, who is herself at risk for Huntington's disease—her mother died of it 30 years ago. Dr. Wexler, president of the Hereditary Disease Foundation, has made numerous trips to study and aid the people of the Venezuelan village of Lake Maracaibo, many of whom suffer from the disease. All are related to one woman who died of the disease in the early 1800s. Wexler took blood and tissue samples and gave neurological tests to the inhabitants of the village. The samples and test results enabled the researchers to find the single gene that causes Huntington's disease.

In October 1993, Wexler received an Albert Lasker Medical Research Award, a prestigious honor that is often a precursor to a Nobel Prize. The award was given in recognition for her contribution to the international effort that culminated in the discovery of the Huntington's disease gene. At the awards ceremony, she explained to first lady Hillary Rodham Clinton that her genetic heritage has made her uninsurable—she would lose her health coverage if she switched jobs. She told Mrs. Clinton that more Americans will be in the same situation as more genetic discoveries are made, unless the health care system is reformed. The first lady incorporated this information into her speech at the awards ceremony: "It is likely that in the next years, every one of us will have a preexisting condition and will be uninsurable. . . . What will happen as we discover those genes for breast cancer, or prostate cancer, or osteoporosis, or any of the thousands of other conditions that affect us as human beings?"

choose to avoid conception, to adopt, to use artificial insemination by donor, or to use pre-natal testing to determine whether a fetus is affected—at which point the additional controversy regarding abortion could enter the picture.

The history of genetic screening programs is not an impressive one. In the 1970s, mass screening of blacks for sickle-cell anemia was instituted. This program caused unwarranted panic; those who were told they had sickle-cell trait feared that they would develop symptoms of the disease and often did not understand the probability that their children would inherit the disease (see Exercises 9 and 10). Some people with sickle-cell trait were denied health insurance and life insurance.

The Row over Sickle-Cell

NEWSWEEK

FEBRUARY 12, 1973

. . . Two years ago, President Nixon listed sickle-cell anemia along with cancer as diseases requiring special Federal attention. . . . Federal spending for sickle-cell anemia programs has risen from a scanty $1 million a year to $15 million for 1973. At the same time, in what can only be described as a headlong rush, at least a dozen states have passed laws requiring sickle-cell screening for blacks.

While all these efforts have been undertaken with the best intentions of both whites and blacks, in recent months the campaign has begun to stir widespread and bitter controversy.

Some of the educational programs have been riddled with misinformation and have unduly frightened the black community. To quite a few Negroes, the state laws are discriminatory—and to the extent that they might inhibit childbearing, even genocidal. . . .

Parents whose children have the trait often misunderstand and assume they have the disease. In some cases, airlines have allegedly refused to hire black stewardesses who have the trait, and some carriers have been turned down by life-insurance companies—or issued policies at high-risk rates.

Because of racial overtones and the stigma that attaches to persons found

to have the sickle-cell trait, many experts seriously object to mandatory screening programs. They note, for example, that there are no laws requiring testing for Cooley's anemia [or other disorders that have a hereditary basis]. . . . Moreover, there is little that a person who knows he has the disease or the trait can do about it. "I don't feel," says Dr. Robert L. Murray, a black geneticist at Washington's Howard University, "that people should be required by law to be tested for something that will provide information that is more negative than positive." . . . Fortunately, some of the mandatory laws are being repealed.

6.2

EXERCISES

1. A spinner is designed for a board game so that the arrow can point to either 1, 2, 3, or 4, as shown in Figure 6.6.
   a. Determine the probability distribution for this spinner.
   b. Find the probability of not spinning a 1.

2. A spinner is designed for a board game so that the arrow can point to either 1, 2, 3, 4, or 5, as shown in Figure 6.7.
   a. Determine the probability distribution for this spinner.
   b. Find the probability of spinning an odd number.

FIGURE 6.6

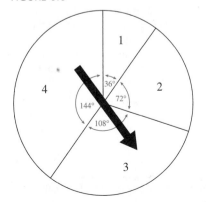

FIGURE 6.7

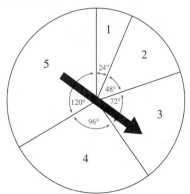

3. A new dartboard is designed by a toy manufacturer, as shown in Figure 6.8.
   a. Determine the probability distribution for this dartboard.
   b. Find the probability of not hitting a white zone.

FIGURE 6.8

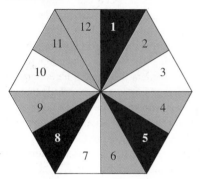

4. A new dartboard is designed by a toy manufacturer, as shown in Figure 6.8.
   a. Determine the probability distribution for this dartboard.
   b. Find the probability of not hitting a black zone.

5. Mendel found that snapdragons have no color dominance; a snapdragon with one red gene and one white gene will have pink flowers. Suppose a snapdragon with red flowers is crossed with one with pink flowers.
   a. Determine the probability distribution for this cross.
   b. Find the probability of a red offspring.
   c. Find the probability of a pink offspring.
   d. Find the probability of a white offspring.

6. Two pink snapdragons are crossed (see Exercise 5).
   a. Determine the probability distribution for this cross.
   b. Find the probability of a red offspring.
   c. Find the probability of a pink offspring.
   d. Find the probability of a white offspring.

7. One parent is a cystic fibrosis carrier and the other has no cystic fibrosis gene.
   a. Determine the probability distribution for their child.
   b. Find the probability that their child would have cystic fibrosis.
   c. Find the probability that their child would be a carrier.
   d. Find the probability that their child would not have cystic fibrosis and not be a carrier.
   e. Find the probability that their child would be healthy (i.e., free of symptoms).

8. One parent is a cystic fibrosis carrier and the other has the disease.
   a. Determine the probability distribution for their child.
   b. Find the probability that their child would have cystic fibrosis.

c. Find the probability that their child would be a carrier.
d. Find the probability that their child would not have cystic fibrosis and not be a carrier.
e. Find the probability that their child would be healthy (i.e., free of symptoms).

9. Carrier detection tests show that two prospective parents have sickle-cell trait.
   a. Determine the probability distribution for their child.
   b. Find the probability that their child would have sickle-cell anemia.
   c. Find the probability that their child would have sickle-cell trait.
   d. Find the probability that their child would be healthy (i.e., free of symptoms).

10. Carrier detection tests show that one prospective parent has sickle-cell trait and the other has no sickle-cell gene.
    a. Determine the probability distribution for their child.
    b. Find the probability that their child would have sickle-cell anemia.
    c. Find the probability that their child would have sickle-cell trait.
    d. Find the probability that their child would be healthy (i.e., free of symptoms).

Tay-Sachs disease is a recessive disease characterized by an abnormal accumulation of certain fat compounds in the spinal cord and brain, resulting in paralysis, severe mental impairment, and blindness. There is no effective treatment, and death usually occurs before the age of 5. The disease occurs once in 3600 births among Ashkenazi Jews (Jews from central and eastern Europe), but only once in 600,000 births in other populations. Carrier detection tests and fetal monitoring tests are available. The successful use of these tests, coupled with an aggressive counseling program, has resulted in a decrease of 90% in the incidence of this disease.

11. Carrier detection tests show that only one prospective parent is a carrier of Tay-Sachs.
    a. Determine the probability distribution for their child.
    b. Find the probability that their child would have the disease.
    c. Find the probability that their child would be a carrier.
    d. Find the probability that their child would be healthy (i.e., free of symptoms).

12. Carrier detection tests show that two prospective parents are carriers of Tay-Sachs.
    a. Determine the probability distribution for their child.
    b. Find the probability that their child would have the disease.
    c. Find the probability that their child would be a carrier.
    d. Find the probability that their child would be healthy (i.e., free of symptoms).

13. A parent started to exhibit the symptoms of Huntington's disease after the birth of his or her child. Assume that this parent carries a single gene for Huntington's disease and that the other carries no such gene.
  a. Determine the probability distribution for their child.
  b. Find the probability that their child would have the disease.
  c. Find the probability that their child would be a carrier.
  d. Find the probability that their child would be healthy (i.e., free of symptoms).

 Answer the following questions using complete sentences.

14. What prompted Dr. Nancy Wexler's interest in Huntington's disease? What resulted from this interest?

15. In the United States, 52% of the babies are boys and 48% are girls. Do these percentages contradict an assumption that boys and girls are equally likely? Why or why not?

16. In the 1970s there was mass screening of blacks for sickle-cell anemia and mass screening of Jews for Tay-Sachs disease. One of these was a successful program; one was not. Write a research paper on these two programs.

6.3

BASIC RULES OF PROBABILITY

One basic rule about probabilities is that they are always between 0 and 1. The smallest possible probability is that of an impossible event (an event equal to the null set); that probability is 0. The largest probability is that of a certain event (an event equal to the sample space); that probability is 1. *If you ever get a negative answer or an answer greater than 1 when you calculate a probability, go back and find your error.*

Probability Rules		
Rule 1	$p(\emptyset) = 0$	The probability of the null set is 0.
Rule 2	$p(S) = 1$	The probability of the sample space is 1.
Rule 3	$0 \leq p(E) \leq 1$	Probabilities are between 0 and 1 (inclusive).

Probability rules 1, 2, and 3 can be formally verified (for a sample space with equally likely outcomes) as follows:

Rule 1: $\quad p(\emptyset) = \dfrac{n(\emptyset)}{n(S)} = \dfrac{0}{n(S)} = 0$

Rule 2: $\quad p(S) = \dfrac{n(S)}{n(S)} = 1$

Rule 3: $\quad E$ is a subset of S; therefore,

$$0 \leq n(E) \leq n(S)$$

$$\dfrac{0}{n(S)} \leq \dfrac{n(E)}{n(S)} \leq \dfrac{n(S)}{n(S)} \qquad \text{Dividing by } n(S)$$

$$0 \leq p(E) \leq 1$$

Mutually Exclusive Events

Two events that cannot both occur at the same time are called **mutually exclusive.** In other words, E and F are mutually exclusive if and only if $E \cap F = \emptyset$.

EXAMPLE 1 A die is rolled. Let E be the event "an even number comes up," F the event "a number greater than 3 comes up," and G the event "an odd number comes up."
  a. Are E and F mutually exclusive?

b. Are E and G mutually exclusive?

Solution
a. $E = \{2, 4, 6\}$, $F = \{4, 5, 6\}$, and $E \cap F = \{4, 6\} \neq \emptyset$ (see Figure 6.9). Therefore, E and F are *not* mutually exclusive; the number that comes up could be *both* even *and* greater than 3. In particular, it could be 4 or 6.

FIGURE 6.9

b. $E = \{2, 4, 6\}$, $G = \{1, 3, 5\}$, and $E \cap G = \emptyset$. Therefore, E and G *are* mutually exclusive; the number that comes up could *not* be both even and odd.

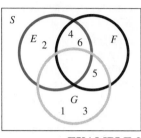

EXAMPLE 2 Let M be the event "being a mother," F the event "being a father," and D the event "being a daughter."

a. Are events M and D mutually exclusive?
b. Are events M and F mutually exclusive?

Solution
a. M and D are mutually exclusive if $M \cap D = \emptyset$. $M \cap D$ is the set of all people who are both mothers and daughters, and that set is not empty. A person can be a mother and a daughter at the same time. M and D are not mutually exclusive because being a mother does not exclude being a daughter.

b. M and F are mutually exclusive if $M \cap F = \emptyset$. $M \cap F$ is the set of all people who are both mothers and fathers, and that set is empty. A person cannot be a mother and a father at the same time. M and F are mutually exclusive because being a mother does exclude the possibility of being a father.

Pair of Dice Probabilities

In order to find probabilities involving the rolling of a pair of dice, we must first determine the sample space. This can be done in either of two ways. If you roll a 3 and a 5, you can consider the outcome to be the ordered pair (3, 5), or you can consider the outcome

FIGURE 6.10
Outcomes of rolling two dice

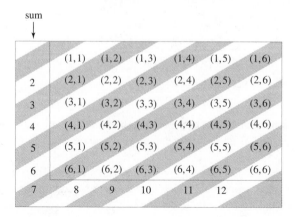

to be 8. We will proceed with the ordered pair approach. (We will explore the other approach in Exercise 47.) Figure 6.10 lists all possible ordered pair outcomes and the resulting sums. Notice that $n(S) = 6 \cdot 6 = 36$.

EXAMPLE 3   A pair of dice is rolled. Find the probability of each of the following events.

    a. The sum is 7.
    b. The sum is greater than 9.
    c. The sum is even.
    d. The sum is not greater than 9.
    e. The sum is greater than 9 and even.
    f. The sum is greater than 9 or even.

Solution   **a.** To find the probability that the sum is 7, let D be the event "the sum is 7." From Figure 6.10, $D = \{(1, 6), (2, 5), (3, 4), (4, 3), (5, 2), (6, 1)\}$, so $n(D) = 6$; therefore,

$$p(D) = \frac{n(D)}{n(S)} = \frac{6}{36} = \frac{1}{6}$$

This means that if we were to roll a pair of dice a large number of times, we should expect to get a sum of 7 approximately one-sixth of the time.

> ✔   Notice that $p(D) = \frac{1}{6}$ is between 0 and 1, as are all probabilities.

b. To find the probability that the sum is greater than 9, let E be the event "the sum is greater than 9."
$E = \{(4, 6), (5, 5), (6, 4), (5, 6), (6, 5), (6, 6)\}$, so $n(E) = 6$; therefore,

$$p(E) = \frac{n(E)}{n(S)} = \frac{6}{36} = \frac{1}{6}$$

This means that if we were to roll a pair of dice a large number of times, we should expect to get a sum greater than 9 approximately one-sixth of the time.

c. Let F be the event "the sum is even."
$F = \{(1, 1), (1, 3), (2, 2), (3, 1), \ldots, (6, 6)\}$, so $n(F) = 18$ (refer to Figure 6.10); therefore,

$$p(F) = \frac{n(F)}{n(S)} = \frac{18}{36} = \frac{1}{2}$$

This means that if we were to roll a pair of dice a large number of times, we should expect to get an even sum approximately half of the time.

d. We could find the probability that the sum is not greater than 9 by counting, as in (a), (b), and (c), but the counting would be rather excessive. It is easier to use one of the cardinal number formulas from Chapter 5 on sets. The event "the sum is not greater than 9" is the complement of event E ("the sum is greater than 9") and can be expressed as E'.

$$n(E') = n(U) - n(E) \qquad \text{Cardinal number formula}$$
$$= n(S) - n(E) \qquad \text{"Universal set" and "sample space"}$$
$$= 36 - 6 = 30 \qquad \text{represent the same idea.}$$
$$p(E') = \frac{n(E')}{n(S)} = \frac{30}{36} = \frac{5}{6}$$

This means that if we were to roll a pair of dice a large number of times, we should expect to get a sum that's not greater than 9 approximately five-sixths of the time.

e. The event "the sum is greater than 9 and even" can be expressed as the event $E \cap F$.
$E \cap F = \{(4, 6), (5, 5), (6, 4), (6, 6)\}$, so $n(E \cap F) = 4$; therefore,

$$p(E \cap F) = \frac{n(E \cap F)}{n(S)} = \frac{4}{36} = \frac{1}{9}$$

This means that if we were to roll a pair of dice a large number of times, we should expect to get a sum that's both greater than 9 and even approximately one-ninth of the time.

f. Finding the probability that the sum is greater than 9 or even by counting would require an excessive amount of counting. It is easier to use one of the cardinal number formulas from Chapter 5. The event "the sum is greater than 9 or even" can be expressed as the event $E \cup F$.

$$n(E \cup F) = n(E) + n(F) - n(E \cap F) \qquad \text{Cardinal number formula}$$
$$= 6 + 18 - 4 \qquad \text{From parts (b), (c), and (e)}$$
$$= 20$$
$$p(E \cup F) = \frac{n(E \cup F)}{n(S)} = \frac{20}{36} = \frac{5}{9}$$

This means that if we were to roll a pair of dice a large number of times, we should expect to get a sum that's either greater than 9 or even approximately five-ninths of the time. ●

More Probability Rules

In Example 3 above, we used some cardinal number formulas from Chapter 5 to avoid excessive counting. Some people find it easier to use these rules to calculate probabilities when they are expressed in the language of probability theory.

More Probability Rules

Rule 4 $p(E \cup F) = p(E) + p(F) - p(E \cap F)$
Rule 5 If E and F are mutually exclusive, then $p(E \cup F) = p(E) + p(F)$.
Rule 6 $p(E) + p(E') = 1$ [or equivalently, $p(E) = 1 - p(E')$ or $p(E') = 1 - p(E)$]

In part (d) of Example 3, finding the probability that the sum is not greater than 9 could be done with probability rule 6 rather than with a cardinal number formula:

$$p(E') = 1 - p(E)$$
$$= 1 - \frac{1}{6} \qquad \text{From (b)}$$
$$= \frac{5}{6}$$

Similarly, in part (f) of Example 3, finding the probability that the sum is greater than 9 or even could be done with probability rule 4 rather than with a cardinal number formula:

$$p(E \cup F) = p(E) + p(F) - p(E \cap F)$$
$$= \frac{1}{6} + \frac{1}{2} - \frac{1}{9} \qquad \text{From (b), (c), and (e)}$$
$$= \frac{3}{18} + \frac{9}{18} - \frac{2}{18} = \frac{5}{9}$$

Probabilities and Venn Diagrams

Venn diagrams can be used to illustrate probabilities in the same way they are used in set theory. In this case, we label each region with its probability rather than its cardinal number.

EXAMPLE 4   Zaptronics manufactures compact discs and their cases for several major record labels. A recent sampling of the product indicated that 5% have defective packaging, 3% have a defective disc, and 7% have at least one of the two defects. Find the probability that a Zaptronics product has the following:

a. both defects
b. neither defect

Solution   **a.** Let P be the event "the packaging is defective" and D the event "the disc is defective." We are given $p(P) = 5\% = 0.05$, $p(D) = 3\% = 0.03$, and $p(P \cup D) = 7\% = 0.07$. We are asked to find $p(P \cap D)$. To do so, substitute into probability rule 4.

$$
\begin{aligned}
p(P \cup D) &= p(P) + p(D) - p(P \cap D) \qquad &\text{Probability rule 4} \\
0.07 &= 0.05 + 0.03 - p(P \cap D) \qquad &\text{Substituting} \\
0.07 &= 0.08 - p(P \cap D) \\
p(P \cap D) &= 0.01 = 1\%
\end{aligned}
$$

This means that 1% of Zaptronics' products have defective packaging *and* a defective disc. The Venn diagram for this problem is shown in Figure 6.11.

FIGURE 6.11

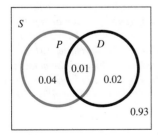

b. According to the Venn diagram, the probability of neither defect is $0.93 = 93\%$.

Alternatively, we are asked to find the probability that the product does not have defective packaging *and* does not have a defective disc—that is, to find $p(P' \cap D')$. This is equal to $p((P \cup D)')$, by De Morgan's law. And $(P \cup D)'$ is the complement of an event whose probability is given to us. Thus,

$$
\begin{aligned}
p(P' \cap D') &= p((P \cup D)') \qquad &\text{De Morgan's law} \\
&= 1 - p(P \cup D) \qquad &\text{Probability rule 6} \\
&= 1 - 0.07 \qquad &\text{Substituting} \\
&= 0.93 = 93\%
\end{aligned}
$$

This means that 93% of Zaptronics' products are defect-free.   •

6.3

EXERCISES

In Exercises 1–10, determine whether E and F are mutually exclusive. Write a sentence justifying your answer.

1. *E* is the event "being a doctor," and *F* is the event "being a woman."

2. *E* is the event "it's raining," and *F* is the event "it's sunny."

3. *E* is the event "being single," and *F* is the event "being married."

4. *E* is the event "having naturally blond hair," and *F* is the event "having naturally black hair."

5. *E* is the event "having brown hair," and *F* is the event "having gray hair."

6. *E* is the event "being a plumber," and *F* is the event "being a stamp collector."

7. *E* is the event "wearing boots," and *F* is the event "wearing sandals."

8. *E* is the event "wearing shoes," and *F* is the event "wearing socks."

9. If a die is rolled once, *E* is the event "getting a four," and *F* is the event "getting an odd number."

10. If a die is rolled once, *E* is the event "getting a four," and *F* is the event "getting an even number."

In Exercises 11–18 a card is dealt from a complete deck of 52 cards (no jokers). Use probability rules (when appropriate) to find the probability that the card is as stated. (Count an ace as high.)

11. **a.** a jack and red   **b.** a jack or red
    c. not a red jack

12. **a.** a jack and a heart   **b.** a jack or a heart
    c. not a jack of hearts

13. **a.** a ten and a spade   **b.** a ten or a spade
    c. not a ten of spades

14. **a.** a five and black   **b.** a five or black
    c. not a black five

15. **a.** under a four   **b.** above a nine
    c. both under a four and above a nine
    d. either under a four or above a nine

16. **a.** above a jack   **b.** below a three
    c. both above a jack and below a three
    d. either above a jack or below a three

17. **a.** above a five   **b.** below a ten
    c. both above a five and below a ten
    d. either above a five or below a ten

18. **a.** above a seven   **b.** below a queen
    c. both above a seven and below a queen
    d. either above a seven or below a queen.

In Exercises 19–26, use complements to find the probability that a card dealt from a full deck (no jokers) is as stated. (Count an ace as high.)

19. not a queen   20. not a seven

21. not a face card   22. not a heart

23. above a three   24. below a queen

25. below a jack   26. above a five

27. If $o(E) = 5{:}9$, find $o(E')$.

28. If $o(E) = 1{:}6$, find $o(E')$.

29. If $p(E) = \frac{2}{7}$, find $o(E)$ and $o(E')$.

30. If $p(E) = \frac{3}{8}$, find $o(E)$ and $o(E')$.

31. If $o(E) = a{:}b$, find $o(E')$.

32. If $p(E) = \frac{a}{b}$, find $o(E')$.

   HINT: Use Exercise 37 from Section 6.1 and Exercise 31 above.

In Exercises 33–38, use Exercise 32 above to find the odds that a card dealt from a full deck (no jokers) is as stated.

33. not a king   34. not an eight

35. not a face card   36. not a club

37. above a four   38. below a king

In Exercises 39–42, use the following information. In order to determine the effect their salespersons have on purchases, a department store polled 700 shoppers regarding whether or not they made a purchase and whether or not they were pleased with the service they received. Of those who made a purchase, 151 were happy with the service and 133 were not. Of those who made no purchase, 201 were happy with the service and 215 were not. Use probability rules (when appropriate) to find the probability of the event stated.

39. **a.** A shopper made a purchase.
    b. A shopper did not make a purchase.

40. **a.** A shopper was happy with the service received.
    b. A shopper was unhappy with the service received.

41. a. A shopper made a purchase and was happy with the service.
   b. A shopper made a purchase or was happy with the service.

42. a. A shopper made no purchase and was unhappy with the service.
   b. A shopper made no purchase or was unhappy with the service.

In Exercises 43–46, use the following information: A supermarket polled 1000 customers regarding the size of their bill. The results were given in Figure 6.12. Use probability rules (when appropriate) to find the relative frequency with which a customer's bill is as stated.

43. a. less than $40.00   **b.** $40.00 or more

44. a. less than $80.00   **b.** $80.00 or more

45. a. between $40.00 and $79.99
   b. not between $40.00 and $79.99

46. a. between $20.00 and $79.99
   b. not between $20.00 and $79.99

FIGURE 6.12
Supermarket bills

Size of Bill	Number of Customers
Below $20.00	208
$20.00–$39.99	112
$40.00–$59.99	183
$60.00–$79.99	177
$80.00–$99.99	198
$100.00 or above	122

47. A pair of dice is rolled. Consider the outcome of rolling a 3 and a 5 to be the number 8.
   a. Find the sample space.
   b. Create a probability distribution for the experiment.
   c. Find the probability that the sum is 8.
   d. Find the probability that the sum is greater than 8.
   e. Find the probability that the sum is 8 or less.
   f. Why can the outcomes of this experiment be considered to be either equally likely (as in Example 3) or not equally likely (as in this exercise)?
   g. Could Exercise 3 in Section 6.2 be approached in either of these two ways? If so, describe the equally likely approach. If not, discuss why.

In Exercises 48–54, find the probability that the sum is as stated when a pair of dice is rolled.

48. a. 2   **b.** 4   **c.** 6

49. a. 7 or 11   **b.** 7 or 11 or doubles

50. a. 8 or 10   **b.** 8 or 10 or doubles

51. a. odd and greater than 7
   b. odd or greater than 7

52. a. even and less than 5   **b.** even or less than 5

53. a. even and doubles   **b.** even or doubles

54. a. odd and doubles   **b.** odd or doubles

55. Maya is taking two courses: photography and economics. Student records indicate that the probability of passing photography is 0.75, that of failing economics is 0.65, and that of passing at least one of the two courses is 0.85. Find the probability of the following.
   a. Maya will pass economics.
   b. Maya will pass both courses.
   c. Maya will fail both courses.
   d. Maya will pass exactly one course.

56. Alex is taking two courses: algebra and U.S. history. Student records indicate that the probability of passing algebra is 0.35, that of failing U.S. history is 0.35, and that of passing at least one of the two courses is 0.80. Find the probability of the following.
   a. Alex will pass history.
   b. Alex will pass both courses.
   c. Alex will fail both courses.
   d. Alex will pass exactly one course.

57. Of all the flashlights in a large shipment, 15% have a defective bulb, 10% have a defective battery, and 5% have both defects. If you purchase one of the flashlights in this shipment, find the probability that it has the following.
   a. a defective bulb or a defective battery
   b. a good bulb or a good battery
   c. a good bulb and a good battery

58. Of all the videotapes in a large shipment, 20% have a defective tape, 15% have a defective case, and 10% have both defects. If you purchase one of the videotapes in this shipment, find the probability that it has the following.
   a. a defective tape or a defective case
   b. a good tape or a good case
   c. a good tape and a good case

In Exercises 59–60, use the following information: Lily Orteaga owns a video rental store called LilyO's Videos. Lily knows both the age and the address of all of her renters because she requires that renters have a copy of their driver's license on file. During the last three months, Lily accumulated data on the age of every renter and the part of town they live in; those data are given in *Figure 6.13*.

FIGURE 6.13

Age	Parkview District	Central City	South of Town
Under 20	1287	1582	2038
20–40	3758	3677	3114
Over 40	2895	2156	2066

59. a. What is the probability that a renter lives in the Parkview district?
  b. What is the probability that a renter is 20–40 years old?
  c. What is the probability that a renter lives in the Parkview district and is 20–40 years old?
  d. What is the probability that a renter lives in the Parkview district or is 20–40 years old?

60. a. What is the probability that a renter lives in the central city?
  b. What is the probability that a renter is over 40 years old?

c. What is the probability that a renter lives in the central city and is over 40 years old?
  d. What is the probability that a renter lives in the central city or is over 40 years old?

61. Verify probability rule 4 for a sample space with equally likely outcomes.
  HINT: Divide the cardinal number formula for the union of sets from Section 5.1 by $n(S)$.

62. Verify probability rule 5 for a sample space with equally likely outcomes.
  HINT: Start with probability rule 4. Then use the fact that E and F are mutually exclusive.

63. Verify probability rule 6 for a sample space with equally likely outcomes.
  HINT: Are E and E' mutually exclusive?

 Answer the following using complete sentences.

64. What does probability rule 2 say about the sum of the probabilities in a probability distribution?

65. What is the complement of a certain event?

66. Write an essay in which you compare and contrast mutually exclusive events and impossible events.

FRACTIONS ON A GRAPHING CALCULATOR

Some graphing calculators (including the TI-82, TI-83, TI-85, and TI-86) will add, subtract, multiply, and divide fractions and will give answers in reduced fractional form.

Reducing Fractions

The fraction 42/70 reduces to 3/5. To do this on your calculator, you must make your screen read "42/70→Frac." The way that you do this varies.

TI-82/83:
- Type 42 $\boxed{\div}$ 70, but do not press $\boxed{\text{ENTER}}$. This causes "42/70" to appear on the screen.
- Press the $\boxed{\text{MATH}}$ button.
- Highlight option 1 "→Frac". (Option 1 is automatically highlighted. If we were selecting a different option, we would use the $\boxed{\uparrow}$ and $\boxed{\downarrow}$ buttons to highlight it.)
- Press $\boxed{\text{ENTER}}$. This causes "42/70→Frac" to appear on the screen.
- Press $\boxed{\text{ENTER}}$. This causes "3/5" to appear on the screen.

TI-85/86:
- Type 42 $\boxed{\div}$ 70, but do not press $\boxed{\text{ENTER}}$. This causes "42/70" to appear on the screen.
- Press $\boxed{\text{2nd}}$ $\boxed{\text{MATH}}$.
- Press $\boxed{\text{MISC}}$ (i.e., $\boxed{\text{F5}}$).
- Press $\boxed{\text{MORE}}$ until "→Frac" appears.
- Press $\boxed{\text{FRAC}}$ (i.e., $\boxed{\text{F1}}$). This causes "42/70→Frac" to appear on the screen.
- Press $\boxed{\text{ENTER}}$. This causes "3/5" to appear on the screen.

EXAMPLE 1   Use your calculator to compute

$$\frac{1}{6} + \frac{1}{2} - \frac{1}{9}$$

and give your answer in reduced fractional form.

Solution   Make your screen read "1/6+1/2−1/9→Frac" by typing

$$1 \boxed{\div} 6 \boxed{+} 1 \boxed{\div} 2 \boxed{-} 1 \boxed{\div} 9$$

and then inserting the "→Frac" command as described above. Once you press $\boxed{\text{ENTER}}$, the screen reads "5/9."

EXERCISES

In Exercises 67 and 68, reduce the given fractions to lowest terms, both (a) by hand and (b) with a calculator. Check your work by comparing the two answers. (Answers are not given in the back of the book.)

67. 18/33

68. −42/72

In Exercises 69–74, perform the indicated operations and reduce the answers to lowest terms, both (a) by hand and (b) with a calculator. Check your work by comparing the two answers. (Answers are not given in the back of the book.)

69. $\dfrac{6}{15} \cdot \dfrac{10}{21}$

70. $\dfrac{6}{15} \div \dfrac{10}{21}$

71. $\dfrac{6}{15} + \dfrac{10}{21}$

72. $\dfrac{6}{15} - \dfrac{10}{21}$

73. $\dfrac{7}{6} - \dfrac{5}{7} + \dfrac{9}{14}$

74. $\dfrac{-8}{5} - \left(\dfrac{-3}{28} + \dfrac{5}{21} \right)$

75. How could you get decimal answers to the above exercises, rather than fractional answers?

6.4

COMBINATORICS AND PROBABILITY

Finding a probability involves finding the number of outcomes in an event and the number of outcomes in the sample space. So far, we have used the Probability Rules as an alternative to excessive counting. Another alternative is combinatorics—that is, permutations, combinations, and the fundamental counting principle—as covered in Chapter 5. The flowchart used in Chapter 5 is summarized below.

Which Counting Technique?

1. If the problem involves more than one category, use the ***fundamental principle of counting*** and multiply the number of choices for each category.
2. Within any one category, if the order of selection is important, use ***permutations:*** r items can be selected from a group of n items in

$$nPr = \frac{n!}{(n-r)!}$$

ways.

3. Within any one category, if the order of selection is *not* important, use ***combinations:*** r items can be selected from a group of n items in

$$_nC_r = \frac{n!}{r! \cdot (n-r)!}$$

ways.

EXAMPLE 1 A group of three people is selected at random. What is the probability that at least two of them will have the same birthday?

Solution We will assume that all birthdays are equally likely, and for the sake of simplicity we will ignore leap year's day (February 29). The experiment is to ask three people their birthdays. One possible outcome is (May 1, May 3, August 23). The sample space is the set of all possible lists of three birthdays. In order to find the number of elements in the sample space, we follow the flowchart in Figure 5.27. The selected items are birthdays. They can be repeated (people can share the same birthday), so we must use the fundamental principle of counting. We make three boxes, one for each birthday. The first birthday may be selected in any of 365 different ways, since there are 365 different days in a year. The second birthday may be selected in any of 365 ways also, as can the third birthday, because people can share the same birthday. Thus, the number of elements in the sample space is

$$n(S) = \boxed{365} \cdot \boxed{365} \cdot \boxed{365}$$

The event E is the set of all possible lists of three birthdays in which at least two of those birthdays are the same. It is rather difficult to compute $n(E)$ directly; instead, we will compute $n(E')$ and use a probability rule. E' is the set of all possible lists of three birthdays in which no two of those birthdays are the same. In order to find the number of elements in

E', we again follow the flowchart in Figure 5.27. The birthdays *cannot* be repeated, there is only one category (birthdays), and the order of selection does matter [(May 1, May 3, August 23) is a different list than (August 23, May 3, May 1)]. Thus, we use permutations, and the number of elements in E' is

$$n(E') = {}_{365}P_3 = \frac{365!}{(365 - 3)!}$$

$$= \frac{365 \cdot 364 \cdot 363 \cdot 362!}{362!}$$

$$= 365 \cdot 364 \cdot 363 \qquad \text{Canceling}$$

We are now ready to compute $p(E)$.

$$p(E) = 1 - p(E') \qquad \text{Probability rule 6}$$

$$= 1 - \frac{n(E')}{n(S)}$$

$$= 1 - \frac{365 \cdot 364 \cdot 363}{365 \cdot 365 \cdot 365}$$

$$= 1 - \frac{364 \cdot 363}{365 \cdot 365} \qquad \text{Canceling}$$

$$= 0.008204 \ldots \approx 0.008$$

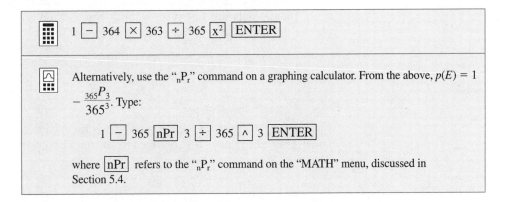

$\boxed{}$ 1 $\boxed{-}$ 364 $\boxed{\times}$ 363 $\boxed{\div}$ 365 $\boxed{x^2}$ $\boxed{\text{ENTER}}$

Alternatively, use the "$_nP_r$" command on a graphing calculator. From the above, $p(E) = 1 - \dfrac{_{365}P_3}{365^3}$. Type:

1 $\boxed{-}$ 365 $\boxed{\text{nPr}}$ 3 $\boxed{\div}$ 365 $\boxed{\wedge}$ 3 $\boxed{\text{ENTER}}$

where $\boxed{\text{nPr}}$ refers to the "$_nP_r$" command on the "MATH" menu, discussed in Section 5.4.

This result is not at all surprising; it means that it is extremely unlikely that two or more people in a group of three share a birthday. You will, however, be surprised in the exercises when the group is increased to 30 members. •

EXAMPLE 2 Connecticut, Louisiana, Oregon, and Virginia all operate 6/44 lotteries; that is, the player selects any six of the numbers from 1 to 44. The state determines six winning numbers, usually by having a mechanical device choose six balls from a container filled with balls numbered from 1 to 44. If a player's six selections match the six winning numbers, the player wins first prize. (Smaller prizes are awarded to players who select five or four of the winning numbers.) Find the probability of winning first prize.

Many states give citizens a chance to win money in lotteries. But how great an opportunity is it?

Solution Let E represent the event "winning first prize." To find the probability, we must first find the number of outcomes in event E and the number of outcomes in the sample space S.

There is only one first-prize-winning combination, so $n(E) = 1$. However, the sample space is huge. One possible selection is

| 1 | 32 | 43 | 4 | 15 | 26 |

Another is

| 41 | 12 | 4 | 24 | 25 | 37 |

Finding $n(S)$ by counting the elements in S is impractical, so we will use combinatorics. Order does not matter, because the player can choose the six numbers in any order. Therefore, we use *combinations*.

$$n(S) = {}_{44}C_6 = \frac{44!}{6! \cdot (44 - 6)!} = 7{,}059{,}052$$

Alternatively, use the "$_nC_r$" command on a graphing calculator. Type

$$44 \;\boxed{\text{nCr}}\; 6 \;\boxed{\text{ENTER}}$$

where $\boxed{\text{nCr}}$ refers to the "$_nC_r$" command on the "MATH" menu, discussed in Section 5.4.

As you can see, the sample space is amazingly huge. The probability of selecting all six winning numbers is

$$p(E) = \frac{n(E)}{n(S)} = \frac{1}{7,059,052} \approx 0.00000014$$

This means that only one out of approximately *seven million* combinations is the first-prize-winning combination. Few events in life are less likely than this. •

One common form of poker is five-card draw, in which each player is dealt five cards. The order in which the cards are dealt is unimportant, so we compute probabilities with combinations rather than permutations.

EXAMPLE 3 Find the probability of being dealt four aces.

Solution The sample space consists of all possible five-card hands that can be dealt from a deck of 52. There are

$$_{52}C_5 = \frac{52!}{5! \cdot 47!} = 2,598,960$$

possible hands.

The event consists of all possible five-card hands that include four aces and one non-ace. This involves two categories (aces and non-aces), so we will use the fundamental counting principle and multiply the number of ways of getting four aces and the number of ways of getting one non-ace. There is

$$_4C_4 = \frac{4!}{4! \cdot 0!} = 1$$

way of getting four aces, and there are

$$_{48}C_1 = \frac{48}{1! \cdot 47!} = 48$$

ways of getting one non-ace. (These numbers could certainly be obtained with common sense rather than combinations.) Thus, the event consists of

$$_4C_4 \cdot {}_{48}C_1 = 1 \cdot 48 = 48$$

elements, and the probability of being dealt four aces is

$$p(E) = \frac{_4C_4 \cdot {}_{48}C_1}{_{52}C_5} = \frac{48}{2,598,960} \approx 0.00001847$$

✔ In the event, there is a distinction between two categories (aces and non-aces); in the sample space, there is no such distinction. Thus, the numerator of

$$p(E) = \frac{_4C_4 \cdot {}_{48}C_1}{_{52}C_5}$$

has two parts (one for each category), and the denominator has one part. Also, the numbers in front of the C's add correctly (4 aces + 48 non-aces = 52 cards to choose from), and the numbers after the C's add correctly (4 aces + 1 non-ace = 5 cards to select). •

Lotteries and Keno

The first lotteries appeared during the 15th century in France and Belgium, when cities used lotteries to raise money to fortify their defenses and to aid the poor. The first lottery that paid cash prizes was probably started in Florence, Italy, in 1530. It was so successful that many other Italian cities began to offer their own lotteries. When Italian cities united to form a nation, the first national lottery was created. *Lotto,* the Italian national lottery, continues today and is regarded as the basis for such modern gambling games as keno, state lotteries, bingo, and the illegal numbers game.

Public lotteries have a long history in the United States. The settlement of Jamestown was financed in part by an English lottery. George Washington managed a lottery that paid for a road through the Cumberland mountains. Several universities, including Harvard, Dartmouth, Yale, and Columbia, were partly financed by lotteries.

The game of keno is a casino version of a lottery. In this game, the casino has a container filled with balls numbered from 1 to 80. The player buys a keno ticket, with which he or she selects anywhere from 1 to 15 (usually 6, 8, 9, or 10) of these 80 numbers; the player's selections are called "spots." The casino chooses 20 winning numbers, using a mechanical device to ensure a fair game. If a sufficient number of the player's spots are winning numbers, the player receives an appropriate payoff.

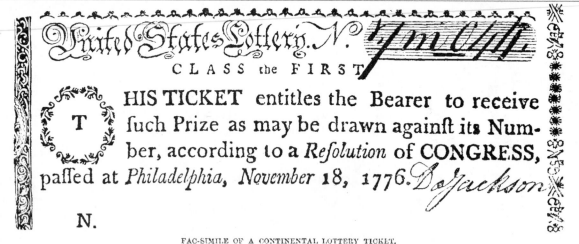

The United States Congress operated a lottery to help fund the Revolutionary War.

EXAMPLE 4   Find the probability of being dealt four of a kind.

Solution   The sample space is the same as in Example 3, and the event is very similar. The number of ways of getting four twos or four kings is the same as the number of ways of getting four aces, and there are 13 denominations (two through ace). Therefore, the number of ways of getting four cards of the same denomination is

$$13 \cdot {}_4C_4$$

and the probability of being dealt four of a kind is

$$\frac{13 \cdot {}_4C_4 \cdot {}_{48}C_1}{{}_{52}C_5} = \frac{13 \cdot 1 \cdot 48}{2{,}598{,}960} = \frac{624}{2{,}598{,}960} \approx 0.0002401$$

See Example 8 in Section 5.4.

✔ The numerator of

$$p(E) = \frac{n(E)}{n(S)} = \frac{13 \cdot {}_4C_4 \cdot {}_{48}C_1}{{}_{52}C_5}$$

has two parts (one for each category) and the denominator has one part. The numbers in front of the C's add correctly (4 of one denomination + 48 of another = 52 cards to choose from), and the numbers after the C's add correctly (4 of a kind + 1 other = 5 cards to select). ●

EXAMPLE 5 Find the probability of being dealt five hearts.

Solution The sample space is the same as in Example 4. The event consists of all possible five-card hands that include five hearts and no non-hearts. This involves two categories (hearts and non-hearts), so we will use the fundamental counting principle and multiply the number of ways of getting five hearts and the number of ways of getting no non-hearts. There are

$${}_{13}C_5 = \frac{13!}{5! \cdot 8!} = 1287$$

ways of getting five hearts, and there is

$${}_{39}C_0 = \frac{39!}{0! \cdot 39!} = 1$$

way of getting no non-hearts. Thus, the probability of being dealt five hearts is

$$p(E) = \frac{{}_{13}C_5 \cdot {}_{39}C_0}{{}_{52}C_5} = \frac{1287 \cdot 1}{2{,}598{,}960} \approx 0.000495198$$

✔ In the event, there is a distinction between two categories (hearts and non-hearts); in the sample space, there is no such distinction. Thus, the numerator of

$$p(E) = \frac{{}_{13}C_5 \cdot {}_{39}C_0}{{}_{52}C_5}$$

has two parts (one for each category), and the denominator has one part. Also, the numbers in front of the C's add correctly (13 hearts + 39 non-hearts = 52 cards to choose from), and the numbers after the C's add correctly (5 hearts + 0 non-hearts = 5 cards to select). ●

Notice that in Example 5 we could argue that since we're only selecting hearts, we can disregard the non-hearts. This would lead to the answer obtained in Example 5.

$$p(E) = \frac{_{13}C_5}{_{52}C_5} = \frac{1287}{2,598,960} \approx 0.000495198$$

However, this approach would not allow us to check our work in the manner described above; the numbers in front of the C's don't add correctly, nor do the numbers after the C's.

6.4

EXERCISES

1. A group of 30 people is selected at random. What is the probability that at least two of them will have the same birthday?

2. A group of 60 people is selected at random. What is the probability that at least two of them will have the same birthday?

3. In 1990, California switched from a 6/49 lottery to a 6/53 lottery. Now the state has a 6/51 lottery.
   a. Find the probability of winning first prize in a 6/49 lottery.
   b. Find the probability of winning first prize in a 6/53 lottery.
   c. Find the probability of winning first prize in a 6/51 lottery.
   d. How much more probable is it that one will win the 6/49 lottery than the 6/53 lottery?
   e. Why do you think California switched from a 6/49 lottery to a 6/53 lottery? And why do you think the state then switched to a 6/51 lottery? (Answer using complete sentences.)

4. Find the probability of winning second prize—that is, picking five of the six winning numbers—with a 6/53 lottery.

5. Find the probability of winning second prize (that is, picking five of the six winning numbers) with a 6/44 lottery, as played in Connecticut, Louisiana, Oregon, and Virginia.

6. Find the probability of winning third prize—that is, picking four of the six winning numbers—with a 6/44 lottery.

7. Currently, the most popular type of lottery is the 5/35 lottery. It is played in Arizona, Illinois, Massachusetts, Connecticut, Iowa, Kentucky, South Dakota, Maine, New Hampshire, and Vermont.
   a. Find the probability of winning first prize.
   b. Find the probability of winning second prize.

8. The second most popular type of lottery is the 6/49 lottery. It is currently played in Massachusetts, Michigan, Florida, Kentucky, Maryland, Washington, and Wisconsin.

   a. Find the probability of winning first prize.
   b. Find the probability of winning second prize.

9. The 5/39 lottery is currently played in California, New York, Michigan, Minnesota, Pennsylvania, and Maryland.
   a. Find the probability of winning first prize.
   b. Find the probability of winning second prize.

10. The 6/42 lottery is currently played in Arizona, Colorado, and Massachusetts.
    a. Find the probability of winning first prize.
    b. Find the probability of winning second prize.

11. There is an amazing variety of lotteries played in the United States. Currently, the following lotteries are played: 5/26, 5/32, 5/34, 5/35, 5/36, 5/37, 5/38, 5/39, 5/40, 5/42, 5/52, 6/25, 6/30, 6/33, 6/36, 6/39, 6/40, 6/41, 6/42, 6/44, 6/46, 6/47, 6/48, 6/49, 6/50, 6/51, and 6/54. Which is the easiest to win? Which is the hardest to win? Explain your reasoning.
    HINT: It isn't necessary to compute every single probability.

12. In the game of keno, if six spots are marked, the player wins if four or more of his or her spots are selected. Complete the following probability distribution.

Outcome	Probability
6 winning spots	
5 winning spots	
4 winning spots	
3 winning spots	
Less than 3 winning spots	

13. In the game of keno, if eight spots are marked, the player wins if five or more of his or her spots are selected. Complete the following probability distribution.

Outcome	Probability
8 winning spots	
7 winning spots	
6 winning spots	
5 winning spots	
4 winning spots	
Less than 4 winning spots	

14. In the game of keno, if nine spots are marked, the player wins if six or more of his or her spots are selected. Complete the following probability distribution.

Outcome	Probability
9 winning spots	
8 winning spots	
7 winning spots	
6 winning spots	
5 winning spots	
Less than 5 winning spots	

15. a. Find the probability of being dealt five spades when playing five-card draw poker.
  b. Find the probability of being dealt five cards of the same suit when playing five-card draw poker.
  c. When you are dealt five cards of the same suit, you have either a *flush* (if the cards are not in sequence) or a *straight flush* (if the cards are in sequence). For each suit, there are ten possible straight flushes ("ace, two, three, four, five," through "ten, jack, queen, king, ace"). Find the probability of being dealt a straight flush.
  d. Find the probability of being dealt a flush.

16. a. Find the probability of being dealt an "aces over kings" full house (three aces and two kings).
  b. Why are there 13 · 12 different types of full houses?
  c. Find the probability of being dealt a full house.

You order twelve burritos to go from a Mexican restaurant, five with hot peppers and seven without. However, the restaurant forgot to label them. If you pick three burritos at random, find the probability of each event in Exercises 17–24.

17. All have hot peppers.

18. None have hot peppers.

19. Exactly one has hot peppers.

20. Exactly two have hot peppers.

21. At most one has hot peppers.

22. At least one has hot peppers.

23. At least two have hot peppers.

24. At most two have hot peppers.

25. Two hundred people apply for two jobs. Sixty of the applicants are women.
  a. If two persons are selected at random, what is the probability that both are women?
  b. If two persons are selected at random, what is the probability that only one is a woman?
  c. If two persons are selected at random, what is the probability that both are men?
  d. If you were an applicant and the two selected people were not of your gender, should the above probabilities have an impact on your situation? Why?

26. Two hundred people apply for three jobs. Sixty of the applicants are women.
  a. If three persons are selected at random, what is the probability that all are women?
  b. If three persons are selected at random, what is the probability that exactly two are women?

c. If three persons are selected at random, what is the probability that exactly one is a woman?

d. If three persons are selected at random, what is the probability that none are women?

e. If you were an applicant, and the three selected people were not of your gender, should the above probabilities have an impact on your situation? Why?

6.5

PROBABILITY DISTRIBUTIONS AND EXPECTED VALUE

Suppose you're at a casino playing roulette, concentrating on the $1 single-number bet. This bet involves your selecting a single number from a set of 38 numbers. If the casino selects that number also (by spinning a ball in a wheel), you win $35. If the casino selects another number, you lose your dollar. How much should you expect to win or lose, on average, if you were to play this game many times?

The probability of your winning is 1/38, because there are 38 equally likely outcomes and only one of them results in your winning. This means that if you play the game a large number of times, it is most likely that you will win once for every 38 times you place the bet (and lose the other 37 times). Your average winnings would be

$$\frac{\$35 + 37 \cdot (-\$1)}{38} = \frac{-\$2}{38} \approx -\$0.053$$

per game. This is called the *expected value* of your bet, because you should expect to lose about a nickel per game if you play the game a long time. If you play a few times, anything could happen—you could win every single bet (though it's not likely). The house makes the bet so many times that it can be certain that its profit will be $0.053 per dollar bet.

The standard way to find the expected value of an experiment is to make a probability distribution for the experiment that lists the value of each outcome and its probability, as shown in Figure 6.14. Once this augmented probability distribution is completed, multiply the value of each outcome by the probability of that outcome and add the results.

Roulette, the oldest casino game played today, has been popular since it was introduced to Paris in 1765. Does this game have any good bets?

FIGURE 6.14

Outcome	Value	Probability
Win	$35	1/38
Lose	−$1	37/38

The expected value is then

$$(\$35)(1/38) + (-\$1)(37/38) = -\$2/38 \approx -\$0.053$$

It is easy to see that this calculation is algebraically equivalent to the calculation done above.

Finding an expected value of a bet is very similar to finding your average test score in a class. Suppose you're a student in a class in which you've taken four tests. If your scores were 80%, 76%, 90%, and 90%, your average test score would be

$$\frac{80 + 76 + 2 \cdot 90}{4} = 84\%$$

or, equivalently,

$$80 \cdot \frac{1}{4} + 76 \cdot \frac{1}{4} + 90 \cdot \frac{2}{4}$$

The difference between finding an average test score and finding the expected value of a bet is that with the average test score you are summarizing what *has* happened, whereas with a bet you are using probabilities to project what *will* happen.

Expected Value

To find the **expected value** (or "long-term average") of an experiment, multiply the value of each outcome of the experiment by its probability and add the results.

EXAMPLE 1 By analyzing her sales records, a saleswoman has found that her weekly commission has the following probability distribution.

Commission	0	$100	$200	$300	$400
Probability	0.05	0.15	0.25	0.45	0.1

Find the saleswoman's expected commission.

Solution To find the expected commission, we multiply each possible commission by its probability and add the results. Therefore,

$$\text{expected commissions} = (0)(0.05) + (100)(0.15) + (200)(0.25)$$
$$+ (300)(0.45) + (400)(0.1)$$
$$= 240$$

On the basis of her history, the saleswoman should expect to average $240 per week in future commissions. Certainly, anything can happen in the future—she could receive a $700 commission (it's not likely, though, because it has never happened before).   •

Why the House Wins

Four of the "best" bets that can be made in a casino game of chance are the pass, don't pass, come, and don't come bets in craps. They all have almost the exact same expected value, −$0.014. In the long run, *there isn't a single bet in any game of chance with which you can expect to break even, let alone make a profit.* After all, the casinos are out to make money. The expected values for $1 bets in the more common games are shown in Figure 6.15.

FIGURE 6.15
Expected values of common games of chance

Game	Expected Value of $1 Bet
Baccarat	−$0.014
Blackjack	−$0.06 to +$0.10 (varies with strategies)
Craps	−$0.014 for pass, don't pass, come, don't come bets *only*
Slot machines	−$0.13 to ? (varies)
Keno (eight-spot ticket)	−$0.29
Many state lotteries	−$0.50

It is possible to achieve a positive expected value in blackjack and other card games in which a number of hands are played from the same deck without reshuffling before each hand. To do this, the player must memorize which cards have been played in previous hands. Many casinos use four decks at once to discourage memorization. Some people try to sneak small homemade computers into the casino to aid them. Naturally, the casinos forbid this.

Decision Theory

Which is the better bet—a $1 single-number bet in roulette or a lottery ticket? Each costs $1. The roulette bet pays $35, but the lottery ticket might pay several million dollars. Lotteries are successful in part because the possibility of winning a large amount of money distracts people from the fact that winning is extremely unlikely. In Example 2 of Section 6.4, we found that the probability of winning first prize in many state lotteries is $\frac{1}{7,059,052} \approx 0.00000014$. At the beginning of this section, we found that the probability of winning the roulette bet is $\frac{1}{38} \approx 0.03$.

A more informed decision would take into account not only the potential winnings and losses but also their probabilities. The expected value of a bet does just that, since its calculation involves both the value and the probability of each outcome. We found that the expected value of a $1 single-number bet in roulette is about −$0.053. The expected value of many state lotteries is −$0.50 (see Figure 6.15). The roulette bet is a much better bet than is the lottery. (Of course, there is a third option, which has an even better expected value of $0.00. Not gambling!)

A decision always involves choosing between various alternatives. If you compare the expected values of the alternatives, then you are taking into account the alternatives' potential winnings and losses as well as their probabilities. This form of decision making is called **decision theory.**

EXAMPLE 2   The saleswoman in Example 1 has been offered a new job that has a fixed weekly salary of $290. Financially, which is the better job?

Solution   In Example 1, we found that her expected weekly commission was $240. The new job has a guaranteed weekly salary of $290. Financial considerations indicate that she should take the new job.   •

6.5

EXERCISES

1. Based on his previous experience, the public librarian at Smallville knows that the number of books checked out by a person visiting the library has the following probability distribution:

Number of Books	0	1	2	3	4	5
Probability	0.15	0.35	0.25	0.15	0.05	0.05

Find the expected number of books checked out by a person visiting this library.

2. Based on his sales records, a salesman knows that his weekly commissions have the following probability distribution:

Commission	0	$1,000	$2,000	$3,000	$4,000
Probability	0.15	0.2	0.45	0.1	0.1

Find the salesman's expected commission.

3. Of all workers at a certain factory, the proportions earning certain hourly wages are as follows:

Hourly Wage	$8.50	$9.00	$9.50	$10.00	$12.50	$15.00
Proportion	20%	15%	25%	20%	15%	5%

Find the expected hourly wage that a worker at this factory makes.

4. Of all students at the University of Metropolis, the proportions taking certain numbers of units are shown across the bottom of the page. Find the expected number of units that a student at U.M. takes.

5. Show why the calculation at the top of page 300 is algebraically equivalent to the calculation on the same page below Figure 6.14.

6. In Example 1, the saleswoman's most likely weekly commission was $300 (since that commission has the highest probability). With her new job (in Example 2), she will always make $290 per week. This implies that she would be better off with the old job. Is this reasoning more or less valid than that used in Example 2? Why?

7. Maria just inherited $10,000. Her bank has a savings account that pays 4.1% interest per year. Some of her friends recommended a new mutual fund, which has been in business for three years. During its first year, the fund went up in value by 10%; during the second year, it went down by 19%; and during its third year, it went up by 14%. She is attracted by the mutual fund's potential for relatively high earnings but concerned by the possibility of actually losing some of her inheritance. The bank's rate is low, but it is insured by the federal government. Use decision theory to find the best investment. (Assume that the fund's past behavior predicts its future behavior.)

8. Trang has saved $8000. It is currently in a bank savings account that pays 3.9% interest per year. He is considering putting the money into a speculative investment that would either earn 20% in one year if the investment succeeds or lose 18% in one year if it fails. At what probability of success would the speculative investment be the better choice?

Units	3	4	5	6	7	8	9	10	11	12	13	14
Proportion	3%	4%	5%	6%	5%	4%	8%	12%	13%	13%	15%	12%

9. Erica has her savings in a bank account that pays 4.5% interest per year. She is considering buying stock in a pharmaceuticals company that is developing a cure for cellulite. Her research indicates that she could earn 50% in one year if the cure is successful or lose 60% in one year if it is not. At what probability of success would the pharmaceuticals stock be the better choice?

10. Debra is buying prizes for a game at her school's fund-raiser. The game has three levels of prizes, and she has already bought the second and third prizes. She wants the first prize to be nice enough to attract people to the game. The game's manufacturer has supplied her with the probabilities of winning first, second, and third prizes. Tickets cost $3 each, and she wants the school to profit an average of $1 per ticket. How much should she spend on each first prize?

Prize	Cost of Prize	Probability
1st	?	.15
2nd	$1.25	.30
3rd	$0.75	.45

11. Few students manage to complete their schooling without taking a standardized admissions test such as the Scholastic Achievement Test, or S.A.T. (used for admission to college); the Law School Admissions Test, or L.S.A.T.; and the Graduate Record Exam, or G.R.E. (used for admission to graduate school). Sometimes, these multiple-choice tests discourage guessing by subtracting points for wrong answers. In particular, a correct answer will be worth +1 point, and an incorrect answer on a question with 5 listed answers (a through e) will be worth $-\frac{1}{4}$ point.
    a. Find the expected value of a random guess.
    b. Find the expected value of eliminating one answer and guessing between the remaining 4 possible answers.
    c. Find the expected value of eliminating three answers and guessing between the remaining 2 possible answers.
    d. Use decision theory and your answers to parts (a), (b), and (c) to create a guessing strategy for standardized tests such as the S.A.T.

12. Find the expected value of a $1 bet in six-spot keno if three winning spots pay $1 (but you pay $1 to play, so you actually break even), four winning spots pay $3 (but you pay $1 to play, so you profit $2), five pay $100, and six pay $2600. (You might want to use the probability distribution computed in Exercise 12 of Section 6.4.)

13. Find the expected value of a $1 bet in eight-spot keno if four winning spots pay $1 (but you pay $1 to play, so you actually break even), five winning spots pay $5 (but you pay $1 to play, so you profit $4), six winning spots pay $100, seven winning spots pay $1480, and eight winning spots pay $19,000. (You might want to use the probability distribution computed in Exercise 13 of Section 6.4.)

14. Find the expected value of a $1 bet in nine-spot keno if five winning spots pay $1 (but you pay $1 to play, so you actually break even), six winning spots pay $50 (but you pay $1 to play, so you profit $49), seven pay $390, eight pay $6000, and nine pay $25,000. (You might want to use the probability distribution computed in Exercise 14 of Section 6.4.)

15. Arizona's "Fantasy Five" is a 5/35 lottery. It differs from many other state lotteries in that its payouts are set; they do not vary with sales. To win first prize, you must select all 5 of the winning numbers. To win second prize, you must select any 4 of the 5 winning numbers; to win third prize, you must select any 3 of the 5 winning numbers. The first prize jackpot is $50,000 (but you pay $1 to play, so you profit $49,999). Second prize pays $500, and third prize pays $5. If you select two or fewer winning numbers, you lose your $1. Find the expected value of the Fantasy Five. (You might want to use the probabilities computed in Exercise 7 of Section 6.4.)

16. Write a paragraph in which you compare the states' fiscal policies concerning their lotteries with the casinos' fiscal policies concerning their keno games. Assume that the expected value of Arizona's Fantasy Five is representative of that of other states' lotteries, and assume that the expected value of a $1 keno bet as described in Exercises 12–14 is representative of that of other keno bets.

17. Trustworthy Insurance Co. estimates that a certain home has a 1% chance of burning down in any one year. They calculate that it would cost $120,000 to rebuild that home. What should Trustworthy's annual premium for fire insurance be on this home?

18. Mr. and Mrs. Trump have applied to the Trustworthy Insurance Co. for insurance on Mrs. Trump's diamond tiara. The tiara is valued at $97,500. Trustworthy estimates that the jewelry has a 2.3% chance of being stolen in any one year. What should Trustworthy's annual premium for insurance be on his home?

19. The Black Gold Oil Co. is considering drilling either in Jed Clampett's back yard or his front yard. After thorough testing and analysis, they estimate that there is a 30% chance of striking oil in the back yard, and a 40% chance in the front yard. They also estimate that the back yard site would either net $60 million (if oil is found) or lose $6 million (if oil is not found), and the front yard site would either net 40 million or lose $6 million. Use decision theory to determine where they should drill.

6.6

CONDITIONAL PROBABILITY

Probabilities and Polls

Public opinion polls, such as those found in newspapers and magazines and on television, frequently categorize the respondents by sex, age, race, or level of education. This is done so that the reader or listener can make comparisons and observe trends, such as "people over 40 are more likely to support the social security system than are people under 40." The tool that enables us to observe such trends is conditional probability.

In a newspaper poll concerning violence on television, 600 people were asked, "What is your opinion of the amount of violence on prime-time television—is there too much violence on television?" Their responses are indicated in Figure 6.16.

FIGURE 6.16
Results of "Violence on Television" poll

	Yes	No	Don't Know	Total
Men	162	95	23	280
Women	256	45	19	320
Total	418	140	42	600

Is there too much violence on TV?

Six hundred people were surveyed in this poll; that is, the sample space consists of 600 responses. Of these, 418 said they thought there was too much violence on television, so the probability of a "yes" response is $\frac{418}{600}$, or about $0.70 = 70\%$. The probability of a "no" response is $\frac{140}{600}$, or about $0.23 = 23\%$.

If we are asked to find the probability that a *woman* responded yes, we do not consider all 600 responses but instead limit the sample space to only the responses from women.

	Yes	No	Don't Know	Total
Women	256	45	19	320

The probability that a woman responded yes is $\frac{256}{320} = 0.80 = 80\%$.

Suppose we label the events in the following manner: W is the event that a response is from a woman, M is the event that a response is from a man, Y is the event that a response is yes, and N is the event that a response is no. Then the event that a woman responded yes would be written as

$$Y \mid W$$

The vertical bar stands for the phrase "given that"; the event $Y \mid W$ is read "a response is yes, given that the response is from a woman." The probability of this event is called a **conditional probability**:

$$p(Y \mid W) = \frac{256}{320} = \frac{4}{5} = 0.80 = 80\%$$

The numerator of this probability, 256, is the number of responses that are yes and are from women; that is, $n(Y \cap W) = 256$. The denominator, 320, is the number of responses that are from women; that is, $n(W) = 320$. A *conditional probability* is a probability whose sample space has been limited to only those outcomes that fulfill a certain condition. Because an event is a subset of the sample space, the event must also fulfill that condition. The numerator of $p(Y \mid W)$ is 256 rather than 418 even though there were 418 yes responses, because many of those 418 responses were made by men; we are interested only in the probability that a woman responded yes.

Conditional Probability of an Event with Equally Likely Outcomes

The *conditional probability* of event A given event B is

$$p(A \mid B) = \frac{n(A \cap B)}{n(B)}$$

if the outcomes are equally likely.

EXAMPLE 1   Using the data in Figure 6.16, find the following.

a. the probability that a response is yes, given that the response is from a man
b. the probability that a response is from a man, given that the response is yes
c. the probability that a response is yes and is from a man

Solution **a.** *Finding* $p(Y \mid M)$: We are told to consider only the male responses—that is, to limit our sample space to men.

	Yes	No	Don't Know	Total
Men	162	95	23	280

$$p(Y \mid M) = \frac{n(Y \cap M)}{n(M)} = \frac{162}{280} \approx 0.58 = 58\%$$

In other words, approximately 58% of the men responded yes. (Recall that 80% of the women responded yes. This poll indicates that men and women do not have the same opinion regarding violence on television and, in particular, that a woman is more likely to oppose the violence.)

b. *Finding* $p(M \mid Y)$: We are told to consider only the yes responses.

	Yes
Men	162
Women	256
Total	418

$$p(M \mid Y) = \frac{n(M \cap Y)}{n(Y)} = \frac{162}{418} \approx 0.39 = 39\%$$

Therefore, of those who responded yes, approximately 39% were male.

c. *Finding* $p(Y \cap M)$: This is *not* a conditional probability (there is no vertical bar), so we do *not* limit our sample space.

$$p(Y \cap M) = \frac{n(Y \cap M)}{n(S)} = \frac{162}{600} = 0.27 = 27\%$$

Therefore, of all those polled, 27% were men who responded yes. ●

Notice that in Example 1, each of the three probabilities has the same numerator [$n(Y \cap M)$, the number of responses that are yes and are from men] but a different denominator. The denominator in (a) is the number of male responses, and in (b) it is the number of yes responses. In (c), the probability is not a conditional probability, so its sample space is not limited, and the denominator is the entire original sample space of 600 responses. *If you calculate a conditional probability incorrectly, check whether you are using the correct limited sample space.*

The conditional probability of A given B is defined as

$$p(A \mid B) = \frac{n(A \cap B)}{n(B)}$$

only if all outcomes are equally likely. The following definition applies regardless of whether outcomes are equally likely.

Conditional Probability Definition

The *conditional probability* of event A given event B is

$$p(A \mid B) = \frac{p(A \cap B)}{p(B)}$$

The above definition differs from the earlier definition in that it defines $p(A \mid B)$ in terms of the *probabilities* $p(A \cap B)$ and $p(B)$, whereas the earlier definition defines $p(A \mid B)$ in terms of the *cardinal numbers* $n(A \cap B)$ and $n(B)$.

We could have used this definition of conditional probability in doing Example 1(a). This would involve first computing $p(Y \cap M)$ and $p(M)$, and then dividing.

$$p(Y \cap M) = 0.27 \qquad \text{From Example 1(c)}$$

$$p(M) = \frac{n(M)}{n(S)} = \frac{280}{600} = 0.466 \ldots$$

$$p(Y \mid M) = \frac{p(Y \cap M)}{p(M)} = \frac{0.27}{0.466 \ldots} \approx 0.58 = 58\%$$

Either the cardinal number method [used in Example 1(a)] or the probabilities method (used above) can be used when outcomes are equally likely. The cardinal number method involves fewer computations and is more intuitive, but it can be used only when outcomes are equally likely.

It is easy to see why these two definitions are equivalent if outcomes are equally likely.

$$p(A \mid B) = \frac{p(A \cap B)}{p(B)} \qquad \text{The second definition of conditional probability}$$

$$= \frac{\dfrac{n(A \cap B)}{n(S)}}{\dfrac{n(B)}{n(S)}} \qquad \text{Definition of probability}$$

$$= \frac{\dfrac{n(A \cap B)}{\cancel{n(S)}}}{\dfrac{n(B)}{\cancel{n(S)}}} \cdot \frac{\cancel{n(S)}}{\cancel{n(S)}} \qquad \text{Multiplying by } \frac{n(S)}{n(S)} = 1 \text{ and canceling}$$

$$= \frac{n(A \cap B)}{n(B)} \qquad \text{The original definition of conditional probability}$$

EXAMPLE 2   A man and a woman have a child. Both parents are cystic fibrosis carriers. They know that their child does not have cystic fibrosis because she shows no symptoms, but they are concerned that she might be a carrier. Find the probability that she is a carrier.

Solution   Figure 6.17 shows the Punnett square and probability distribution for the child.

FIGURE 6.17

	C	**c**
C	CC	Cc
c	Cc	cc

Outcome	Probability
CC	1/4
Cc	2/4 5 1/2
cc	1/4

We are asked to find $p(\text{carrier} \mid \text{no symptoms})$. We cannot divide cardinal numbers and use

$$p(A \mid B) = \frac{n(A \cap B)}{n(B)}$$

because the outcomes are not equally likely. Instead, we must divide probabilities:

$$p(A \mid B) = \frac{p(A \cap B)}{p(B)}$$

$$p(\text{carrier} \mid \text{no symptoms}) = \frac{p(\text{carrier and no symptoms})}{p(\text{no symptoms})}$$

The CC child has no cystic fibrosis gene. That child is not a carrier and has no symptoms. The Cc child has one cystic fibrosis gene. That child is a carrier and has no symptoms. Thus, the numerator ("carrier and no symptoms") refers to the Cc child, and the denominator ("no symptoms") refers to the CC child as well as the Cc child.

$$p(\text{carrier} \mid \text{no symptoms}) = \frac{p(\text{carrier and no symptoms})}{p(\text{no symptoms})}$$

$$= \frac{p(\text{Cc})}{p(\text{CC}) + p(\text{Cc})}$$

$$= \frac{2/4}{1/4 + 2/4}$$

$$= \frac{2/4}{3/4} = \frac{2}{3}$$

The Product Rule

If two cards are dealt from a full deck (no jokers), how would you find the probability that both are hearts? The probability that the first card is a heart is easy to find—it's $\frac{13}{52}$, because there are 52 cards in the deck and 13 of them are hearts. The probability that the second card is a heart is more difficult to find. There are only 51 cards left in the deck (one was already dealt), but how many of these are hearts? The number of hearts left in the deck depends on the first card that was dealt. If it was a heart, then there are 12 hearts left in the deck; if it was not a heart, then there are 13 hearts left. We could certainly say that the probability that the second card is a heart *given that the first card was a heart* is $\frac{12}{51}$.

Therefore, the probability that the first card is a heart is $\frac{13}{52}$, and the probability that the second card is a heart given that the first was a heart is $\frac{12}{51}$. How do we put these two probabilities together to find the probability that *both* the first and the second cards are hearts? Should we add them? Subtract them? Multiply them? Divide them?

The answer is obtained by algebraically rewriting the conditional probability definition:

$$p(A \mid B) = \frac{p(A \cap B)}{p(B)}$$

$$p(A \mid B) \cdot p(B) = \frac{p(A \cap B)}{p(B)} \cdot p(B) \qquad \text{Multiplying by } p(B)$$

$$p(A \mid B) \cdot p(B) = p(A \cap B) \qquad \text{Canceling}$$

Product Rule

For any events A and B, the probability of A and B is

$$p(A \cap B) = p(A \mid B) \cdot p(B)$$

EXAMPLE 3   If two cards are dealt from a full deck, find the probability that both are hearts.

Solution

$$
\begin{array}{ccccc}
p(A \cap B) & = & p(A \mid B) & \cdot & p(B) \\
\end{array}
$$

$$p(\text{2nd heart and 1st heart}) = p(\text{2nd heart} \mid \text{1st heart}) \cdot p(\text{1st heart})$$

$$
\begin{array}{ccccc}
 & = & \dfrac{12}{51} & \cdot & \dfrac{13}{52} \\[2ex]
 & = & \dfrac{4}{17} & \cdot & \dfrac{1}{4} \\
\end{array}
$$

$$= \frac{1}{17} \approx 0.06 = 6\%$$

Therefore, there is a 6% probability that both cards are hearts.   •

Tree Diagrams

Many people find that a *tree diagram* helps them understand problems like the one in Example 3, in which an experiment is performed in stages over time. Figure 6.18 on page 311 shows the tree diagram for Example 3. The first column gives a list of the possible outcomes of the first stage of the experiment; in Example 3, the first stage is dealing the first card, and its outcomes are "heart," and "not a heart." The branches leading to those outcomes represent their probabilities. The second column gives a list of the possible outcomes of the second stage of the experiment; in Example 3, the second stage is dealing the second card. A branch leading from a first-stage outcome to a second-stage outcome is the conditional probability $p(\text{2nd stage outcome} \mid \text{1st stage outcome})$.

Looking at the top pair of branches, we see that the first branch stops at "first card is a heart" and the probability is $p(\text{1st heart}) = \frac{13}{52}$. The second branch starts at "first card is a heart" and stops at "second card is a heart" and gives the conditional probability $p(\text{2nd heart} \mid \text{1st heart}) = \frac{12}{51}$. The probability we were asked to calculate in Example 3, $p(\text{1st heart and 2nd heart})$, is that of the top limb:

$$p(\text{1st heart and 2nd heart}) = p(\text{2nd heart} \mid \text{1st heart}) \cdot p(\text{1st heart})$$

$$= \frac{12}{51} \cdot \frac{13}{52}$$

(We use the word **limb** to refer to a sequence of branches that starts at the beginning of the tree.) Notice that the sum of the probabilities of the four limbs is 1.00. Because the four limbs are the only four possible outcomes of the experiment, they must add up to 1.

Conditional probabilities always start at their condition, never at the beginning of the tree. For example $p(\text{2nd heart} \mid \text{1st heart})$ is a conditional probability; its condition is that the first card is a heart. Thus, its branch starts at the box "first card is a heart." However,

FIGURE 6.18

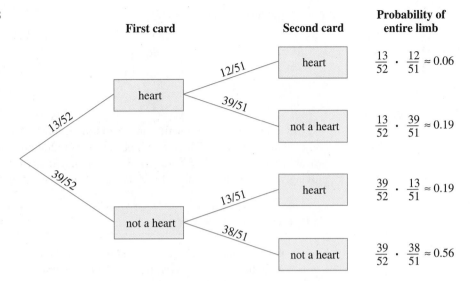

| | **First card** | **Second card** | **Probability of entire limb** |

p(1st heart) is not a conditional probability, so it starts at the beginning of the tree. Similarly, p(1st heart and 2nd heart) is not a conditional probability, so it too starts at the beginning of the tree. The product rule tells us that

$$p(\text{1st heart and 2nd heart}) = p(\text{2nd heart} \mid \text{1st heart}) \cdot p(\text{1st heart})$$

That is, the product rule tells us to multiply the branches that make up the top horizontal limb. In fact, "Multiply when moving horizontally across a limb" is a restatement of the product rule.

EXAMPLE 4   Two cards are drawn from a full deck. Use the tree diagram in Figure 6.18 to find the probability that the second card is a heart.

Solution   The second card can be a heart if the first card is a heart *or* if it is not. The event "the second card is a heart" is the union of the following two mutually exclusive events:

E = 1st heart and 2nd heart

F = 1st not heart and 2nd heart

We previously used the tree diagram to find that

$$p(E) = \frac{13}{52} \cdot \frac{12}{51}$$

Similarly,

$$p(F) = \frac{39}{52} \cdot \frac{13}{51}$$

Thus, we add the probabilities of limbs that result in the second card being a heart:

$$p(\text{2nd heart}) = p(E \cup F)$$
$$= p(E) + p(F) \qquad \text{Probability rule 5}$$
$$= \frac{13}{52} \cdot \frac{12}{51} + \frac{39}{52} \cdot \frac{13}{51} = 0.25$$

In Example 4 above, the first and third limbs represent the only two ways that the second card can be a heart. These two limbs represent mutually exclusive events, so we used probability rule 5 [$p(E \cup F) = p(E) + p(F)$] to add their probabilities. In fact, "add when moving vertically from limb to limb" is a good restatement of probability rule 5.

Tree Diagram Summary

- Conditional probabilities start at their condition.
- Nonconditional probabilities start at the beginning of the tree.
- Multiply when moving horizontally across a limb.
- Add when moving vertically from limb to limb.

EXAMPLE 5 Big Fun Bicycles manufactures its product at two plants, one in Korea and one in Peoria. The Korea plant manufactures 60% of the bicycles; 4% of the Korean bikes are defective; and 5% of the Peorian bikes are defective.

a. Draw a tree diagram that shows this information.
b. Use the tree diagram to find the probability that a bike is defective and came from Korea.
c. Use the tree diagram to find the probability that a bike is defective.
d. Use the tree diagram to find the probability that a bike is defect-free.

Solution a. First, we need to determine which probabilities have been given and find their complements, as shown in Figure 6.19.

FIGURE 6.19

Probabilities Given	Complements of These Probabilities
$p(\text{Korea}) = 60\% = 0.60$	$p(\text{Peoria}) = p(\text{not Korea}) = 1 - 0.60 = 0.40$
$p(\text{defective} \mid \text{Korea}) = 4\% = 0.04$	$p(\text{not defective} \mid \text{Korea}) = 1 - 0.04 = 0.96$
$p(\text{defective} \mid \text{Peoria}) = 5\% = 0.05$	$p(\text{not defective} \mid \text{Peoria}) = 1 - 0.05 = 0.95$

The first two of these probabilities [$p(\text{Korea})$ and $p(\text{Peoria})$] are not conditional, so they start at the beginning of the tree. The next two probabilities [$p(\text{defective} \mid \text{Korea})$ and $p(\text{not defective} \mid \text{Korea})$] are conditional, so they start at their condition (Korea). Similarly, the last two probabilities are conditional, so they start at their condition (Peoria). This placement of the probabilities yields the tree diagram in Figure 6.20.

FIGURE 6.20
Tree diagram for Example 5

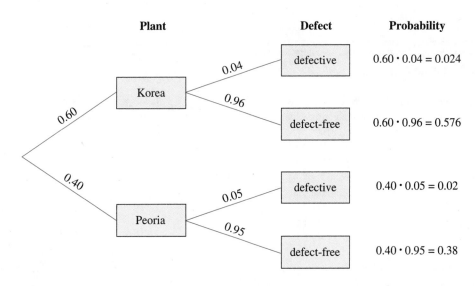

b. The probability that a bike is defective and came from Korea is a nonconditional probability, so it starts at the beginning of the tree. Do not confuse it with the conditional probability that a bike is defective *given that* it came from Korea, which starts at its condition (Korea). The former is the limb that goes through "Korea" and stops at "defective"; the latter is one branch of that limb. We use the product rule to multiply when moving horizontally across a limb.

p(defective and Korea) = p(defective | Korea) · p(Korea)

$\qquad$ = 0.04 · 0.60 = 0.024 $\qquad$ Product rule

This means that 2.4% of all of Big Fun's bikes are defective bikes manufactured in Korea.

c. The event that a bike is defective is the union of two mutually exclusive events:

The bike is defective and came from Korea.

The bike is defective and came from Peoria.

These two events are represented by the first and third limbs of the tree. We use probability rule 5 to add when moving vertically from limb to limb.

p(defective) = p(defective and Korea $\cup$ defective and Peoria)

$\qquad$ = p(defective and Korea) + p(defective and Peoria)

$\qquad$ = 0.024 + 0.02 = 0.044

This means that 4.4% of Big Fun's bicycles are defective.

d. The probability that a bike is defect-free is the complement of (c).

p(defect-free) = p(not defective) = 1 − 0.044 = 0.956

Alternatively, we can find the sum of all the limbs that stop at "defect-free."

p(defect-free) = 0.576 + 0.38 = 0.956

This means that 95.6% of Big Fun's bicycles are defect-free. $\qquad$ •

6.6

EXERCISES

1. Use the data in Figure 6.16 to find the given probabilities. Also, write a sentence explaining what each means.
   - **a.** $p(N)$
   - **b.** $p(W)$
   - **c.** $p(N \mid W)$
   - **d.** $p(W \mid N)$
   - **e.** $p(N \cap W)$
   - **f.** $p(W \cap N)$

2. Use the data in Figure 6.16 to find the given probabilities. Also, write a sentence explaining what each means.
   - **a.** $p(Y)$
   - **b.** $p(M)$
   - **c.** $p(Y \mid M)$
   - **d.** $p(M \mid Y)$
   - **e.** $p(Y \cap M)$
   - **f.** $p(M \cap Y)$

In Exercises 3–6, use Figure 6.21, which gives information on the number of drivers and the number of accidents in 1987. Assume that no driver had more than one accident. (Round off to the nearest hundredth.)

3. Find the probability that a driver had an accident.

4. Find the probability that a driver had an accident, given that the driver was under twenty.

5. Find the probability that a driver had an accident, given that the driver was twenty to twenty-four.

6. Find the probability that a driver had an accident, given that the driver was forty-five to fifty-four.

FIGURE 6.21

Age Group	Number of Drivers	Number of Accidents
Under 20	14,100,000	5,200,000
20–24	16,900,000	5,800,000
25–34	39,800,000	9,100,000
35–44	33,200,000	5,400,000
45–54	23,400,000	2,900,000
55–64	18,500,000	2,200,000
65–74	12,500,000	1,500,000
75 and over	3,600,000	900,000
Total	162,000,000	33,000,000

Source: National Safety Council's *Accident Facts,* 1988 edition.

7. A spinner is designed for a board game so that the arrow can point to either 1, 2, 3, or 4, as shown in Figure 6.22.

FIGURE 6.22

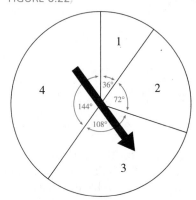

   - **a.** Determine the probability distribution for this spinner. (See Section 6.2, Exercise 1.)
   - **b.** Find the probability of spinning a 1.
   - **c.** Find the probability of spinning a 1, given that an odd number is spun.

8. A spinner is designed for a board game so that the arrow can point to either 1, 2, 3, 4, or 5, as shown in Figure 6.23.

FIGURE 6.23

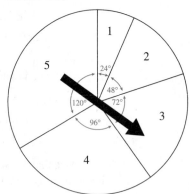

   - **a.** Determine the probability distribution for this spinner. (See Section 6.2, Exercise 2.)
   - **b.** Find the probability of spinning a 3.
   - **c.** Find the probability of spinning a 3, given that an odd number is spun.

9. A new dartboard is designed by a toy manufacturer, as shown in Figure 6.24. (See Section 6.2, Exercise 3.)

FIGURE 6.24

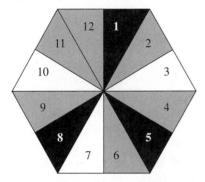

a. Find the probability of hitting a black zone.
b. Find the probability of hitting a black odd-numbered zone.
c. Find the probability of hitting a black zone, given that an odd-numbered zone is hit.

10. A new dartboard is designed by a toy manufacturer, as shown in Figure 6.24. (See Section 6.2, Exercise 4.)
a. Find the probability of hitting a white zone.
b. Find the probability of hitting a white odd-numbered zone.
c. Find the probability of hitting a white zone, given that an odd-numbered zone was hit.

In Exercises 11–14, cards are dealt from a full deck of 52. Find the probabilities of the given events.

11. a. The first card is a club.
b. The second card is a club, given that the first was a club.
c. The first and second cards are both clubs.
d. Draw a tree diagram illustrating this.

12. a. The first card is a king.
b. The second card is a king, given that the first was a king.
c. The first and second cards are both kings.
d. Draw a tree diagram illustrating this.

13. a. The first card is a diamond.
b. The second card is a spade, given that the first was a diamond.
c. The first card is a diamond and the second is a spade.
d. Draw a tree diagram illustrating this.

14. a. The first card is a jack.
b. The second card is an ace, given that the first card was a jack.
c. The first card is a jack and the second is an ace.
d. Draw a tree diagram illustrating this.

*In Exercises 15 and 16, determine which probability the indicated branch in Figure 6.25 refers to. For example, the branch labeled * refers to the probability p(A).*

FIGURE 6.25

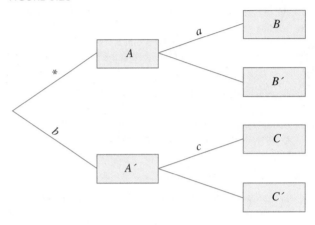

15. a. the branch labeled *a*
b. the branch labeled *b*
c. the branch labeled *c*

16. a. Should probabilities (*) and (*a*) be added or multiplied? What rule tells us that? What is the result of combining them?
b. Should probabilities (*b*) and (*c*) be added or multiplied? What rule tells us that? What is the result of combining them?
c. Should the probabilities that result from parts (*a*) and (*b*) of this exercise be added or multiplied? What rule tells us that? What is the result of combining them?

In Exercises 17 and 18, a single die is rolled. Find the probabilities of the given events.

17. a. rolling a 6
b. rolling a 6, given that the number rolled is even
c. rolling a 6, given that the number rolled is odd
d. rolling an even number, given that a 6 was rolled

18. a. rolling a 5
b. rolling a 5, given that the number rolled is even
c. rolling a 5, given that the number rolled is odd
d. rolling an odd number, given that a 5 was rolled

In Exercises 19–22, a pair of dice is rolled. Find the probabilities of the given events.

19. a. The sum is 6.
b. The sum is 6, given that the sum is even.

   c. The sum is 6, given that the sum is odd.

   d. The sum is even, given that the sum is 6.

20. a. The sum is 12.

   b. The sum is 12, given that the sum is even.

   c. The sum is 12, given that the sum is odd.

   d. The sum is even, given that the sum is 12.

21. a. The sum is 4.

   b. The sum is 4, given that the sum is less than 6.

   c. The sum is less than 6, given that the sum is 4.

22. a. The sum is 11.

   b. The sum is 11, given that the sum is greater than 10.

   c. The sum is greater than 10, given that the sum is 11.

23. A single die is rolled. Determine which of the following events is least likely, and which is most likely. Do so without making any calculations. Explain your reasoning.

   E_1 is the event "rolling a 4."

   E_2 is the event "rolling a 4, given that the number rolled is even."

   E_3 is the event "rolling a 4, given that the number rolled is odd."

24. A pair of dice is rolled. Determine which of the following events is least likely, and which is most likely. Do so without making any calculations. Explain your reasoning.

   E_1 is the event "rolling a 7."

   E_2 is the event "rolling a 7, given that the number rolled is even."

   E_3 is the event "rolling a 7, given that the number rolled is odd."

In Exercises 25 and 26, use the following information. In order to determine the effect their salespeople have on purchases, a department store polled 700 shoppers regarding whether or not they made a purchase and whether or not they were pleased with the service they received. Of those who made a purchase, 125 were happy with the service they received and 111 were not. Of those who made no purchase, 148 were happy with the service they received and 316 were not.

25. Find the probability that a shopper who was happy with the service made a purchase (round off to the nearest hundredth). What can you conclude?

26. Find the probability that a shopper who was unhappy with the service did not make a purchase. (Round off to the nearest hundredth.) What can you conclude?

In Exercises 27–30, five cards are dealt from a full deck. Find the probabilities of the given events. (Round off to four decimal places.)

27. All are spades.

28. The fifth is a spade, given that the first four were spades.

29. The last four are spades, given that the first was a spade.

30. All are the same suit.

In Exercises 31–36, round off to the nearest hundredth.

31. If three cards are dealt from a full deck, use a tree diagram to find the probability that exactly two are spades.

32. If three cards are dealt from a full deck, use a tree diagram to find the probability that exactly one is a spade.

33. If three cards are dealt from a full deck, use a tree diagram to find the probability that exactly one is an ace.

34. If three cards are dealt from a full deck, use a tree diagram to find the probability that exactly two are aces.

35. If a pair of dice is rolled three times, use a tree diagram to find the probability that exactly two throws result in sevens.

36. If a pair of dice is rolled three times, use a tree diagram to find the probability that all three throws result in sevens.

In Exercises 37–40, use the following information: a personal computer manufacturer buys 38% of its chips from Japan and the rest from America. 1.7% of the Japanese chips are defective, and 1.1% of the American chips are defective.

37. Find the probability that a chip is defective and made in Japan.

38. Find the probability that a chip is defective and made in America.

39. Find the probability that a chip is defective.

40. Find the probability that a chip is defect-free.

In Exercises 41–44, use the following information: The University of Metropolis requires its students to pass an examination in college-level mathematics before they can graduate. The students are given three chances to pass the exam; 61% pass it on their first attempt; 63% of those that take it a second time pass it then; and 42% of those that take it a third time pass it then.

41. What percent of the students pass the exam?

42. What percent of the students are not allowed to graduate because of their performance on the exam?

43. What percent of the students take the exam at least twice?

44. What percent of the students take the test three times?

45. In the game of blackjack, if the first two cards dealt to a player are an ace and either a ten, jack, queen, or king, then the player has a "blackjack," and he or she wins. Find the probability that a player is dealt a blackjack out of a full deck (no jokers).

46. In blackjack, the dealer's first card is dealt face up. If that card is an ace, then the player has the option of "taking insurance." ("Insurance" is a side bet. If the dealer has a blackjack, the player wins the insurance bet and is paid 2 to 1 odds. If the dealer does not have a blackjack, the player loses the insurance bet.) Find the probability that the dealer gets a blackjack if his or her first card is an ace.

In Exercises 47 and 48, use the following information: Lily Orteaga owns a video rental store called LilyO's Videos. Lily knows both the age and the address of all of her renters because she requires that renters have a copy of their driver's license on file. During the last three months she accumulated data on the age of every renter and the part of town they live in; those data are given in Figure 6.26.

FIGURE 6.26

Age	Parkview District	Central City	South of Town
Under 20	1287	1582	2038
20–40	3758	3677	3114
Over 40	2895	2156	2066

47. a. A Parkview district renter is most likely to be in what age bracket?
   b. What is the probability of part (a)?
   c. Is the probability found in part (b) conditional? If so, what is given?

48. a. A 20- to 40-year-old renter is most likely to live in what part of town?
   b. What is the probability of part (a)?
   c. Is the probability found in part (b) conditional? If so, what is given?

49. In 1973, the University of California at Berkeley admitted 1494 of 4321 female applicants for graduate study, and 3738 of 8442 male applicants. (*Source:* P. J. Bickel, E. A. Hammel, and J. W. O'Connell, "Sex Bias in Graduate Admissions: Data from Berkeley," *Science,* vol. 187, 7 February 1975.)

a. Find the probability that an applicant was admitted.
b. Find the probability that an applicant was admitted, given that he was male.
c. Find the probability that an applicant was admitted, given that she was female.
d. Do these numbers indicate a possible bias against women?
e. Berkeley's graduate students are admitted by the department to which they apply, rather than by a campus-wide admissions panel. When $p(\text{admission} \mid \text{male})$ and $p(\text{admission} \mid \text{female})$ were computed for each of the school's more than 100 departments, it was found that in four departments, $p(\text{admission} \mid \text{male})$ was greater than $p(\text{admission} \mid \text{female})$ by a significant amount, and that in six departments, $p(\text{admission} \mid \text{male})$ was less than $p(\text{admission} \mid \text{female})$ by a significant amount. Do these data indicate a possible bias against women?
f. What conclusions would you make, and what further information would you obtain, if you were an affirmative action officer for the campus?

50. The authors of "Sex Bias in Graduate Admissions: Data from Berkeley" attempt to explain the paradox in Exercise 49 by discussing an imaginary school with only two departments: "machismatics" and "social warfare." Machismatics admitted 200 of 400 male applicants for graduate study and 100 of 200 female applicants, while social warfare admitted 50 of 150 male applicants for graduate study and 150 of 450 female applicants. For the school as a whole, and for each of the two departments, answer the following questions.
a. What is the probability that an applicant was admitted?
b. What is the probability that an applicant was admitted, given that he was male?
c. What is the probability that an applicant was admitted, given that she was female?
d. Do these numbers indicate a bias against women?
e. How would you explain the paradox illustrated in this problem and in Exercise 49?

51. Figure 6.27 gives information of the incidence of tuberculosis in New York City and in Richmond, Virginia in 1910.

FIGURE 6.27

Source: Morris R. Cohen and Ernest Nagel, *An Introduction to Logic and Scientific Method* (New York: Harcourt Brace & Co., 1934).

| | New York City | | Richmond | |
	Population	TB Deaths	Population	TB Deaths
Caucasian	4,675,000	8400	81,000	130
Non-Caucasian	92,000	500	47,000	160

a. Find the probability that a New York City resident died of tuberculosis.

b. Find the probability that a Caucasian New York City resident died of tuberculosis.

c. Find the probability that a non-Caucasian New York City resident died of tuberculosis.

d. Find the probability that a Richmond resident died of tuberculosis.

e. Find the probability that a Caucasian Richmond resident died of tuberculosis.

f. Find the probability that a non-Caucasian Richmond resident died of tuberculosis.

g. Which city had a more severe problem with tuberculosis?

52. In Exercise 2, find the following probabilities.

a. $p(Y' \mid M)$

b. $p(Y \mid M')$

c. $p(Y' \mid M')$

d. Which event, $Y' \mid M$, $Y \mid M'$, or $Y' \mid M'$, is the complement of the event $Y \mid M$? Why?

53. In Exercise 1, find the following probabilities.

a. $p(N' \mid W)$

b. $p(N \mid W')$

c. $p(N' \mid W')$

d. Which event, $N' \mid W$, $N \mid W'$, or $N' \mid W'$, is the complement of the event $N \mid W$? Why?

54. If A and B are arbitrary events, what is the complement of the event $A \mid B$?

55. Write a paper discussing why it is necessary to assume that no driver had more than one accident in Exercises 3–6.

HINT: In 1987, there were fewer accidents than drivers. In some future year it may be that there will be more accidents than drivers. What would this do to your answer to Exercise 3?

In your paper, explain what further information would be needed to answer Exercises 3–6, and how that information would affect the answers. Would they be larger or smaller than your original answers? Would they differ from your original answers by a relatively large or small amount?

6.7

INDEPENDENCE

Dependent and Independent Events

Consider the dealing of two cards from a full deck. An observer who saw that the first card was a heart would be better able to predict whether the second card will be a heart than another observer who did not see the first card. If the first card was a heart, there is one less heart in the deck, so it is slightly less likely that the second card will be a heart. In particular,

$$p(\text{2nd heart} \mid \text{1st heart}) = \frac{12}{51} \approx 0.24$$

Whereas, as we saw in Example 4 of Section 6.6,

$$p(\text{2nd heart}) = 0.25$$

These two probabilities are different because of the effect the first card drawn has on the second. We say that the two events "first card is a heart" and "second card is a heart" are *dependent;* the result of dealing the second card depends, to some extent, on the result of dealing the first card. In general, two events E and F are **dependent** if $p(E \mid F) \neq p(E)$.

Consider two successive tosses of a single die. An observer who saw that the first toss resulted in a three would be *no better able* to predict whether the second toss will result in a three than another observer who did not observe the first toss. The die does not remember. In particular,

$$p(\text{2nd toss is a three}) = \frac{1}{6}$$

and

$$p(\text{2nd toss is a three} \mid \text{1st toss was a three}) = \frac{1}{6}$$

These two probabilities are the same, because the first toss has no effect on the second toss. We say that the two events "first toss is a three" and "second toss is a three" are **independent;** the result of the second toss does *not* depend on the result of the first toss. In general, two events E and F are *independent* if $p(E \mid F) = p(E)$.

Independence-Dependence Definitions

Two events E and F are *independent* if $p(E \mid F) = p(E)$.
  (*Think: Knowing F does not affect E's probability.*)
Two events E and F are *dependent* if $p(E \mid F) \neq p(E)$.
  (*Think: Knowing F does affect E's probability.*)

Many people have difficulty distinguishing between *independent* and *mutually exclusive*. (Recall that two events E and F are mutually exclusive if $E \cap F = \emptyset$; that is, if one event excludes the other.) This is probably because the relationship between mutually exclusive events and the relationship between independent events both could be described, in a very loose sort of way, by saying that "the two events have nothing to do with each other." *Never think this way;* mentally replacing "mutually exclusive" or "independent" with "having nothing to do with each other" only obscures the distinction between these two concepts. E and F are independent if knowing that F has occurred does not affect the probability that E will occur. E and F are dependent if knowing that F has occurred does affect the probability that E will occur. E and F are mutually exclusive if E and F cannot occur simultaneously.

EXAMPLE 1 Let F be the event "a person has freckles" and R the event "a person has red hair."

a. Are F and R independent?
b. Are F and R mutually exclusive?

Solution **a.** F and R are independent if $p(F \mid R) = p(F)$. With $p(F \mid R)$, we are given that a person has red hair; with $p(F)$, we are not given that information. Does knowing that a person has red hair affect the probability that the person has freckles? Yes, it does; $p(F \mid R) > p(F)$. Therefore, F and R are not independent; they are dependent.

b. F and R are mutually exclusive if $F \cap R = \emptyset$. Many people have both freckles and red hair, so $F \cap R \neq \emptyset$, and F and R are not mutually exclusive. In other words, having freckles does not exclude the possibility of having red hair; freckles and red hair can occur simultaneously. ●

EXAMPLE 2 Let T be the event "a person is tall" and R the event "a person has red hair."

a. Are T and R independent?
b. Are T and R mutually exclusive?

Solution a. T and R are independent if $p(T\,|\,R) = p(T)$. With $p(T\,|\,R)$, we are given that a person has red hair; with $p(T)$, we are not given that information. Does knowing that a person has red hair affect the probability that the person is tall? No, it does not; $p(T\,|\,R) = p(T)$, so T and R are independent.

b. T and R are mutually exclusive if $T \cap R = \varnothing$. $T \cap R$ is the event "a person is tall and has red hair." There are tall people who have red hair, so $T \cap R \neq \varnothing$, and T and R are not mutually exclusive. In other words, being tall does not exclude the possibility of having red hair; being tall and having red hair can occur simultaneously. ●

In Examples 1 and 2, we had to rely on our personal experience in concluding that knowledge that a person has red hair does affect the probability that he or she has freckles and does not affect the probability that he or she is tall. It may be the case that you have seen only one red-haired person, and she was short and without freckles. Independence is better determined by computing the appropriate probabilities than by relying on one's own personal experiences. This is especially crucial in determining the effectiveness of an experimental drug. *Double-blind* experiments, in which neither the patient nor the doctor knows whether the given medication is the experimental drug or an inert substance, are often done to assure reliable, unbiased results.

Independence is an important tool in determining whether an experimental drug is an effective vaccine. Let D be the event that the experimental drug was administered to a patient and R the event that the patient recovered. It is hoped that $p(R\,|\,D) > p(R)$, that is, that the rate of recovery is greater among those who were given the drug. In this case, R and D are dependent. Independence is also an important tool in determining whether an advertisement effectively promotes a product. An ad is effective if p(consumer purchases product $|$ consumer saw ad) $> p$(consumer purchases product).

EXAMPLE 3 Use probabilities to determine whether the events "thinking there is too much violence in television" and "being a man" in Example 1 of Section 6.6 are independent.

Solution Two events E and F are independent if $p(E\,|\,F) = p(E)$. The events "responding yes to the question on violence in television" and "being a man" are independent if $p(Y\,|\,M) = p(Y)$. We need to compute these two probabilities and compare them.

In Example 1 of Section 6.6, we found $p(Y\,|\,M) \approx 0.58$. We can use the data from the poll in Figure 6.16 to find $p(Y)$.

$$p(Y) = \frac{418}{600} \approx 0.70$$

$$p(Y\,|\,M) \neq p(Y)$$

The events "responding yes to the question on violence in television" and "being a man" are dependent. According to the poll, men are less likely to think that there is too much violence on television. ●

In Example 3, what should we conclude if we found that $p(Y) = 0.69$, and $p(Y\,|\,M) = 0.67$? Should we conclude that $p(Y\,|\,M) \neq p(Y)$ and that the events "thinking that there is too much violence on television" and "being a man" are dependent? Or should we conclude that $p(Y \mid M) \approx p(Y)$ and that the events are (probably) independent? In this particular case, the probabilities are relative frequencies rather than theoretical probabilities, and relative frequencies can vary. A group of 600 people was polled in order to determine the opinions of the entire viewing public; if the same question was asked of a different group, a somewhat different set of relative frequencies could result. While it would be reasonable to conclude that the events are (probably) independent, it would be most appropriate to include more people in the poll and make a new comparison.

Product Rule for Independent Events

The product rule says that $p(A \cap B) = p(A \mid B) \cdot p(B)$. If A and B are independent, then $p(A \mid B) = p(A)$. Combining these two equations, we get the following rule:

Product Rule for Independent Events

If A and B are independent events, then the probability of A and B is

$$p(A \cap B) = p(A) \cdot p(B)$$

A common error made in computing probabilities, by both students and professionals, is using the formula $p(A \cap B) = p(A) \cdot p(B)$ without verifying that A and B are independent. In fact, the Federal Aviation Administration (FAA) has stated that this is the most frequently encountered error in probabilistic analyses of airplane component failures. If it is not known that A and B are independent, you must use the product rule $p(A \cap B) = p(A \mid B) \cdot p(B)$.

EXAMPLE 4   If a pair of dice is rolled twice, find the probability that each roll results in a seven.

Solution   In Example 3 of Section 6.3, we found that the probability of a seven is $\frac{1}{6}$. The two rolls are independent (one roll has no influence on the next), so the probability of a seven is $\frac{1}{6}$ regardless of what might have happened on an earlier roll; we can use the product rule for independent events:

$$p(A \cap B) \quad = \quad p(A) \quad \cdot \quad p(B)$$
$$p(\text{1st is 7 and 2nd is 7}) = p(\text{1st is 7}) \cdot p(\text{2nd is 7})$$
$$= \quad \frac{1}{6} \quad \cdot \quad \frac{1}{6}$$
$$= \frac{1}{36}$$

See Figure 6.28 on page 322.

FIGURE 6.28

1st roll   **2nd roll**

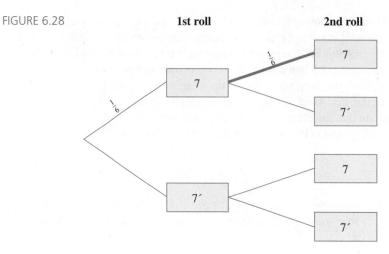

The thicker branch of the tree diagram starts at the event "1st roll is 7" and ends at the event "2nd roll is 7," so it is the conditional probability $p(\text{2nd is 7} \mid \text{1st is 7})$. However, the two rolls are independent, so $p(\text{2nd is 7} \mid \text{1st is 7}) = p(\text{2nd is 7})$. We are free to label this branch as either of these two equivalent probabilities.　•

6.7

EXERCISES

In Exercises 1–8, use your own personal experience with the events described to determine whether (a) E and F are independent and (b) E and F are mutually exclusive. (Where appropriate, these events are meant to be simultaneous; for example, in Exercise 2, "E and F" would mean "it's simultaneously raining and sunny.") Write a sentence justifying each of your answers.

1. *E* is the event "being a doctor," and *F* is the event "being a woman."

2. *E* is the event "it's raining," and *F* is the event "it's sunny."

3. *E* is the event "being single," and *F* is the event "being married."

4. *E* is the event "having naturally blond hair," and *F* is the event "having naturally black hair."

5. *E* is the event "having brown hair," and *F* is the event "having gray hair."

6. *E* is the event "being a plumber," and *F* is the event "being a stamp collector."

7. *E* is the event "wearing shoes," and *F* is the event "wearing sandals."

8. *E* is the event "wearing shoes," and *F* is the event "wearing socks."

In Exercises 9–10, use probabilities, rather than your own personal experience, to determine whether (a) E and F are independent and (b) E and F are mutually exclusive.

9. If a die is rolled once, *E* is the event "getting a 4," and *F* is the event "getting an odd number."

10. If a die is rolled once, and *E* is the event "getting a 4," and *F* is the event "getting an even number."

11. Use probabilities to determine whether the events "responding yes to the question on violence in television"

and "being a woman" in Example 1 of Section 6.6 are independent.

12. Use probabilities to determine if the events "having a defect" and "being manufactured in Peoria" in Example 5 of Section 6.6 are independent.

In Exercises 13 and 14, use the following information: In order to determine the effect their salespeople have on purchases, a department store polled 700 shoppers regarding whether or not they made a purchase, and whether or not they were pleased with the service they received. Of those who made a purchase, 125 were happy with the service they received and 111 were not. Of those who made no purchase, 148 were happy with the service they received and 316 were not.

13. Are the events "being happy with the service" and "making a purchase" independent? What conclusion can you make?

14. Are the events "being unhappy with the service" and "not making a purchase" independent? What conclusion can you make?

15. A personal computer manufacturer buys 38% of its chips from Japan and the rest from America. Of the Japanese chips, 1.7% are defective, whereas 1.1% of the American chips are defective. Are the events "defective" and "Japanese-made" independent? What conclusion can you draw" (See Exercises 37–40 in Section 6.6.)

16. A skateboard manufacturer buys 23% of its ball bearings from a supplier in Akron, 38% from one in Atlanta, and the rest from a supplier in Los Angeles. Of the ball bearings from Akron, 4% are defective; 6.5% of those from Atlanta are defective; and 8.1% of those from Los Angeles are defective.
    a. Find the probability that a ball bearing is defective.
    b. Are the events "defective" and "from the Los Angeles supplier" independent?
    c. Are the events "defective" and "from the Atlanta supplier" independent?
    d. What conclusion can you draw?

17. Suppose that the space shuttle has three separate computer control systems—the main system and two backup duplicates of it. The first backup would monitor the main system and kick in if the main system failed. Similarly, the second backup would monitor the first. We can assume that a failure of one system is independent of a failure of another system, since the systems are separate. The probability of failure for any one system on any one mission is known to be 0.01.
    a. Find the probability that the shuttle is left with no computer control system on a mission.

b. How many backup systems does the space shuttle need if the probability that the shuttle is left with no computer control system on a mission must be $\frac{1}{1 \text{ billion}}$?

18. Recall from Section 6.1 that Antoine Gombauld, the Chevalier de Méré, had lost money over the years by betting with even odds that he could roll at least one pair of 6's in 24 rolls of a pair of dice, and that he could not understand why. This problem finds the probability of winning that bet, and the expected value of the bet.
    a. Find the probability of rolling a double 6 in one roll of a pair of dice.
    b. Find the probability of not rolling a double 6 in one roll of a pair of dice.
    c. Find the probability of never rolling a double 6 in 24 rolls of a pair of dice.
       HINT: This would mean that the first roll is not a double 6 *and* the second roll is not a double 6 *and* the third is not a double 6 . . .
    d. Find the probability of rolling at least one double 6 in 24 rolls of a pair of dice.
       HINT: Use complements.
    e. Find the expected value of this bet if $1 is wagered.

19. Recall from Section 6.1 that Antoine Gombauld, the Chevalier de Méré, had made money over the years by betting with even odds that he could roll at least one 6 in four rolls of a single die. This problem finds the probability of winning that bet, and the expected value of the bet.
    a. Find the probability of rolling a 6 in one roll of one die.
    b. Find the probability of not rolling a 6 in one roll of one die.
    c. Find the probability of never rolling a 6 in four rolls of one die.
       HINT: This would mean that the first roll is not a 6 *and* the second roll is not a 6 *and* the third is not a 6 *and* the fourth is not a 6.
    d. Find the probability of rolling at least one 6 in four rolls of a die.
       HINT: Use complements.
    e. Find the expected value of this bet if $1 is wagered.

20. Probability theory began when Antoine Gombauld, the Chevalier de Méré, asked his friend Blaise Pascal why he had made money over the years betting that he could roll at least one 6 in four rolls of a single die, but he had lost money betting that he could roll at least one pair of 6's in 24 rolls of a pair of dice. Use decision theory and the results of Exercises 18 and 19 to answer Gombauld.

21. Write a paper in which you compare and contrast the concepts of independence and mutual exclusivity.

6.8

BAYES' THEOREM

Trees in Medicine

Usually, medical diagnostic tests are not 100% accurate. A test might indicate the presence of a disease when the patient is in fact healthy (this is called a **false positive**), or it might indicate the absence of a disease when the patient does in fact have the disease (a **false negative**). Probability trees can be used to determine the probability that a person whose test results were positive actually has the disease.

EXAMPLE 1   Medical researchers have recently devised a diagnostic test for "white lung" (an imaginary disease caused by the inhalation of chalk dust). Teachers are particularly susceptible to this disease; studies have shown that half of all teachers are afflicted with it. The test correctly diagnoses the presence of white lung in 99% of the persons who have it and correctly diagnoses its absence in 98% of the persons who do not have it. Find the probability that a teacher whose test results are positive actually has white lung and the probability that a teacher whose test results are negative does not have white lung.

Solution   First, we determine which probabilities have been given and find their complements, as shown in Figure 6.29. We use $+$ to denote the event that a person receives a positive diagnosis and $-$ to denote the event that a person receives a negative diagnosis.

FIGURE 6.29

Probabilities Given	Complements of Those Probabilities
$p(\text{ill}) = 0.50$	$p(\text{healthy}) = p(\text{not ill}) = 1 - 0.50 = 0.50$
$p(- \mid \text{healthy}) = 98\% = 0.98$	$p(+ \mid \text{healthy}) = 1 - 0.98 = 0.02$
$p(+ \mid \text{ill}) = 99\% = 0.99$	$p(- \mid \text{ill}) = 1 - 0.99 = 0.01$

The first two of these probabilities [$p(\text{ill})$ and $p(\text{healthy})$] are not conditional, so they start at the beginning of the tree. The next two probabilities [$p(- \mid$ healthy) and $p(+ \mid$ healthy)] are conditional, so they start at their condition (healthy). Similarly, the last two probabilities are conditional, so they start at their condition (ill). This placement of the probabilities yields the tree diagram in Figure 6.30.

The four probabilities to the right of the tree are:

$p(\text{ill and } +) = 0.495$

$p(\text{ill and } -) = 0.005$

$p(\text{healthy and } +) = 0.01$

$p(\text{healthy and } -) = 0.49$

FIGURE 6.30

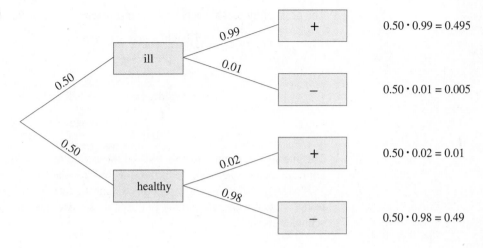

How can we combine these probabilities to find $p(\text{ill} \mid +)$? First, we can rewrite this conditional probability with the conditional probability definition.

$$p(A \mid B) = \frac{p(A \cap B)}{p(B)}$$

$$p(\text{ill} \mid +) = \frac{p(\text{ill and } +)}{p(+)}$$

We have already found the numerator; $p(\text{ill and } +) = 0.495$. In the denominator, $+$ denotes the event that a person receives a positive diagnosis. That event is the union of two mutually exclusive events:

The diagnosis is positive and the person is ill

The diagnosis is positive and the person is healthy

These two events are represented by the first and third limbs of the tree. We use probability rule 5 to add when moving vertically from limb to limb.

$$p(+) = p(+ \text{ and ill} \cup + \text{ and healthy})$$
$$= p(+ \text{ and ill}) + p(+ \text{ and healthy})$$
$$= 0.495 + 0.01 = 0.505$$

Thus

$$p(\text{ill} \mid +) = \frac{p(\text{ill and } +)}{p(+)}$$

$$= \frac{0.495}{0.505} = 0.9801 \ldots \approx 98\%$$

This means that approximately 98% of those who test positive for white lung actually have the disease (and 2% of those who test positive don't have the disease).

The probability that a teacher whose test results are negative does not have white lung is

$$p(\text{healthy} \mid -) = \frac{p(\text{healthy and } -)}{p(-)}$$

$$= \frac{p(\text{healthy and } -)}{p(\text{healthy and } -) + p(\text{ill and } -)}$$

$$= \frac{0.49}{0.49 + 0.005} = 0.98989\ldots \approx 99\%$$

This means that approximately 99% of those who test negative for white lung actually do not have the disease (and 1% of those who test negative do have the disease).

The probabilities show that this diagnostic test works well in determining whether a teacher actually has white lung. In the exercises, we will see how well it would work with schoolchildren. ●

Let's look more carefully at the technique used in Example 1. It involves finding $p(A \mid C)$ using a tree of the form

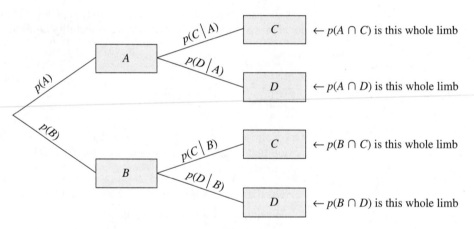

where A and B are mutually exclusive events whose union forms the sample space S. Since we are finding $p(A \mid C)$, we are given C, and we can disregard any branches that do not involve C. Thus, we are concerned only with the following portion of the above tree:

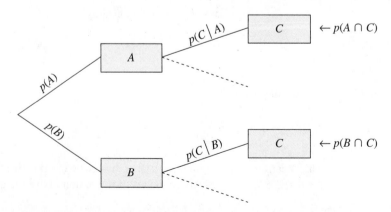

We can rewrite the conditional probability $p(A \mid C)$ with the conditional probability definition.

$$p(A \mid C) = \frac{p(A \cap C)}{p(C)}$$

The numerator, $p(A \cap C)$, is the top limb of our tree. That limb is composed of two branches, $p(A)$ and $p(C \mid A)$. The product rule tells us to multiply when moving horizontally across a limb, so

$$p(A \cap C) = p(C \mid A) \cdot p(A).$$

In the denominator, C is the union of two mutually exclusive events:

$$C = (A \cap C) \cup (B \cap C)$$

These two events are represented by the first and third limbs of the tree. We use probability rule 5 to add when moving vertically from limb to limb.

$$\begin{aligned} p(C) &= p((A \cap C) \cup (B \cap C)) \\ &= p(A \cap C) + p(B \cap C) \end{aligned}$$

Furthermore,

$$p(A \cap C) = p(C \mid A) \cdot p(A)$$

as discussed above. Similarly,

$$p(B \cap C) = p(C \mid B) \cdot p(B)$$

Thus

$$\begin{aligned} p(A \mid C) &= \frac{p(A \cap C)}{p(C)} \\ &= \frac{p(A \cap C)}{p(A \cap C) + p(B \cap C)} \\ &= \frac{p(C \mid A) \cdot p(A)}{p(C \mid A) \cdot p(A) + p(C \mid B) \cdot p(B)} \end{aligned}$$

This result is called **Bayes' theorem.**

Bayes' Theorem

If A and B are mutually exclusive events whose union forms the sample space S, then

$$p(A \mid C) = \frac{p(C \mid A) \cdot p(A)}{p(C \mid A) \cdot p(A) + p(C \mid B) \cdot p(B)}$$

It is not necessary to remember the above formula in order to apply Bayes' theorem. Many people find it easier to make a tree diagram and apply the technique used in Example 1.

6.8

EXERCISES

1. Overwhelmed with their success in diagnosing white lung in teachers, public health officials decided to administer the test to all schoolchildren, even though only one child in 1000 has contracted the disease. Recall from Example 1 that the test correctly diagnoses the presence of white lung in 99% of the persons who have it and correctly diagnoses its absence in 98% of the persons who do not have it.
   a. Find the probability that a schoolchild whose test results are positive actually has white lung.
   b. Find the probability that a schoolchild whose test results are negative does not have white lung.
   c. Find the probability that a schoolchild whose test results are positive does not have white lung.
   d. Find the probability that a schoolchild whose test results are negative actually has white lung.
   e. Which of these events is a false positive? Which is a false negative?
   f. Which of these probabilities would you be interested in if you or one of your family members tested positive?
   g. Discuss the usefulness of this diagnostic test, both for teachers (as in Example 1) and for schoolchildren.

2. In 1996 the Centers for Disease Control estimated that 1,000,000 of the 261,000,000 residents of the United States are HIV-positive. The SUDS diagnostic test correctly diagnoses the presence of AIDS/HIV 99.9% of the time and correctly diagnoses its absence 99.6% of the time.
   a. Find the probability that a person whose test results are positive actually has HIV.
   b. Find the probability that a person whose test results are negative does not have HIV.
   c. Find the probability that a person whose test results are positive does not have HIV.
   d. Find the probability that a person whose test results are negative actually has HIV.
   e. Which of these probabilities would you be interested in if you or someone close to you tested positive?
   f. Which of these events is a false positive? Which is a false negative?
   g. Discuss the usefulness of this diagnostic test.
   h. It has been proposed that all immigrants to the United States should be tested for HIV before being allowed into the country. Discuss this proposal.

3. Assuming that the SUDS test cannot be made more accurate, what changes in the circumstances described in Exercise 2 would increase the usefulness of the SUDS diagnostic test? Give a specific example of this change in circumstances, and demonstrate how it would increase the test's usefulness by computing appropriate probabilities.

4. Compare and contrast the circumstances and the probabilities in Example 1 and in Exercises 1, 2, and 3. Discuss the difficulties in using a diagnostic test when that test is not 100% accurate.

5. Big Fun Bicycles manufactures its product at two plants: one in Korea and one in Peoria. Sixty percent of their bicycles are manufactured at the Korean plant, 4% of the Korean bikes are defective, and 5% of the Peorian bikes are defective. (See Example 5 in Section 6.6.)
   a. What percent of the defective bikes are Korean?
   b. What percent of the defective bikes are Peorian?
   c. What percent of the nondefective bikes are Peorian?

6. A personal computer manufacturer buys 38% of its chips from Japan and the rest from America. 1.7% of the Japanese chips are defective, while 1.1% of the American chips are defective. (See Exercises 37–40 in Section 6.6.)
   a. What percent of the defective bikes are Japanese?
   b. What percent of the defective bikes are American?
   c. What percent of the nondefective bikes are Japanese?

7. The Toyonda Automobile Co. buys turn-signal indicators from three different manufacturers: one in Ohio, one in Kentucky, and one in California. Twenty-five percent of its turn-signal indicators are purchased from the Kentucky manufacturer, 35% are purchased from the Ohio manufacturer, and the rest are purchased from the California manufacturer. Two percent of the Kentucky turn-signal indicators are defective, 4.3% of the Ohio indicators are defective, and 1.7% of the California indicators are defective.
   a. What percent of the defective indicators are made in Kentucky?
   b. What percent of the defective indicators are made in Ohio?
   c. What percent of the defective indicators are made in California?

8. A cable television firm buys television remote control devices from three different manufacturers: one in Korea, one in Singapore, and one in California. Forty-two percent of its remote controls are purchased from the Korean manufacturer, 23% are purchased from the Singapore manufacturer, and the rest are purchased from the California manufacturer. Three percent of the Korean remote controls are defective, 7.1% of the Singapore remote controls are defective, and 1.9% of the California remote controls are defective.

a. What percent of the defective remote controls are made in Korea?

b. What percent of the defective remote controls are made in Singapore?

c. What percent of the defective remote controls are made in California?

9. What would Bayes' theorem state for the tree diagram shown in Figure 6.31?

10. What would Bayes' theorem state for a tree diagram that has four initial possibilities (A_1, A_2, A_3, and A_4) and is otherwise similar to that shown in Figure 6.31?

11. The personnel director of Moon Microcomputers found that 68% of the technical support staff prove to be unsatisfactory because of insufficient mathematics and English skills. She has written a test that hopefully will screen out unqualified applicants. She administers the test to all of the currently employed technical support staff. Eighty-five percent of the satisfactory staff pass the test, and 40% of the unsatisfactory staff pass the test.

a. On the basis of these data, what is the probability that a new applicant who passes the test will turn out to be a satisfactory employee?

b. What is the probability that a new applicant who fails the test would turn out to be an unsatisfactory employee?

12. The admissions officer of the University of Metropolis found that 55% of its students fail to complete their degree. A faculty committee has written a test that hopefully will screen out unqualified applicants. The test is administered to all entering freshmen. Four years later, 63% of those who graduated had passed the test, and 45% of those who hadn't graduated had passed the test.

a. On the basis of these data, what is the probability that a new applicant who passes the test would complete his or her degree?

b. What is the probability that a new applicant who fails the test would not complete his or her degree?

13. Thirty-eight percent of the registered voters of the state of Confusion are Republicans, 51% are Democrats, and the rest are independents. Thirty-one percent of the Republicans voted for a proposition to increase funding for the schools, 76% of the Democrats voted for the proposition, and 79% of the independents voted for the proposition.

a. What percent of those in favor of the proposition were Republicans?

b. What percent of those in favor of the proposition were Democrats?

c. What percent of those opposed to the proposition were Republicans?

 Answer the following questions using complete sentences.

14. Under what circumstances is it appropriate to use Bayes' theorem?

FIGURE 6.31

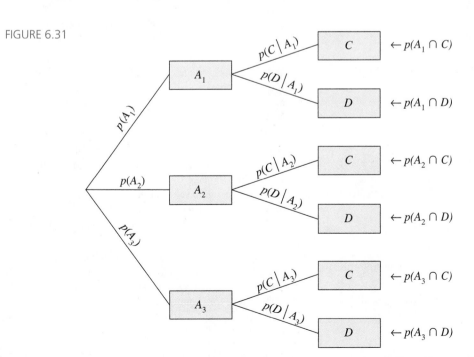

CHAPTER 6 REVIEW

Terms

carrier	dominant gene	impossible event	probability distribution
certain event	event	independent events	Punnett square
codominant	expected value	mutually exclusive events	recessive gene
conditional probability	experiment	odds	relative frequency
decision theory	false positive and false	outcome	sample space
dependent event	negative	probability	

Review Exercises

1. A card is dealt from a well-shuffled deck of 52 cards. Find and interpret the probability and the odds of being dealt each of the following.
   a. a red card
   b. a queen
   c. a club
   d. the queen of clubs
   e. a queen or a club
   f. not a queen

2. Three coins are tossed. Find each of the following.
   a. the experiment
   b. the sample space
   c. the event E that exactly two are tails
   d. the event F that two or more are tails
   e. the probability of E and the odds of E
   f. the probability of F and the odds of F

3. A pair of dice is tossed. Find the probability of rolling each of the following.
   a. a 7
   b. an 11
   c. a 7, an 11, or doubles
   d. a number that's both odd and greater than 8
   e. a number that's either odd or greater than 8
   f. a number that's neither odd nor greater than 8

4. Three cards are dealt from a deck of 52. Find the probability of each of the following.
   a. All three are hearts.
   b. Exactly two are hearts.
   c. At least two are hearts.
   d. The first is an ace of hearts, the second a two of hearts, and the third a three of hearts.

5. A pair of dice is rolled three times. Find the probability of each of the following.
   a. All three are 7's.
   b. Exactly two are 7's.
   c. At least two are 7's.

6. Gregor's Garden Corner buys 40% of its plants from the Green Growery and the balance from Herb's Herbs. Of the plants from the Green Growery, 20% must be returned, and 10% of those from Herb's Herbs must be returned.
   a. Find the probability that a plant must be returned.
   b. Are the events "a plant must be returned" and "a plant was from Herb's Herbs" independent?

7. A long-stemmed pea is dominant over a short-stemmed pea. A pea with one long-stemmed gene and one short-stemmed gene is crossed with a pea with two short-stemmed genes. Find the following.
   a. the probability that the offspring will be long-stemmed
   b. the probability that the offspring will be short-stemmed

8. Cystic fibrosis is caused by a recessive gene. If two cystic fibrosis carriers produce a child, find the probability of each of the following.
   a. The child will have cystic fibrosis.
   b. The child will be a carrier.
   c. The child will neither have the disease nor be a carrier.

9. Sickle-cell anemia is caused by a codominant gene. If two sickle-cell carriers produce a child, find the probability of each of the following.
   a. The child will have the disease.
   b. The child will have sickle-cell trait.
   c. The child will neither have the disease nor have sickle-cell trait.

10. Huntington's disease is caused by a dominant gene. If one parent has Huntington's disease, find the probability of each of the following. (Assume that the affected parent inherited a single gene for Huntington's disease.)
    a. The child will have the disease.
    b. The child will be healthy.

11. Tay-Sachs disease is caused by a recessive gene. If one parent is a Tay-Sachs carrier and the other has no Tay-Sachs gene, find the probability of each of the following.

   a. The child will have the disease.

   b. The child will be a carrier.

   c. The child will neither have the disease nor be a carrier.

12. Jock O'Neill, a sportscaster, and Trudy Bell, a member of the state assembly, are both running for governor of the state of Erehwon. A recent telephone poll asked 800 randomly selected voters for whom they planned to vote. The results of this poll are shown in Figure 6.32.

FIGURE 6.32

	Jock O'Neill	**Trudy Bell**	**Undecided**
Urban Residents	266	184	22
Rural Residents	131	181	16

   a. Find the probability that an urban resident supports O'Neill and the probability that an urban resident supports Bell.

   b. Find the probability that a rural resident supports O'Neill and the probability that a rural resident supports Bell.

   c. Find the probability that an O'Neill supporter lives in an urban area and the probability that an O'Neill supporter lives in a rural area.

   d. Find the probability that a Bell supporter lives in an urban area and the probability that a Bell supporter lives in a rural area.

   e. Where are O'Neill supporters more likely to live? Where are Bell supporters more likely to live?

   f. Which candidate do the urban residents tend to prefer? The rural residents?

   g. Are the events "supporting O'Neill" and "living in an urban area" independent?

   h. Based on the poll, who is ahead in the gubernatorial race?

13. Are the following events independent or dependent? Are they mutually exclusive?

   a. "It's springtime" and "it's sunny."

   b. "It's springtime" and "it's Monday."

   c. "It's springtime" and "it's autumn."

   d. "The first card dealt is an ace" and "the second card dealt is an ace."

   e. "The first roll of the dice results in a 7" and "the second roll results in a 7."

14. The Nissota Automobile Co. buys emergency flashers from two different manufacturers: one in Arkansas and one in Nevada. Thirty-nine percent of its turn-signal indicators are purchased from the Arkansas manufacturer, and the rest are purchased from the Nevada manufacturer. Two percent of the Arkansas turn-signal indicators are defective, and 1.7% of the Nevada indicators are defective.

   a. What percent of the defective indicators are made in Arkansas?

   b. What percent of the defective indicators are made in Nevada?

15. In nine-spot keno, five winning spots break even, six winning spots pay $50, seven pay $390, eight pay $6000, and nine pay $25,000.

   a. Create a probability distribution for this game.

   b. Find the expected value of a $1 bet.

16. Some $1 bets in craps (specifically, a pass, don't pass, come, or don't come bet) have expected values of −$0.014. Use decision theory to compare these bets with a $1 bet in nine-spot keno. (See Exercise 15.)

 Answer the following questions using complete sentences.

17. What is a conditional probability?

18. What is meant by independence?

19. Why are probabilities always between 0 and 1 (inclusive)?

20. Give an example of a permutation and a similar example of a combination.

21. Give an example of two events that are mutually exclusive and an example of two events that are not mutually exclusive.

22. What three mathematicians invented probability theory? Why?

MARKOV CHAINS 7

Markov chains are used to analyze trends and predict the future. They have many applications in business, sociology, the physical sciences, and biology. For example, in business, Markov chains can be used to predict the future success of a product. In sociology, Markov chains can be used to predict the eventual outcome of trends such as the shrinking of the American middle class and the growth of the suburbs at the expense of central cities and rural areas. In this chapter, we investigate these two sociological problems and also see how businesses use Markov chains.

The method of Markov chains consists of using probabilistic information in the analysis of current trends in order to predict their outcomes. Probability trees are used to analyze trends, and matrices are used to summarize the probability trees and simplify the calculations. As an alternative to probability trees, matrices make the work easier.

Andy Warhol's *100 Cans* illustrates America's fascination with products. Markov chains can be used to predict the success of those products.

7.1

INTRODUCTION TO MARKOV CHAINS

Suppose we repeatedly observe a certain changeable quality at successive points in time. For example, a college student might observe her own class standing at the beginning of each school year. These observations vary among elements of the finite set {freshman, sophomore, junior, senior, quit, graduated}. If the student is a freshman in 1998, then she could be a freshman or a sophomore or have quit in 1999, each with a certain probability. If the probability of making a certain observation at a certain time depends only on the immediately preceding observation, and if the observations' outcomes are elements of a finite set, then these observations and their probabilities form a **Markov chain.** Our college student's observations form a Markov chain; the probability that she graduates in 2002 depends only on her status in 2001 and not on her status in any previous year.

Markov chains were developed in the early 1900s by Andrei Markov, a Russian mathematician. They have many applications in the physical sciences, business, sociology, and biology. In business, Markov chains are used to analyze data on customer satisfaction with a product and the effect of the product's advertising and to predict what portion of the market the product will eventually command. In sociology, Markov chains are used to analyze sociological trends, such as the shrinking of the American middle class and the growth of the suburbs at the expense of central cities and rural areas, and to predict the eventual outcome of such trends. Markov chains are also used to predict the weather, to analyze genetic inheritance, and to predict the daily fluctuation in a stock's price.

EXAMPLE 1   Bif is a laundry detergent. The company that makes Bif has just launched a new advertising campaign. A market analysis indicates that, as a result of this campaign, 40% of the consumers who currently *do not* use Bif will buy it the next time they buy a detergent. Another market analysis has studied customer loyalty toward Bif. It indicates that 80% of the consumers who currently *do* use Bif will buy it again the next time they buy a detergent. Rewrite these data in probability form and find their complements.

Solution   We are given

p(next purchase is Bif | current purchase is Bif) = 0.8

p(next purchase is Bif | current purchase is not Bif) = 0.4

We can compute the complements:

p(next purchase is not Bif | current purchase is Bif) = $1 - 0.8 = 0.2$

p(next purchase is not Bif | current purchase is not Bif) = $1 - 0.4 = 0.6$

These relationships are summarized in Figure 7.1.

FIGURE 7.1

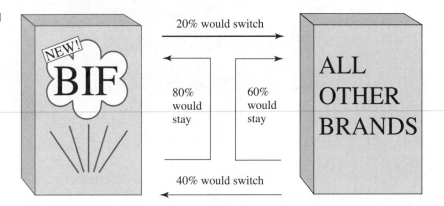

Notice that Example 1 describes a Markov chain because:

- the observations are made at successive points in time (shopping trips at which a detergent is purchased)
- the observations' outcomes are elements of a finite set ("Bif is purchased" and "something other than Bif is purchased")
- the probability of making a certain observation at a certain time depends only on the outcome of the immediately preceding observation (the probability that the next purchase is Bif depends only on the current purchase and not on any previous purchases)

The observations' outcomes are called **states.** A **state** is a condition, form, or stage. The states in Example 1 are "a consumer purchases Bif" ("Bif" or B for short) and "a consumer purchases something other than Bif" ("not Bif" or B′ for short). All states manifest themselves as both **current states** and **following states.** The current states in Example 1 are "a consumer *currently* purchases Bif" and "a consumer *currently* purchases something other than Bif," and the next following states are "a consumer will purchase Bif *the next* *time he or she purchases a detergent"* and "a consumer will purchase something other than Bif *the next time he or she purchases a detergent."*

Transition Matrices

A **transition matrix** is a matrix whose entries are the probabilities of passing from current states to following states. A transition matrix has a row and a column for each state. The rows refer to current manifestations of those states, while the columns refer to later (or following) manifestations of those states. A transition matrix is read like a chart; an entry in a certain row and column represents the probability of making a transition from the current state represented by that row to the following state represented by that column.

EXAMPLE 2   Write the transition matrix T for the data in Example 1.

Solution   Our states are "Bif" and "not Bif" (or just B and B'); therefore, our transition matrix T will be the following 2×2 matrix:

First following
state

$$T = \begin{array}{cc} & \begin{array}{cc} B & B' \end{array} \\ \begin{bmatrix} 0.8 & 0.2 \\ 0.4 & 0.6 \end{bmatrix} & \begin{array}{c} B \\ B' \end{array} \end{array} \quad \text{Current state}$$

The entry 0.4 is the probability that a consumer makes a transition from buying something other than Bif (current state is B') to buying Bif (first following state is B).   •

We can make several observations about transition matrices.

1. *A transition matrix must be square,* because there is one row and one column for each state. If you create a transition matrix that is not square, go back and see which state you did not list as both a current state and a following state.
2. *Each entry in a transition matrix must be between 0 and 1 (inclusive),* because the entries are probabilities. If you create a transition matrix that has an entry that is less than 0 or greater than 1, go back and find your error.
3. *The sum of the entries of any row must be* 1, because the entries of a row are the probabilities of changing from the state represented by that row to *all* of the possible following states. If you create a transition matrix that has a row that does not add to 1, go back and find your error.

Transition Matrix Observations

1. A transition matrix must be square.
2. Each entry in a transition matrix must be between 0 and 1 (inclusive).
3. The sum of the entries of any row must be 1.

The transition matrix in Example 2 shows the transitional trends in the consumers' selection of their next detergent purchases. What effect will these trends have on Bif's success as a product? Will Bif's market share increase or decrease?

Probability Matrices

EXAMPLE 3   A marketing analysis for Bif shows that Bif currently commands 25% of the market. Write this, and its complement, in probability form.

Solution   We are given:

$$p(\text{current purchase is Bif}) = .25$$

Its complement is:

$$p(\text{current purchase is not Bif}) = 1 - .25 = .75 \qquad \bullet$$

A **probability matrix** is a row matrix in which each entry is the probability of a possible state. The columns of a probability matrix must be labeled in the same way as the rows and columns of the transition matrices.

EXAMPLE 4   Write the probability matrix P for the data in Example 3.

Solution   A probability matrix has one row, with a column for each state; our states are "Bif" and "not Bif," so our probability matrix will be a 1×2 matrix.

$$\begin{array}{cc} \text{B} & \text{B}' \end{array}$$
$$P = [.25 \quad .75] \qquad \bullet$$

We can make several observations about probability matrices, which parallel the observations made earlier about transition matrices.

Probability Matrix Observations

1. A probability matrix is a row matrix.
2. Each entry in a probability matrix must be between 0 and 1 (inclusive).
3. The sum of the entries of the row must be 1.

EXAMPLE 5   Use a probability tree to predict Bif's market share after the first following purchase. In other words, find $p(\text{1st following purchase is Bif})$ and $p(\text{1st following purchase is not Bif})$.

Solution   This can be computed with the tree in Figure 7.2 on page 337. We want $p(\text{1st following purchase is Bif})$, which is the sum of the probabilities of the limbs that stop at Bif. Similarly, $p(\text{1st following purchase is not Bif})$ is the sum of the probabilities of the limbs that stop at not Bif.

$$p(\text{1st following purchase is Bif}) = (0.25 \cdot 0.8) + (0.75 \cdot 0.4)$$
$$= 0.2 + 0.3 = 0.5$$
$$p(\text{1st following purchase is not Bif}) = (0.25 \cdot 0.2) + (0.75 \cdot 0.6)$$
$$= 0.05 + 0.45 = 0.5$$

In other words, after the first following purchase, Bif will command 50% of the market (up from a previous 25%). Remember, though, that this is only a prediction, based on the assumption that current trends continue unchanged.

FIGURE 7.2

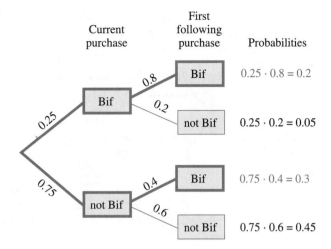

Notice that the calculations done in Example 5 are the same ones that are done in computing the product of the probability matrix P and the transition matrix T:

$$PT = [0.25 \quad 0.75] \cdot \begin{bmatrix} 0.8 & 0.2 \\ 0.4 & 0.6 \end{bmatrix}$$

$$= [0.25 \cdot 0.8 + 0.75 \cdot 0.4 \quad 0.25 \cdot 0.2 + 0.75 \cdot 0.6]$$

$$= [0.2 + 0.3 \quad 0.05 + 0.45]$$

$$= [0.5 \quad 0.5]$$

This parallelism between trees and matrices is illustrated in Figure 7.3.

FIGURE 7.3

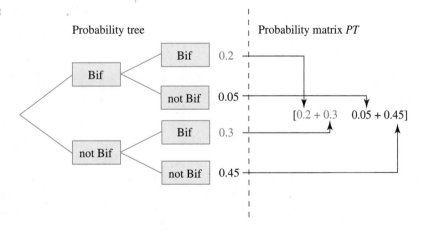

EXAMPLE 6 Use a probability tree to find Bif's market share after the second following purchase.

Solution This can be computed with the tree shown in Figure 7.4 on page 338. This tree is the tree from Example 5, altered to include one more set of branches. We want p(2nd following purchase is Bif), which is the sum of the probabilities of the limbs that stop at Bif under "2nd following purchase."

FIGURE 7.4

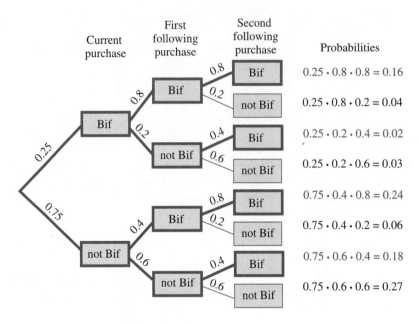

Therefore,

$$p(\text{2nd following purchase is Bif}) = (0.25 \cdot 0.8 \cdot 0.8) + (0.25 \cdot 0.2 \cdot 0.4)$$
$$+ (0.75 \cdot 0.4 \cdot 0.8) + (0.75 \cdot 0.6 \cdot 0.4)$$
$$= 0.6$$

In other words, after the second following purchase, Bif will command 60% of the market. Remember, though, that this is only a prediction, based on the assumption that current trends continue unchanged. •

Notice that the calculation done in Example 6 could be rewritten in a factored form:

$$(\boxed{0.25 \cdot 0.8} \cdot 0.8) + (\boxed{0.25 \cdot 0.2} \cdot 0.4) + (\boxed{0.75 \cdot 0.4} \cdot 0.8) + (\boxed{0.75 \cdot 0.6} \cdot 0.4)$$
$$= (\boxed{0.25 \cdot 0.8} + \boxed{0.75 \cdot 0.4})(0.8) + (\boxed{0.25 \cdot 0.2} + \boxed{0.75 \cdot 0.6})(0.4) \qquad \text{Factoring}$$
$$= (\qquad \boxed{0.5} \qquad)(0.8) + (\qquad \boxed{0.5} \qquad)(0.4)$$

From PT   From PT

Part of this ($\boxed{0.25 \cdot 0.8} + \boxed{0.75 \cdot 0.4} = \boxed{0.5}$ and $\boxed{0.25 \cdot 0.2} + \boxed{0.75 \cdot 0.6} = \boxed{0.5}$) is the same as the calculation we did in finding *PT*. And the whole thing is the same calculation that is done in computing row one, column one of the product of the probability matrix *PT* and the transition matrix *T*:

$$PT^2 = \quad PT \quad \cdot \quad T$$

$$= [0.5 \quad 0.5] \cdot \begin{bmatrix} 0.8 & 0.2 \\ 0.4 & 0.6 \end{bmatrix}$$

$$= [0.5 \cdot 0.8 + 0.5 \cdot 0.4 \quad 0.5 \cdot 0.2 + 0.5 \cdot 0.6]$$

$$= [0.6 \quad 0.4]$$

If this parallelism between probability trees and matrices continues, then p(2nd following purchase is not Bif), when computed from the above tree, should match the second entry in PT^2. From the tree:

p(2nd following purchase is not Bif)

$(\boxed{0.25 \cdot 0.8} \cdot 0.2) + (\boxed{0.25 \cdot 0.2} \cdot 0.6) + (\boxed{0.75 \cdot 0.4} \cdot 0.2) + (\boxed{0.75 \cdot 0.6} \cdot 0.6)$

$= (\boxed{0.25 \cdot 0.8} + \boxed{0.75 \cdot 0.4})(0.2) + (\boxed{0.25 \cdot 0.2} + \boxed{0.75 \cdot 0.6})(0.6)$ Factoring

$= (\qquad \boxed{0.5} \qquad)(0.2) + (\qquad \boxed{0.5} \qquad)(0.6)$

This does indeed match the second entry of PT^2. Predictions of Bif's future market shares can be made either with trees or with matrices. The matrix work is certainly simpler, both to set up and to calculate.

EXAMPLE 7   Use matrices to find Bif's market share after the third following purchase.

Solution   Bif's market share will be an entry in PT^3. One way to do this calculation is to observe that

$$PT^3 = P(T^2 T) = (PT^2)T \qquad \text{Associative property}$$

and to use the previously calculated PT^2.

$$PT^3 = (PT^2)T = [0.6 \quad 0.4] \cdot \begin{bmatrix} 0.8 & 0.2 \\ 0.4 & 0.6 \end{bmatrix} = [0.64 \quad 0.36]$$

Thus, p(3rd following purchase is Bif) $= 0.64$. [Also, p(3rd following purchase is not Bif) $= 0.36$.] This means that three purchases after the time when the market analysis was done, 64% of the detergents purchased will be Bif (*if current trends continue*). In other words, Bif's market share will be 64%. Bif will be doing quite well—recall that it had a market share of only 25% at the beginning of the new advertising campaign.   •

Predictions with Markov Chains

1. *Create the probability matrix P.* P is a row matrix whose entries are the initial probabilities of the states.
2. *Create the transition matrix T.* T is the square matrix whose entries are the probabilities of passing from current states to first following states. The rows refer to the current states, and the columns refer to the first following states.
3. *Calculate PT^n.* The matrix PT^n is a row matrix whose entries are the probabilities of the nth following states. Be careful that you multiply in the correct order: $PT^n \neq T^n P$.

EXAMPLE 8   In sociology, a family is considered to belong to the lower class, the middle class, or the upper class, depending on the family's total annual income. Sociologists have found that the strongest determinant of an individual's class is the class of his or her parents.

a. Convert the data given below (U.S. Bureau of the Census; *Statistical Abstract of the United States, 1996*) on family incomes in 1994 into a probability matrix.

Class	Family Income	Percent of Population
Lower class	Under $15,000	16%
Middle class	$15,000–$74,999	67%
Upper class	$75,000 or more	17%

b. Census data suggest the following (illustrated in Figure 7.5):

- Of those individuals whose parents belong to the lower class, 21% will become members of the middle class, and 1% will become members of the upper class.
- Of those individuals whose parents belong to the middle class, 6% will become members of the lower class, and 4% will become members of the upper class.
- Of those individuals whose parents belong to the upper class, 1% will become members of the lower class, and 10% will become members of the middle class.

FIGURE 7.5

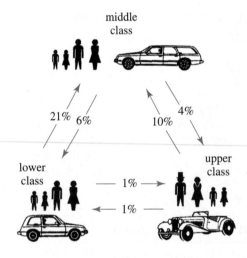

Convert this information into a transition matrix.

c. Predict the percent of U.S. families in the lower, middle, and upper classes after one generation.
d. Predict the percent of U.S. families in the lower, middle, and upper classes after two generations.

Solution
a. *Creating the probability matrix P:* Recall that a probability matrix has one row, with a column for each state. Our states are "lower," "middle," and "upper"; our probability matrix will be a 1×3 matrix.

$$\begin{matrix} \text{low} & \text{mid} & \text{up} \end{matrix}$$
$$p = [0.16 \quad 0.67 \quad 0.17]$$

b. *Creating the transition matrix T:* Because we have three classes, the transition matrix T will be a 3×3 matrix, with rows referring to the current state, or class of the parents, and columns referring to the first following state, or class of the child.

We are given the following portion of T:

Class of child

$$
T = \begin{array}{c c c}
 & \text{low} & \text{mid} & \text{up} \\
\begin{bmatrix} ? & 0.21 & 0.01 \\ 0.06 & ? & 0.04 \\ 0.01 & 0.10 & ? \end{bmatrix} & \begin{array}{l} \text{low} \\ \text{mid} \\ \text{up} \end{array} & \text{Class of parents}
\end{array}
$$

Filling in the blanks is easy. If 21% of the children of lower-class parents become members of the middle class and 1% become members of the upper class, that leaves 78% to remain members of the lower class. [Alternatively, the entries in any row must add to 1, so the missing entry is $1 - (0.21 + 0.01) = 0.78$.] Similar calculations fill in the other two blanks. Thus, we have

Class of child

$$
T = \begin{array}{c c c}
 & \text{low} & \text{mid} & \text{up} \\
\begin{bmatrix} 0.78 & 0.21 & 0.01 \\ 0.06 & 0.90 & 0.04 \\ 0.01 & 0.10 & 0.89 \end{bmatrix} & \begin{array}{l} \text{low} \\ \text{mid} \\ \text{up} \end{array} & \text{Class of parents}
\end{array}
$$

c. *Predicting the distribution after one generation:* We are asked to calculate PT^1, the probabilities of the first following generation.

$$
\begin{aligned}
PT^1 = PT &= [0.16 \quad 0.67 \quad 0.17] \cdot \begin{bmatrix} 0.78 & 0.21 & 0.01 \\ 0.06 & 0.90 & 0.04 \\ 0.01 & 0.10 & 0.89 \end{bmatrix} \\
&= [0.1667 \quad 0.6536 \quad 0.1797] \\
&\quad\;\; \text{low} \quad\;\; \text{mid} \quad\;\; \text{up} \\
&\approx [0.17 \quad 0.65 \quad 0.18]
\end{aligned}
$$

This means that, if current trends continue, in one generation's time the lower class will grow from 16% of the population to approximately 17%, the middle class will shrink from 67% to 65%, and the upper class will grow from 17% to 18%.

> ✔ Notice that the row adds to 1, as it should.

d. *Predicting the distribution after two generations:* We are asked to calculate PT^2, the probabilities of the second following generation. The easiest way to do this calculation is to observe that $PT^2 = (PT)T$ and to use the previously calculated PT. Our answers will be more accurate if we do round off PT before multiplying it by T.

$$
\begin{aligned}
PT^2 = (PT)T &= [0.1667 \quad 0.6536 \quad 0.1797] \cdot \begin{bmatrix} 0.78 & 0.21 & 0.01 \\ 0.06 & 0.90 & 0.04 \\ 0.01 & 0.10 & 0.89 \end{bmatrix} \\
&= [0.171039 \quad 0.641217 \quad 0.187744] \\
&\quad\;\; \text{low} \quad\;\; \text{mid} \quad\;\; \text{up} \\
&\approx [0.17 \quad 0.64 \quad 0.19]
\end{aligned}
$$

Andrei Andreevich Markov

1856–1922

Andrei Markov lived most of his life in St. Petersburg, Russia. During his lifetime, St. Petersburg (once Leningrad) was the capital of czarist Russia, a major seaport and commercial center, as well as an international center of literature, theater, music, and ballet. Markov's family belonged to the upper class; his father worked for the forestry department and managed a private estate.

In high school, Markov showed a talent for mathematics but generally was not a good student. He studied mathematics at St. Petersburg University, where he received his bachelor's, master's, and doctor's degrees. He then went on to teach at St. Petersburg University. The head of

the mathematics department, P. L. Chebyshev, was a famous mathematician and statistician. Markov became a consistent follower of Chebyshev's ideas. Nominated by Chebyshev, Markov was elected to the prestigious St. Petersburg Academy of Sciences.

Markov's research was primarily in the areas of statistics, probability theory, calculus, and number theory. His most famous work, on Markov chains, was motivated solely by theoretical concerns. In fact, he never wrote about their applications other than in a linguistic analysis of Pushkin's *Eugene Onegin*.

During the early 20th century, Markov participated in the liberal movement that climaxed in the Russian Revolution. When the czar overruled the election of author and revolutionary Maxim Gorky to the St. Petersburg Academy of Sciences, Markov wrote letters of protest to academic and state officials. When the czar dissolved the duma (an elected assembly), Markov denounced the czarist government. When the government celebrated the 300th anniversary of the House of Romanov (the czars' house), Markov organized a

celebration of the 200th anniversary of the publishing of Jacob Bernoulli's book on probabilities, *Ars Conjectandi*. After the czar had finally abdicated, Markov asked the academy to send him to teach mathematics at a secondary school in a small country town in the center of Russia. He returned to St. Petersburg after a winter of famine. Soon after his return, his health declined rapidly, and he died.

This means that, if current trends continue, in two generations' time the lower class will grow from 16% of the population to approximately 17%, the middle class will shrink from 67% to 64%, and the upper class will grow from 17% to 19%.

✔ Notice that the row adds to 1, as it should. •

7.1

EXERCISES

In Exercises 1–6, (a) rewrite the given data (and, if appropriate, their complements) in probability form and (b) convert these probabilities into a probability matrix.

1. A marketing analysis shows that KickKola currently commands 14% of the cola market.

2. A marketing analysis shows that SoftNWash currently commands 26% of the fabric softener market.

3. A census report shows that currently 32% of the residents of Metropolis own their own home and that 68% rent.

4. A survey shows that 23% of the shoppers in seven midwestern states regularly buy their groceries at Safe Shop, 29% regularly shop at PayNEat, and the balance shop at any one of several smaller markets.

5. Silver's Gym currently commands 48% of the health club market in Metropolis, Fitness Lab commands 37%, and ThinNFit commands the balance of the market.

6. Smallville has three Chinese restaurants: Asia Gardens, Chef Chao's, and Chung King Village. Currently, Asia Gardens gets 41% of the business, Chef Chao's gets 33%, and Chung King Village gets the balance.

In Exercises 7–12, (a) rewrite the given data in probability form and (b) convert the data into a transition matrix.

7. A marketing analysis for KickKola indicates that 12% of the consumers who do not currently drink KickKola will purchase KickKola the next time they buy a cola (in response to a new advertising campaign) and that 63% of the consumers who currently drink KickKola will purchase it the next time they buy a cola.

8. A marketing analysis for SoftNWash indicates that 9% of the consumers who do not currently use SoftNWash will purchase SoftNWash the next time they buy a fabric softener (in response to a free sample sent to selected consumers) and that 29% of the consumers who currently use SoftNWash will purchase it the next time they buy a fabric softener.

9. The Metropolis census report shows that 12% of the renters plan to buy a home in the next 12 months and that 3% of the homeowners plan to sell their home and rent instead.

10. The survey for Safe Shop indicates that 8% of the consumers who currently shop at Safe Shop will purchase their groceries at PayNEat the next time they shop and that 5% will switch to some other store. Also, 12% of the consumers who currently shop at PayNEat will purchase their groceries at Safe Shop the next time they shop, and 2% will switch to some other

store. In addition, 13% of the consumers who currently shop at neither store will purchase their groceries at Safe Shop the next time they shop, and 10% will shop at PayNEat.

11. An extensive survey of Metropolis's gym users indicates that 71% of the current members of Silver's will continue their annual membership when it expires, 12% will quit and join Fitness Lab, and the rest will quit and join ThinNFit. Fitness Lab has been unable to keep its equipment in good shape, and as a result 32% of its members will defect to Silver's and 34% will leave for ThinNFit. ThinNFit's members are quite happy, and as a result 96% plan to renew their annual membership, with half of the balance planning to move to Silver's and half to Fitness Lab.

12. Chung King Village recently mailed a coupon to all residents of Smallville offering two dinners for the price of one. As a result, 67% of those who normally eat at Asia Gardens plan on trying Chung King Village within the next month when they eat Chinese food, and 59% of those who normally eat at Chef Chao's plan on trying Chung King Village. Also, all of Chung King Village's normal customers will return to take advantage of the special. And 30% of those who normally eat at Asia Gardens will eat there within the next month because of its convenient location. Chef Chao's has a new chef who isn't doing very well, and as a result only 15% of those who normally eat there are planning on returning the next time they eat Chinese food.

13. Use matrices and the information in Exercises 1 and 7 to predict KickKola's market share at each purchase given.
    a. the next following purchase
    b. the second following purchase

14. Use matrices and the information in Exercises 2 and 8 to predict SoftNWash's market share at each purchase given.
    a. the next following purchase
    b. the second following purchase

15. Use matrices and the information in Exercises 5 and 11 to predict the market shares of Silver's Gym, Fitness Lab, and ThinNFit at the following times.
    a. in one year
    b. in two years
    c. in three years

16. Use matrices and the information in Exercises 6 and 12 to predict the market shares of Asia Gardens, Chef Chao's, and Chung King Village at the following times.
    a. in one month
    b. in two months
    c. in three months

17. Sierra Cruiser currently commands 41% of the mountain bike market. A nationwide survey performed by *Get Out of My Way* magazine indicates that 31% of the bike owners who do not currently own a Sierra Cruiser will purchase a Sierra Cruiser the next time they buy a mountain bike and that 12% of the bikers who currently own a Sierra Cruiser will purchase one the next time they buy a mountain bike. If the average customer buys a new mountain bike every two years, predict Sierra Cruiser's market share in four years.

18. A marketing analysis shows that Clicker Pens currently commands 46% of the pen market. The analysis also indicates that 46% of the consumers who do not currently own a Clicker pen will purchase a Clicker the next time they buy a pen and that 37% of the consumers who currently own a Clicker will not purchase one the next time they buy a pen. If the average consumer buys a new pen every three weeks, predict Clicker's market share after six weeks.

19. a. The Census Bureau classifies all residents of the United States as residents of a central city, or a suburb, or a nonmetropolitan area. In the 1990 census, the bureau reported that the central cities held 77.8 million people, the suburbs 114.9 million people, and the nonmetropolitan areas 56.0 million people. Compute the proportion of U.S. residents for each category and present those proportions as a probability matrix. (Round off to 3 decimal places.)

   b. The Census Bureau (in *Current Population Reports*, P20-463, *Geographic Mobility, March 1990 to March 1991*) also reported the information in the table below and in Figure 7.6 regarding migration between these three areas from 1990 to 1991. (Numbers are in thousands.)

Moved from	**Moved to**		
	Central City	Suburb	Nonmetropolitan Area
Central city	(*x*)	4946	736
Suburb	2482	(*x*)	964
Nonmetropolitan area	741	1075	(*x*)

Note: Persons moving from one central city to another were not considered. This is represented in the above chart by "(*x*)".

Use the above data and the data in part (a) to compute the probabilities that a central city resident will move to a suburb or to a nonmetropolitan area, the probabilities that a suburban resident will move to a central city or to a nonmetropolitan area, and the probabilities that a nonmetropolitan resident will move to a central city or to

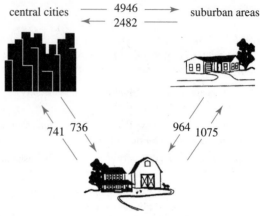

FIGURE 7.6 Movers between cities, suburbs, and nonmetropolitan areas, and net change due to migration, 1990–1991 (numbers in thousands)

a suburb. Present these probabilities in the form of a transition matrix. (Round off to 3 decimal places.)

   c. Predict the percent of U.S. residents who will reside in a central city, a suburb, and a nonmetropolitan area in 1991.

   d. Predict the percent of U.S. residents residing in a central city, a suburb, and a nonmetropolitan area in 1992.

   e. Which prediction is the stronger one—that in part (c) or that in part (d)? Why?

20. a. The Census Bureau (in *Statistical Abstract of the United States: 1996*) reported the following 1993 national and regional populations.

Region	**Population (in thousands)**
U.S.	257,800
Northeast	51,275
Midwest	61,040
South	89,426
West	56,059

Compute the proportion of U.S. residents of each region and present those proportions as a probability matrix. (Round off to 3 decimal places.)

   b. The Census Bureau (in *Current Population Reports*, P20-465, *Geographic Mobility: March 1993 to March 1994*) also reported the information in the table below and in Figure 7.7 regarding movement between regions from 1993 to 1994. (Numbers are in thousands.)

Region Moved from	Region Moved to			
	Northeast	Midwest	South	West
Northeast	(*x*)	94	449	143
Midwest	53	(*x*)	468	215
South	201	371	(*x*)	388
West	93	251	419	(*x*)

Note: The Census data involved interregional movement; thus, persons moving from one part of a region to another part of the same region were not considered. This is represented in the above chart by "(*x*)."

FIGURE 7.7
Movers between regions, 1990–1991 (numbers in thousands)

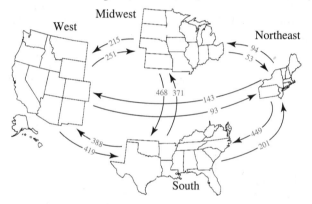

Use the data contained in the chart and the data in part (a) to compute the probabilities that a resident of any one region will move to any of the other regions. Present these probabilities in the form of a transition matrix. (Round off to 3 decimal places.)

c. Predict the percent of U.S. residents who will reside in the Northeast, the Midwest, the South, and the West in 1994.

d. Predict the percent of U.S. residents who will reside in the Northeast, the Midwest, the South, and the West in 1995.

e. Which prediction is the stronger one—that in part (c) or that in part (d)? Why?

21. a. In 1981, Neil Sampson, executive vice president of the National Association of Conservation Districts, compiled data on the shifting land use pattern in the United States. [*Source:* R. Neil Sampson, *Farmland or Wasteland: A Time to Choose* (Rodale Press, 1981).] He stated that in 1977 there were 413 million acres of cropland, 127 million acres of land that had a high or medium potential for conversion to cropland (for example, grasslands or forests), and 856 million acres of land that had little or no potential for

conversion to cropland (such as urban land or land with heavily deteriorated soil). Compute the proportion of land in each category and present those proportions as a probability matrix. (Round off to 3 decimal places.)

b. He also estimated that between 1967 and 1977, the following shifts in land use occurred (in millions of acres):

Previous Use of Land	New Use of Land		
	Cropland	Potential Cropland	Noncropland
Cropland	(*X*)	17	35
Potential cropland	34	(*X*)	0[1]
Noncropland	0	0[1]	(*X*)

[1]Unknown, but assumed to cancel each other out

FIGURE 7.8

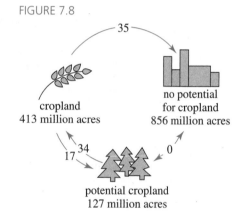

Use the data contained in the chart and the data in part (a) to compute the probabilities that an acre of land of any of the three categories will shift to another use (in ten years time). Present these probabilities in the form of a transition matrix. (Round off to 3 decimal places.)

c. Predict the amount of land that will be cropland, potential cropland, and noncropland in 1987, assuming the trend described above continues.

d. Predict the amount of land that will be cropland, potential cropland, and noncropland in 1997, assuming the trend described above continues.

 Answer the following questions using complete sentences.

22. In creating Markov chains, was Markov motivated by theoretical concerns or by specific applications?

23. Was Markov a supporter of the czar or of the revolution?

24. How did Markov celebrate the 300th anniversary of the House of Romanov?

25. What did Markov do when the czar abdicated?

26. Are PT, PT^2, and PT^3 transition matrices or probability matrices? Why?

27. What are the differences between a transition matrix and a probability matrix?

28. What information would a marketing analyst need in order to predict the future market share of a product? How would he or she obtain such information?

29. When a prediction is made using Markov chains, that prediction is based on a number of assumptions, one involving the trend, one involving the data summarized in the probability matrix P, and one involving the data summarized in the transition matrix T. What are these assumptions? What would make the assumptions invalid?

30. Compare and contrast the tree method of Markov chains with the matrix method of Markov chains. What are the advantages of the two different methods?

7.2

REGULAR MARKOV CHAINS

In the previous section we were given Bif's current market share (25%), and we predicted Bif's market share after one, two, and three purchases (that is, we found the probabilities of the first, second, and third following states) by computing PT, PT^2, and PT^3. In a similar way, we could predict Bif's market share after four purchases by computing PT^4. These predictions are shown in Figure 7.9. It appears that the trend in increasing market share created by Bif's new advertising campaign and customer satisfaction with the product will stabilize and that Bif will ultimately command close to 70% of the market.

FIGURE 7.9

Bif's Projected Market Share	Bif	Not Bif	Probability Matrix
Current purchase	25%	75%	P
1st following purchase	50%	50%	$PT = PT^1$
2nd following purchase	60%	40%	PT^2
3rd following purchase	64%	36%	PT^3
4th following purchase	65.6%	34.4%	PT^4

If Bif had started out with a much smaller initial market share, say only 10% rather than 25%, a similar pattern would emerge, as shown in Figure 7.10.

$$P = [0.1 \quad 0.9]$$

$$T = \begin{bmatrix} 0.8 & 0.2 \\ 0.4 & 0.6 \end{bmatrix} \quad \text{As before}$$

FIGURE 7.10

Bif's Projected Market Share	Bif	Not Bif	Probability Matrix
Current purchase	10%	90%	P
1st following purchase	44%	56%	$PT = PT^1$
2nd following purchase	57.6%	42.4%	PT^2
3rd following purchase	63.04%	36.96%	PT^3
4th following purchase	65.22%	34.78%	PT^4

As a result of a new advertising campaign and customers' satisfaction with the product, Bif is predicted to increase its market share from 25% . . .

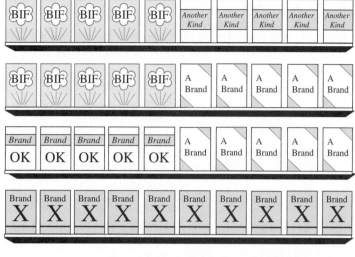

. . . to 66% after the fourth following purchase.

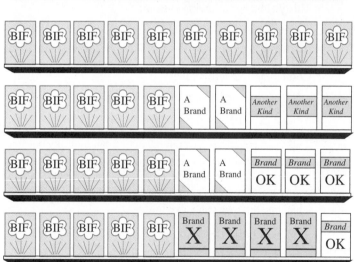

Surprisingly, it appears that Bif's new advertising campaign and customer satisfaction with the product would have the same ultimate effect regardless of Bif's initial market share. Under certain circumstances, PT^n will stabilize, and it will do so in a way that is unaffected by the value of P. In this case, the Markov chain is said to be **regular.** This occurs when it is possible to go from any given state to each other state. (*Note:* It is also necessary that all states are "aperiodic." This requirement is discussed in more advanced courses; we will ignore it.)

Notice that it is possible to go from Bif to not Bif (20% of Bif users will switch to not Bif) and to go from not Bif to Bif (40% of not Bif users will switch to Bif). Thus, the Bif Markov chain is regular, and PT^n will stabilize.

Our goal in this section is to find the level at which the trend will stabilize. The **equilibrium matrix** L, the matrix at which the trend stabilizes, is a probability matrix $L = PT^n$ such that all following probability matrices are equal. In other words, multiplying the equilibrium matrix L by T would have no effect, and LT would equal L. In fact, we find the equilibrium matrix by solving the matrix equation $LT = L$.

EXAMPLE 1   Make a long-range forecast for Bif's market share given the transition data from Section 7.1.

Solution   The long-range forecast is the probability matrix at which the trend in Bif's changing market share stabilizes; that is, it is the equilibrium matrix L. We find L by solving the matrix equation $LT = L$.

$$L = [x, \quad y] \qquad \text{A row matrix because } L \text{ is a probability matrix}$$

$$T = \begin{bmatrix} 0.8 & 0.2 \\ 0.4 & 0.6 \end{bmatrix} \qquad \text{From Section 7.1}$$

$$\begin{array}{ccc} L & \cdot & T \end{array} = L$$

$$[x \quad y] \cdot \begin{bmatrix} 0.8 & 0.2 \\ 0.4 & 0.6 \end{bmatrix} = [x \quad y]$$

$$[0.8x + 0.4y \quad 0.2x + 0.6y] = [x \quad y] \qquad \text{Multiplying}$$

This yields the following system:

$$0.8x + 0.4y = x$$
$$0.2x + 0.6y = y$$

Combining like terms gives

$$-0.2x + 0.4y = 0$$
$$0.2x - 0.4y = 0$$

These two equations are equivalent (we can multiply one equation by -1 to get the other), so we can discard one. This leaves us in a difficult spot—our single equation has an infinite number of solutions, and we are looking for a single solution. However, recall that $L = [x \quad y]$ is a probability matrix, so its entries must add to 1. Thus, $x + y = 1$ is our needed second equation. Now we can solve the system

$$0.2x - 0.4y = 0$$
$$x + y = 1$$

Because this is a small, simple system, solving it with elimination would be easiest. If it were a larger, messier system, we would want to use the Gauss-Jordan method.

Proceeding with the elimination method, we multiply the second equation by 0.4. The resulting system is

$$0.2x - 0.4y = 0$$
$$0.4x + 0.4y = 0.4$$

Adding gives

$$0.6x = 0.4$$
$$\frac{0.6x}{0.6} = \frac{0.4}{0.6}$$
$$x = \frac{2}{3}$$

Substituting into $x + y = 1$ gives

$$\frac{2}{3} + y = 1$$

$$y = \frac{1}{3}$$

Thus, $L = [x \quad y] = \begin{bmatrix} \frac{2}{3} & \frac{1}{3} \end{bmatrix}$, and Bif's market share should stabilize at $\frac{2}{3}$, or approximately 67%. As a check, multiply L by T and see if you get L:

✔ $LT = \begin{bmatrix} \frac{2}{3} & \frac{1}{3} \end{bmatrix} \cdot \begin{bmatrix} 0.8 & 0.2 \\ 0.4 & 0.6 \end{bmatrix}$

$\approx [0.6667 \quad 0.3333]$

$\approx \begin{bmatrix} \frac{2}{3} & \frac{1}{3} \end{bmatrix}$ •

Long-Range Predictions with Regular Markov Chains

1. *Create the transition matrix T. T* is the square matrix, discussed in Section 7.1, whose entries are the probabilities of passing from current states to next following states. The rows refer to the current states, and the columns refer to the next following states.
2. *Determine if the chain is regular.* This occurs when it is possible to go from each state to each other state. (*Note:* It is also necessary that all states are aperiodic. We will ignore this requirement.)
3. *Create the equilibrium matrix L.* The equilibrium matrix L is the long-range prediction. L is the matrix that solves the matrix equation $LT = L$. If there are two states, then $L = [x \quad y]$. If there are three states, then $L = [x \quad y \quad z]$.
4. *Find and simplify the system of equations described by $LT = L$.*
5. *Discard any redundant equations and include the equation $x + y = 1$ (or $x + y + z = 1$, and so on).*
6. *Solve the resulting system.* Use elimination if the system is small; use the Gauss-Jordan method and possibly a graphing calculator or computer if the system is big.
7. ✔ *Check your work by verifying that $LT = L$.*

EXAMPLE 2   In sociology, a family is considered to belong to the lower class, the middle class, or the upper class, depending on the family's total annual income. Sociologists have found that the strongest determinant of an individual's class is the class of his or her parents. Census data suggest the following (see Figure 7.11).

- Of those individuals whose parents belong to the lower class, 21% will become members of the middle class, and 1% will become members of the upper class.
- Of those individuals whose parents belong to the middle class, 6% will become members of the lower class, and 4% will become members of the upper class.

FIGURE 7.11

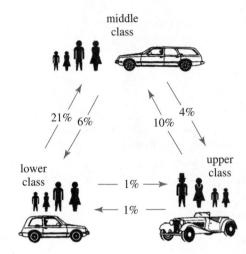

middle
class

21% 6%

10% 4%

lower
class

upper
class

— 1% →

← 1% —

- Of those individuals whose parents belong to the upper class, 1% will become members of the lower class, and 10% will become members of the middle class.

Make a long-range prediction of the levels at which the lower, middle, and upper classes will stabilize.

Solution It is possible to go from any given state to each other state, so the Markov chain is regular and will reach an equilibrium. We are asked to find the equilibrium matrix L.

Step 1 *Create the transition matrix T.* We did this in Section 7.1.

$$T = \begin{bmatrix} 0.78 & 0.21 & 0.01 \\ 0.06 & 0.90 & 0.04 \\ 0.01 & 0.10 & 0.89 \end{bmatrix}$$

Step 2 *Create the equilibrium matrix L. L* is a row matrix, and in this problem L has three entries—one for lower class, one for middle class, and one for upper class.

$$L = [x \quad y \quad z]$$

Step 3 *Find and simplify the system of equations described by $LT = L$.*

$$L \quad \cdot \quad T \quad = \quad L$$

$$[x \quad y \quad z] \cdot \begin{bmatrix} 0.78 & 0.21 & 0.01 \\ 0.06 & 0.90 & 0.04 \\ 0.01 & 0.10 & 0.89 \end{bmatrix} = [x \quad y \quad z]$$

$$[0.78x + 0.06y + 0.01z \quad 0.21x + 0.90y + 0.10z \quad 0.01x + 0.04y + 0.89z]$$
$$= [x \quad y \quad z]$$

This matrix equation describes the following system:

$$0.78x + 0.06y + 0.01z = x$$
$$0.21x + 0.90y + 0.10z = y$$
$$0.01x + 0.04y + 0.89z = z$$

We can simplify this system by combining like terms.

$$-0.22x + 0.06y + 0.01z = 0$$
$$0.21x - 0.10y + 0.10z = 0$$
$$0.01x + 0.04y - 0.11z = 0$$

We can further simplify the system by multiplying by 100.

$$-22x + 6y + 1z = 0$$
$$21x - 10y + 10z = 0$$
$$1x + 4y - 11z = 0$$

Step 4 *Discard any redundant equations and include the equation $x + y + z = 1$.* The third equation is the negative of the sum of the first two equations, so its presence is redundant. Thus, any one of the three equations can be dropped. $L = [x \quad y \quad z]$ is a probability matrix, so the sum of its entries must be 1; that is, $x + y + z = 1$.

$$-22x + 6y + 1z = 0 \qquad\qquad\qquad\qquad\qquad \textbf{(1)}$$
$$21x - 10y + 10z = 0 \qquad\qquad\qquad\qquad\qquad \textbf{(2)}$$
$$x + y + z = 1 \qquad\qquad\qquad\qquad\qquad\qquad \textbf{(3)}$$

Step 5 *Solve the resulting system.* Because this is a larger system, we'll use the Gauss-Jordan method and round off all calculations to four decimal places.

$$\begin{bmatrix} 1 & 1 & 1 & 1 \\ -22 & 6 & 1 & 0 \\ 21 & -10 & 10 & 0 \end{bmatrix}$$
Putting equation (3) first, to take advantage of the 1

$$\begin{array}{l} 22R1 + R2 : R2 \\ -21R1 + R3 : R3 \end{array} \quad \begin{bmatrix} 1 & 1 & 1 & 1 \\ 0 & 28 & 23 & 22 \\ 0 & -31 & -11 & -21 \end{bmatrix}$$

$$R2 \div 28 : R2 \quad \begin{bmatrix} 1 & 1 & 1 & 1 \\ 0 & 1 & 0.8214 & 0.7857 \\ 0 & -31 & -11 & -21 \end{bmatrix}$$

$$\begin{array}{l} R1 - R2 : R1 \\ 31R2 + R3 : R3 \end{array} \quad \begin{bmatrix} 1 & 0 & 0.1786 & 0.2143 \\ 0 & 1 & 0.8214 & 0.7857 \\ 0 & 0 & 14.4643 & 3.3571 \end{bmatrix}$$

$$R3 \div 14.4643 : R3 \quad \begin{bmatrix} 1 & 0 & 0.1786 & 0.2143 \\ 0 & 1 & 0.8214 & 0.7857 \\ 0 & 0 & 1 & 0.2321 \end{bmatrix}$$

R1 − 0.1786R3 : R1
R2 − 0.8214R3 : R2

$$\begin{bmatrix} 1 & 0 & 0 & 0.1728 \\ 0 & 1 & 0 & 0.5951 \\ 0 & 0 & 1 & 0.2321 \end{bmatrix}$$

The resulting solution is

$$L [x \quad y \quad z] = [0.1728 \quad 0.5951 \quad 0.2321] \approx [0.17 \quad 0.60 \quad 0.23]$$

Alternatively, we could use a TI graphing calculator's "reduced row echelon form" (or "rref") feature and go directly from

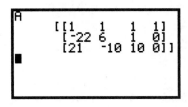

to

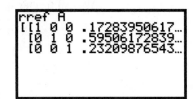

Step 6 ✔ *Check your work by verifying that LT = L.*

$$\begin{array}{ccc} L & \cdot & T & = & L \end{array}$$

$$[0.1728 \quad 0.5951 \quad 0.2321] \cdot \begin{bmatrix} 0.78 & 0.21 & 0.01 \\ 0.06 & 0.90 & 0.04 \\ 0.01 & 0.10 & 0.89 \end{bmatrix}$$

$$= [0.172811 \quad 0.595088 \quad 0.232101]$$

$$\approx [0.1728 \quad 0.5951 \quad 0.2321]$$

This means that, if current trends continue, the lower class will eventually stabilize at 17% of the population, the middle class will stabilize at 60%, and the upper class will stabilize at 23%. (Recall that in 1994 the lower class was 16%, the middle class was 67%, and the upper class was 17%, according to the Census Bureau.) It is important to remember that this prediction is based on the assumption that current trends continue. , ●

7.2

EXERCISES

In Exercises 1–4, find the equilibrium matrix L by solving LT = L for L.

1. $T = \begin{bmatrix} 0.1 & 0.9 \\ 0.2 & 0.8 \end{bmatrix}$

2. $T = \begin{bmatrix} 0.5 & 0.5 \\ 0.6 & 0.4 \end{bmatrix}$

3. $T = \begin{bmatrix} 0.3 & 0.2 & 0.5 \\ 0.1 & 0.8 & 0.1 \\ 0.4 & 0.3 & 0.3 \end{bmatrix}$

4. $T = \begin{bmatrix} 0.4 & 0.3 & 0.3 \\ 0.2 & 0.7 & 0.1 \\ 0.3 & 0.3 & 0.4 \end{bmatrix}$

5. A marketing analysis shows that 12% of the consumers who do not currently drink KickKola will purchase KickKola the next time they buy a cola and that 63% of the consumers who currently drink KickKola will purchase it the next time they buy a cola. Make a long-range prediction of KickKola's ultimate market share, assuming that current trends continue. (See Exercise 13 in Section 7.1.)

6. A marketing analysis shows that 9% of the consumers who do not currently use SoftNWash will purchase SoftNWash the next time they buy a fabric softener and that 29% of the consumers who currently use SoftNWash will purchase it the next time they buy a fabric softener. Make a long-range prediction of SoftNWash's ultimate market share, assuming that current trends continue. (See Exercise 14 in Section 7.1.)

7. An extensive survey of Metropolis's gym users indicates that 71% of the current members of Silver's Gym will continue their annual membership when it expires, 12% will quit and join Fitness Lab, and the rest will quit and join ThinNFit. Fitness Lab has been unable to keep its equipment in good shape, and as a result 32% of its members will defect to Silver's and 34% will leave for ThinNFit. ThinNFit's members are quite happy, and as a result 96% plan on renewing their annual memberships, with half of the balance planning on moving to Silver's and half to Fitness Lab. Make a long-range prediction of the ultimate market shares of the three health clubs, assuming that current trends continue. (See Exercise 15 in Section 7.1.)

8. Smallville has three Chinese restaurants: Asia Gardens, Chef Chao's, and Chung King Village. Chung King Village recently mailed a coupon to all residents of Smallville offering two dinners for the price of one. As a result, 67% of those who normally eat at Asia Gardens plan to try Chung King Village within the next month, and 59% of those who normally eat at Chef Chao's plan to try Chung King Village. Also, all of Chung King Village's normal customers will return to take advantage of the special. And 30% of those who normally eat at Asia Gardens will eat there within the next month because of its convenient location. Chef Chao's has a new chef who isn't doing very well, and as a result only 15% of those who normally eat there are planning to return next time. Make a long-range prediction of the ultimate market shares of the three restaurants, assuming that current trends continue. (See Exercise 16 in Section 7.1.)

9. A census report shows that 32% of the residents of Metropolis own their own home and that 68% rent. The report shows that 12% of the renters plan to buy a home in the next 12 months and that 3% of the home owners plan to sell their home and rent instead. Make a long-range prediction of the percent of Metropolis residents who will own their own home and the percent who will rent. Give two assumptions on which this prediction is based. (See Exercises 3 and 9 in Section 7.1.)

10. A survey shows that 23% of the shoppers in seven midwestern states regularly buy their groceries at Safe Shop, 29% regularly shop at PayNEat, and the balance shop at any one of several smaller markets. The survey indicates that 8% of the consumers who currently shop at Safe Shop will purchase their groceries at PayNEat the next time they shop and that 5% will switch to some other store. Also, 12% of the consumers who currently shop at PayNEat will purchase their groceries at Safe Shop the next time they shop, and 2% will switch to some other store. In addition, 13% of the consumers who currently shop at neither store will purchase their groceries at Safe Shop the next time they shop, and 10% will shop at PayNEat. Predict the percent of midwestern shoppers who will be regular Safe Shop customers and the percent who will be regular PayNEat customers as a result of this trend. (See Exercises 4 and 10 in Section 7.1.)

11. a. The Census Bureau classifies all residents of the United States as residents of a central city, or a suburb, or a nonmetropolitan area. In the 1990 census, the bureau reported that the central cities held 77.8 million people, the suburbs 114.9 million people, and the nonmetropolitan areas 56.0 million people. The Census Bureau (in Current Population Reports, P20–463, *Geographic Mobility: March 1990 to March 1991*) also reported the information in the table below and in Figure 7.12 regarding migration between these three areas from 1990 to 1991. (Numbers are in thousands.) Compute the probabilities that a central city resident will move to a suburb or to a nonmetropolitan area, the probabilities that a suburban resident will move to a central city or to a

	Moved to		
Moved from	Central City	Suburb	Nonmetropolitan Area
Central city	(x)	4946	736
Suburb	2482	(x)	964
Nonmetropolitan area	741	1075	(x)

Note: Persons moving from one central city to another were not considered. This is represented in the above chart by "(x)."

FIGURE 7.12 Movers between cities, suburbs, and nonmetropolitan areas, and net change due to migration, 1990–1991 (numbers in thousands)

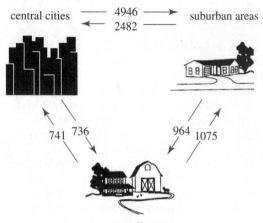

nonmetropolitan areas

nonmetropolitan area, and the probabilities that a nonmetropolitan resident will move to a central city or to a suburb. Present these probabilities in the form of a transition matrix. (Round off to 3 decimal places.) (See Exercise 19(b) in Section 7.1.)

b. Make a long-range prediction of the ultimate percent of U.S. residents who will reside in a central city, a suburb, and a nonmetropolitan area, assuming that the trend indicated by the above data continues.

12. a. The Census Bureau (*Statistical Abstract of the United States, 1996*) reported the following 1993 national and regional populations:

Region	Population (in thousands)
U.S.	257,800
Northeast	51,275
Midwest	61,040
South	89,426
West	56,059

In Current Population Reports, P20–485, *Geographic Mobility: March 1993 to March 1994*, the Census Bureau also reported the information in the table below and in Figure 7.13 regarding movement between the regions from 1993 to 1994. (Numbers are in thousands.) Compute the probabilities that a resident of

Region Moved from	Region Moved to			
	Northeast	Midwest	South	West
Northeast	(x)	94	449	143
Midwest	53	(x)	468	215
South	201	371	(x)	388
West	93	251	419	(x)

any one region will move to any of the other regions. Present these probabilities in the form of a transition matrix. (Round off to 3 decimal places.) (See Exercises 20(b) in Section 7.1.)

b. Make a long-range prediction of the ultimate percent of U.S. residents who will reside in the Northeast, the Midwest, the South, and the West, assuming that the trend described by the above data continues.

FIGURE 7.13

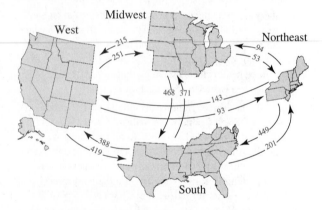

13. Seven different makes of automobile are available for purchase in the country of Outer Moldavia: the Asian-manufactured Toyonda and Nissota, the American Reo, Henry J, and DeSoto, and the European Hugo. The Moldavian Department of Motor Vehicles kept records of the sale of all new and used autos. These records show that 60% of those people who sold their old Toyonda replaced it with a new Toyonda, 10% replaced it with a Nissota, 12% switched to a Reo, 8% switched to a Henry J, 7% bought a DeSoto, and 3% actually bought a Hugo. Similar statistics on the other brands are shown in Figure 7.14. If this trend continues, what percent of the Moldavian market will each brand of automobile command?

FIGURE 7.14

	Car Purchased					
Car Sold	Toyonda	Nissota	Reo	Henry J	DeSoto	Hugo
Toyonda	60%	10%	12%	8%	7%	3%
Nissota	9%	64%	7%	8%	9%	3%
Reo	19%	10%	54%	9%	7%	1%
Henry J	5%	6%	5%	69%	12%	3%
DeSoto	3%	5%	13%	16%	61%	2%
Hugo	18%	15%	18%	16%	17%	16%

14. a. Use the information on U.S. class structure in Example 8 in Section 7.1 to calculate PT, PT^2, PT^3, etc., and list each of those matrices, rounded to 2 decimal places.

Continue in this fashion until your result agrees with the result of Example 2 in this section. (*Note:* Use some form of technology.)

b. For which value of n did PT^n approximate the result of Example 2 with 2-decimal-place accuracy?

c. Is it reasonable to approximate the equilibrium matrix L with 2-decimal-place accuracy by calculating PT, PT^2, PT^3, etc. until you get two successive answers that round to the same result? Why or why not?

 Answer the following questions using complete sentences.

15. What information would a marketing analyst need in order to make a long-range prediction of the future market share of a product? How would he or she obtain such information?

16. How could a marketing analyst predict when the equilibrium matrix will be achieved?

7.3

ABSORBING MARKOV CHAINS

Not all Markov chains are regular, and thus not all Markov chains have an equilibrium matrix. In this section we will investigate another type of Markov chain for which long-range predictions can be made; these are called *absorbing chains.*

Regular Chains versus Absorbing Chains

EXAMPLE 1 The University of Erehwon has studied its students' paths through college. The university has discovered that:

- 70% of all freshmen return the next year as sophomores, 15% return as freshmen, and the rest quit
- 75% of all sophomores return the next year as juniors, 15% return as sophomores, and the rest quit
- 80% of all juniors return the next year as seniors, 10% return as juniors, and the rest quit
- 85% of all seniors graduate, 10% return as seniors, and the rest quit

(For the sake of this analysis, we will assume that the university does not allow a student who has quit to reenroll, and that the university does not change a student's class standing midyear.)

What is the probability that an entering freshman will graduate?

a. Write the transition matrix T for this data.

b. Determine if this transition matrix is regular.

Solution **a.** There are 5 states: freshman, sophomore, junior, senior, graduated, and quit. The transition matrix is

$$
T = \begin{array}{c} \\ \\ F \\ So \\ J \\ Sr \\ G \\ Q \end{array}
\begin{array}{cccccc} F & So & J & Sr & G & Q \\ \end{array}
\left[\begin{array}{cccccc}
0.15 & 0.7 & 0 & 0 & 0 & 0.15 \\
0 & 0.15 & 0.75 & 0 & 0 & 0.1 \\
0 & 0 & 0.1 & 0.8 & 0 & 0.1 \\
0 & 0 & 0 & 0.1 & 0.85 & 0.05 \\
0 & 0 & 0 & 0 & 1 & 0 \\
0 & 0 & 0 & 0 & 0 & 1
\end{array}\right]
\begin{array}{c}
F \\ So \\ J \\ Sr \\ G \\ Q
\end{array}
$$

b. This is not a regular transition matrix because it is not possible to go from Sr to either F, So, or J; nor is it possible to go from G or Q to any other state. ●

In Example 1, the states G and Q are called **absorbing states,** because it is not possible to leave those states. Once a student has graduated, that student cannot become a freshman, sophomore, junior, or senior or quit. A Markov chain is called an **absorbing chain** if:

- the chain has at least one absorbing state, and
- it is possible to go from any nonabsorbing state to an absorbing state

EXAMPLE 2   Determine if the Markov chain in Example 1 is absorbing.

Solution   The chain has 2 absorbing states: G and Q.
The nonabsorbing states are F, So, J, and Sr. It is possible to go from each of these nonabsorbing states to either G or Q. For example, you can go from F to G by going first to So, then to J, then to Sr, and finally to G. Thus, the chain is absorbing.   •

Long-Range Predictions with Absorbing Chains

The University of Erehwon would like to use its data to answer two questions:

- How many years can an entering freshman expect to spend at the university?
- What is the probability that an entering freshman will graduate?

We will discuss the first question now; we will discuss the second question later in this section. The first question asks for an expected value, so we need probabilities and values. We are given that the student is an entering freshman, so we need conditional probabilities such as those found in T and T^2.

EXAMPLE 3   Use T and T^2 to estimate the number of years an entering freshman can expect to spend as a freshman.

Solution   The university does not change a student's class standing midyear, so 100% of the entering freshmen spend their first year as freshmen. As discussed in Sections 7.1 and 7.2, the transformation matrix T gives the probabilities of passing from current states to first following states, and the transformation matrix T^2 gives the probabilities of passing from current states to second following states. Row 1, column 1 of T tells us that 15% of entering freshmen return as freshmen. The matrix T^2 is given below; it tells us that 2.25% of the entering freshmen spend their third year as freshmen.

$$
T^2 = \begin{array}{c} \\ \\ \\ \\ \\ \\ \end{array}
\begin{array}{c c c c c c}
F & So & J & Sr & G & Q \\
\end{array}
\left[
\begin{array}{c c c c c c}
0.0225 & 0.21 & 0.525 & 0 & 0 & 0.2425 \\
0 & 0.0225 & 0.1875 & 0.6 & 0 & 0.19 \\
0 & 0 & 0.01 & 0.16 & 0.68 & 0.15 \\
0 & 0 & 0 & 0.01 & 0.935 & 0.055 \\
0 & 0 & 0 & 0 & 1 & 0 \\
0 & 0 & 0 & 0 & 0 & 1 \\
\end{array}
\right]
\begin{array}{c}
F \\ So \\ J \\ Sr \\ G \\ Q
\end{array}
$$

We have found that:

- 100% of the entering freshmen spend their first year as freshmen
- 15% of the entering freshmen spend their second year as freshmen
- 2.25% of the entering freshmen spend their third year as freshmen

Thus, the expected value of the number of years a freshman spends as a freshman is approximately

$$(1.00)(1) + (0.15)(1) + (0.0225)(1) = 1.1725$$

This is only an approximation because some could spend more years as freshmen. (This is not the standard way to compute expected values; see Exercise 7.)   •

Notice that we could have used matrices to solve Example 3 by adding $I + T + T^2$ and then looking in row 1, column 1. In fact, $I + T + T^2 + T^3 + T^4 + T^5 + \cdots$ would give us the exact number of years a freshman spends as a freshman; it would, however, be a very tedious calculation. There is an easy way around this dilemma. Let the matrix X be this sum of matrices.

$$X = I + T + T^2 + T^3 + T^4 + T^5 + \cdots$$

If we multiply each side of this equation by $I - T$, simplify, and solve the result for X, we will get a quick way of calculating X.

$$(I - T)X = (I - T)(I + T + T^2 + T^3 + T^4 + T^5 + \cdots) \qquad \text{Multiplying by } I - T$$

$$(I - T)X = (I + T + T^2 + T^3 + T^4 + T^5 + \cdots) - (T + T^2 + T^3 + T^4 + T^5 + T^6 + \cdots)$$

$$\qquad\qquad\qquad \text{Distributing the } I \qquad\qquad\qquad\qquad \text{Distributing the } T$$

$$(I - T)X = I \qquad\qquad\qquad \text{Canceling}$$

$$(I - T)^{-1}(I - T)X = (I - T)^{-1}I \qquad \text{Solving for } X$$

$$X = (I - T)^{-1}$$

Thus, we could have used matrices to solve Example 3 by finding $(I - T)^{-1}$ and looking in row 1, column 1. This calculation can be made simpler by using only a part of T. The part of T that is about absorbing states is of no value to us in this work, since we are investigating the length of time spent in a nonabsorbing state. Thus, we will discard T's last 2 rows and columns (the G and Q rows and columns). We will call the resulting matrix N, because it contains only the nonabsorbing states.

$$N = \begin{array}{c} \\ \\ \\ \\ \\ \end{array} \begin{array}{cccc} \text{F} & \text{So} & \text{J} & \text{Sr} \\ \end{array}$$

$$N = \begin{bmatrix} 0.15 & 0.7 & 0 & 0 \\ 0 & 0.15 & 0.75 & 0 \\ 0 & 0 & 0.1 & 0.8 \\ 0 & 0 & 0 & 0.1 \end{bmatrix} \begin{array}{c} \text{F} \\ \text{So} \\ \text{J} \\ \text{Sr} \end{array}$$

The matrix N is an abbreviated version of T that gives only nonabsorbing data. Similarly, $(I - N)^{-1}$ is an abbreviated version of $(I - T)^{-1}$ that gives only nonabsorbing data. •

EXAMPLE 4 Use $(I - N)^{-1}$ to find the average number of years a freshman spends as a freshman.

Solution

$$I - N = \begin{bmatrix} 1 & 0 & 0 & 0 \\ 0 & 1 & 0 & 0 \\ 0 & 0 & 1 & 0 \\ 0 & 0 & 0 & 1 \end{bmatrix} - \begin{bmatrix} 0.15 & 0.7 & 0 & 0 \\ 0 & 0.15 & 0.75 & 0 \\ 0 & 0 & 0.1 & 0.8 \\ 0 & 0 & 0 & 0.1 \end{bmatrix}$$

$$= \begin{bmatrix} 0.85 & -0.7 & 0 & 0 \\ 0 & 0.85 & -0.75 & 0 \\ 0 & 0 & 0.9 & -0.8 \\ 0 & 0 & 0 & 0.9 \end{bmatrix}$$

We can find $(I - N)^{-1}$ with the Gauss-Jordan method, as discussed in Section 4.2, or by entering $I - N$ into a graphing calculator and using the $\boxed{x^{-1}}$ button to compute its inverse, as discussed in Section 4.1. Either way, we get

$$(I - N)^{-1} = \begin{bmatrix} 1.1764\ldots & 0.9688\ldots & 0.8073\ldots & 0.7176\ldots \\ 0 & 1.1764\ldots & 0.9803\ldots & 0.8714\ldots \\ 0 & 0 & 1.1111\ldots & 0.9876\ldots \\ 0 & 0 & 0 & 1.1111\ldots \end{bmatrix}$$

The matrix $(I - N)^{-1}$ is derived from nonabsorbing state information, so it is labeled with the nonabsorbing states.

$$(I - N)^{-1} \approx \begin{array}{c} \\ \\ \\ \\ \\ \end{array} \begin{matrix} \text{F} & \text{So} & \text{J} & \text{Sr} \\ \begin{bmatrix} 1.18 & 0.97 & 0.81 & 0.72 \\ 0 & 1.18 & 0.98 & 0.87 \\ 0 & 0 & 1.11 & 0.99 \\ 0 & 0 & 0 & 1.11 \end{bmatrix} & \begin{matrix} \text{F} \\ \text{So} \\ \text{J} \\ \text{Sr} \end{matrix} \end{matrix}$$

Rounding to 2 decimal places

By looking in row 1, column 1 of this matrix, we can see that, on the average, a freshman spends 1.18 years as a freshman. Notice that this answer was much easier to obtain than was that of Example 4, where we estimated that a freshman spends slightly more than 1.1725 years as a freshman. •

Some graphing calculators (including the TI-82, TI-83, TI-85, and TI-86) can generate identity matrices. To generate a 4×4 identity matrix, make the screen read "identity 4" ["identity (4)" on a TI-83].

TI-82:	• Press $\boxed{\text{MATRX}}$ • Scroll to "MATH" • Select "identity"	• Press 4 $\boxed{\text{ENTER}}$
TI-83:	• Press $\boxed{\text{MATRX}}$ • Scroll to "MATH" • Select "identity"	• Press 4 $\boxed{)}$ $\boxed{\text{ENTER}}$
TI-85/86:	• Press $\boxed{\text{2nd}}$ $\boxed{\text{MATRX}}$ • Select the "OPS" menu (i. e., $\boxed{\text{F4}}$) • Select "ident" (i. e., $\boxed{\text{F3}}$)	• Press 4 $\boxed{\text{ENTER}}$
	The above makes "identity" appear on the screen	*The above makes the identity matrix a 4×4 matrix*

Thus in Example 4, $(I - N)^{-1}$ can be calculated by storing the 4×4 matrix N (as matrix "[A]" on a TI-82 or TI-83, or matrix "N" on a TI-85 or TI-86) and then making the screen read:

TI-82:	$(\text{identity } 4 - [\text{A}])^{-1}$
TI-83:	$(\text{identity } (4) - [\text{A}])^{-1}$
TI-85/86:	$(\text{identity } 4 - \text{N})^{-1}$

We are now in a position to answer the university's first question.

EXAMPLE 5   How many years can an entering freshman expect to spend at the university?

Solution   By inspecting $(I - N)^{-1}$, we can tell that the expected time a freshman stays at the university is:

freshman time + sophomore time + junior time + senior time

$$= 1.1764\ldots + 0.9688\ldots + 0.8073\ldots + 0.7176\ldots$$

$$\approx 3.67 \text{ years}$$

This is an expected value, as discussed in Chapter 6 on probability, so it represents a long-term average.

Realize that this is the expected time to *stay* at the university, not the expected time to *graduate,* and thus reflects the fact that a number of students quit before graduating.   •

Finding the Probability of Being Absorbed

Common sense tells us that if we begin in a nonabsorbing state, we will eventually end up in an absorbing state. All freshmen eventually either graduate or quit. This conclusion is supported by T^5, which gives transitional information for a 5-year time interval.

$$T^5 \approx \begin{array}{c} \\ \\ \\ \\ \\ \\ \\ \end{array} \begin{bmatrix} 0.00 & 0.00 & 0.01 & 0.07 & 0.54 & 0.38 \\ 0 & 0.00 & 0.00 & 0.01 & 0.73 & 0.26 \\ 0 & 0 & 0.00 & 0.00 & 0.84 & 0.16 \\ 0 & 0 & 0 & 0.00 & 0.94 & 0.06 \\ 0 & 0 & 0 & 0 & 1 & 0 \\ 0 & 0 & 0 & 0 & 0 & 1 \end{bmatrix} \begin{array}{l} \text{F} \\ \text{So} \\ \text{J} \\ \text{Sr} \\ \text{G} \\ \text{Q} \end{array}$$

with columns labeled F, So, J, Sr, G, Q — Rounded to 2 decimal places

After 5 years, all but 8% of the freshman have been absorbed (that is, have graduated or quit), all but 1% of the sophomores have been absorbed, and all of the juniors and seniors have been absorbed (at 2-decimal-place accuracy). After a few more years, all of the freshmen and sophomores should be absorbed too.

EXAMPLE 6   Use T^5 to estimate the probability that an entering freshman will graduate.

Solution   By inspecting T^5, we can tell that 54% of the freshmen have graduated after 5 years and that an additional 7% + 1% could still graduate, given enough time. Thus, between 54% and 54% + 7% + 1% = 62% of the freshmen will graduate.   •

It is doubtful that the university would be satisfied with this estimate; a 54% graduation rate is much worse than a 62% rate. The information in N is not sufficient to answer this question, because N contains only nonabsorbing information and the question is about being absorbed. Thus, in addition to forming the matrix N out of the matrix T, we will also form the matrix A, which gives information about making transitions from nonabsorbing states to absorbing states. For reasons that will not be discussed here, the matrix $(I - N)^{-1}A$ gives the probability of being absorbed into each of the absorbing states.

The matrix $(I - N)^{-1}$ was used to answer the first question about the number of years an average student spends at the university; it is also used to answer the second question about the probability of graduation. The matrix $(I - N)^{-1}$ is called the **fundamental matrix** because it is fundamental in making long-range predictions with absorbing chains.

EXAMPLE 7   What is the probability that an entering freshman will graduate?

Solution   After forming the matrices N and A, we will calculate $(I - N)^{-1}A$; this gives the probability of being absorbed into each of the absorbing states.

Step 1   *Discard the rows of T that correspond to absorbing states.* We will discard the last 2 rows: the G row and the Q row.

$$\begin{array}{ccccccc} \text{F} & \text{So} & \text{J} & \text{Sr} & \text{G} & \text{Q} & \\ \begin{bmatrix} 0.15 & 0.7 & 0 & 0 & 0 & 0.15 \\ 0 & 0.15 & 0.75 & 0 & 0 & 0.1 \\ 0 & 0 & 0.1 & 0.8 & 0 & 0.1 \\ 0 & 0 & 0 & 0.1 & 0.85 & 0.05 \end{bmatrix} & \begin{array}{l} \text{F} \\ \text{So} \\ \text{J} \\ \text{Sr} \end{array} \end{array}$$

Nonabsorbing columns   Absorbing columns

Step 2   *Create the matrix N from the nonabsorbing columns of the matrix in Step 1, and create the matrix A from the absorbing columns.*

$$N = \begin{bmatrix} 0.15 & 0.7 & 0 & 0 \\ 0 & 0.15 & 0.75 & 0 \\ 0 & 0 & 0.1 & 0.8 \\ 0 & 0 & 0 & 0.1 \end{bmatrix} \qquad A = \begin{bmatrix} 0 & 0.15 \\ 0 & 0.1 \\ 0 & 0.1 \\ 0.85 & 0.05 \end{bmatrix}$$

Step 3   *Find the label $(I - N)^{-1}A$.*

$$(I - N)^{-1}A = \begin{bmatrix} 1.1764\ldots & 0.9688\ldots & 0.8073\ldots & 0.7176\ldots \\ 0 & 1.1764\ldots & 0.9803\ldots & 0.8714\ldots \\ 0 & 0 & 1.1111\ldots & 0.9876\ldots \\ 0 & 0 & 0 & 1.1111\ldots \end{bmatrix} \begin{bmatrix} 0 & 0.15 \\ 0 & 0.1 \\ 0 & 0.1 \\ 0.85 & 0.05 \end{bmatrix}$$

$$\approx \begin{array}{cc} \text{G} & \text{Q} \\ \begin{bmatrix} 0.61 & 0.39 \\ 0.74 & 0.26 \\ 0.84 & 0.16 \\ 0.94 & 0.06 \end{bmatrix} & \begin{array}{l} \text{F} \\ \text{So} \\ \text{J} \\ \text{Sr} \end{array} \end{array}$$

Rounded to 2 decimal places

The rows of $(I - N)^{-1}A$ are labeled with the nonabsorbing states, and the columns are labeled with the absorbing states.

Step 4   *The matrix $(I - N)^{-1}A$ gives the probability of being absorbed into each of the absorbing states.* We are now in a position to answer the university's other question: What is the probability that an entering freshman will graduate? From the above matrix, we can see that that probability is 0.61.

Notice that this solution is in agreement with that of Example 6, where we estimated that between 54% and 62% of the freshmen will graduate. •

Long-Range Predictions with Absorbing Markov Chains

1. *Determine if the chain is absorbing.* This occurs if:

   - the chain has at least one absorbing state, and
   - it is possible to go from any nonabsorbing state to an absorbing state

   If the chain is absorbing, everything will eventually end up in an absorbing state, with predictable probabilities, after a predictable number of periods.

2. *Create the transition matrix T.* List the absorbing states last. Be certain that the rows and columns are in the same order.
3. *Discard the rows that correspond to absorbing states.* This information is no longer needed.
4. *Create the matrix N from the nonabsorbing columns of the matrix in Step 3, and create the matrix A from the absorbing columns.*
5. *Find and label the fundamental matrix $(I - N)^{-1}$.* This matrix gives the expected number of periods that will be spent in the nonabsorbing states before being absorbed. The matrix $(I - N)^{-1}$ is derived from nonabsorbing state information, so it is labeled with the nonabsorbing states.
6. *Find and label $(I - N)^{-1}A$.* This matrix gives the probability of being absorbed into each of the absorbing states. The rows of $(I - N)^{-1}A$ are labeled with the nonabsorbing states, and the columns are labeled with the absorbing states.

7.3

EXERCISES

1. The university of Utopia has studied its students' paths through college. The university has discovered that:

   - 85% of all freshmen return the next year as sophomores, 8% return as freshmen, and the rest quit
   - 87% of all sophomores return the next year as juniors, 7% return as sophomores, and the rest quit
   - 89% of all juniors return the next year as seniors, 6% return as juniors, and the rest quit
   - 91% of all seniors graduate, 5% return as seniors, and the rest quit

   (Assume that the university does not allow a student who has quit to reenroll, and that the university does not change a student's class standing midyear.)

   a. Write the transition matrix T for these data.
   b. Determine if this transition matrix is regular, absorbing, or neither. Justify your answer.
   c. Use T and T^2 to estimate the average number of years a freshman spends as a freshman.
   d. Use the fundamental matrix to find the average number of years a freshman spends as a freshman.
   e. How many years can a beginning freshman expect to spend at the university?
   f. How many additional years can a junior expect to spend at the university?
   g. Use T^5 to estimate the probability that an entering freshman will graduate.

h. Use the fundamental matrix to find the probability that an entering freshman will graduate.

i. What is the probability that a junior will graduate?

2. In Example 4, we found that

$$(I - N)^{-1} \approx \begin{array}{c} \\ \\ \\ \\ \\ \end{array} \begin{array}{cccc} \text{F} & \text{So} & \text{J} & \text{Sr} \\ \begin{bmatrix} 1.18 & 0.97 & 0.81 & 0.72 \\ 0 & 1.18 & 0.98 & 0.87 \\ 0 & 0 & 1.11 & 0.99 \\ 0 & 0 & 0 & 1.11 \end{bmatrix} & \begin{array}{c} \text{F} \\ \text{So} \\ \text{J} \\ \text{Sr} \end{array} \end{array}$$

The first entry of row 1 is greater than 1, and all of the other entries are less than 1. Explain why this makes sense, given the circumstances at the University of Erehwon.

3. In 1981, Neil Sampson, executive vice president of the National Association of Conservation Districts, compiled data on the shifting land use pattern in the United States. [*Source:* R. Neil Sampson, *Farmland or Wasteland: A Time to Choose* (Rodale Press, 1981).] He estimated that between 1967 and 1977, the following shifts in land use occurred (in millions of acres):

Previous Use of Land	New Use of Land		
	Cropland	Potential Cropland	Noncropland
Cropland	(x)	17	35
Potential cropland	34	(x)	0[1]
Noncropland	0	0[1]	(x)

[1]Unknown, but assumed to cancel each other out

He also stated that in 1977 there were 413 million acres of cropland, 127 million acres of potential cropland, and 856 million acres of noncropland.

a. Predict the number of years that potential cropland will stay as potential cropland.
HINT: You found the matrix *T* in Section 7.1, Exercise 21.

b. Predict the number of years that cropland will stay as cropland.

c. What percent of the potential cropland will eventually become noncropland?

d. What percent of the cropland will eventually become noncropland?

4. The law firm of Cochran and Shapiro employs three types of attorneys: junior attorneys, senior attorneys, and partners. During any given year, 10% of the junior attorneys are promoted to senior attorneys, and 10% are asked to leave the firm. During any given year, 5% of the senior attorneys are

promoted to partners, and 13% are asked to leave. (Junior attorneys must be promoted to senior attorneys before they can become partners.) Attorneys who perform unsatisfactorily are never demoted; they are either not advanced or asked to leave.

a. Write the transition matrix *T* for these data.

b. Determine if this transition matrix is regular, absorbing, or neither. Justify your answer.

c. What is the probability that a junior attorney will eventually become partner?

d. How long should a newly hired junior attorney expect to stay in that position?

e. How long should a newly hired junior attorney expect to stay with the firm?

f. What is the probability that a senior attorney will eventually become partner?

5. Bay Fleet Supply sells auto and truck parts to firms that own fleets of vehicles. When a firm buys parts from Bay Fleet, it is given three months to pay. If the account is not paid in that amount of time, Bay Fleet closes the account, turns it over to a collection agency, and no further transactions occur. Thus, Bay Fleet classifies its accounts as new, one month overdue, two months overdue, three months overdue, paid, or a bad debt. Bay Fleet has studied their past records and have discovered that:

- 70% of all new accounts are paid off in one month
- 60% of all accounts that are one month overdue are paid off at the end of that month
- 50% of all accounts that are two months overdue are paid off at the end of that month
- 60% of all accounts that are three months overdue are turned over to a collection agency

a. Write the transition matrix *T* for these data.

b. Determine if this transition matrix is regular, absorbing, or neither. Justify your answer.

c. What is the probability that a new account will eventually be paid?

d. What is the probability that a 1-month-overdue account will eventually become a bad debt?

e. How many months should Bay Fleet Supply expect to wait for payment from an average new customer?

f. If Bay Fleet's sales average $125,000 per month, how much money will be written off as a bad debt each month? Each year?

6. Woody's Christmas Tree Farm has two types of trees: short trees (5 ft and under) and big trees (over 5 ft). Each year, 16% of all the short trees die, 19% are sold for $30 each, and 10% grow to be over 5 ft. Each year, 5% of the big trees die, and 45% are sold for $45 each.

a. What percent of all seedlings will eventually be sold for $30?

b. What percent of all seedlings will eventually be sold for $45?

c. What percent of all big trees will eventually be sold?

d. What is the expected value of a seedling?

HINT: Use two categories, "sold for $30" and "sold for $45," rather than just "sold."

7. In Example 3, we found that 100% of the entering freshmen spend their first year as freshmen, 15% of the entering freshmen spend their second year as freshmen, and 2.25% of the entering freshmen spend their third year as freshmen. We used these data to compute the expected value of the number of years a freshman spends as a freshman. However, that computation was quite different from the normal way that we compute expected values.

a. Find the percent of entering freshmen that spend *only* their first year as freshmen. (Assume that no one spends more than 3 years as a freshman.)

b. Find the percent of entering freshmen that spend *only* their first 2 years as freshmen. (Assume that no one spends more than 3 years as a freshman.)

c. Find the percent of entering freshmen that spend their first 3 years as freshmen.

d. Use parts (a), (b), and (c) to complete the following chart, and use the chart to calculate the expected value.

Total Years as Freshman	Probability
1	
2	
3	

e. Why does the expected value calculation in Example 3 give the same result?

8. Create a transition matrix T that is neither regular nor absorbing. Explain why your matrix lacks these two qualities.

CHAPTER 7 REVIEW

Terms

absorbing Markov chains
absorbing states
current versus following states
equilibrium matrix

fundamental matrix
Markov chain
probability matrix

regular Markov chains
states
transition matrix

Review Exercises

1. Department of Motor Vehicles records indicate that 12% of the automobile owners in the state of Jefferson own a Toyonda. A recent survey of Jefferson automobile owners, commissioned by Toyonda Motors, shows that 62% of the Toyonda owners would buy a Toyonda for their next car and that 16% of those automobile owners who do not own a Toyonda would buy a Toyonda for their next car. The same survey indicates that automobile owners buy a new car an average of once every three years.

a. Predict Toyonda Motors' market share after three years.

b. Predict Toyonda Motors' market share after six years.

c. Make a long-range prediction of Toyonda Motors' market share.

d. What assumptions are these predictions based on?

2. Currently, 23% of the residences in Foxtail County are apartments, 18% are condominiums or townhouses, and the balance are single-family houses. The Foxtail County Contractors' Association (FCCA) has commissioned a survey that shows that 3% of the apartment residents plan to move to a condominium or townhouse within the next two years and that 5% plan to move to a single-family house. The same survey indicates that 1% of those who currently reside in a condominium or townhouse plan to move to an apartment and that 11% plan to move to a single-family house; also, 2% of the single-family house dwellers plan to move to an apartment, and 4% plan to move to a condominium or townhouse.

  a. What recommendation should the FCCA make regarding the construction of apartments, condominiums, and single-family houses in the next four years?

  b. What long-term recommendations should the FCCA make?

  c. What important factors does the survey ignore?

3. *U.S.A. Yesterday,* a nationwide newspaper, has found that 25% of its subscribers cancel their subscriptions in their first year, 32% of its second-year subscribers cancel their subscriptions in their second year, and 8% of its long-term subscribers cancel their subscriptions during any given year.

  a. On the average, how long will a new subscriber subscribe to *U.S.A. Yesterday*?

  b. On the average, how long will a second-year subscriber continue to subscribe?

4. Lucky Larry is in Las Vegas. He needs $60 to get home, but he has only $30. He has decided to make $10 "pass" bets in the game of craps until he either has $60 or is broke. The "pass" bet is one of the best bets in a casino; it has a probability of winning of 0.49. It has 1-to-1 house odds, so if Larry wins a $10 bet he gains $10, and if he loses a $10 bet he loses $10. Find the probability that Larry will make enough to get home.

GAME THEORY   8

Game theory, invented by the German-American mathematician John von Neumann in 1926, is the study of how to make decisions when competing with an aggressive opponent in a gamelike situation. Game theory can be applied to each of the following three situations.

- Three chains of fast food outlets are competing for sales. Each chain chooses an advertising program and a pricing policy; one chain's choices affect not only that chain's sales but also the competing chains' sales.
- Two countries are at war. Each country's leaders make decisions involving tactics, personnel, equipment and financing; each country's decisions affect the outcome of the war.
- A group of people are playing poker. Each player makes decisions involving discarding cards, betting, and bluffing, and each player's decisions affect not only that player but the others as well.

While only the last of these three scenarios is normally considered to be a game, the other two also could be thought of as games. Each involves two or more parties in competition for a reward, where each party's decisions affect all of the competitors' rewards.

8.1

INTRODUCTION TO GAME THEORY

Reward Matrices

Reggie and Claire are playing a classic game called *two-finger morra*. This game involves two players, each of whom simultaneously puts out a hand with either one or two fingers showing. Reggie and Claire agreed to the betting structure shown in Figure 8.1.

Claire pays Reggie $2, since both Reggie and Claire are showing two fingers.

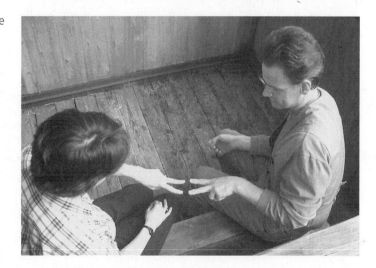

FIGURE 8.1

	Claire Shows 1 Finger	**Claire Shows 2 Fingers**
Reggie Shows 1 Finger	Claire pays Reggie $6	Reggie pays Claire $11
Reggie Shows 2 Fingers	Claire pays Reggie $3	Claire pays Reggie $2

If, for example, Reggie and Claire each show one finger, then Claire must pay Reggie $6.

To analyze the game, the betting structure is summarized in a matrix, called the **reward matrix.** If we let positive matrix entries indicate money paid to Reggie and negative entries indicate money paid to Claire, we obtain the following matrix:

$$R \left\{ \begin{array}{c} 1 \\ 2 \end{array} \overset{\overset{\displaystyle C}{\overbrace{\hspace{2em}}}}{\overset{\begin{array}{cc} 1 & 2 \end{array}}{\begin{bmatrix} 6 & -11 \\ 3 & 2 \end{bmatrix}}} \right.$$

Clearly, this is a two-person game. The two players are traditionally referred to as the **row player** and the **column player.** Here, Reggie is the row player and Claire is the column player.

The 6 in row 1, column 1 indicates that Claire pays Reggie $6. In this event, Reggie's reward is $+6$ and Claire's is -6. Thus, for this outcome as well as the other outcomes, the sum of the rewards is 0, and the game is referred to as a **zero-sum game.**

Reggie and Claire assured each other that the game is fair, because if the above four possibilities are equally likely, then the expected value of a bet is

$$(1/4)(6) + (1/4)(-11) + (1/4)(3) + (1/4)(2) = 0$$

This means that Reggie and Claire's winnings would probably average to $0 per game if the game is played a large number of times. Secretly, however, Reggie suspects that he has an advantage because he has more ways of winning than does Claire, and Claire suspects that she has an advantage because when she wins her winnings are bigger than Reggie's.

Pure Strategies

The simplest of all strategies is one in which the player consistently chooses the same row (or column). Such strategies are called **pure strategies.** As simple as they are, there are circumstances in which pure strategies are the most effective strategies.

EXAMPLE 1   Use the reward matrix

$$\begin{array}{cc} & \begin{array}{cc} 1 & 2 \end{array} \\ \begin{array}{c} 1 \\ 2 \end{array} & \begin{bmatrix} 6 & -11 \\ 3 & 2 \end{bmatrix} \end{array}$$

to find Reggie's best pure strategy and Claire's best pure strategy. Also determine if either should ever deviate from his or her best pure strategy.

Solution   Reggie's highest winnings ($6) occur when both he and Claire show one finger. However, if he consistently shows one finger, Claire will soon learn Reggie's strategy and will consistently show two. This would result in Reggie paying Claire $11 per game. On the other hand if Reggie consistently shows two, Claire will learn Reggie's strategy and will consistently show two. This would result in Claire paying Reggie $2 per game. Reggie would be better off receiving $2 than paying $11, so his best pure strategy is to consistently show two fingers.

If Claire consistently shows one finger, she will have to pay Reggie either $6 or $3. Reggie would respond to this pure strategy by consistently showing one and would win $6 per game. If Claire consistently shows two, she will either be paid $11 by Reggie or she will have to pay Reggie $2. Reggie would respond by showing two and would win $2 per game. Claire would be better off paying $2 than paying $6, so her best pure strategy is to consistently show two fingers.

Claire should not deviate from her best pure strategy; if she does, she will have to pay Reggie $3 per game. Similarly, Reggie should not deviate from his best pure strategy; if he does, he will have to pay Claire $11 per game.   •

If Reggie and Claire wisely adhere to their best pure strategies, then each will always choose two fingers and Claire will pay Reggie $2 per game. (Clearly, it would be even wiser of Claire to quit the game.) This $2 is called the **value** of the game. A game with a value of 0 is said to be **fair** because such a game favors neither player. This version of two-finger morra is not fair; it favors Reggie.

In Example 1 we determined Reggie's best pure strategy essentially by first selecting the minimal entry in each row and then selecting the largest of those row minimums, as shown in Figure 8.2. The row minimums are -11 and 2; the largest of these row minimums is 2.

We determined Claire's best pure strategy by first selecting the maximal entry in each column and then selecting the smallest of those column maximums, as shown in Figure 8.2. The column maximums are 6 and 2; the smallest of these column maximums is 2.

FIGURE 8.2

	Column 1 (1 Finger)	Column 2 (2 Fingers)	Row Minimum
Row 1 (1 Finger)	6	-11	-11
Row 2 (2 Fingers)	3	2	$\boxed{2}$
Column Maximum	6	$\boxed{2}$	

Notice that both Reggie's and Claire's earlier analyses of the game were wrong. Both claimed that the game was fair because if the four possible outcomes are equally likely, then the expected value of the game is 0. In fact, the possibilities are not necessarily equally likely, and the game is not fair; its value is 2, not 0. Furthermore, Reggie suspected that he had an advantage because he had more ways of winning than did Claire, and Claire suspected that she had an advantage because her reward was larger than Reggie's; these aspects of the game actually have no impact.

EXAMPLE 2   Find the row player's and the column player's best pure strategies for the game whose reward matrix is

$$\begin{bmatrix} 2 & 1 & 2 \\ 4 & 3 & 4 \\ 2 & 1 & 2 \end{bmatrix}$$

Also, find the value of the game. If the game is not fair, determine whom it favors.

Solution   For this reward matrix, the row minimums are 1, 3, and 1. These represent the worst outcomes of each of the row player's pure strategies (they are also the most likely outcomes, assuming an aggressive opponent). The largest of those row minimums is 3. This represents the best that the row player can achieve.

FIGURE 8.3

	Column 1	Column 2	Column 3	Row Minimum
Row 1	2	1	2	1
Row 2	4	3	4	3
Row 3	2	1	2	1
Column Maximum	4	3	4	

The column maximums are 4, 3, and 4. These represent the worst (and most likely) outcomes of each of the column player's pure strategies. The smallest of those column maximums is 3. This represents the best that the column player can achieve.

The row player's best pure strategy is to consistently select row 2, and the column player's best pure strategy is to consistently select column 2. The value of the game is +3 (i.e., the column player will consistently lose 3 and the row player will consistently gain 3). The game is not fair; it favors the row player.

Strictly Determined Games

Examples 1 and 2 have an important characteristic in common: there is an entry in the reward matrix that is simultaneously the largest of the row minimums and the smallest of the column maximums. This entry is called a **saddle point,** because (as shown in Figure 8.4) if we draw a surface above the matrix, where the distances from the matrix to the surface are given by the matrix entries, the surface will resemble a saddle.

FIGURE 8.4

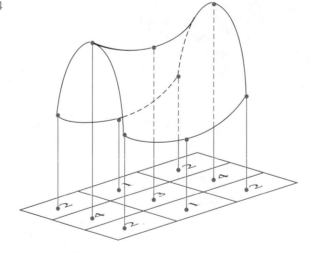

If a game contains a saddle point, the row player's best strategy is to consistently choose the row containing the saddle point, and the column player's best strategy is to consistently choose the column containing the saddle point. Such a game is said to be **strictly determined,** and the **value** of the game is the value of the saddle point itself. The games in Examples 1 and 2 are strictly determined.

Two-Person Constant-Sum Games

EXAMPLE 3

Two television networks, MBC and DBS, are competing for an audience of 150 million viewers during the Tuesday 8 P.M. time slot. Each network will choose to air a medical drama, a legal drama, or a comedy. The projected number of MBC viewers resulting from each possible choice are shown in the following reward matrix.

$$\begin{array}{c} & & \text{DBS} \\ & & \overbrace{\begin{array}{ccc} \text{med} & \text{law} & \text{com} \end{array}} \\ \text{MBC} \left\{ \begin{array}{c} \text{med} \\ \text{law} \\ \text{com} \end{array} \right. & \begin{bmatrix} 52 & 23 & 90 \\ 68 & 87 & 75 \\ 57 & 21 & 95 \end{bmatrix} \end{array}$$

If, for example, both networks select medical dramas, then 52 million viewers are projected to watch MBC, and $150 - 52 = 98$ million are projected to watch DBS.

Determine if this game is strictly determined. If so, find the networks' best pure strategies, the saddle point, and the value of the game; and determine if either network should deviate from its best pure strategy.

Solution

As shown in Figure 8.5, the row minimums are 23, 68, and 21. These represent the worst and most likely outcomes of each of MBC's possible strategies. The largest of those row minimums, 68, represents MBC's best possible market share.

FIGURE 8.5

	Medical Drama	**Legal Drama**	**Comedy**	**Row Minimum**
Medical Drama	52	23	90	23
Legal Drama	68	87	75	68
Comedy	57	21	95	21
Column Maximum	68	87	95	

The column maximums are 68, 87, and 95. These represent the worst and most likely outcomes of each of DBS's strategies. The smallest of those column maximums, 68, corresponds to DBS's best possible market share ($150 - 68 = 82$ million viewers).

MBC's best strategy is to air a legal drama, and DBS's best strategy is to air a medical drama. This combination of shows is a saddle point, and the value of the game is $+68$ (i.e., MBC will have 68 million viewers, and DBS will have $150 - 68 = 82$ million viewers).

Neither MBC nor DBS would benefit from a midseason change of strategies; if MBC replaced its legal drama with a medical drama or comedy, the number of MBC viewers would drop from 68 million to either 52 million or 57 million, respectively. If DBS replaced its medical drama with a legal drama or comedy, the number of MBC viewers would rise from 68 million to either 87 million or 75 million, respectively, and the number of DBS viewers would drop accordingly.   •

Notice that the game in Example 3 is not a zero-sum game. The sum of the rewards for the saddle point is 68 million + 82 million = 150 million. Clearly, the sum of the rewards for other outcomes is also 150 million. Thus, the game is referred to as a **constant-sum game.**

Games That Are Not Strictly Determined

Not all games are strictly determined. If in Example 3 the projected number of MBC viewers was 69 million rather than 57 million for an MBC comedy competing against a DBS medical drama, we would have the following reward matrix:

$$
\text{MBC} \begin{cases} \\ \\ \end{cases}
\begin{array}{c}
 \\ \text{med} \\ \text{law} \\ \text{com}
\end{array}
\overset{\displaystyle \text{DBS}}{\overset{\text{med \quad law \quad com}}{\begin{bmatrix} 52 & 23 & 90 \\ 68 & 87 & 75 \\ 69 & 21 & 95 \end{bmatrix}}}
$$

As shown in Figure 8.6, the row minimums are as they were in Example 3, and the largest of the row minimums is still 68. The column maximums, however, are now 69, 87, and 95, and the smallest of the column maximums is now 69.

FIGURE 8.6

	Medical Drama	**Legal Drama**	**Comedy**	**Row Minimum**
Medical Drama	52	23	90	23
Legal Drama	68	87	75	68
Comedy	69	21	95	21
Column Maximum	69	87	95	

The game does not have a saddle point because the largest of the row minimums is not the same as the smallest of the column maximums; thus, the game is not strictly determined. If MBC were to air a legal drama, DBS would respond with a medical drama. But then MBC would be tempted to cancel its legal drama midseason and replace it with a comedy. It would then be in DBS's best interest to replace its medical drama with a legal drama.

Nonstrictly determined games do have optimal strategies. We have just seen that a pure strategy is not effective if a game is not strictly determined; instead, a **mixed strategy,** where the player does not consistently choose the same row, is most effective. Mixed strategies for games without saddle points are discussed in the next section.

Historical Note

John von Neumann

1903–1957

Born in Hungary in 1903, John von Neumann lived in Germany until he was 27. His grade school teachers recognized his mathematical abilities and advised his father that a conventional education in mathematics would waste his talent. His father, a wealthy banker, arranged for him to receive private tutoring in mathematics from university professors. By the time von Neumann was 19, he had published his first paper and was recognized as a professional mathematician. He received a Ph.D. in mathematics from the University of Berlin; his doctoral thesis was on set theory, discussed in Chapter 5.

Von Neumann emigrated to the United States in 1930 and became a lecturer at Princeton University. After three years there, he was invited to join Princeton's new Institute for Advanced Study, where he remained for the rest of his life.

When World War II broke out, von Neumann, a Jew from Germany, participated in several scientific projects related to the war effort, most notably the development of the hydrogen bomb at Los Alamos.

His game theory amazed the scientific community because it provided a strategic analysis of a subject that seemed to be beyond analysis—games of skill. Furthermore, game theory had a significant influence on economics, where it was applied to gamelike competitive situations. In fact, von Neumann and Oskar Morgenstern, a Princeton economist, wrote a book on game theory and its economic applications, entitled *Theory of Games and Economic Behavior.*

Von Neumann wrote about 150 papers in mathematics, physics, and computer science. He died in 1957.

THEORY OF
GAMES
AND ECONOMIC
BEHAVIOR

By JOHN VON NEUMANN, and
OSKAR MORGENSTERN

PRINCETON
PRINCETON UNIVERSITY PRESS
1944

8.1

EXERCISES

In Exercises 1–10, determine whether the given zero-sum game is strictly determined. If it is, find the row player's best pure strategy, the column player's best pure strategy, the value of the game, and whether the game favors either player. If the game is not strictly de-termined, describe the most likely sequence of four moves if the game started with the row player pursuing his or her highest reward.

1. $\begin{bmatrix} 2 & 3 \\ 1 & 4 \end{bmatrix}$
2. $\begin{bmatrix} 3 & -4 \\ -2 & 2 \end{bmatrix}$

3. $\begin{bmatrix} 10 & -8 \\ -11 & 9 \end{bmatrix}$ 4. $\begin{bmatrix} 5 & -2 \\ 4 & 0 \end{bmatrix}$

5. $\begin{bmatrix} 4 & -2 & 5 \\ -6 & 3 & 0 \\ 1 & -1 & 1 \end{bmatrix}$

6. $\begin{bmatrix} -3 & 2 & 5 \\ 6 & -1 & 0 \\ 4 & 3 & 4 \end{bmatrix}$

7. $\begin{bmatrix} 2 & -3 & -6 \\ -7 & 0 & -1 \\ -5 & -4 & -5 \end{bmatrix}$

8. $\begin{bmatrix} 5 & -7 & 2 & -2 \\ -4 & 3 & 8 & 1 \end{bmatrix}$

9. $\begin{bmatrix} -2 & 3 & 7 \\ 0 & 1 & 2 \end{bmatrix}$

10. $\begin{bmatrix} 3 & 1 & 8 & 7 \\ 4 & 2 & 6 & 9 \end{bmatrix}$

11. For what values of x is the zero-sum game $\begin{bmatrix} x & -2 \\ 0 & 2 \end{bmatrix}$ strictly determined?

12. For what values of x is the zero-sum game $\begin{bmatrix} -2 & x \\ 0 & 2 \end{bmatrix}$ strictly determined?

In Exercises 13–17:

    a. Determine if the game is a zero-sum game or a constant-sum game.

    b. Determine if this game is strictly determined.

    c. If the game is strictly determined, find the players' best pure strategies, the value of the game, whether the game favors either player, and the effect on each player of following the best pure strategies. If it is not strictly determined, describe the most likely sequence of four moves if the game started with the row player pursuing his or her highest reward.

13. At the Main Street freeway offramp, the RunzGood and Petrolia gas stations are in competition. The RunzGood station is an independent station and tends to have lower

prices. Petrolia, however, is a national brand and thus has better advertising and a better image. Their managers are allowed to change prices on a daily basis. Figure 8.7, below, describes the result of this competition. For example, if both stations raise prices, then the RunzGood station gains 3% of the two stations' combined business.

14. Marcos and Jean are dating. This Friday, they are going to dinner and a movie. Jean would prefer to see a movie about relationships, and Marcos would prefer to see an action movie. At dinner, Jean could either be honest and tell Marcos that she wants to see the movie about relationships or be altruistic and tell Marcos that she wants to see the action movie. Similarly, Marcos could be honest or altruistic. If Jean is honest and Marcos is altruistic, then they would see the relationship movie; Jean would be pleased and Marcos would be displeased. If Marcos is honest and Jean is altruistic, then they would see the action movie; Marcos would be pleased and Jean would be displeased. If both are honest or if both are altruistic, then they would argue. Assign a reward of +1 to "being pleased," −1 to "being displeased," and 0 to "arguing."

15. The tri-city area currently has no cellular phone stores, but two chains (Global Fon and U-Call) are planning to open branches in the area. Twenty percent of the tri-city residents live in the town of Backwater, 30% live in Podunk, and 50% live in River City. If both stores locate in the same town, they will each get 50% of the tri-city area business. Otherwise, a store will get all of the business of the town it's located in, none of the business of the town its competitor is located in, and half of the business of the third town.

16. All of the 18- to 25-year-old residents of the city of Erehwon who listen to the radio while commuting listen to either KEWB or KYA. Currently, KEWB has the more popular morning program, and KYA has the more popular evening program. Currently, KEWB gets 55% of the two stations' combined advertising revenue, and KYA gets 45%. A market research survey indicates that if KYA switches its morning and evening shows, its market share will increase to 70%, and if KEWB also switches its shows, KYA's market share will increase to 75%. Finally, if KEWB switches its two shows and KYA does not, KEWB's market share is projected to decrease to 45%.

FIGURE 8.7

	Petrolia Raises Prices	Petrolia Doesn't Change Prices	Petrolia Lowers Prices
RunzGood Raises Prices	RunzGood gains 3%	Petrolia gains 2%	Petrolia gains 5%
RunzGood Doesn't Change Prices	RunzGood gains 1%	no change	Petrolia gains 3%
RunzGood Lowers Prices	RunzGood gains 4%	RunzGood gains 1%	RunzGood gains 2%

17. Willie Jordan and Frank Brown are the only candidates for governor of the state of Confusion. One week before the election, a poll shows that Jordan has the lead in the southern part of the state and Brown has the lead in the north, but overall the two are tied. The fate of the campaign hinges on how Jordan and Brown spend their remaining advertising money in the remaining week. Each can decide to spend the bulk of his money in the north or in the south, or to split it equally. Each candidate is free to change his advertising tactics on a daily basis. The campaign managers' projected results of these decisions are given in Figure 8.8.

 Answer the following questions using complete sentences.

18. Select one strictly determined game from Exercises 13–17 and discuss what would happen if either player deviated from his or her best pure strategy.

19. If a competitor uses game theory to analyze her competition, does the competitor acquire an optimistic or pessimistic view of her circumstances? What assumption does game theory make that forces this view?

20. Why is the row player's best pure strategy found by selecting the minimal entry of each row and then selecting the largest of the row minimums? Why is the column player's best pure strategy found by selecting the maximal entry of each column and then selecting the smallest of the column maximums?

21. Why is it that, in a strictly determined game, either player would suffer by deviating from his or her best pure strategy?

22. What did von Neumann do to assist the United States during World War II? What aspect of von Neumann's background could well have motivated this action?

FIGURE .8.8

	Jordan Emphasizes North	**Jordan Splits Money Equally**	**Jordan Emphasizes South**
Brown Emphasizes North	Brown gains 3%	Brown gains 5%	No net change
Brown Splits Money Equally	Jordan gains 4%	No net change	Brown gains 2%
Brown Emphasizes South	No net change	Jordan gains 6%	Jordan gains 3%

8.2

MIXED STRATEGIES

In the previous section we found that some games have saddle points and are strictly determined; that is, both the row player and the column player are better off if each uses a pure strategy. If either player deviates from his or her best pure strategy, that player's reward will decrease. We also found that some games are not strictly determined.

EXAMPLE 1   Two people are playing two-finger morra. In this variation of the game one player, whom we will call the Odd player (or just "Odd"), wins if the sum of the fingers is odd, and the other player ("Even") wins if the sum is even. The dollar amount of the reward is the sum of the fingers. Is this game strictly determined?

Solution   If each player puts out one finger, the sum of the fingers is 2, and Odd pays Even $2. If Odd puts out one finger and Even puts out two, then the sum is 3, and Even pays Odd $3. If we make the Odd player the row player, we obtain the following reward matrix:

$$\text{Odd} \begin{cases} 1 \\ 2 \end{cases} \overset{\overset{\text{Even}}{\overbrace{\quad 1 \qquad 2 \quad}}}{\begin{bmatrix} -2 & 3 \\ 3 & -4 \end{bmatrix}}$$

As shown in Figure 8.9, the row minimums are -2 and -4. The largest of the row minimums is -2, since $-2 > -4$. The column maximums are both 3.

FIGURE 8.9

	Column 1 (1 Finger)	Column 2 (2 Fingers)	Row Minimum
Row 1 (1 Finger)	-2	3	$\boxed{-2}$
Row 2 (2 Fingers)	3	-4	-4
Column Maximum	$\boxed{3}$	$\boxed{3}$	

There is no entry in the reward matrix that is simultaneously the largest of the row minimums (-2) and the smallest of the column maximums (3). Thus there is no saddle point, and the game is not strictly determined. •

Randomized Strategies

Notice that if in Example 1 Odd adopts the pure strategy of always putting out one finger, Even would respond by always putting out one finger, and Odd must pay Even $2. This would prompt Odd to always put out two fingers, and Even would pay Odd $3. Of course Even would then shift strategies. Clearly, a pure strategy would not be effective, and the players would have to adopt some other form of strategy (i.e., a **mixed strategy**). Von Neumann discovered that the best form of mixed strategy is a **randomized strategy,** where the row player selects a row at random, according to a certain probability distribution. For example, the row player might use the following probability distribution:

FIGURE 8.10

Outcome	Probability
Row 1	2/3
Row 2	1/3

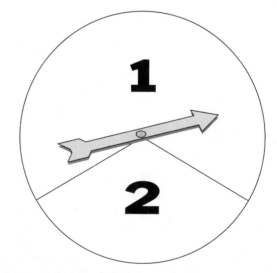

If the row player used this distribution in a predictable manner, such as repeatedly selecting row 1 twice and then row 2 once, the column player could well learn this strategy and could respond in a way that minimizes its effectiveness. Thus, the row player must use some chance device, such as the spinner shown in Figure 8.10, that would select row 1 two-thirds of the time.

Games with saddle points are strictly determined; that is, the players adopt pure strategies, and the course of the game is determined in advance (assuming that the players are aggressive and able). Games without saddle points are not strictly determined; if a player uses a randomized strategy, the course of the game is subject to chance and anything can happen. There is no fixed value of the game; there is only a most likely or *expected* value. (Expected values are discussed in Chapter 6.)

Randomized Strategies versus Pure Strategies

If the row player adopts a randomized strategy, the column player can respond with a pure strategy or a randomized strategy. At first, we will discuss randomized strategies applied against pure strategies, where each competitor has two strategies. Later in this section, we will discuss randomized strategies applied against randomized strategies, and games in which the competitors have more than two strategies.

EXAMPLE 2   In the two-finger morra game described in Example 1, Odd adopts a randomized strategy in which Odd shows one finger two-thirds of the time and two fingers one-third of the time.

a. Compute the expected value of this randomized strategy if Even always responds with one finger.
b. Compute the expected value of this randomized strategy if Even always responds with two fingers.
c. Will Odd's randomized strategy work to his advantage?

Solution   In Example 1, we found the reward matrix:

$$
\begin{array}{cc}
 & \begin{array}{cc} 1 & 2 \end{array} \\
\begin{array}{c} 1 \\ 2 \end{array} & \begin{bmatrix} -2 & 3 \\ 3 & -4 \end{bmatrix}
\end{array}
$$

a. If Even always responds with one finger, then the possible outcomes of a game have the following values (from the reward matrix) and probabilities:

Outcome	Value	Probability
Odds shows 1	−2	2/3
Odd shows 2	3	1/3

The expected value of the randomized strategy is then

$$(-2)(2/3) + (3)(1/3) = -1/3 \approx -\$0.33$$

This means that Odd will most probably lose an average of $0.33 per game (and Even will most probably win $0.33 per game) if the game is played a large number of times.

b. If Even always responds with two fingers, then we get

Outcome	Value	Probability
Odd shows 1	3	2/3
Odd shows 2	−4	1/3

The expected value of the randomized strategy is then

$$(3)(2/3) + (−4)(1/3) = 2/3 \approx \$0.67$$

This means that Odd will most probably win an average of $0.67 per game (and Even will most probably lose $0.67 per game) if the game is played a large number of times.

c. This particular randomized strategy will not work to Odd's advantage, because (assuming that Even is an able opponent) Even would always respond with one finger and Odd would tend to lose $0.33 per game. ●

In Example 2, we found that the expected value of Odd's randomized strategy is:

* −$0.33 if Even always responds with one finger
* $0.67 if Even always responds with two fingers

We also found that the higher expected value ($0.67) is not attainable, and that the lower expected value (−$0.33) is more realistic (assuming that Even is an aggressive and able opponent). For this reason, −$0.33 is called the **expected value of the randomized strategy.**

Don't let the terminology confuse you. When the row player adopts a randomized strategy, the column player can respond with either of his or her pure strategies, and there are three different expected values:

* the expected value of the randomized strategy applied against the column player's *first* pure strategy
* the expected value of the randomized strategy applied against the column player's *second* pure strategy
* the expected value of the randomized strategy

The latter expected value assumes that the column player will adopt the pure strategy that works to his own advantage.

Probability Matrices and Expected Value

The above expected value computations could be performed with matrices. If we list the probabilities of either of the rows being selected in a row matrix P_r, then we obtain

$$P_r = [2/3 \quad 1/3]$$

P_r is called the **row player's probability matrix.** The expected value calculations done in Example 2 are the same ones done in computing the product of the row player's probability matrix P_r and the reward matrix R:

$$P_r \cdot R = [2/3 \quad 1/3] \cdot \begin{bmatrix} -2 & 3 \\ 3 & -4 \end{bmatrix}$$

$$= [(2/3)(-2) + (1/3)(3) \quad (2/3)(3) + (1/3)(-4)]$$

$$= [-1/3 \quad 2/3]$$

Expected Value: Randomized Strategy versus Pure Strategy

If in a game with reward matrix $R = \begin{bmatrix} a & b \\ c & d \end{bmatrix}$, the row player adopts a randomized strategy where the probabilities of selecting rows 1 and 2 are p_1 and p_2 respectively, and the column player adopts a pure strategy, then the expected values of the randomized strategy against the pure strategies are contained in the matrix

$$E(\text{pure}) = P_r \cdot R = [p_1 \quad p_2] \begin{bmatrix} a & b \\ c & d \end{bmatrix}$$

The expected value of the randomized strategy is the smaller entry in the matrix $E(\text{pure})$.

EXAMPLE 3   In the two-finger morra game described in Example 1, Odd adopts a randomized strategy in which Odd shows one finger one-fourth of the time.

  a. Compute the expected value of this randomized strategy if Even always responds with one finger, and the expected value if Even always responds with two fingers.
  b. Compute the expected value of the randomized strategy.
  c. Will the randomized strategy work to Odd's advantage?

Solution   a. The expected values of the randomized strategy against the pure strategies are contained in the matrix

$$E(\text{pure}) = P_r \cdot R = [1/4 \quad 3/4] \cdot \begin{bmatrix} -2 & 3 \\ 3 & -4 \end{bmatrix}$$

$$= [(1/4)(-2) + (3/4)(3) \quad (1/4)(3) + (3/4)(-4)]$$

$$= [7/4 \quad -9/4]$$

$$= [1.75 \quad -2.25]$$

This means that Odd's expected winnings will be $1.75 per game if Even always responds with one finger, or $-$2.25 if Even always responds with two fingers, if the game is played a large number of times.

  b. Even should always select two fingers, so that Odd would tend to lose $2.25 per game. The expected value of the randomized strategy is $-$2.25.
  c. The randomized strategy will not work to Odd's advantage.   •

Optimal Randomized Strategies: The Geometric Method

In Examples 2 and 3 the Odd player had different randomized strategies, but in each case the Even player was able to find a pure strategy that would keep Odd from winning, in the long run. Odd needs a randomized strategy that does not give Even this option. Odd needs to find his **optimal randomized strategy,** the strategy with the highest expected value.

Let p_1 be the probability with which Odd shows one finger. Then p_2 is the probability that Odd shows two fingers, and $p_1 + p_2 = 1$.

$$E(\text{pure}) = P_r \cdot R = [p_1 \quad p_2] \cdot \begin{bmatrix} -2 & 3 \\ 3 & -4 \end{bmatrix}$$

$$= [-2p_1 + 3p_2 \quad 3p_1 - 4p_2]$$

$$= [-2p_1 + 3(1 - p_1) \quad 3p_1 - 4(1 - p_1)] \qquad \text{Since } p_2 = 1 - p_1$$

$$= [-2p_1 + 3 - 3p_1 \quad 3p_1 - 4 + 4p_1]$$

$$= [-5p_1 + 3 \quad 7p_1 - 4]$$

Thus, the expected value is $-5p_1 + 3$ if Even always responds with one finger, and it is $7p_1 - 4$ if Even always responds with two fingers. These expected values are graphed in Figure 8.11, with p_1 on the horizontal axis and the resulting expected values on the vertical axis. Only that part of the graphs where p_1 is between 0 and 1 is shown; p_1 is a probability, so p_1 must be between 0 and 1.

FIGURE 8.11

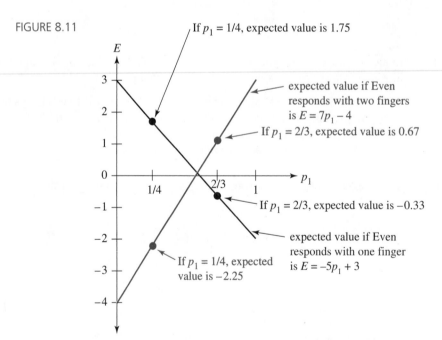

If $p_1 = 1/4$, expected value is 1.75

expected value if Even responds with two fingers is $E = 7p_1 - 4$

If $p_1 = 2/3$, expected value is 0.67

If $p_1 = 2/3$, expected value is −0.33

expected value if Even responds with one finger is $E = -5p_1 + 3$

If $p_1 = 1/4$, expected value is −2.25

The results of Examples 2 and 3 can be seen in the graph. In Example 2, we found that if $p_1 = 2/3$, the expected value would be −$0.33 if Even responds with one finger, or $0.67 if Even responds with two fingers. And in Example 3, we found that if $p_1 = 1/4$, the expected value would be $1.75 if Even responds with one finger, or −$2.25 if Even responds with two fingers.

Naturally, Even will always respond in a way that eliminates the higher expected value to Odd. This effectively removes the top portion of the graphs (as shown in Figure 8.12), because the top portion contains the higher expected values. The graph makes it clear that for most values of p_1, the randomized strategy has a negative expected value (that is, Odd would be expected to lose in the long run). The graph also makes it clear that the expected value is positive near the intersection point, and it is highest at the intersection point. To find that intersection point, set the two expected values equal to each other.

FIGURE 8.12

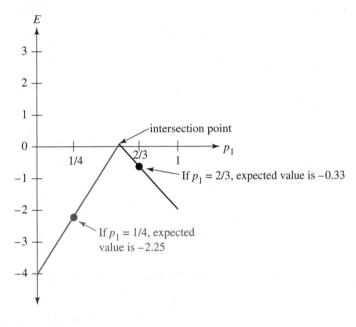

$$-5p_1 + 3 = 7p_1 - 4$$
$$7 = 12p_1$$
$$p_1 = 7/12$$

Odd's optimal randomized strategy is the one given by the intersection point; that is, Odd should show one finger $p_1 = 7/12$ of the time, and two fingers $p_2 = 1 - 7/12 = 5/12$ of the time. If Even always responds with one finger, the resulting expected value will be

$$-5p_1 + 3 = -5(7/12) + 3 = 1/12 \approx \$0.08$$

And if Even always responds with two fingers, the resulting expected value will be

$$7p_1 - 4 = 7(7/12) - 4 = 1/12 \approx \$0.08$$

(Of course, it would suffice to compute just one of these expected values; the graph makes it clear that the two expected values are equal at the point of intersection.) This means that Odd will tend to win $0.08 per game regardless of whether Even always responds with one finger or two fingers, if the game is played a large number of times and if Odd follows his or her optimal randomized strategy. This $1/12 \approx \$0.08$ is called the **expected value of the game.**

We have now discussed four different expected values:

- the expected value of a specific randomized strategy applied against the column player's *first* pure strategy
- the expected value of a specific randomized strategy applied against the column player's *second* pure strategy
- the expected value of the specific randomized strategy
- the expected value of the game

The latter expected value assumes that the row player adopts his or her optimal randomized strategy.

Randomized Strategies versus Randomized Strategies

Neither of Even's pure strategies will give Even an advantage over Odd. Perhaps Even could gain an advantage if Even were to respond with a randomized strategy. To investigate this possibility, let q_1 be the probability that Even shows one finger, and let $q_2 = 1 - q_1$ be the probability that Even shows two fingers. To find the expected value of the game, we must first find the probability

p(Odd shows 1 finger ∩ Even shows 1 finger)

as well as the probabilities of the other outcomes of the game. It is true that

p(Odd shows 1 │ Even shows 1) = p(Odd shows 1 │ Even shows 2)

because Odd's choice is not affected by Even's choice. Thus Odd's choices and Even's choices are independent and we can use the product rule for independent events.

$p(A \cap B) = p(A) \cdot p(B)$

p(Odd shows 1 ∩ Even shows 1) = p(Odd shows 1) · p(Even shows 1)

$$= p_1 \cdot q_1$$

Similar computations yield the probabilities in the following chart.

Outcome	Probability	Value
Odd shows 1, Even shows 1	$p_1 \cdot q_1$	-2
Odd shows 1, Even shows 2	$p_1 \cdot q_2$	3
Odd shows 2, Even shows 1	$p_2 \cdot q_1$	3
Odd shows 2, Even shows 2	$p_2 \cdot q_2$	-4

The expected value is then:

$$(-2)(p_1 q_1) + (3)(p_1 q_2) + (3)(p_2 q_1) + (-4)(p_2 q_2)$$

$$= -2(7/12)q_1 + 3(7/12)q_2 + 3(5/12)q_1 - 4(5/12)q_2 \qquad \text{Since } p_1 = 7/12 \text{ and}$$

$$= (-14/12)q_1 + (21/12)q_2 + (15/12)q_1 - (20/12)q_2 \qquad p_2 = 5/12$$

$$= (1/12)q_1 + (1/12)q_2$$

$$= (1/12)q_1 + (1/12)(1 - q_1) \qquad \text{Since } q_2 = 1 - q_1$$

$$= (1/12)q_1 + (1/12) - (1/12 q_1)$$

$$= 1/12$$

This means that if Odd is playing his or her optimal and randomized strategy, all of Even's randomized strategies have the same result; they all have an expected value of 1/12. We previously found that both of Even's pure strategies resulted in an expected value of 1/12. Thus, Even is helpless against Odd's optimal randomized strategy.

The above expected value computation could be performed with matrices. If we list the probabilities of either of the rows being selected in the probability row matrix P_r, and the probabilities of either of the columns being selected in the **column player's probability matrix** P_c, then we obtain

$$P_r = [p_1 \quad p_2] \qquad P_c = \begin{bmatrix} q_1 \\ q_2 \end{bmatrix}$$

The expected value of the game is

$$
\begin{aligned}
E(\text{random}) &= P_r \cdot R \cdot P_c \\
&= [p_1 \quad p_2] \begin{bmatrix} -2 & 3 \\ 3 & -4 \end{bmatrix} \begin{bmatrix} q_1 \\ q_2 \end{bmatrix} \\
&= [-2p_1 + 3p_2 \quad 3p_1 - 4p_2] \begin{bmatrix} q_1 \\ q_2 \end{bmatrix} \\
&= [(-2p_1 + 3p_2)(q_1) + (3p_1 - 4p_2)(q_2)] \\
&= [-2p_1q_1 + 3p_2q_1 + 3p_1q_2 - 4p_2q_2]
\end{aligned}
$$

Clearly, this leads to the same result as does the expected value computation on page 382.

If you use matrices, you can cut down on the computations by using the previously computed $E(\text{pure}) = [p_1 \quad p_2] \begin{bmatrix} -2 & 3 \\ 3 & -4 \end{bmatrix}$.

$$
\begin{aligned}
E(\text{random}) &= P_r \cdot R \cdot P_c \\
&= [p_1 \quad p_2] \begin{bmatrix} -2 & 3 \\ 3 & -4 \end{bmatrix} \begin{bmatrix} q_1 \\ q_2 \end{bmatrix} \\
&= E(\text{pure}) \begin{bmatrix} q_1 \\ q_2 \end{bmatrix} \\
&= [-5p_1 + 3 \quad 7p_1 - 4] \begin{bmatrix} q_1 \\ q_2 \end{bmatrix} && \text{From Example 3} \\
&= [(-5p_1 + 3)q_1 + (7p_1 - 4)q_2] \\
&= [(-5p_1 + 3)q_1 + (7p_1 - 4)(1 - q_1)] && \text{Since } q_2 = 1 - q_1 \\
&= [-5p_1q_1 + 3q_1 + 7p_1 - 4 - 7p_1q_1 + 4q_1] \\
&= [-12p_1q_1 + 7q_1 + 7p_1 - 4] \\
&= [-12(7/12)q_1 + 7q_1 + 7(7/12) - 4] && \text{Since } p_1 = 7/12 \\
&= [-7q_1 + 7q_1 + 7(7/12) - 4] \\
&= [49/12 - 48/12] \\
&= [1/12]
\end{aligned}
$$

Expected Value: Randomized Strategy versus Randomized Strategy

If in a game with reward matrix $R = \begin{bmatrix} a & b \\ c & d \end{bmatrix}$, both the row player and the column player adopt randomized strategies, where:

- the row player's strategy is given by the probability row matrix

$$P_r = [\,p_1 \quad p_2\,] \quad \text{and}$$

- the column player's strategy is given by the probability column matrix

$$P_c = \begin{bmatrix} q_1 \\ q_2 \end{bmatrix}$$

(where row selections and column selections are independent)

then the expected value of the game is

$$E(\text{random}) = P_r \cdot R \cdot P_c = [\,p_1 \quad p_2\,]\begin{bmatrix} a & b \\ c & d \end{bmatrix}\begin{bmatrix} q_1 \\ q_2 \end{bmatrix}$$

Steps for the Geometric Method of Finding the Optimal Randomized Strategy

1. Determine the reward matrix $R = \begin{bmatrix} a & b \\ c & d \end{bmatrix}$.
2. If the row player adopts a randomized strategy where the probabilities of selecting rows 1 and 2 are p_1 and p_2, respectively, and the column player adopts a pure strategy, then the expected values of the randomized strategy against the pure strategies are contained in the matrix

$$E(\text{pure}) = P_r \cdot R = [\,p_1 \quad p_2\,]\begin{bmatrix} a & b \\ c & d \end{bmatrix}$$

3. Substitute $1 - p_1$ for p_2 and simplify.
4. Graph the expected values of the column player's pure strategies on one set of axes. Remember that p_1 is a probability, so $0 \le p_1 \le 1$.
5. Remove the portions of the graphs that contain the higher expected values, because the column player will not allow the row player to achieve those values.
6. Find the row player's optimal randomized strategy (i.e., the strategy with the highest achievable expected value) and the expected value of the game (i.e., the highest achievable expected value).
7. Describe the row player's optimal randomized strategy by giving the probability with which he or she should select row 1 and the probability with which he or she should select row 2.
8. Determine if the column player could do better with a randomized strategy by computing $E(\text{random}) = P_r \cdot R \cdot P_c = [\,p_1 \quad p_2\,]\begin{bmatrix} a & b \\ c & d \end{bmatrix}\begin{bmatrix} q_1 \\ q_2 \end{bmatrix}$.

Removing Dominated Strategies

The method of finding a game's optimal randomized strategy discussed earlier in this section involves graphing two line segments, removing parts of those line segments, and selecting the highest point on the resulting graph. Thus, it applies only to games with two variables—that is, to games where the row player and the column player each have two strategies. Some games with more than two strategies per player can be simplified by removing ineffective strategies. Ineffective strategies that can be removed from consideration are called **dominated strategies**, and the reward matrix that results from the removal of dominated strategies is called the **reduced reward matrix.**

EXAMPLE 4   Consider the game described by the reward matrix

$$R = \begin{bmatrix} 5 & -7 & 6 \\ -4 & 4 & 0 \\ -3 & 5 & 0 \end{bmatrix}$$

    a. Identify any dominated strategies and remove them.
    b. Find the row player's optimal randomized strategy.
    c. Discuss the result if the column player responds with a pure strategy.
    d. Discuss the result if the column player responds with a randomized strategy.

Solution   a. *Identifying and removing dominated strategies.* Positive entries indicate payments made by the column player to the row player, so the larger the number, the better for the row player. Each of the entries in row 3 is greater than or equal to each of the corresponding entries in row 2 ($-3 \geq -4$, $5 \geq 4$, $0 \geq 0$). The row player should never play row 2 because, for each of the column player's responses, row 3 results in a higher reward than does row 2. Thus, row 3 dominates row 2.

        Negative entries indicate payments made by the row player to the column player, so the smaller the number, the better for the column player. Each of the entries in column 1 is less than or equal to each of the corresponding entries in column 3. The column player should never play column 3, because for each of the row player's responses, column 1 results in a lower (i.e., better) reward than does column 3. Thus, column 1 dominates column 3. If we remove row 2 and column 3, we obtain the reduced reward matrix R'.

$$R' = \begin{bmatrix} 5 & -7 \\ -3 & 5 \end{bmatrix}$$

    b. *Finding the row player's optimal randomized strategy.*

$$E = P_r \cdot R' = \begin{bmatrix} p_1 & p_2 \end{bmatrix} \cdot \begin{bmatrix} 5 & -7 \\ -3 & 5 \end{bmatrix}$$

$$= \begin{bmatrix} 5p_1 - 3p_2 & -7p_1 + 5p_2 \end{bmatrix}$$

$$= \begin{bmatrix} 5p_1 - 3(1 - p_1) & -7p_1 + 5(1 - p_1) \end{bmatrix} \qquad \text{Since } p_2 = 1 - p_1$$

$$= \begin{bmatrix} 8p_1 - 3 & -12p_1 + 5 \end{bmatrix}$$

Thus, the expected value is $8p_1 - 3$ if the column player always responds with column 1, and it is $-12p_1 + 5$ if the column player always responds with column 2. These expected values are graphed in Figure 8.13. Only that part of the graph where p_1 is between 0 and 1 is shown, since p_1 is a probability.

FIGURE 8.13

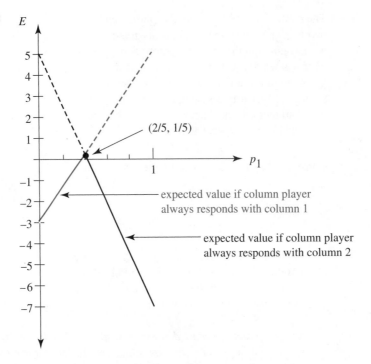

Since the column player will always respond in a way that results in a lower expected value to the row player, the top portion of the graphs have been removed. The graph makes it clear that the expected value is highest at the intersection point. To find that intersection point, set the two expected values equal to each other.

$$8p_1 - 3 = -12p_1 + 5$$
$$20p_1 = 8$$
$$p_1 = 8/20 = 2/5$$
$$p_2 = 1 - p_1 = 3/5$$

In reference to the reduced reward matrix R', the row player should play row 1 two-fifths of the time and row 2 three-fifths of the time. In reference to the original reward matrix R, the row player should play row 1 two-fifths of the time, play row 3 three-fifths of the time, and never play row 2.

c. If the column player always selects column 1, the expected value of the game is given by $8p_1 - 3$; if the column player always selects column 2, the expected value of the game is given by $-12p_1 + 5$. The row player's best randomized strategy requires $p_1 = 2/5$ and, since this is the intersection point, the column player's two pure strategies have the same expected value:

$$8p_1 - 3 = 8(2/5) - 3 = 1/5.$$

This means that the row player will tend to win 1/5 per game if the column player responds with a pure strategy (other than column 3, the dominated strategy) if the game is played a large number of times. This is not a fair game; it favors the row player.

d. If the column player responds with a randomized strategy, the expected value of the game is

$$P_r \cdot R' \cdot P_c = \begin{bmatrix} p_1 & p_2 \end{bmatrix} \begin{bmatrix} 5 & -7 \\ -3 & 5 \end{bmatrix} \begin{bmatrix} q_1 \\ q_2 \end{bmatrix}$$

$$= \begin{bmatrix} 2/5 & 3/5 \end{bmatrix} \begin{bmatrix} 5 & -7 \\ -3 & 5 \end{bmatrix} \begin{bmatrix} q_1 \\ q_2 \end{bmatrix}$$

$$= \begin{bmatrix} 1/5 & 1/5 \end{bmatrix} \begin{bmatrix} q_1 \\ q_2 \end{bmatrix}$$

$$= \left[(1/5)q_1 + (1/5)q_2 \right]$$

$$= \left[(1/5)q_1 + (1/5)(1 - q_1) \right]$$

$$= \left[(1/5)q_1 + (1/5) - (1/5)q_1 \right]$$

$$= \left[1/5 \right]$$

This means that if the row player is playing his or her optimal randomized strategy, all of the column player's randomized strategies have the same result; they all have a value of 1/5. In part (c) we found that both of the column player's pure strategies resulted in an expected value of 1/5. Thus, the column player is helpless against the row player's optimal randomized strategy.

The following flowchart includes the methods of Sections 8.1 and 8.2.

Game Theory Flowchart

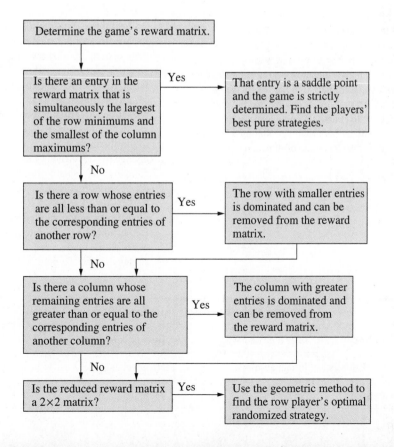

8.2

EXERCISES

1. Two people are playing two-finger morra. In this variation of the game, the odd player wins $1 if the sum of the fingers is odd, and the even player wins $1 if the sum is even.
   a. Show that the game is not strictly determined.
   b. Should Odd adopt a randomized strategy in which Odd shows one finger one-fourth of the time and two fingers three-fourths of the time? (Assume that Even responds with a pure strategy.)
   c. Find E(pure), and graph the expected values of the column player's pure strategies on one set of axes. Show the results of part (b) on your graph.
   d. Remove the portion of the graphs that contains the higher expected values and discuss what assumptions imply that these portions should be removed.
   e. Find the game's expected value.
   f. Describe the row player's optimal randomized strategy.
   g. Is the game fair?

In Exercises 2–10:

   a. Determine if the game is strictly determined.
   b. If it is strictly determined, find the row player's best pure strategy, the column player's best pure strategy, the value of the game, and whether the game favors either player.
   c. If it is not strictly determined, find the row player's optimal randomized strategy, assuming that the column player responds with a pure strategy. Determine if the column player could do better with a randomized strategy. Find the expected value of the game.

2. $\begin{bmatrix} 3 & -4 \\ -2 & 2 \end{bmatrix}$

3. $\begin{bmatrix} 10 & -8 \\ -11 & 9 \end{bmatrix}$

4. $\begin{bmatrix} 4 & -5 \\ -1 & 1 \end{bmatrix}$

5. $\begin{bmatrix} 5 & 2 \\ 7 & 3 \end{bmatrix}$

6. $\begin{bmatrix} -6 & 9 \\ 4 & -1 \end{bmatrix}$

7. $\begin{bmatrix} 3 & -6 & 2 \\ 1 & 0 & -1 \\ -9 & -7 & 1 \end{bmatrix}$

8. $\begin{bmatrix} 3 & 1 \\ -2 & 2 \\ 2 & -1 \end{bmatrix}$

9. $\begin{bmatrix} 5 & 4 \\ 2 & 0 \\ -1 & 3 \end{bmatrix}$

10. $\begin{bmatrix} 9 & 10 & -9 \\ 0 & 2 & -10 \\ -12 & 0 & -8 \end{bmatrix}$

11. Willie Jordan and Frank Brown are the only candidates for governor of the state of Confusion. One week before the election, a poll shows that Jordan has the lead in the southern part of the state and Brown has the lead in the north, but overall the two are tied. The fate of the campaign hinges on how Jordan and Brown spend their remaining advertising money in the remaining week. Each can decide to spend the bulk of his money in the north or in the south, or to split it equally. Each candidate is free to change his advertising tactics on a daily basis. The campaign managers' projected results of these decisions are given in Figure 8.14, below.
   NOTE: This is Exercise 17 from Section 8.1.
   Find Brown's optimal randomized strategy, assuming that Jordan adopts a pure strategy. Determine if Jordan could do better with a randomized strategy. Discuss the impact of this strategy on the election.

FIGURE 8.14

	Jordan Emphasizes North	Jordan Splits Money Equally	Jordan Emphasizes South
Brown Emphasizes North	Brown gains 3%	Brown gains 5%	No net change
Brown Splits Money Equally	Jordan gains 4%	No net change	Brown gains 2%
Brown Emphasizes South	No net change	Jordan gains 6%	Jordan gains 3%

FIGURE 8.15

	Standard's Production Is Low	Standard's Production Is Medium	Standard's Production Is High
Petrolia's Production Is Low	Petrolia's revenue is $5000	Petrolia's revenue is $6000	Petrolia's revenue is $4000
Petrolia's Production Is Medium	Petrolia's revenue is $4000	Petrolia's revenue is $5000	Petrolia's revenue is $3000
Petrolia's Production Is High	Petrolia's revenue is $3000	Petrolia's revenue is $7000	Petrolia's revenue is $6000

12. The Petrolia Oil Co. and the Standard Oil Co. produce all of the gasoline sold in the state of Erehwon. The total revenue earned by these two firms is currently $10,000 per day. Once a week, each of the firms determines how much gasoline to produce. When both firms' production levels are low, they each earn $5,000 in revenue. Similar data for other production levels are given in Figure 8.15. Find Petrolia's optimal randomized strategy, assuming that Standard adopts a pure strategy. Determine if Standard could do better with a randomized strategy. Discuss the impact of these strategies on revenues.

13. The Chavez Agricultural Corporation owns a large tract of land that can be used successfully to grow wheat or rice, depending on the amount of rain. If the entire tract of land is planted with rice, a profit of $10 million will be generated in a wet year, a profit of $5 million will be generated in a normal year, and $5 million will be lost in a dry year. If the entire tract of land is planted with wheat, a profit of $12 million will be generated in a normal year, a profit of $3 million will be generated in a dry year, and $3 million will be lost in a wet year.
  a. According to the above information, what climatic conditions are better for wheat? For rice?
  b. Use game theory to determine the proportion of the tract of land that should be planted with wheat and the proportion that should be planted with rice.
  c. What would result if the land is planted as discussed in part (b) during a wet year? A normal year? A dry year?

14. The Round Valley area has two home improvement stores: Household Depot and the Handyman's Emporium. Each week, each store advertises either through a mailer, an ad in the local newspaper, a television ad, or a radio ad. If Household Depot advertises through a mailer and Handyman's Emporium advertises through the newspaper, then Household Depot's portion of the two firms' combined revenues drops by 1% (and Handyman's Emporium's share increases by 1%). Figure 8.16 gives similar data for other media choices.

FIGURE 8.16

Household Depot's Advertisement	Handyman's Emporium's Advertisement			
	Mail	Newspaper	TV	Radio
Mail	2%	−1%	1%	5%
Newspaper	1%	3%	4%	1%
TV	1%	−1%	0%	2%
Radio	−1%	2%	3%	−1%

  a. What is Household Depot's best strategy?
  b. What is Handyman's Emporium's best strategy?
    HINT: The reward matrix differs from that of part (a) in two ways: the determination of which store is the row player and the entries of the matrix.
  c. What would result if Household Depot followed its best strategy and Handyman's Emporium didn't?
  d. What would result if Handyman's Emporium followed its best strategy and Household Depot didn't?
  e. What would result if both stores followed their best strategies?

Exercise 15 involves a strictly determined game that was solved in Section 8.1. Use the methods of this section to find the row player's optimal randomized strategy. Discuss the application of the two different methods to the problem and compare the results.

15. Marcos and Jean are dating. This Friday, they are going to dinner and a movie. Jean would prefer to see a movie about relationships, and Marcos would prefer to see an action movie. At dinner, Jean could either be honest and tell Marcos that she wants to see the movie about relationships,

or she could be altruistic and tell Marcos that she wants to see the action movie. Similarly, Marcos could be honest or altruistic. If Jean is honest and Marcos is altruistic, then they would see the relationship movie; Jean would be pleased and Marcos would be displeased. If Marcos is honest and Jean is altruistic, then they would see the action movie; Marcos would be pleased and Jean would be displeased. If both are honest or if both are altruistic, then they would argue. Assign a reward of $+1$ to "being pleased," -1 to "being displeased," and 0 to "arguing."
NOTE: This is Exercise 14 from Section 8.1.

 Answer the following questions using complete sentences.

16. What is the difference between the value of a strictly determined game and the expected value of a nonstrictly determined game?

17. What is the difference between the expected value of a randomized strategy against a pure strategy, the expected value of a randomized strategy against a randomized strategy, and the expected value of a game?

8.3

GAME THEORY AND LINEAR PROGRAMMING

In Section 8.2 we discussed the geometric method of finding the optimal random strategy of a game. That method involves graphing two line segments and selecting the optimal point on the graph. Thus, it applies only to games with two variables—that is, to games where the row player and the column player each have two nondominated strategies. Many games involve more than two strategies. When a player has *two or more* strategies, the optimal random strategy can be found using linear programming.

We will use linear programming to find the optimal randomized strategy for the two-finger morra game from Section 8.2. In that section we found that the game's reward matrix is

$$R = \begin{bmatrix} -2 & 3 \\ 3 & -4 \end{bmatrix}$$

The simplex method requires that none of the variables have negative values. In order to apply the simplex method, the reward matrix must contain only positive entries, and our reward matrix contains two negative entries. This can be remedied by requiring that the column player pay the row player an extra $5 per game, in addition to the payments of $2, $3, or $4 described by the above reward matrix. While this would certainly be to the column player's disadvantage and would artificially inflate the value of the game by $5, it does not affect the probabilities for the row player's optimal randomized strategy. Once we compute these probabilities and the artificially inflated value of the game, we can compute the true value by subtracting $5.

If the column player pays the row player $5 per game in addition to the rewards in the above reward matrix R, then the reward matrix is changed to

$$R' = \begin{bmatrix} -2+5 & 3+5 \\ 3+5 & -4+5 \end{bmatrix} = \begin{bmatrix} 3 & 8 \\ 8 & 1 \end{bmatrix}$$

If E is the expected value of the game associated with the original reward matrix R, and E' is the expected value of the game associated with the new reward matrix R', then $E' = E + 5$.

Let p_1 be the probability with which Odd shows one finger. Then $p_2 = 1 - p_1$ is the probability that Odd shows two fingers, and the expected values of the column player's pure strategies are contained in the matrix

$$E(\text{pure}) = P_r \cdot R' = [p_1 \quad p_2] \cdot \begin{bmatrix} 3 & 8 \\ 8 & 1 \end{bmatrix}$$

$$= [3p_1 + 8p_2 \quad 8p_1 + 1p_2]$$

$$= [3p_1 + 8(1 - p_1) \quad 8p_1 + 1(1 - p_1)] \qquad \text{Since } p_2 = 1 - p_1$$

$$= [3p_1 + 8 - 8p_1 \quad 8p_1 + 1 - p_1]$$

$$= [-5p_1 + 8 \quad 7p_1 + 1]$$

Each value of p_1 gives the row player a different randomized strategy. The first entry of the above matrix is the expected value of the game if the column player responds with column 1; the second entry is the expected value if the column player responds with column 2. Naturally, the column player will select the column that results in the smaller expected value, because smaller values mean smaller rewards to the row player (and hence larger rewards to the column player). Thus, the expected value of the game is the smaller of $-5p_1 + 8$ and $7p_1 + 1$.

If $-5p_1 + 8$ is smaller than $7p_1 + 1$, then $E' = -5p_1 + 8$ and $E' < 7p_1 + 1$. If, on the other hand, $7p_1 + 1$ is smaller than $5p_1 + 8$, then $E' = 7p_1 + 1$ and $E' < -5p_1 + 8$. Either way, we have that

$$E' \le -5p_1 + 8 \text{ and } E' \le 7p_1 + 1$$

The row player wants to maximize his reward, so the objective of the linear programming problem is to maximize E'. Thus, our linear programming problem is:

Constraints

$C_1\colon E' \le -5p_1 + 8$

$C_2\colon E' \le 7p_1 + 1$

$C_3\colon 0 \le p_1 \le 1$ Since p_1 is a probability

Objective Function

Maximize $z = E'$

We must rewrite the constraints and the objective function with the variables on the left side and introduce slack variables. Also, the requirement that $p_1 \ge 0$ in constraint C_3 can be dropped because the simplex method requires that *all* of the variables are nonnegative.

$C_1\colon E' \le -5p_1 + 8 \quad \rightarrow \quad E' + 5p_1 \le 8 \quad \rightarrow \quad E' + 5p_1 + s_1 = 8$

$C_2\colon E' \le 7p_1 + 1 \quad \rightarrow \quad E' - 7p_1 \le 1 \quad \rightarrow \quad E' - 7p_1 + s_2 = 1$

$C_3\colon 0 \le p_1 \le 1 \quad \rightarrow \quad p_1 \le 1 \quad \rightarrow \quad p_1 + s_3 = 1$

Objective Function

$-E' + z = 0$

Thus, the first simplex matrix is

$$
\begin{array}{cccccccc}
E' & p_1 & s_1 & s_2 & s_3 & z & \\
\left[\begin{array}{cccccc|c}
1 & 5 & 1 & 0 & 0 & 0 & 8 \\
\boxed{1} & -7 & 0 & 1 & 0 & 0 & 1 \\
0 & 1 & 0 & 0 & 1 & 0 & 1 \\
-1 & 0 & 0 & 0 & 0 & 1 & 0
\end{array}\right] & \begin{array}{l} 8/1 \\ 1/1 \leftarrow \\ 1 \\ \ \end{array}
\end{array}
$$

$\qquad\qquad\uparrow$

R1 − R2 : R1
R2 + R4 : R4

$$
\left[\begin{array}{cccccc|c}
0 & \boxed{12} & 1 & -1 & 0 & 0 & 7 \\
1 & -7 & 0 & 1 & 0 & 0 & 1 \\
0 & 1 & 0 & 0 & 1 & 0 & 1 \\
0 & -7 & 0 & 1 & 0 & 1 & 1
\end{array}\right]
\begin{array}{l} 7/12 \leftarrow \\ -1/7 \\ \ \\ \ \end{array}
$$

$\qquad\qquad\uparrow$

R1 ÷ 12

$$
\left[\begin{array}{cccccc|c}
0 & \boxed{1} & 1/12 & -1/12 & 0 & 0 & 7/12 \\
1 & -7 & 0 & 1 & 0 & 0 & 1 \\
0 & 1 & 0 & 0 & 1 & 0 & 1 \\
0 & -7 & 0 & 1 & 0 & 1 & 1
\end{array}\right]
$$

7R1 + R2 : R2
−R1 + R3 : R3
7R1 + R4 : R4

$$
\begin{array}{cccccc}
E' & p_1 & s_1 & s_2 & s_3 & z \\
\left[\begin{array}{cccccc|c}
0 & 1 & 1/12 & -1/12 & 0 & 0 & 7/12 \\
1 & 0 & 7/12 & 5/12 & 0 & 0 & 61/12 \\
0 & 0 & -1/12 & 1/12 & 1 & 0 & 5/12 \\
0 & 0 & 7/12 & 5/12 & 0 & 1 & 61/12
\end{array}\right]
\end{array}
$$

$$E' = 61/12, \qquad p_1 = 7/12 \quad \rightarrow \quad p_2 = 1 - p_1 = 1 - 7/12 = 5/12$$

Recall that E' is the artificially inflated expected value of the game; the true expected value is E, where $E = E' - 5 = 61/12 - 5 = 1/12$.

In Section 8.2 we solved the above game with the geometric method and obtained the same solution with less effort. The following game could not be solved with the geometric method because each player has three nondominated strategies.

EXAMPLE 1 In the children's game rock-paper-scissors, two players simultaneously put out a hand in the form of a rock (a closed fist), a paper (an open hand), or scissors (with two extended fingers). If both players make the same choice, the game is a tie. Otherwise, rock beats scissors (because a rock can dull scissors), scissors beats paper (because scissors can cut paper), and paper beats rock (because paper can cover a rock).

a. Show that the game is not strictly determined.
b. Find and remove any dominated strategies.
c. Find the row player's optimal randomized strategy.
d. Is the game fair?

THE FAR SIDE   By GARY LARSON

**Tension mounts in the final heat of the
paper-rock-scissors event.**

Solution   **a.** The reward matrix R is

$$
\begin{array}{c c}
 & \begin{array}{ccc} R & P & S \end{array} \\
\begin{array}{c} R \\ P \\ S \end{array} &
\left[\begin{array}{ccc}
0 & -1 & 1 \\
1 & 0 & -1 \\
-1 & 1 & 0
\end{array} \right]
\end{array}
$$

As shown in Figure 8.17, the row minimums are all -1, and the column maximums are all 1.

FIGURE 8.17

	Column 1 (Rock)	Column 2 (Paper)	Column 3 (Scissors)	Row Minimum
Row 1 (Rock)	0	−1	1	−1
Row 2 (Paper)	1	0	−1	−1
Row 3 (Scissors)	−1	1	0	−1
Column Maximum	1	1	1	

There is no entry in the reward matrix that is simultaneously the largest of the row minimums (-1) and the smallest of the column maximums (1). Thus, there is no saddle point, and the game is not strictly determined.

b. There is no row whose entries are all less than or equal to the corresponding entries of another row, so there is no dominated row. There is no column whose entries are all greater than or equal to the corresponding entries of another column, so there is no dominated column.

c. To find the row player's optimal randomized strategy, we must rewrite the reward matrix so that it contains only positive entries. Adding 2 to each entry, we obtain

$$R' = \begin{bmatrix} 2 & 1 & 3 \\ 3 & 2 & 1 \\ 1 & 3 & 2 \end{bmatrix}$$

If we let p_1 be the probability with which the row player shows rock, p_2 the probability with which the row player shows paper, and p_3 the probability with which the row player shows scissors, then $p_1 + p_2 + p_3 = 1$ and the row player's strategy is given by the probability row matrix

$$P_r = [p_1 \quad p_2 \quad p_3]$$

The expected values of the column player's pure strategies are contained in the matrix

$$E(\text{pure}) = P_r \cdot R$$

$$= [p_1 \quad p_2 \quad p_3] \begin{bmatrix} 2 & 1 & 3 \\ 3 & 2 & 1 \\ 1 & 3 & 2 \end{bmatrix}$$

$$= [2p_1 + 3p_2 + p_3 \quad p_1 + 2p_2 + 3p_3 \quad 3p_1 + p_2 + 2p_3]$$

$$= [2p_1 + 3p_2 + (1 - p_1 - p_2) \quad p_1 + 2p_2 + 3(1 - p_1 - p_2)$$
$$3p_1 + p_2 + 2(1 - p_1 - p_2)] \qquad \text{Substituting } 1 - p_1 - p_2 \text{ for } p_3$$

$$= [p_1 + 2p_2 + 1 \quad -2p_1 - p_2 + 3 \quad p_1 - p_2 + 2] \qquad \text{Simplifying}$$

The expected value E' of the game is equal to the smallest of the three entries in the above matrix, and it is less than the other two entries. Thus, E' is less than or equal to each of these three entries.

C_1: $E' \le p_1 + 2p_2 + 1 \quad \rightarrow \quad E' - p_1 - 2p_2 \le 1 \quad \rightarrow \quad E' - p_1 - 2p_2 + s_1 = 1$

C_2: $E' \le -2p_1 - p_2 + 3 \quad \rightarrow \quad E' + 2p_1 + p_2 \le 3 \quad \rightarrow \quad E' + 2p_1 + p_2 + s_2 = 3$

C_3: $E' \le p_1 - p_2 + 2 \quad \rightarrow \quad E' - p_1 + p_2 \le 2 \quad \rightarrow \quad E' - p_1 + p_2 + s_3 = 2$

C_4: $p_1 \le 1 \qquad\qquad\qquad \rightarrow \quad p_1 + s_4 = 1$

C_4: $p_2 \le 1 \qquad\qquad\qquad \rightarrow \quad p_2 + s_5 = 1$

Maximize $z = E' \quad \rightarrow \quad -E' + z = 0$

$$
\begin{array}{cccccccccc}
E' & p_1 & p_2 & s_1 & s_2 & s_3 & s_4 & s_5 & z \\
\end{array}
$$

$$
\left[\begin{array}{cccccccccc}
\boxed{1} & -1 & -2 & 1 & 0 & 0 & 0 & 0 & 0 & 1 \\
1 & 2 & 1 & 0 & 1 & 0 & 0 & 0 & 0 & 3 \\
1 & -1 & 1 & 0 & 0 & 1 & 0 & 0 & 0 & 2 \\
0 & 1 & 0 & 0 & 0 & 0 & 1 & 0 & 0 & 1 \\
0 & 0 & 1 & 0 & 0 & 0 & 0 & 1 & 0 & 1 \\
-1 & 0 & 0 & 0 & 0 & 0 & 0 & 0 & 1 & 0 \\
\end{array}\right]
\begin{array}{l}
1/1 \leftarrow \\
3/1 \\
2/1 \\
1/0 \\
1/0 \\
\\
\end{array}
$$

$\uparrow$

R2 − R1 : R2
R3 − R1 : R3
R1 + R6 : R6

$$
\left[\begin{array}{cccccccccc}
1 & -1 & -2 & 1 & 0 & 0 & 0 & 0 & 0 & 1 \\
0 & 3 & 3 & -1 & 1 & 0 & 0 & 0 & 0 & 2 \\
0 & 0 & \boxed{3} & -1 & 0 & 1 & 0 & 0 & 0 & 1 \\
0 & 1 & 0 & 0 & 0 & 0 & 1 & 0 & 0 & 1 \\
0 & 0 & 1 & 0 & 0 & 0 & 0 & 1 & 0 & 1 \\
0 & -1 & -2 & 1 & 0 & 0 & 0 & 0 & 1 & 1 \\
\end{array}\right]
\begin{array}{l}
-1/2 \\
2/3 \\
1/3 \leftarrow \\
1/0 \\
1/1 \\
\\
\end{array}
$$

$\uparrow$

R3 ÷ 3

$$
\left[\begin{array}{cccccccccc}
1 & -1 & -2 & 1 & 0 & 0 & 0 & 0 & 0 & 1 \\
0 & 3 & 3 & -1 & 1 & 0 & 0 & 0 & 0 & 2 \\
0 & 0 & \boxed{1} & -1/3 & 0 & 1/3 & 0 & 0 & 0 & 1/3 \\
0 & 1 & 0 & 0 & 0 & 0 & 1 & 0 & 0 & 1 \\
0 & 0 & 1 & 0 & 0 & 0 & 0 & 1 & 0 & 1 \\
0 & -1 & -2 & 1 & 0 & 0 & 0 & 0 & 1 & 1 \\
\end{array}\right]
$$

R1 + 2R3 : R1
R2 − 3R3 : R2
R5 − R3 : R5
R6 + 2R3 : R6

$$
\left[\begin{array}{cccccccccc}
1 & -1 & 0 & 1/3 & 0 & 2/3 & 0 & 0 & 0 & 5/3 \\
0 & \boxed{3} & 0 & 0 & 1 & -1 & 0 & 0 & 0 & 1 \\
0 & 0 & 1 & -1/3 & 0 & 1/3 & 0 & 0 & 0 & 1/3 \\
0 & 1 & 0 & 0 & 0 & 0 & 1 & 0 & 0 & 1 \\
0 & 0 & 0 & 1/3 & 0 & -1/3 & 0 & 1 & 0 & 2/3 \\
0 & -1 & 0 & 1/3 & 0 & 2/3 & 0 & 0 & 1 & 5/3 \\
\end{array}\right]
\begin{array}{l}
-5/3 \\
1/3 \leftarrow \\
(1/3)/0 \\
1/1 \\
(2/3)/0 \\
\\
\end{array}
$$

$\uparrow$

R2 ÷ 3

$$
\left[\begin{array}{cccccccccc}
1 & -1 & 0 & 1/3 & 0 & 2/3 & 0 & 0 & 0 & 5/3 \\
0 & \boxed{1} & 0 & 0 & 1/3 & -1/3 & 0 & 0 & 0 & 1/3 \\
0 & 0 & 1 & -1/3 & 0 & 1/3 & 0 & 0 & 0 & 1/3 \\
0 & 1 & 0 & 0 & 0 & 0 & 1 & 0 & 0 & 1 \\
0 & 0 & 0 & 1/3 & 0 & -1/3 & 0 & 1 & 0 & 2/3 \\
0 & -1 & 0 & 1/3 & 0 & 2/3 & 0 & 0 & 1 & 5/3 \\
\end{array}\right]
$$

	E'	p_1	p_2	s_1	s_2	s_3	s_4	s_5	z	
R1 + R2 : R1	1	0	0	1/3	1/3	1/3	0	0	0	2
R4 − R2 : R4	0	1	0	0	1/3	−1/3	0	0	0	1/3
R2 + R6 : R6	0	0	1	−1/3	0	1/3	0	0	0	1/3
	0	0	0	0	−1/3	1/3	1	0	0	2/3
	0	0	0	1/3	0	−1/3	0	1	0	2/3
	0	0	0	1/3	1/3	1/3	0	0	1	2

$$E' = 2, \quad p_1 = 1/3, \quad p_2 = 1/3 \quad \rightarrow \quad p_3 = 1 - p_1 - p_2 = 1 - 1/3 - 1/3 = 1/3$$

The row player's optimal randomized strategy is to play rock, paper, and scissors 1/3 of the time each.

Recall that E' is the artificially inflated value of the game; the true value is $E = E' - 2 = 2 - 2 = 0$.

d. The game is fair, since its value is 0. •

The following flowchart expands the flowchart given in Section 8.2 to include the methods of Section 8.3 as well.

Game Theory Flowchart

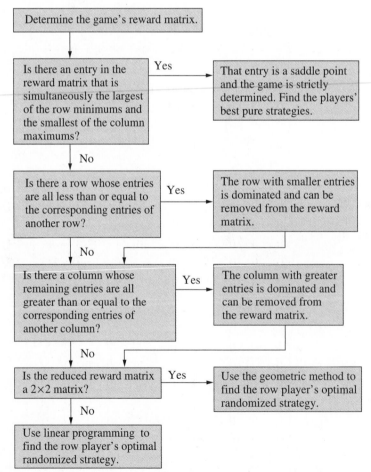

> ### Steps for the Linear Programming Method of Finding the Optimal Randomized Strategy
>
> 1. Determine the reward matrix R.
> 2. If R has nonpositive entries, create R' by adding a constant to each entry of R. If E and E' are the expected values of the game associated with reward matrices R and R' respectively, then $E' = E + $ (the added constant).
> 3. If the row player adopts a randomized strategy where the probabilities of selecting rows 1 through n are p_1 through p_n, respectively, then
>
> $$P_r = [\,p_1 \quad p_2 \cdots p_n\,] \quad \text{where } p_1 + p_2 + \cdots + p_n = 1$$
>
> If the column player adopts a pure strategy, then the expected values of those pure strategies are contained in the matrix
>
> $$E(\text{pure}) = P_r \cdot R'$$
>
> Calculate $E(\text{pure})$.
> 4. Solve $p_1 + p_2 + \cdots + p_n = 1$ for p_n and substitute the result into $E(\text{pure})$.
> 5. The expected value E' of the game is less than or equal to each of the entries of the matrix in Step 4. Each of these resulting inequalities is a constraint in the linear programming problem. List these constraints.
> 6. Each of the probabilities $p_1, p_2, p_3, \ldots$ must be less than or equal to 1. List these constraints.
> 7. State the objective, which is to maximize $z = E'$.
> 8. Solve the linear programming problem, and find the values of $p_1, p_2, \ldots, p_n$, and E'.
> 9. Describe the row player's optimal randomized strategy by giving the probability with which he or she should select each row.
> 10. Subtract the constant from Step 2 from E' to obtain E, the expected value of the game.

8.3

EXERCISES

In Exercises 1–4, determine if the game is strictly determined. If not, use linear programming to find the row player's optimal randomized strategy. [Note: These are Exercises 3–6 from Section 8.2.]

1. $\begin{bmatrix} 10 & -8 \\ -11 & 9 \end{bmatrix}$

2. $\begin{bmatrix} 4 & -5 \\ -1 & 1 \end{bmatrix}$

3. $\begin{bmatrix} 5 & 2 \\ 7 & 3 \end{bmatrix}$

4. $\begin{bmatrix} -6 & 9 \\ 4 & -1 \end{bmatrix}$

In Exercises 5–8, determine if the game is strictly determined. If not, remove dominated strategies and use linear programming to find the row player's optimal randomized strategy.

5. $\begin{bmatrix} 1 & 2 & -2 & 1 \\ -1 & -2 & 3 & 0 \\ -2 & -3 & -3 & -1 \end{bmatrix}$

6. $\begin{bmatrix} -1 & 1 & 3 \\ 0 & -2 & 2 \end{bmatrix}$

7. $\begin{bmatrix} 1 & -2 & 3 \\ -1 & 2 & -3 \\ -3 & -1 & -2 \end{bmatrix}$

8. $\begin{bmatrix} 3 & -1 & 0 \\ -2 & 0 & 1 \\ -3 & 1 & 1 \end{bmatrix}$

Use linear programming to solve Exercises 9–12.

9. Willie Jordan and Frank Brown are the only candidates for governor of the state of Confusion. One week before the

FIGURE 8.18

	Jordan Emphasizes North	**Jordan Splits Money Equally**	**Jordan Emphasizes South**
Brown Emphasizes North	Brown gains 3%	Brown gains 5%	No net change
Brown Splits Money Equally	Jordan gains 4%	No net change	Brown gains 2%
Brown Emphasizes South	No net change	Jordan gains 6%	Jordan gains 3%

FIGURE 8.19

	Standard's Production Is Low	**Standard's Production Is Medium**	**Standard's Production Is High**
Petrolia's Production Is Low	Petrolia's revenue is $5000	Petrolia's revenue is $6000	Petrolia's revenue is $4000
Petrolia's Production Is Medium	Petrolia's revenue is $4000	Petrolia's revenue is $5000	Petrolia's revenue is $3000
Petrolia's Production Is High	Petrolia's revenue is $3000	Petrolia's revenue is $7000	Petrolia's revenue is $6000

election, a poll shows that Jordan has the lead in the southern part of the state and Brown has the lead in the north, but overall the two are tied. The fate of the campaign hinges on how Jordan and Brown spend their remaining advertising money in the remaining week. Each can decide to spend the bulk of his money in the north or in the south, or to split it equally. Each candidate is free to change his advertising tactics on a daily basis. The campaign managers' projected results of these decisions are given in Figure 8.18, above. Find Brown's optimal randomized strategy.
NOTE: This is Exercise 11 from Section 8.2.

10. The Petrolia Oil Co. and the Standard Oil Co. produce all of the gasoline sold in the state of Erehwon. The total revenue earned by these two firms is currently $10,000 per day. Once a week, each of the firms determines how much gasoline to produce. When both firms' production levels are low, they each earn $5000 in revenue. Similar data for other production levels are given in Figure 8.19, above. Find Petrolia's optimal randomized strategy.
NOTE: This is Exercise 12 from Section 8.2.

11. Two people are playing three-finger morra. In this variation of the game, each player simultaneously puts out a hand with either one, two, or three fingers showing. One player, whom we will call the odd player, wins if the sum of the fingers is odd, and the other player, the even player, wins if the sum is even. The dollar amount of the reward is the sum of the fingers. Find the odd player's best strategy. Is the game fair?

12. Two television networks, MBC and DBS, are competing for an audience of 150 million viewers during the Tuesday 8 P.M.

time slot. Each network will choose to air a medical drama, a legal drama, or a comedy. The projected number of MBC viewers (in millions) resulting from each possible choice are shown in the following reward matrix.

$$
\text{MBC} \left\{
\begin{array}{c}
 \\
\text{med} \\
\text{law} \\
\text{com}
\end{array}
\overbrace{
\begin{array}{ccc}
\text{med} & \text{law} & \text{com} \\
\begin{bmatrix} 52 & 23 & 90 \\ 68 & 87 & 75 \\ 69 & 21 & 95 \end{bmatrix}
\end{array}
}^{\text{DBS}}
\right.
$$

If, for example, both networks select medical dramas, then 52 million viewers are projected to watch MBC, and $150 - 52 = 98$ million are projected to watch DBS. Find MBC's best strategy.

Exercises 13 and 14 involve strictly determined games that were solved in Section 8.1. Use the methods of this section to find the row player's optimal randomized strategy. Discuss the application of the two different methods to the problem and compare the results.

13. The tri-city area currently has no cellular phone stores, but two chains (Global Fon and U-Call) are planning to open branches in the area. Twenty percent of the tri-city residents live in the town of Backwater, 30% live in Podunk, and 50% live in River City. If both stores locate in the same town, they will each get 50% of the tri-city area business. Otherwise, a store will get all of the business of the town it's located in, none of the business of the town its competitor is located in, and half of the business of the third town.

14. All of the 18- to 25-year-old residents of the city of Erehwon who listen to the radio while commuting listen to either KEWB or KYA. Currently, KEWB has the more popular morning program, and KYA has the more popular evening program. Currently, KEWB gets 55% of the two stations' combined advertising revenue, and KYA gets 45%.

A market research survey indicates that if KYA switches its morning and evening shows, its market share will increase to 70%; and if KEWB also switches its shows, KYA's market share will increase to 75%. Finally, if KEWB switches its two shows and KYA does not, KEWB's market share is projected to decrease to 45%.

CHAPTER 8 REVIEW

Terms

column player	expected value of a randomized strategy against a pure strategy	reward matrix
column player's probability matrix		row player
constant-sum game	fair game	row player's probability matrix
dominated strategy	game theory	saddle point
expected value of a game	mixed strategy	strictly determined game
expected value of a randomized strategy	pure strategy	value of a strictly determined game
	randomized strategy	zero-sum game
	reduced reward matrix	

Review Exercises

In each of the following Exercises, do either part (a), (b), or (c), as appropriate.

    a. If the game is strictly determined, find the players' best pure strategies, the value of the game, whether the game favors either player, and the effect on each player of following their best pure strategies.

    b. If the game is not strictly determined and the geometric method applies, find the row player's optimal mixed strategy using that method, assuming that the competitor adopts a pure strategy. Determine if the competitor could do better with a randomized strategy. Discuss the impact of this strategy on the game.

    c. If the game is not strictly determined and the geometric method doesn't apply, find the row player's optimal mixed strategy and the value of the game, using linear programming.

1. The central valley area currently has no television disc receiver stores, but two chains (the row player Discs-R-Us and the column player 500 Channels!) are planning to open branches in the area. Twenty percent of the area residents live in the suburb of Smallville, 35% live in Gotham, and 45% live in Elmwood. If both stores locate in the same town, they will each get 50% of the central valley area business. Otherwise, it is projected that a store will get 90% of the business of the town it's located in, 10% of the business of the town its competitor is located in, and half of the business of the third town.

2. Discs-R-Us (from Exercise 1) has discovered that it will have slightly less expensive prices than its competitor. Company planners think that this will mean that Discs-R-Us will get 60% of the central valley area business if they locate in the same town as 500 Channels!, rather than 50%. All other data from Exercise 1 are unchanged.

3. The Central State Mall has two department stores: Bloomworthy's and Mercy's. Each week, their managers decide whether or not to have a sale. Figure 8.20 describes the result of this competition. (Bloomworthy's is the row player.)

FIGURE 8.20

	Mercy's Has a Sale	Mercy's Doesn't Change Prices
Bloomworthy's Has a Sale	Bloomworthy's gains 4%	Bloomworthy's gains 5%
Bloomworthy's Doesn't Change Prices	Mercy's gains 2%	No change

STATISTICS 9

Statistics are everywhere. The news, whether reported in a newspaper, on television, or over the radio, includes statistics of every kind. Shopping for a new car, you will certainly examine the statistics (average miles per gallon, acceleration times, braking distances, and so on) of the various makes and models you're considering. Statistics abound in government studies, and their interpretation affects us all. Industry is driven by statistics— they are essential to the direction of quality control, marketing research, productivity, and many other factors. Sporting events are laden with statistics concerning the past performance of the teams and players.

A person who understands the nature of statistics is equipped to see beyond the short-term and individual perspective. He or she is also better prepared to deal with those who use statistics in misleading ways. To many people, the word *statistics* conjures up an image of an endless list of facts and figures. Where do they come from? What do they mean? In this chapter you will learn to handle basic statistical problems and expand your knowledge of the meanings, uses, and misuses of statistics.

9.1

FREQUENCY DISTRIBUTIONS

The field of **statistics** can be defined as the science of collecting, organizing, and summarizing data in such a way that valid conclusions can be drawn from them. **Descriptive statistics** refers to collecting, organizing, and summarizing data, and **inferential statistics** refers to drawing conclusions from those data.

If a clothing manufacturer was considering marketing a line of clothes for tall people, it might collect data to determine the heights of a group of shoppers. A **data point** is one particular item in the collected data; for example, one shopper's height is a data point.

One way to organize data is to create a **frequency distribution,** a table that lists each data point along with the number of times it occurs (its **frequency**). Usually, a frequency distribution is easier to understand if it also lists each data point's **relative frequency,** which is the frequency expressed as a percent of the total number of data points (that is, made *relative* to the total).

EXAMPLE 1   While bargaining for their new contract, the employees of 2 Dye 4 Clothing asked their employers to provide day care service as an employee benefit. Examining the personnel files of the company's 50 employees, the management recorded the number of children under 6 years of age that each employee was caring for. The following results were obtained:

```
0  2  1  0  3  2  0  1  1  0
0  1  1  2  4  1  0  1  1  0
2  1  0  0  3  0  0  1  2  1
0  0  2  4  1  1  0  1  2  0
1  1  0  3  5  1  2  1  3  2
```

Organize the data by creating a frequency distribution.

Solution   First, we list each different data point in a column, putting them in order. Then we use tally marks to count the number of times each data point occurs, and we compute the frequencies and relative frequencies, as shown in Figure 9.1.

FIGURE 9.1

Number of Children Under 6	Tally	Frequency	Relative Frequency
0	|||| |||| ||||		16	$\frac{16}{50} = 0.32 = 32\%$
1	|||| |||| |||| |||	18	$\frac{18}{50} = 0.36 = 36\%$
2	|||| ||||	9	$\frac{9}{50} = 0.18 = 18\%$
3	||||	4	$\frac{4}{50} = 0.08 = 8\%$
4	||	2	$\frac{2}{50} = 0.04 = 4\%$
5			1	$\frac{1}{50} = 0.02 = 2\%$
		$n = 50$	Total $= 100\%$

Notice the two totals listed at the bottom of Figure 9.1. The total of the frequencies (n) is needed to compute the relative frequencies. Furthermore, the two totals help check the calculations—n should equal the number of data points, and the total of the relative frequencies should be 100%.

The raw data have now been organized and summarized. At this point, we can see that about one-third of the employees have no need for child care (32%), while the remaining two-thirds (68%) have at least one child under 6 years of age and would benefit from company-sponsored day care. The most common trend (that is, the data point with the highest relative frequency for the 50 employees) is having one child (36%). •

Grouped Data

When raw data consist of only a few distinct values (for instance, the data in Example 1, which consisted of only the numbers 0, 1, 2, 3, 4, and 5), we can easily organize the data and determine any trends by listing each data point along with its frequency and relative frequency. However, when the raw data consist of many nonrepeated data points, listing each one separately does not help us see any trends the data set might contain. In such cases it is useful to group the data and determine the frequency and relative frequency of each group, rather than of each data point.

EXAMPLE 2   Keith Reed is an instructor for an acting class offered through a local arts academy. The class is open to anyone who is at least 16 years old. Forty-two people are enrolled; their ages are as follows:

$$26 \quad 16 \quad 21 \quad 34 \quad 45 \quad 18 \quad 41 \quad 38 \quad 22$$
$$48 \quad 27 \quad 22 \quad 30 \quad 39 \quad 62 \quad 25 \quad 25 \quad 38$$
$$29 \quad 31 \quad 28 \quad 20 \quad 56 \quad 60 \quad 24 \quad 61 \quad 28$$
$$32 \quad 33 \quad 18 \quad 23 \quad 27 \quad 46 \quad 30 \quad 34 \quad 62$$
$$49 \quad 59 \quad 19 \quad 20 \quad 23 \quad 24$$

Organize the data by creating a frequency distribution.

Solution   This example is quite different from Example 1. Example 1 had only six different data values, whereas this example has many. Listing each distinct data point and its frequency might not summarize the data well enough for us to draw conclusions. Instead, we will work with grouped data.

First, we find the largest and smallest values (62 and 16). Subtracting, we find the range of ages to be $62 - 16 = 46$ years. In working with grouped data, it is customary to create between four and eight groups of data points. We arbitrarily choose six groups, with the first group beginning at the smallest data point, 16. To find the beginning of the second group (and hence the end of the first group), divide the range by the number of groups, round off this answer to be consistent with the data, and then add the result to the smallest data point:

$$46 \div 6 = 7.6666666 \approx 8 \qquad \text{(this is the width of each group)}$$

The beginning of the second group is $16 + 8 = 24$, so the first group consists of people from 16 up to (but not including) 24 years of age.

In a similar manner, find the remaining **group boundaries.** The second group consists of people from 24 up to but not including 32 ($24 + 8 = 32$) years of age. The third group consists of people from 32 up to but not including 40 years of age. The frequency distribution is shown in Figure 9.2.

FIGURE 9.2

$x =$ Age	Tally	Frequency	Relative Frequency
$16 \leq x < 24$	⊮ ⊮ I	11	$\frac{11}{42} \approx 26\%$
$24 \leq x < 32$	⊮ ⊮ III	13	$\frac{13}{42} \approx 31\%$
$32 \leq x < 40$	⊮ II	7	$\frac{7}{42} \approx 17\%$
$40 \leq x < 48$	III	3	$\frac{3}{42} \approx 7\%$
$48 \leq x < 56$	II	2	$\frac{2}{42} \approx 5\%$
$56 \leq x < 64$	⊮ I	6	$\frac{6}{42} \approx 14\%$
		$n = 42$	Total $= 100\%$

Now that the data have been organized, we can observe various trends: Ages 24 to 32 are most common (31% is the highest relative frequency), and ages 48 to 56 are least common (5% is the lowest). Also, over half the people enrolled (57%) are 16 to 32 years old.•

When we are working with grouped data, we can choose the groups in any desired fashion. The method used in Example 2 might not be appropriate in all situations. For example, we used the smallest data point as the beginning of the first group, but we could have begun the first group at an even smaller number.

Constructing a Frequency Distribution

1. If the data do not consist of many different values, do not group the data. Instead, list each different value.

If the data consist of many different values, group the data:

- Determine the number of groups (usually four to eight).
- Find the width of the groups by dividing the range (high minus low) by the number of groups. Round off this answer so that the groups have reasonable boundaries.
- Find the group boundaries. Add the group width to the lowest data point (or a lower point) to find the beginning of the second group. Add the group width to the beginning of the second group to find the beginning of the third group. Continue this process until all group boundaries have been found.

2. Tally the number of times each individual value occurs (in the case of ungrouped data) or the number of data points in each group of values (in the case of grouped data).

3. List the frequency for each individual value (in the case of ungrouped data) or for each group of values (in the case of grouped data).

4. Find the relative frequencies by dividing the frequencies by the total number of data points in the distribution. Express the resulting decimal as a percent.

Histograms

A frequency distribution can be graphically depicted by a **histogram,** a bar chart that shows how the data are distributed. A horizontal axis is used to display the group boundaries (or the different individual values in the case of ungrouped data), and two vertical axes are used for the frequencies and the relative frequencies, respectively. (Histograms are slightly different if the groups don't have equal widths; this will be discussed later in this section.) The histogram depicting the distribution of the ages of the people in Keith Reed's acting class (Example 2) is shown in Figure 9.3. Notice how the three axes are labeled and how the markings on the relative frequency axis are consistent with those on the frequency axis.

FIGURE 9.3
Ages of the people in Keith Reed's acting class

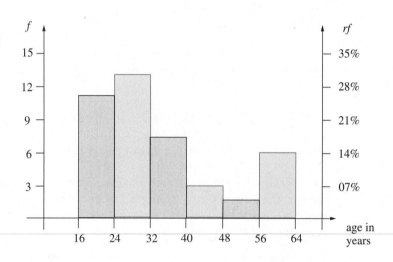

Relative Frequency Density

What happens if the intervals do not have equal width? For instance, suppose the ages of the people in Keith Reed's acting class are those given in the frequency distribution shown in Figure 9.4.

FIGURE 9.4

x = Age	Frequency	Relative Frequency	Group Width
$15 \le x < 20$	4	$\frac{4}{42} \approx 10\%$	5
$20 \le x < 25$	9	$\frac{9}{42} \approx 21\%$	5
$25 \le x < 30$	8	$\frac{8}{42} \approx 19\%$	5
$30 \le x < 45$	11	$\frac{11}{42} \approx 26\%$	15
$45 \le x < 65$	10	$\frac{10}{42} \approx 24\%$	20
	$n = 42$	Total $= 100\%$	

Using frequency and relative frequency as the vertical scales, the histogram depicting this new distribution is given in Figure 9.5.

FIGURE 9.5
Why is this histogram
misleading?

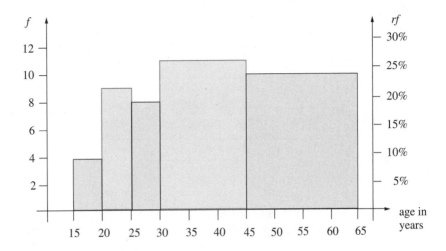

Does the histogram in Figure 9.5 give a truthful representation of the distribution? No; the rectangle over the interval from 45 to 65 appears to be larger than the rectangle over the interval from 30 to 45, yet the interval from 45 to 65 contains fewer data points than the interval from 30 to 45. This is misleading; rather than comparing the heights of the rectangles, our eyes naturally compare the areas of the rectangles. Therefore, in order to make an accurate comparison, *the areas of the rectangles must correspond to the relative frequencies of the intervals.*

If the rectangles' areas should correspond to the relative frequencies, then what would the heights correspond to?

area = relative frequency

length · width = relative frequency

$$\text{length} = \frac{\text{relative frequency}}{\text{group width}}$$

Thus, the area of a rectangle will correspond to the relative frequency of its interval if the rectangle's height or length is that relative frequency divided by the group width. This quantity is called the interval's **relative frequency density.**

The word *density* is used here because of the similarity of relative frequency density to weight density. In statistics, density is used to determine the concentration of data in a given interval: Relative frequency density = relative frequency/group width. In science, weight density is used to determine the concentration of weight in a given volume: weight density = weight/volume. For example, 1 cubic foot of water weighs 62.4 pounds, so the density of water is 62.4 pounds/cubic foot.

Adding a new column to the frequency distribution given in Figure 9.4, we obtain the relative frequency densities shown in Figure 9.6.

FIGURE 9.6

x = Age	Frequency	Relative Frequency	Group Width	Relative Frequency Density
$15 \leq x < 20$	4	$\frac{4}{42} \approx 10\%$	5	$\frac{4}{42} \div 5 \approx 0.020$
$20 \leq x < 25$	9	$\frac{9}{42} \approx 21\%$	5	$\frac{9}{42} \div 5 \approx 0.043$
$25 \leq x < 30$	8	$\frac{8}{42} \approx 19\%$	5	$\frac{8}{42} \div 5 \approx 0.038$
$30 \leq x < 45$	11	$\frac{11}{42} \approx 26\%$	15	$\frac{11}{42} \div 15 \approx 0.017$
$45 \leq x < 65$	10	$\frac{10}{42} \approx 24\%$	20	$\frac{10}{42} \div 20 \approx 0.012$
	$n = 42$	Total = 100%		

We now construct a histogram using relative frequency density as the vertical scale. The histogram depicting the distribution of the ages of the people in Keith Reed's acting class (using the frequency distribution in Figure 9.6) is shown in Figure 9.7.

FIGURE 9.7
Ages of the people in Keith Reed's acting class

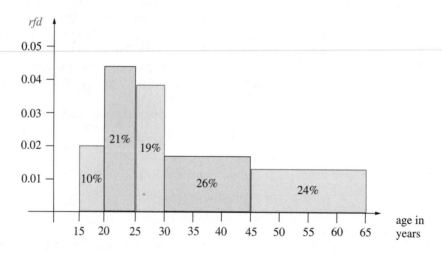

Notice that the groups' relative frequency densities are given on the vertical axis, since relative frequency densities are the heights of the rectangles, whereas the groups' relative frequencies are given inside the rectangles, since relative frequencies are the areas of the rectangles.

Comparing the histograms in Figures 9.5 and 9.7, we see that using relative frequency density as the vertical scale (rather than frequency) gives a more truthful representation of a distribution when the group widths are unequal.

Constructing a Histogram

1. Create the horizontal axis. Mark the group boundaries on the axis. Label the axis appropriately.
2. If the groups have equal width, create two vertical axes.

   - One vertical axis is for the frequencies and is labeled "*f*."
   - One vertical axis is for the relative frequencies and is labeled "*rf*."
   - Mark the frequency axis in a way that is consistent with the data.
   - Mark the relative frequency axis in a way that is consistent with the frequency axis.

   If the intervals do not have equal width, create one vertical axis.

   - This vertical axis is for relative frequency densities and is labeled "*rfd*."
   - Mark this axis in a way that is consistent with the data.

3. Draw the rectangles. If the intervals do not have equal width, give the relative frequencies inside the rectangles.

EXAMPLE 3   In order to study the output of a machine that fills bags with corn chips, a quality control engineer randomly selected and weighed a sample of 200 bags of chips. Figure 9.8 summarizes the data. Construct a complete frequency distribution and a histogram for the weights of the bags of corn chips.

FIGURE 9.8

x = Weight (ounces)	f = Number of Bags
$15.3 \leq x < 15.5$	10
$15.5 \leq x < 15.7$	24
$15.7 \leq x < 15.9$	36
$15.9 \leq x < 16.1$	58
$16.1 \leq x < 16.3$	40
$16.3 \leq x < 16.5$	20
$16.5 \leq x < 16.7$	12

Solution   Because each interval has the same width (0.2), we construct a combined frequency and relative frequency histogram. The complete frequency distribution is given in Figure 9.9.

FIGURE 9.9

x	f	$rf = f/n$
$15.3 \leq x < 15.5$	10	0.05
$15.5 \leq x < 15.7$	24	0.12
$15.7 \leq x < 15.9$	36	0.18
$15.9 \leq x < 16.1$	58	0.29
$16.1 \leq x < 16.3$	40	0.20
$16.3 \leq x < 16.5$	20	0.10
$16.5 \leq x < 16.7$	12	0.06
	$n = 200$	Sum = 1.00

We now draw axes with appropriate scales and rectangles (Figure 9.10).

FIGURE 9.10
Weights of bags of corn chips

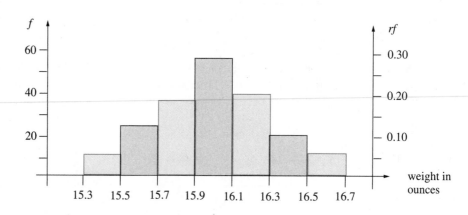

Notice the (near) symmetry of the histogram. We will study this type of distribution in more detail in Section 9.4.

9.1

EXERCISES

1. In order to study the composition of families in Manistee, Michigan, 40 randomly selected married couples were surveyed to determine the number of children in each family. The following results were obtained:

```
2  1  3  0  1  0  2  5  1  2  0  2  2  1
4  3  1  1  3  4  1  1  0  2  0  0  2  2
1  0  3  1  1  3  4  2  1  3  0  1
```

Organize the given data by creating a frequency distribution.

2. In order to study the spending habits of shoppers in Orlando, Florida, 50 randomly selected shoppers at a mall were surveyed to determine the number of credit cards they carried. The following results were obtained:

2 5 0 4 2 1 0 6 3 5 4 3 4 0
5 2 5 2 0 2 5 0 2 5 2 5 4 3
5 6 1 0 6 3 5 3 4 0 5 2 2 5
2 0 2 0 4 2 1 0

Organize the given data by creating a frequency distribution.

3. The speeds, in miles per hour, of 40 randomly monitored cars on Interstate 40 near Winona, Arizona, were as follows:

66   71   76   61   73   78   74   67   80   63   69
78   66   70   77   60   72   58   65   70   64   75
80   75   62   67   72   59   74   65   54   69   73
79   64   68   57   51   68   79

   a. Organize the given data by creating a frequency distribution. (Group the data into six intervals.)
   b. Construct a histogram to represent the data.

4. The weights, in pounds, of 35 packages of ground beef at the Cut Above Market were as follows:

1.0   1.9   2.5   1.2   2.0   0.7   1.3   2.4   1.1
3.3   2.4   0.8   2.3   1.7   1.0   2.8   1.4   3.0
0.9   1.1   1.4   2.2   1.5   3.2   2.1   2.7   1.8
1.6   2.3   2.6   1.3   2.9   1.9   1.2   0.5

   a. Organize the given data by creating a frequency distribution. (Group the data into six intervals.)
   b. Construct a histogram to represent the data.

5. In order to examine the effects of a new registration system, a campus newspaper asked freshmen how long they had to wait in a registration line. Figure 9.11 summarizes the responses. Construct a complete frequency distribution and a histogram to represent the data.

FIGURE 9.11

x = Time in Minutes	Number of Freshmen
$0 \le x < 10$	101
$10 \le x < 20$	237
$20 \le x < 30$	169
$30 \le x < 40$	79
$40 \le x < 50$	51
$50 \le x < 60$	63

6. Figure 9.12 lists the annual salaries of the managers at Universal Manufacturing of Melonville. Construct a complete frequency distribution and a histogram to represent the data.

FIGURE 9.12

x = Salary (in thousands)	Number of Managers
$\$30 \le x < 40$	6
$40 \le x < 50$	12
$50 \le x < 60$	10
$60 \le x < 70$	5
$70 \le x < 80$	7
$80 \le x < 90$	3

7. Figure 9.13 lists the hourly wages of the workers at Universal Manufacturing of Melonville. Construct a complete frequency distribution and a histogram to represent the data.

FIGURE 9.13

x = Hourly Wage	Number of Employees
$\$4.00 \le x < 5.50$	21
$5.50 \le x < 7.00$	35
$7.00 \le x < 8.50$	42
$8.50 \le x < 10.00$	27
$10.00 \le x < 11.50$	18
$11.50 \le x < 13.00$	9

8. In order to study the output of a machine that fills boxes with cereal, a quality control engineer weighed 150 boxes of Brand X cereal. Figure 9.14 summarizes her findings. Construct a complete frequency distribution and a histogram to represent the data.

FIGURE 9.14

x = Weight (in ounces)	Number of Boxes
$15.3 \le x < 15.6$	13
$15.6 \le x < 15.9$	24
$15.9 \le x < 16.2$	84
$16.2 \le x < 16.5$	19
$16.5 \le x < 16.8$	10

9. The ages of more than 52 million women who gave birth in the United States between June 1987 and June 1988 are given in Figure 9.15. Construct a complete frequency distribution and a histogram to represent the data.

FIGURE 9.15
Ages of women giving birth

Age	Number of Women
$18 \leq x < 25$	13,167,000
$25 \leq x < 30$	10,839,000
$30 \leq x < 35$	10,838,000
$35 \leq x < 40$	9,586,000
$40 \leq x < 45$	8,155,000

Source: U.S. Bureau of the Census

10. The age composition of the population of the United States in 1990 is given in Figure 9.16. Replace the interval "85 and over" with the interval $85 \leq x \leq 100$ and construct a complete frequency distribution and a histogram to represent the data.

FIGURE 9.16
Age composition of the population of the United States

Age	Number of People
$0 < x < 5$	18,354,000
$5 \leq x < 18$	45,250,000
$18 \leq x < 21$	11,727,000
$21 \leq x < 25$	15,011,000
$25 \leq x < 45$	80,755,000
$45 \leq x < 55$	25,223,000
$55 \leq x < 60$	10,532,000
$60 \leq x < 65$	10,616,000
$65 \leq x < 75$	18,107,000
$75 \leq x < 85$	10,055,000
85 and over	3,080,000

Source: U.S. Bureau of the Census

In Exercises 11 and 12, use the estimated age composition of the 12,935,000 students enrolled in institutions of higher education in the United States during 1990, as given in Figure 9.17.

FIGURE 9.17
Age composition of students in higher education

Age of Males	Number of Students
$14 \leq x < 18$	111,000
$18 \leq x < 20$	1,380,000
$20 \leq x < 22$	1,158,000
$22 \leq x < 25$	928,000
$25 \leq x < 30$	847,000
$30 \leq x < 35$	591,000
35 and over	878,000
total	5,893,000

Age of Females	Number of Students
$14 \leq x < 18$	126,000
$18 \leq x < 20$	1,502,000
$20 \leq x < 22$	1,253,000
$22 \leq x < 25$	958,000
$25 \leq x < 30$	840,000
$30 \leq x < 35$	743,000
35 and over	1,620,000
total	7,042,000

Source: U.S. National Center for Education Statistics

11. Using the data in Figure 9.17, replace the interval "35 and over" with the interval $35 \leq x \leq 100$ and construct a complete frequency distribution and a histogram to represent the male data.

12. Using the data in Figure 9.17, replace the interval "35 and over" with the interval $35 \leq x \leq 100$ and construct a complete frequency distribution and a histogram to represent the female data.

13. The frequency distribution shown in Figure 9.18 lists the ages of 200 randomly selected students who received a

bachelor's degree at State University last year. Where possible, determine what percent of the graduates had the following ages:

a. less than 23
b. at least 31
c. at most 20
d. not less than 19
e. at least 19 but less than 27
f. not between 23 and 35

FIGURE 9.18

Age	Number of Students
$10 \leq x < 15$	1
$15 \leq x < 19$	4
$19 \leq x < 23$	52
$23 \leq x < 27$	48
$27 \leq x < 31$	31
$31 \leq x < 35$	16
$35 \leq x < 39$	29
39 and over	19

14. The frequency distribution shown in Figure 9.19 lists the number of hours per day a randomly selected sample of teenagers spent watching television. Where possible, determine what percent of the teenagers spent the following number of hours watching television:

a. less than 4 hours
b. at least 5 hours

FIGURE 9.19

Hours per Day	Number of Teenagers
$0 \leq x < 1$	18
$1 \leq x < 2$	31
$2 \leq x < 3$	24
$3 \leq x < 4$	38
$4 \leq x < 5$	27
$5 \leq x < 6$	12
$6 \leq x < 7$	15

c. at least 1 hour
d. less than 2 hours
e. at least 2 hours but less than 4 hours
f. more than 3.5 hours

 Answer the following questions using complete sentences.

15. Explain the difference between frequency, relative frequency, and relative frequency density. What does each measure?

16. When is relative frequency density used as the vertical scale in constructing a histogram? Why?

17. In some frequency distributions, data are grouped in intervals; in others, they are not.
    a. When should data be grouped in intervals?
    b. What are the advantages and disadvantages of using grouped data?

HISTOGRAMS ON A GRAPHING CALCULATOR

In Example 2 of this section we created a frequency distribution, and later a histogram, for the ages of the students in an acting class. Much of this work can be done on a graphing calculator. After entering the data, the histogram can be drawn.

Entering the Data

To enter the data from Example 2, do the following.

Entering the Data on a TI-82/83/86:

* *Put the calculator into statistics mode* by pressing $\boxed{\text{STAT}}$ (TI-86: $\boxed{\text{2nd}}$ $\boxed{\text{STAT}}$).
* *Set the calculator up for entering the data* by selecting "Edit" from the EDIT menu (TI-86: select "Edit" by pressing $\boxed{\text{F2}}$), and the "list screen" appears, as shown in Figure 9.20. (A TI-86's list screen says "xStat," "yStat" and "fStat" instead of "L_1," "L_2," and "L_3," respectively.) If data already appear in a list (as they do in list L_1 in Figure 9.20), use the arrow buttons to highlight the name of the list (i.e., "L_1" or "xStat") and press $\boxed{\text{CLEAR}}$ $\boxed{\text{ENTER}}$.

FIGURE 9.20
A TI-82's list screen

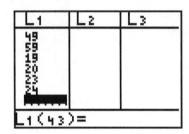

* Use the arrow buttons and the $\boxed{\text{ENTER}}$ button to enter the students' ages in list L_1 (TI-86: in list xStat), in any order. When done, your screen should look similar to that in Figure 9.20. Notice the "$L_1(43)=$" at the bottom of the screen; this indicates that 42 entries have been made, and the calculator is ready to receive the 43rd. This allows you to check whether you have left any entries out. *Note:* If some data points frequently recur, you can enter the data points in list L_1 and their frequencies in list L_2 (TI-86: in list fStat), rather than reentering a data point each time it recurs.
* Press $\boxed{\text{2nd}}$ $\boxed{\text{QUIT}}$.

Entering the Data on a TI-85:

* *Put the calculator into statistics mode* by pressing $\boxed{\text{STAT}}$.
* *Set the calculator up for entering the data* by pressing $\boxed{\text{EDIT}}$ (i.e., $\boxed{\text{F2}}$). Under most circumstances, the calculator will respond by automatically naming the list of *x*-coordinates, "xStat," and the list of *y*-coordinates, "yStat," as shown in Figure 9.21. (You can rename these lists by typing over the calculator-selected names.) Press $\boxed{\text{ENTER}}$ twice to indicate that these two lists are named to your satisfaction.

FIGURE 9.21

- Your calculator is now ready for entering the data. The screen should look like Figure 9.22. [If your screen is messier, you can clear it by pressing $\boxed{\text{CLRxy}}$ (i.e., $\boxed{\text{F5}}$)]. Enter the students' ages (in any order) as x_1, x_2, etc. The corresponding y's are the frequencies; leave them as 1's. When done, your screen should have a blank following "$x_{43} =$"; this indicates that 42 entries have been made, and the calculator is ready to receive the 43rd. This allows you to check whether you have left any entries out.
  Note: If some data points frequently recur, you can enter the data points as x's and their frequencies as y's, rather than reentering a data point each time it recurs.
- Press $\boxed{\text{2nd}}$ $\boxed{\text{QUIT}}$.

FIGURE 9.22

Drawing a Histogram

Once the data are entered, you can draw a histogram.

On a TI-82/83:
- Press $\boxed{\text{Y=}}$ and clear any functions that may appear.
- Press $\boxed{\text{WINDOW}}$ and enter the group boundaries by entering the left boundary of the first group as xmin (16 for this problem), the right boundary of the last group plus 1 as xmax ($64 + 1 = 65$ for this problem), and the group width as xscl (8 for this problem). (The calculator will create only histograms with equal group widths.) Enter 0 for ymin and the largest frequency for ymax (you may guess—it's easy to change it later if you guess wrong).
- Set the calculator up to draw a histogram by pressing $\boxed{\text{2nd}}$ $\boxed{\text{STAT PLOT}}$ and selecting "Plot 1." Turn the plot on and select the histogram icon.
- Tell the calculator to put the data entered in list L_1 on the x-axis by selecting "L_1" for "Xlist," and to consider each entered data point as having a frequency of 1 by selecting "1" for "Freq."
  Note: If some data points frequently recur and you entered their frequencies in list L_2, then select "L_2" rather than "1" for "Freq" by typing $\boxed{\text{2nd}}$ $\boxed{\text{L}_2}$.
- Press $\boxed{\text{GRAPH}}$ to obtain the histogram. If some of the bars are too long or too short for the screen, alter ymax accordingly.
- Press $\boxed{\text{TRACE}}$ to find out the left and right boundaries and the frequency of the bars, as shown in Figure 9.23 on page 414. Use the arrow buttons to move from bar to bar.
- Press $\boxed{\text{2nd}}$ $\boxed{\text{STAT PLOT}}$, select "Plot 1" and turn the plot off, or else the histogram will appear on future graphs.

FIGURE 9.23
The first bar's BOUNDARIES
are 16 and 24; its frequency
is 11.

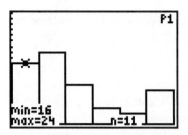

Drawing a Histogram on a TI-85:

- Put the calculator into graphing mode, press $\boxed{\text{y(x)=}}$ and clear any functions that may appear.
- Press $\boxed{\text{2nd}}$ $\boxed{\text{RANGE}}$ and enter the group boundaries by entering the left boundary of the first group as xmin (16 for this problem), the right boundary of the last group plus 1 as xmax (64 + 1 = 65 for this problem), and the group width as xscl (8 for this problem). (The calculator will create only histograms with equal group widths.) Enter 0 for ymin and the largest frequency for ymax (you may guess—it's easy to change it later if you guess wrong). Quit the graphing mode by pressing $\boxed{\text{2nd}}$ $\boxed{\text{QUIT}}$.
- Put the calculator back into statistics mode by pressing $\boxed{\text{STAT}}$, and then press $\boxed{\text{DRAW}}$. If this immediately produces a drawing, clear it by pressing $\boxed{\text{CLDRW}}$.
- To obtain the histogram, press $\boxed{\text{HIST}}$. If some of the bars are too long or too short for the screen, alter ymax accordingly. If the button labels obscure the histogram, press $\boxed{\text{EXIT}}$ once.
- Press $\boxed{\text{CLDRW}}$, or else the histogram will appear on future graphs.

Drawing a Histogram on a TI-86:

- Put the calculator into graphing mode, press $\boxed{\text{y(x)=}}$, and clear any functions that may appear.
- Press $\boxed{\text{2nd}}$ $\boxed{\text{WIND}}$ and enter the group boundaries by entering the left boundary of the first group as xmin (16 for this problem), the right boundary of the last group plus 1 as xmax (64 + 1 = 65 for this problem), and the group width as xscl (8 for this problem). (The calculator will create only histograms with equal group widths.) Enter 0 for ymin and the largest frequency for ymax (you may guess—it's easy to change it later if you guess wrong). Quit the graphing mode by pressing $\boxed{\text{2nd}}$ $\boxed{\text{QUIT}}$.
- Put the calculator back into statistics mode by pressing $\boxed{\text{2nd}}$ $\boxed{\text{STAT}}$.
- Set the calculator up to draw a histogram by pressing $\boxed{\text{PLOT}}$ (i.e., $\boxed{\text{F3}}$) and then $\boxed{\text{Plot 1}}$ (i.e., $\boxed{\text{F1}}$). Turn the plot on, press the down arrow until the symbol after "Type=" is flashing, and press $\boxed{\text{HIST}}$ (i.e., $\boxed{\text{F4}}$).
- Tell the calculator to put the data entered in xStat on the *x*-axis by selecting "xStat" for "Xlist," and to consider each entered data point as having a frequency of 1 by typing "1" for "Freq."
  Note: If some data points frequently recur and you entered their frequencies in list fStat, then enter press $\boxed{\text{fStat}}$ (i.e., $\boxed{\text{F3}}$) rather than "1" for "Freq."
- Press $\boxed{\text{2nd}}$ $\boxed{\text{QUIT}}$ and put the calculator back into statistics mode by pressing $\boxed{\text{2nd}}$ $\boxed{\text{STAT}}$.

- Press ⎡DRAW⎤ to obtain the histogram. If some of the bars are too long or too short for the screen, alter ymax accordingly. If the button labels obscure the histogram, press ⎡EXIT⎤ once.
- Press ⎡PLOT⎤ and then ⎡Plot 1⎤ and turn the plot off, or else the histogram will appear on future graphs.

EXERCISES

18. In August of 1996, *Money* magazine reported that the best savings yields in the country for money-market accounts, bond funds, CD's, and U.S. Treasury Securities were as follows:

5.31%	5.27%	5.25%	5.42%	5.31%	5.28%
4.76%	4.70%	4.61%	5.17%	5.17%	5.16%
7.58%	7.42%	6.94%	7.07%	5.40%	5.38%
5.79%	5.76%	5.73%	6.07%	6.05%	6.03%
6.91%	6.82%	6.82%	4.85%	5.23%	5.46%
5.77%	6.74%				

Construct a histogram to represent the data. (Group the data into six intervals.)

19. In a 1996 article, *Money* magazine (August 1996) claimed that "state regulation (of the insurance industry) is a joke." The article "ranks the states by the percentage of legislators with links to insurers who sit on committees focusing on insurance issues." Those rankings are as follows:

Mississippi 40%	Louisiana 38%	Arkansas 37%
Virginia 35%	North Carolina 33%	Missouri 32%
Florida 31%	Alabama 26%	Georgia 25%
Texas 25%	West Virginia 24%	Indiana 21%
Ohio 18%	Minnesota 18%	Iowa 17%
North Dakota 17%	Pennsylvania 15%	Arizona 14%
Wyoming 14%	Illinois 14%	Kentucky 13%
Idaho 13%	Delaware 13%	Utah 12%
Kansas 12%	Wisconsin 10%	Hawaii 10%
New Mexico 10%	Rhode Island 9%	Maryland 9%
New Jersey 8%	New York 8%	South Carolina 8%
Washington 8%	Tennessee 8%	Maine 8%
Oklahoma 8%	New Hampshire 7%	Connecticut 5%

(Unlisted states were ranked 0%.)

Construct a histogram to represent the data from all 50 states. (Group the data into five intervals.)

9.2

MEASURES OF CENTRAL TENDENCY

Who is the best running back in professional football? How much does a typical house cost? What is the most popular television program? The answers to questions like these have one thing in common: they are based on averages. In order to compare the capabilities of athletes, we compute their average performances. This computation usually involves the ratio of two totals, such as (total yards gained)/(total number of carries) = average gain per carry. In real estate, the average house price is found by listing the prices of all houses for sale (from lowest to highest) and selecting the price in the middle. Television programs are rated by the average number of households tuned in to each particular program.

Rather than listing every data point in a large distribution of numbers, people tend to summarize the data by selecting a representative number, calling it the average. Three figures—the *mean,* the *median,* and the *mode*—describe the "average" or "center" of a distribution of numbers. These averages are known collectively as the **measures of central tendency.**

Population versus Sample

Who will become the next president of the United States? During election years, political analysts spend a lot of time and money trying to determine what percent of the vote each candidate will receive. However, because there are over 175 million registered voters in the United

States, it would be virtually impossible to contact each and every one of them and ask, "Who do you plan to vote for?" Consequently, analysts select a smaller group of people, determine their intended voting patterns, and project their results onto the entire body of all voters.

Who will become the next president of the United States? Political analysts conduct surveys to measure the popularity of the candidates and predict the outcome of an election. What factors would contribute to an erroneous prediction?

FIGURE 9.24

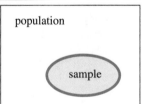

Because of time and money constraints, it is very common for researchers to study the characteristics of a small group in order to estimate the characteristics of a larger group. In this context, the set of all objects under study is called the **population,** and any subset of the population is called a **sample** (see Figure 9.24).

Once we have collected data from the sample, we can summarize it by calculating various descriptive statistics, such as the average value. Inferential statistics then deals with drawing conclusions (hopefully, valid ones!) about the population, based on the descriptive statistics of the sample data.

Sample data are collected and summarized in order to help us draw conclusions about the population. A good sample is representative of the population from which it was taken. Obviously, if the sample is not representative, the conclusions concerning the population might not be valid. The most difficult aspect of inferential statistics is obtaining a representative sample. Remember that conclusions are only as reliable as the sampling process and that information will usually change from sample to sample.

The Mean

Webster's Dictionary defines "average" as "the numerical result obtained by dividing the sum of two or more quantities by the number of quantities; an arithmetical mean." Generally, mathematicians and other scientists call this quantity the *mean* rather than the *average;* they use the word *average* to collectively refer to all three averages (the mean, the median, and the mode).

The **mean** is the average people are most familiar with; it can also be the most misleading. Given a collection of n data points $x_1, x_2, \ldots, x_n$, the mean is found by adding up the data and dividing by the number of data points:

$$\text{the mean of } n \text{ data points} = \frac{x_1 + x_2 + \cdots + x_n}{n}$$

If the data form a sample, the mean is denoted by $\bar{x}$ (read "x bar"); if the data form a population, the mean is denoted by μ (lowercase Greek letter "mu"). The different letters are used so that the distinction between sample data and population data can be emphasized. The distinction is important; samples are used to approximate populations, but the accuracy of that approximation must be questioned.

Mathematicians have developed a type of shorthand, called **summation notation,** to represent the sum of a collection of numbers. The Greek letter Σ ("sigma") corresponds to the letter S and represents the word *sum*. Given a group of data points $x_1, x_2, \ldots, x_n$, we use the symbol Σx to represent their sum; that is $\Sigma x = x_1 + x_2 + \cdots + x_n$.

Definition of the Mean

Given a set of n data points $x_1, x_2, \ldots, x_n$, the *mean*, denoted by either $\bar{x}$ (if the data points form a sample) or μ (if the data points form a population), is

$$\bar{x} \text{ or } \mu = \frac{\Sigma x}{n}$$

Known as the "Sultan of Swat," Babe Ruth hit 714 home runs during his professional career. Although he ranks second in the total number of home runs hit (Hank Aaron is first with 755), Ruth has the highest home run percentage in the history of professional baseball: relative to his 8399 times at bat, "The Bambino" homered 8.5% of the time. (Aaron's percentage is 6.1%.)

If a sample of 460 families have 690 children altogether, then the mean number of children per family is $\bar{x} = 1.5$.

THE FAR SIDE   By GARY LARSON

"Bob and Ruth! Come on in .... Have you met Russell and Bill, our 1.5 children?"

EXAMPLE 1   Rich Bobian wanted to investigate the price of gasoline at various service stations in his area. He surveyed seven stations and recorded the price per gallon of self-serve premium unleaded (92 octane) fuel. The raw data are as follows:

$1.399   1.349   1.299   1.429   1.399   1.379   1.259

Find the mean price per gallon of self-serve premium unleaded gasoline.

Solution

$$\bar{x} = \frac{\Sigma}{n}$$

$$= \frac{\$1.399 + 1.349 + 1.299 + 1.429 + 1.399 + 1.379 + 1.259}{7}$$

$$= \frac{\$9.513}{7}$$

$$= \$1.359 \text{ per gallon}$$

Here, $\bar{x}$ is used rather than μ because the seven stations form a sample of all of the stations in Rick's area. •

EXAMPLE 2   Ten college students were comparing their wages earned at part-time jobs. Nine earned $6 per hour working at jobs ranging from waiting tables to working at a bookstore. One student earned $100 per hour modeling for a major fashion magazine. Find the mean wage of the students.

Solution   Rather than literally adding nine 6's, we'll multiply 6 by 9.

$$\mu = \frac{\Sigma x}{n} = \frac{6 \cdot 9 + 100}{10} = \frac{154}{10} = 15.4$$

The mean wage of the students if $15.40 per hour. Here, μ is used rather than $\bar{x}$ because we found the mean wage *of the students,* not of all college students. •

In Example 2, when we multiplied 6 by 9, we were multiplying the wage ($x = 6$) by its frequency ($f = 9$). This way of calculating the mean is appropriate whenever data points occur with some frequency.

Mean Formula

Given a sample of data points $x_1, x_2, \ldots, x_n$, and corresponding frequencies $f_1, f_2, \ldots, f_n$, the mean is

$$\bar{x} \text{ or } \mu = \frac{\Sigma(fx)}{n}$$

Example 2 seems to indicate that a wage of $15.40 per hour is somehow representative of the wages of the ten students. Is that reasonable? If nine out of ten students earn $6 per hour, can we justify saying that a wage of $15.40 is representative? Of course not! Even though the mean wage *is* $15.40, it is not a convincing "average" for this specific group of data. The mean is inflated because one student made $100 per hour. The wage is called an

extreme value, or **outlier,** because it is significantly different from the rest of the data. Whenever a collection of data has extreme values, the mean can be greatly affected and might not be a representative average.

EXAMPLE 3   In 1992 the U.S. Bureau of Labor Statistics tabulated a survey of workers' ages and wages. The frequency distribution in Figure 9.25 summarizes the age distribution of workers who received minimum wage (then, $4.25 per hour). Find the mean age of a worker receiving minimum wage.

FIGURE 9.25

y = Age	Number of Workers
$16 \leq y < 20$	1660
$20 \leq y < 25$	1254
$25 \leq y < 35$	1118
$35 \leq y < 45$	709
$45 \leq y < 55$	424
$55 \leq y < 65$	337
	$n = 5502$

Source: Bureau of Labor Statistics, U.S. Department of Labor

Solution   To find the mean age of the workers, we must multiply the ages by their frequencies, add the results, and divide by 5502. However, because we are given grouped data, the ages of the individual workers are unknown to us. In this situation we use the midpoint of each interval as representative of the interval; consequently, our answer is an approximation.

To find the midpoint of an interval, add the endpoints and divide by 2. For instance, the midpoint of the interval $16 \leq y < 20$ is $\frac{16 + 20}{2} = 18$. Similar calculations yield the data in Figure 9.26.

FIGURE 9.26

y = Age	f = Frequency	x = Midpoint	f · x
$16 \leq y < 20$	1660	18	29,880
$20 \leq y < 25$	1254	22.5	28,215
$25 \leq y < 35$	1118	30	33,540
$35 \leq y < 45$	709	40	28,360
$45 \leq y < 55$	424	50	21,200
$55 \leq y < 65$	337	60	20,220
	$n = 5502$		$\Sigma(f \cdot x) = 161,415$

The mean age is found by dividing the sum of the ages of all the workers by the number of workers:

$$\bar{x} = \frac{\Sigma(f \cdot x)}{n}$$

$$= \frac{161,415}{5,502}$$

$$= 29.33751 \text{ years}$$

The mean age of the workers earning minimum wage is approximately 29.3 years. We must assume that the workers surveyed form a sample of all U.S. workers because there were only 5502 workers in the survey. Thus, $\bar{x}$ is used rather than μ. ●

> ✔ One common mistake made when working with grouped data is to forget to multiply the midpoint of an interval by the frequency of the interval. Another common mistake is to divide by the number of intervals instead of by the total number of data points.

Suppose you're a student in a class in which you've taken four tests. If your scores were 80%, 76%, 90%, and 90%, your mean test score would be

$$\bar{x} = \frac{\Sigma(fx)}{n} = \frac{1 \cdot 80 + 1 \cdot 76 + 2 \cdot 90}{4} = 84\%$$

Furthermore, the relative frequencies of the scores are $\frac{1}{4}, \frac{1}{4},$ and $\frac{2}{4},$ so the expected value of the test scores is

$$80 \cdot \frac{1}{4} + 76 \cdot \frac{1}{4} + 90 \cdot \frac{2}{4} = 84\%$$

Clearly, these are identical: "expected value" and "mean" are synonyms. When we use the term *expected value* rather than *mean,* we are emphasizing that this number represents what we should expect to occur, over the long term.

The Median

The **median** is the middle value of a distribution of numbers. To find it, we first put the data in numerical order. (If a data point appears more than once, include it as many times as it occurs.) If there is an odd number of data points, the median is the middle data point; if there is an even number of data points, the median is defined to be the mean of the two middle values. In either case, the median separates the distribution into two equal parts. Thus, the median can be viewed as an "average." (The word *median* is also used to describe the strip that runs down the middle of a freeway; half the freeway is on one side, and half is on the other. This common usage is in keeping with the statistical meaning.)

EXAMPLE 4   Find the median of the following sets of raw data.

a. 4   6   1   8   3   10   3
b. 4   6   1   8   3   10   3   9

Solution   **a.** First, we put the data in order. Because there is an odd number of data points ($n = 7$), we pick the middle one:

1   3   3   4   6   8   10
↑
Middle value

The median is 4.

b. We arrange the data first. Because there is an even number of data points ($n = 8$), we pick the two middle values and find their mean:

1   3   3   4   6   8   9   10
↑   ↑

$$\frac{4 + 6}{2} = 5$$

Therefore, the median is 5.   •

EXAMPLE 5   Find the median wage for the ten students in Example 2.

Solution   First, we put the ten wages in order. Because there is an even number of data points ($n = 10$), we pick the two middle values and find their mean:

5   5   5   5   5   5   5   5   5   100
↑   ↑

$$\frac{5 + 5}{2} = 5$$

Therefore, the median wage is $5.00. This is a much more meaningful average than the mean of $15.40.   •

If a collection of data contains extreme values, the median, rather than the mean, is a better indicator of the "average" value. For instance, in discussions of real estate, the median is usually used to express the "average" price of a house. (Why?) In a similar manner, when the incomes of professionals are compared, the median is a more meaningful representation. The discrepancy between mean (average) income and median income is illustrated in the news article shown in Figure 9.27 on page 422. The mean (average) income of physicians in 1992 was $177,400, whereas the median was $148,000.

FIGURE 9.27

Doctors' Average Income in U.S. Is Now $177,000

ASSOCIATED PRESS

WASHINGTON
The average income of the nation's physicians rose to $177,400 in 1992, up 4 percent from the year before, the American Medical Association said yesterday.

Such incomes are apparently helping make careers in medicine more desirable than ever, despite doctors' anxieties about how they would fare under health reform. Medical colleges say the number of applicants for this fall's class has already hit a record.

The average income figures, compiled annually by the AMA and based on a telephone survey of more than 4,100 physicians, ranged from a low of $111,800 for general practitioners and family practice doctors to a high of $253,300 for radiologists.

Physicians' median income was $148,000 in 1992, or 6.5 percent more than a year earlier. Half the physicians earned more than that and half earned less.

The average is pulled higher than the median by the earnings of the highest paid surgeons, anesthesiologists and other specialists at the top end of the scale.

The Association of American Medical Colleges says 43,134 students have already applied for the 16,000 slots available in medical schools this fall. The final number is expected to top 45,000. The old record, set last year, was 42,80.

As recently as 1988, only 26,000 people applied for the same number of seats.

Here are the AMA's average and median net income figures by specialty for 1992:

General/family practice: $111,800, $100,000.
Internal medicine: $159,300, $130,000.
Surgery: $244,600, $207,000.
Pediatrics: $121,700, $112,000.
Obstetrics/gynecology: $215,100, $190,000.
Radiology, $253,300, $240,000.
Psychiatry: $130,700, $120,000.
Anesthesiology: $228,500, $220,000.
Pathology: $189,800, $170,000.
Other: $165,400, $150,000.

The Mode

The third measure of central tendency is the **mode.** The mode is the most frequent number in a collection of data; that is, it is the data point with the highest frequency. Because it represents the most common number, the mode can be viewed as an average. A distribution of data can have more than one mode or none at all.

EXAMPLE 6   Find the mode(s) of the following sets of raw data:
a. 4 10 1 8 5 10 5 10;   **b.** 4 9 1 10 1 10 4 9;   **c.** 9 6 1 8 3 10 3 9.

Solution   a. The mode is 10, because it has the highest frequency (3).
b. There is no mode, because each number has the same frequency (2).
c. The distribution has two modes—namely, 3 and 9—each with a frequency of 2. A distribution that has two modes is called *bimodal.*   ●

In summarizing a distribution of numbers, it is most informative to list all three measures of central tendency. It helps avoid any confusion or misunderstanding in situations in which the word *average* is used. In the hands of someone with questionable intentions, numbers can be manipulated to mislead people. In his book *How to Lie with Statistics,* Darrell Huff states,

The secret language of statistics, so appealing in a fact-minded culture, is employed to sensationalize, inflate, confuse, and oversimplify. Statistical methods and statistical terms are necessary in reporting the mass data of social and economic trends, business conditions, "opinion" polls, the census. But without writers who use the words with honesty and understanding, and readers who know what they mean, the result can only be semantic nonsense.

An educated public should always be on the alert for oversimplification of data via statistics. The next time someone mentions an "average," ask "Which one?" Although people might not intentionally try to mislead you, their findings can be misinterpreted if you do not know the meaning of their statistics and the method by which the statistics were calculated.

9.2

EXERCISES

In Exercises 1–4, find the mean, median, and mode of the given set of raw data.

1. 9   12   8   10   9   11   12
        15   20   9   14   15   21   10

2. 20   25   18   30   21   25   32   27
        32   35   19   26   38   31   20   23

3. 1.2   1.8   0.7   1.5   1.0   0.7   1.9   1.7   1.2
        0.8   1.7   1.3   2.3   0.9   2.0   1.7   1.5   2.2

4. 0.07   0.02   0.09   0.04   0.10   0.08   0.07   0.13
        0.05   0.04   0.10   0.07   0.04   0.01   0.11   0.08

5. Find the mean, median, and mode of each set of data.
   a. 9   9   10   11   12   15
   b. 9   9   10   11   12   102
   c. How do your answers for parts (a) and (b) differ (or agree)? Why?

6. Find the mean, median, and mode for each set of data.
   a. 80   90   100   110   110   140
   b. 10   90   100   110   110   210
   c. How do your answers for parts (a) and (b) differ (or agree)? Why?

7. Find the mean, median, and mode of each set of data.
   a. 2   4   6   8   10   12
   b. 102   104   106   108   110   112
   c. How are the data in (b) related to the data in (a)?
   d. How do your answers for (a) and (b) compare?

8. Find the mean, median, and mode of each set of data.
   a. 12   16   20   24   28   32
   b. 600   800   1,000   1,200   1,400   1,600
   c. How are the data in (b) related to the data in (a)?
   d. How do your answers for (a) and (b) compare?

9. The number of home runs that Babe Ruth hit during each season he played for the New York Yankees is given in Figure 9.28 (*source: The Baseball Encyclopedia,* 1988). Find the mean, median, and mode of the number of home runs per year that Babe Ruth hit while playing for the Yankees.

FIGURE 9.28

Year	HRs		Year	HRs
1920	54		1928	54
1921	59		1929	46
1922	35		1930	49
1923	41		1931	46
1924	46		1932	41
1925	25		1933	34
1926	47		1934	22
1927	60			

10. Fran Tarkenton holds the all-time record in professional football for throwing touchdown passes. The number of touchdown passes for each of his seasons is given in Figure 9.29 on page 424 (*source: The Official NFL Encyclopedia,* 1986). Find the mean, median, and mode of the number of touchdown passes per year thrown by Tarkenton.

FIGURE 9.29
Fran Tarkenton's touchdown passes

Year	TDs	Year	TDs
1961	18	1970	19
1962	22	1971	11
1963	15	1972	18
1964	22	1973	15
1965	19	1974	17
1966	17	1975	25
1967	29	1976	17
1968	21	1977	9
1969	23	1978	25

FIGURE 9.31
Miles run by Todd Booth

Miles Run per Week	Number of Weeks
0	5
1	4
2	10
3	9
4	10
5	7
6	3
7	4

11. The frequency distribution in Figure 9.30 lists the results of a quiz given in Professor Gilbert's statistics class. Find the mean, median, and mode of the scores.

FIGURE 9.30
Quiz scores in Professor Gilbert's statistics class

Score	Number of Students
10	3
9	10
8	9
7	8
6	10
5	2

12. Todd Booth, an avid jogger, kept detailed records of the number of miles he ran per week during the past year. The frequency distribution in Figure 9.31 summarizes his records. Find the mean, median, and mode of the number of miles per week that Todd ran.

13. To study the output of a machine that fills boxes with cereal, a quality control engineer weighed 150 boxes of Brand X cereal. The frequency distribution in Figure 9.32 summarizes his findings. Find the mean weight of the boxes of cereal.

FIGURE 9.32
Amount of Brand X cereal per box

x = Weight (in ounces)	Number of Boxes
$15.3 \leq x < 15.6$	13
$15.6 \leq x < 15.9$	24
$15.9 \leq x < 16.2$	84
$16.2 \leq x < 16.5$	19
$16.5 \leq x \leq 16.8$	10

14. To study the efficiency of its new price-scanning equipment, a local supermarket monitored the amount of time its customers had to wait in line. The frequency distribution in Figure 9.33 on page 425 summarizes the findings. Find the mean amount of time spent in line.

15. If your scores on the first four exams (in this class) are 73, 67, 83, and 81, what score do you need on the next exam for your overall mean to be at least 80?

16. The mean salary of 12 men is $48,000 and the mean salary of 8 women is $29,000. Find the mean salary of all 20 people.

FIGURE 9.33
Time spent waiting in a supermarket checkout line

x = Time (in minutes)	Number of Customers
$0 \le x < 1$	79
$1 \le x < 2$	58
$2 \le x < 3$	64
$3 \le x < 4$	40
$4 \le x \le 5$	35

17. Maria drove from Chicago, Illinois, to Milwaukee, Wisconsin (90 miles), at a mean speed of 60 miles per hour. On her return trip, the traffic was much heavier, and her mean speed was 45 miles per hour. Find Maria's mean speed for the round trip.
HINT: Divide the total distance by the total time.

18. The mean age of a class of 25 students is 23.4 years. How old would a 26th student have to be in order for the mean age of the class to be 24.0 years?

19. The number of civilians holding various federal government jobs, and their mean monthly earnings for May 1992, are given in Figure 9.34.
  a. Find the mean monthly earnings of all civilians employed by the Navy, Air Force, and Army.

FIGURE 9.34
Monthly earnings for civilian jobs (May 1992)

Department	Number of Civilian Workers	Mean Monthly Earnings
State Department	26,164	$5,387.44
Justice Department	95,708	$4,609.08
Congress	20,867	$3,126.66
Department of the Navy	312,303	$2,810.73
Department of the Air Force	212,695	$2,749.66
Department of the Army	344,068	$1,265.71

Source: U.S. Office of Personnel Management

  b. Find the mean monthly earnings of all civilians employed by the State Department, Justice Department, and Congress.

20. The ages of more than 52 million women who gave birth in the United States between June 1987 and June 1988 are given in Figure 9.35. Find the mean age of these women.

FIGURE 9.35
Ages of women giving birth

Age	Number of Women
$18 \le x < 25$	13,167,000
$25 \le x < 30$	10,839,000
$30 \le x < 35$	10,838,000
$35 \le x < 40$	9,586,000
$40 \le x < 45$	8,155,000

Source: U.S. Bureau of the Census

21. The age composition of the population of the United States in 1990 is given in Figure 9.36.

FIGURE 9.36
Age composition of the population of the United States

Age	Number of People
$0 < x < 5$	18,354,000
$5 \le x < 18$	42,250,000
$18 \le x < 21$	11,727,000
$21 \le x < 25$	15,011,000
$25 \le x < 45$	80,755,000
$45 \le x < 55$	25,223,000
$55 \le x < 60$	10,532,000
$60 \le x < 65$	10,616,000
$65 \le x < 75$	18,107,000
$75 \le x < 85$	10,055,000
85 and over	3,080,000

Source: U.S. Bureau of the Census

a. Find the mean age of all people in the United States under the age of 85.

b. Replace the interval "85 and over" with the interval $85 \leq x \leq 100$ and find the mean age of all people in the United States.

In Exercises 22 and 23, use the estimated age composition of the 12,935,000 students enrolled in institutions of higher education in the United States during 1990, as given in Figure 9.37.

22. a. Find the mean age of all male students in higher education under 35.

b. Replace the interval "35 and over" with the interval $35 \leq x \leq 100$ and find the mean age of all male students in higher education.

23. a. Find the mean age of all female students in higher education under 35.

b. Replace the interval "35 and over" with the interval $35 \leq x \leq 100$ and find the mean age of all female students in higher education.

 Answer the following questions using complete sentences.

24. Suppose the mean of Group I is A and the mean of Group II is B. We combine Groups I and II to form Group III. Is the mean of Group III equal to $\frac{A + B}{2}$? Explain.

25. Why do we use the midpoint of an interval when calculating the mean of grouped data?

26. The mean salary of ten employees is $32,000, and the median is $30,000. The highest-paid employee gets a $5,000 raise.
  a. What is the new mean salary of the ten employees?
  b. What is the new median salary of the ten employees?

27. In Exercises 9–14 and 19–22, determine which means are population means and which are sample means, and explain why. In some exercises, either choice could be justified.

FIGURE 9.37
Age composition of students in higher education

Age of Males	Number of Students
$14 \leq x < 18$	111,000
$18 \leq x < 20$	1,380,000
$20 \leq x < 22$	1,158,000
$22 \leq x < 25$	928,000
$25 \leq x < 30$	847,000
$30 \leq x < 35$	591,000
35 and over	878,000
total	5,893,000

Age of Females	Number of Students
$14 \leq x < 18$	126,000
$18 \leq x < 20$	1,502,000
$20 \leq x < 22$	1,253,000
$22 \leq x < 25$	958,000
$25 \leq x < 30$	840,000
$30 \leq x < 35$	743,000
35 and over	1,620,000
total	7,042,000

Source: U.S. National Center for Education Statistics

9.3

MEASURES OF DISPERSION

In order to settle an argument over who was the better bowler, George and Danny agreed to bowl six games, and whoever had the highest "average" would be considered best. Their scores were as follows:

George	185	135	200	185	250	155
Danny	182	185	188	185	180	190

Each bowler then arranged his scores from lowest to highest and computed the mean, median, and mode:

George $\begin{cases} \end{cases}$

135 155 185 185 200 250

$$\text{mean} = \frac{\text{sum of scores}}{6} = \frac{1110}{6} = 185$$

$$\text{median} = \text{middle score} = \frac{185 + 185}{2} = 185$$

$$\text{mode} = \text{most common score} = 185$$

Danny $\begin{cases} \end{cases}$

180 182 185 185 188 190

$$\text{mean} = \frac{1110}{6} = 185$$

$$\text{median} = \frac{185 + 185}{2} = 185$$

$$\text{mode} = 185$$

Much to their surprise, George's mean, median, and mode were exactly the same as Danny's! Using the measures of central tendency alone to summarize their performances, the bowlers appear identical. Even though their averages were identical, however, their performances were not; George was very erratic, while Danny was very consistent. Who is the better bowler? Based on high score, George is better. Based on consistency, Danny is better.

George and Danny's situation points out a fundamental weakness in using only the measures of central tendency to summarize data. In addition to finding the averages of a set of data, the consistency, or spread, of the data should also be taken into account. This is accomplished by using *measures of dispersion,* which determine how the data points differ from the average.

Deviations

It is clear from George and Danny's bowling scores that it is sometimes desirable to measure the relative consistency of a set of data. Are the numbers consistently bunched up? Are they erratically spread out? In order to measure the dispersion of a set of data, we need to identify an average or typical distance between the data points and the mean. The difference between a single data point x and the mean $\bar{x}$ is called the **deviation from the mean** (or simply the **deviation**) and is given by $(x - \bar{x})$. A data point that is close to the mean will have a small deviation, whereas data points far from the mean will have large deviations, as shown in Figure 9.38.

FIGURE 9.38

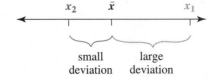

To find the typical deviation of the data points, you might be tempted to add up all the deviations and divide by the total number of data points, thus finding the "average" deviation. Unfortunately, this process leads nowhere. To see why, we will find the mean of the deviations of George's bowling scores.

EXAMPLE 1 George bowled six games, and his scores were 185, 135, 200, 185, 250, and 155. Find the mean of the scores, the deviation of each score, and the mean of the deviations.

Solution $$\bar{x} = \frac{\text{sum of scores}}{6} = \frac{1110}{6} = 185$$

The mean score is 185.

To find the deviations, subtract the mean from each score, as shown in Figure 9.39.

FIGURE 9.39

Score (x)	Deviation ($x - 185$)
135	−50
155	−30
185	0
185	0
200	15
250	65
	sum = 0

$$\text{mean of the deviations} = \frac{\text{sum of deviations}}{6} = \frac{0}{6} = 0$$

The mean of the deviations is zero. ●

In Example 1, the mean of the deviations of the data is zero. This *always* happens; that is, $\frac{\Sigma(x - \bar{x})}{n} = 0$ for any set of data. The negative deviations are the "culprits"—they will always cancel out the positive deviations. Therefore, in order to use deviations to study the spread of the data, we must modify our approach and convert the negatives into positives. We do this by computing the **variance,** which is the mean of the *squares* of the deviations, rather than the mean of the deviations.

Variance and Standard Deviation

Before proceeding, we must be reminded of the difference between a sample and a population. A *population* is the universal set of all possible items under study; a *sample* is any group or subset of items selected from the population. Samples are used when the population is too big to study, and sample data are used to approximate population data. George's six bowling scores represent a sample, not a population; because we do not know the scores of *all* the games George has ever bowled, we are limited to a sample.

A sample mean $\bar{x}$ and a population mean μ are calculated with the same formula. However, a sample variance, denoted by s^2, and a population variance, denoted by σ^2 (σ is the lowercase Greek letter "sigma"), are calculated with slightly different formulas.

Definition of the Variance

Given a *population* of n data points $x_1, x_2, \ldots, x_n$, the **population variance,** denoted by σ^2, is

$$\sigma^2 = \frac{\Sigma(x - \mu)^2}{n}$$

Given a *sample* of n data points $x_1, x_2, \ldots, x_n$, the **sample variance,** denoted by s^2, is

$$s^2 = \frac{\Sigma(x - \bar{x})^2}{n - 1}$$

Because we are working with n data points, you might wonder why we divide by $n - 1$ rather than by n in computing the sample variance. The answer lies in the study of inferential statistics. Recall that inferential statistics deal with the drawing of conclusions concerning the nature of a population based on observations made within a sample. Hence, the variance of a sample can be viewed as an estimate of the variance of the population. However, because the population will vary more than the sample (a population has more data points), dividing the sum of the squares of the sample deviations by n would underestimate the true variance of the entire population. In order to compensate for this underestimation, statisticians have determined that dividing the sum of the squares of the deviations by $n - 1$ rather than by n produces the best estimate of the true population variance.

EXAMPLE 2   George bowled six games, and his scores were 185, 135, 200, 185, 250, and 155. Find the sample variance of his scores.

Solution   The mean of the six data points is 185. The necessary calculations for finding variance are shown in Figure 9.40.

FIGURE 9.40

Data (x)	Deviation $(x - 185)$	Deviation Squared $(x - 185)^2$
135	-50	$(-50)^2 = 2500$
155	-30	$(-30)^2 = 900$
185	0	$(0)^2 = 0$
185	0	$(0)^2 = 0$
200	15	$(15)^2 = 225$
250	65	$(65)^2 = 4225$
		Sum $= 7850$

$$\text{variance} = \frac{\text{sum of the squares of the deviations}}{n - 1}$$

$$s^2 = \frac{7850}{6 - 1}$$

$$= \frac{7850}{5}$$

$$= 1570 \qquad \bullet$$

In Example 2, the variance of George's bowling scores was 1570. What units should be used here—1570 what? The first calculation performed was subtracting the mean score of 185 points from one particular score of 135 points, to obtain a deviation of -50 points. That quantity was squared, to obtain 2500 "square points" (just as the area of a square with 2-inch sides is 4 "square inches"). The sum of this and other similar quantities was computed, so the variance is 1570 "square points." Usually the square root of this quantity is taken, to make the units more reasonable. The result is called the **standard deviation,** and is denoted by either σ (if population data are used) or s (if sample data are used). The standard deviation of George's bowling scores is

$$s = \sqrt{1570} \text{ square points} = 39.6232 \ldots \text{ points} \approx 39.6 \text{ points.}$$

Because they give us information concerning the spread of data, variance and standard deviation are called **measures of dispersion.** Standard deviation (and variance) is a relative measure of the dispersion of a set of data; the larger the standard deviation, the more spread out the data. Consider George's standard deviation of 39.6. This appears to be high, but what exactly constitutes a "high" standard deviation? Unfortunately, because it is a relative measure, there is no hard-and-fast distinction between a "high" and a "low" standard deviation.

By itself, the standard deviation of a set of data might not be very informative, but standard deviations are very useful in comparing the relative consistencies of two sets of data. Given two groups of numbers of the same type (for example, two sets of bowling scores, two sets of heights, two sets of prices, and so on), the set with the lower standard deviation contains data that are more consistent, whereas the data with the higher standard deviation are more spread out. Calculating the standard deviation of Danny's six bowling scores, we find $s = 3.7$. Since Danny's standard deviation is less than George's, we infer that Danny is more consistent. If George's standard deviation is less than 39.6 the next time he bowls six games, we would infer that his game has become more consistent (the scores would not be spread out as far).

Alternate Methods for Finding Variance

The procedure for calculating sample variance is very direct: First find the mean of the data, then find the deviation of each data point, and finally divide the sum of the squares of the deviations by $(n - 1)$. However, using the variance definition to find the variance can be rather tedious. Instead of using the definition of variance (as in Example 2), we can use an alternate formula that contains the two sums Σx and Σx^2, where Σx represents the sum of the data and Σx^2 represents the sum of the squares of the data, as shown in the box on the next page.

Although we will not prove it, the alternate formulas for variance are algebraically equivalent to those given in the variance definition; given any set of data, either method will produce the same answer. At first glance, the alternate formula might appear to be more difficult to use than the definition. Don't be fooled by its appearance! As we will see, the alternate formula is relatively quick and easy to apply.

Alternate Formula for Sample Variance

Given n data points $x_1, x_2, \ldots, x_n$, the *population variance* of the data can be found by

$$\sigma^2 = \frac{1}{n}\left[\Sigma x^2 - \frac{(\Sigma x)^2}{n}\right]$$

and the *sample variance* can be found by

$$s^2 = \frac{1}{n-1}\left[\Sigma x^2 - \frac{(\Sigma x)^2}{n}\right]$$

EXAMPLE 3   Using the alternate formula for sample variance, find the sample standard deviation of George's bowling scores as given in Example 2.

Solution   Recall that George's scores were 185, 135, 200, 185, 250, and 155. To find the standard deviation, we must first find the variance.

The alternate formula for sample variance requires that we find the sum of the data and the sum of the squares of the data. These calculations are shown in Figure 9.41.

FIGURE 9.41

x	x^2
135	18,225
155	24,025
185	34,225
185	34,225
200	40,000
250	62,500
$\Sigma x = 1110$	$\Sigma x^2 = 213{,}200$

Applying the alternate formula for sample variance, we have

$$s^2 = \frac{1}{n-1}\left[\Sigma x^2 - \frac{(\Sigma x)^2}{n}\right]$$

$$= \frac{1}{6-1}\left[213{,}200 - \frac{(1110)^2}{6}\right]$$

$$= \frac{1}{5}[213{,}200 - 205{,}350]$$

$$= \frac{7850}{5} = 1570$$

The variance is $s^2 = 1570$. (Note that this is the same as the variance calculated in Example 2 using the definition.)

Taking the square root, we have

$$s = \sqrt{1570}$$
$$= 39.62322551\ldots$$

Rounded off, the standard deviation of George's bowling scores is $s = 39.6$ points. •

Grouped Data

When we are working with grouped data, the individual data points are unknown. In such cases, the midpoint of each interval should be used as the representative value of the interval.

EXAMPLE 4 In 1992 the U.S. Bureau of Labor Statistics tabulated a survey of workers' ages and wages. The frequency distribution in Figure 9.42 summarizes the age distribution of workers who received minimum wages ($4.25 per hour). Find the standard deviation of the ages of these workers.

FIGURE 9.42

$y = $ Age	Number of Workers
$16 \leq y < 20$	1660
$20 \leq y < 25$	1254
$25 \leq y < 35$	1118
$35 \leq y < 45$	709
$45 \leq y < 55$	424
$55 \leq y < 65$	337
	$n = 5502$

Source: Bureau of Labor Statistics, U.S. Department of Labor

Solution Because we are given grouped data, the first step is to determine the midpoint of each interval. We do this by adding the endpoints and dividing by 2.

To utilize the alternate formula for sample variance, we must find the sum of the data and the sum of the squares of the data. The sum of the data is found by multiplying each midpoint by the frequency of the interval and adding the results. The sum of the squares of the data is found by squaring each midpoint, multiplying by the corresponding frequency, and adding. The calculations are shown in Figure 9.43.

FIGURE 9.43

y = Age	Frequency f	Midpoint x	f · x	f · x²
16 ≤ y < 20	1660	18	29,880	537,840
20 ≤ y < 25	1254	22.5	28,215	634,837.5
25 ≤ y < 35	1118	30	33,540	1,006,200
35 ≤ y < 45	709	40	28,360	1,134,400
45 ≤ y < 55	424	50	21,200	1,060,000
55 ≤ y < 65	337	60	20,220	1,213,200
	n = 5502		Σ(f · x) = 161,415	Σ(f · x²) = 5,586,477.5

Applying the alternate formula for sample variance, we have

$$s^2 = \frac{1}{n-1}\left[\Sigma(f \cdot x^2) - \frac{(\Sigma f \cdot x)^2}{n}\right]$$

$$= \frac{1}{5502 - 1}\left[5,586,477.5 - \frac{(161,415)^2}{5502}\right]$$

$$= \frac{1}{5501}[5,586,477.5 - 4,735,514.763]$$

$$= \frac{850,962.7372}{5501}$$

The variance is $s^2 = 154.6923718$. Taking the square root, we have

$$s = \sqrt{154.6923718}$$

$$= 12.43753882$$

Rounded off, the standard deviation of the ages of the workers receiving minimum wage is $s = 12.4$ years.   ●

In order to obtain the best analysis of a collection of data, we should use the measures of central tendency and the measures of dispersion in conjunction with each other. The most common way to combine these measures is to determine what percent of the data lies within a specified number of standard deviations of the mean. The phrase "one standard deviation of the mean" refers to all numbers within the interval $[\bar{x} - s, \bar{x} + s]$, that is, all numbers that differ from $\bar{x}$ by at most s. Likewise, "two standard deviations of the mean" refers to all numbers within the interval $[\bar{x} - 2s, \bar{x} + 2s]$. One, two, and three standard deviations of the mean are shown in Figure 9.44.

FIGURE 9.44

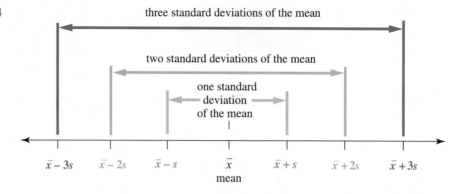

EXAMPLE 5   Jim Courtright surveyed the price of unleaded gasoline at several of the gas stations in his neighborhood. He found the following prices, in dollars per gallon:

$$1.269 \quad 1.249 \quad 1.199 \quad 1.299 \quad 1.269 \quad 1.239 \quad 1.349 \quad 1.289 \quad 1.259$$

What percent of the data lies within one standard deviation of the mean?

Solution   We must first find the mean $\bar{x}$ and the standard deviation s.

Summing the nine data points, we have $\Sigma x = 11.421$. Summing the squares of the data points, we have $\Sigma x^2 = 14.507249$. The mean is $\bar{x} = \frac{(11.421)}{9} = 1.269$.

Using the alternate formula for sample variance, we have

$$s^2 = \frac{1}{n-1}\left[\Sigma x^2 - \frac{(\Sigma x)^2}{n}\right]$$

$$= \frac{1}{9-1}\left[14.507249 - \frac{(11.421)^2}{9}\right]$$

$$= \frac{1}{8}[14.507249 - 14.493249]$$

$$= \frac{0.014}{8}$$

The variance is $s^2 = 0.00175$. Taking the square root, we have

$$s = \sqrt{0.00175}$$

$$= 0.041833001$$

The standard deviation of the prices is 0.0418 (dollars).

To find one standard deviation of the mean, we add and subtract the standard deviation to and from the mean:

$$[\bar{x} - s, \bar{x} + s] = [1.269 - 0.0418, 1.269 + 0.0418]$$

$$= [1.2272, 1.3108]$$

Arranging the data from smallest to largest, we see that seven of the nine data points are between 1.2272 and 1.3108.

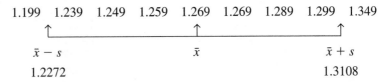

$$1.199 \quad 1.239 \quad 1.249 \quad 1.259 \quad 1.269 \quad 1.269 \quad 1.289 \quad 1.299 \quad 1.349$$

Therefore, $\frac{7}{9} = 0.77777777$ or 78% of the data lie within one standard deviation of the mean.

9.3

EXERCISES

The data in each exercise are sample *data, unless otherwise noted.*

1. Perform each task given the following sample data:

   3 8 5 3 10 13

   a. Use the sample variance definition to find the variance and standard deviation of the data.
   b. Use the alternate formula for sample variance to find the variance and standard deviation of the data.

2. Perform each task given the following sample data:

   6 10 12 12 11 17 9

   a. Use the sample variance definition to find the variance and standard deviation of the data.
   b. Use the alternate formula for sample variance to find the variance and standard deviation of the data.

3. Perform each task given the following sample data:

   10 10 10 10 10 10

   a. Find the variance of the data.
   b. Find the standard deviation of the data.

4. Find the mean and standard deviation of each set of data.
   a. 2   4   6   8   10   12
   b. 102 104 106 108 110 112
   c. How are the data in (b) related to the data in (a)?
   d. How do your answers for (a) and (b) compare?

5. Find the mean and standard deviation of each set of data.
   a. 12   16   20   24   28   32
   b. 600 800 1000 1200 1400 1600
   c. How are the data in (b) related to the data in (a)?
   d. How do your answers for (a) and (b) compare?

6. Find the mean and standard deviation of each set of data.
   a. 50 50 50 50 50
   b. 46 50 50 50 54
   c. 5 50 50 50 95
   d. How do your answers for (a), (b), and (c) compare?

7. Joey and Dee Dee bowled five games at the Rock 'n' Bowl Lanes. Their scores are given in Figure 9.45.
   a. Find the mean score of each bowler. Who has the highest mean?
   b. Find the standard deviation of each bowler's scores.
   c. Who is the more consistent bowler? Why?

FIGURE 9.45
Bowling scores

Joey	144	171	220	158	147
Dee Dee	182	165	187	142	159

8. Paki surveyed the price of unleaded gasoline at several of the gas stations in Novato and Lafayette. The raw data, in dollars per gallon, are given in Figure 9.46.
   a. Find the mean price in each city. Which city has the lowest mean?
   b. Find the standard deviation of prices in each city.
   c. Which city has more consistently priced gasoline? Why?

9. The normal monthly rainfall in Seattle, Washington is given in Figure 9.47 on page 436 (*source:* U.S. Department of Commerce). Find the population mean and standard deviation of the monthly rainfall in Seattle.

FIGURE 9.46
Price (in dollars) of one gallon of unleaded gasoline

Novato	1.309	1.289	1.339	1.309	1.259	1.239
Lafayette	1.329	1.269	1.189	1.349	1.289	1.229

FIGURE 9.47
Rainfall in Seattle

Month	Jan.	Feb.	Mar.	Apr.	May	June
Inches	6.0	4.2	3.6	2.4	1.6	1.4
Month	July	Aug.	Sept.	Oct.	Nov.	Dec.
Inches	0.7	1.3	2.0	3.4	5.6	6.3

10. The normal monthly rainfall in Phoenix, Arizona is given in Figure 9.48 (*source:* U.S. Department of Commerce). Find the population mean and standard deviation of the monthly rainfall in Phoenix.

FIGURE 9.48
Rainfall in Phoenix

Month	Jan.	Feb.	Mar.	Apr.	May	June
Inches	0.7	0.6	0.8	0.3	0.1	0.2
Month	July	Aug.	Sept.	Oct.	Nov.	Dec.
Inches	0.7	1.0	0.6	0.6	0.5	0.8

11. The frequency distribution in Figure 9.49 lists the results of a quiz given in Professor Gilbert's statistics class.
  a. Find the mean and standard deviation of the scores.
  b. What percent of the data lies within one standard deviation of the mean?
  c. What percent of the data lies within two standard deviations of the mean?
  d. What percent of the data lies within three standard deviations of the mean?

FIGURE 9.49
Quiz scores in Professor Gilbert's statistics class

Score	Number of Students	Score	Number of Students
10	5	7	8
9	10	6	3
8	6	5	2

12. Amy surveyed the prices for a quart of a certain brand of motor oil. The sample data, in dollars per quart, is summarized in Figure 9.50.
  a. Find the mean and the standard deviation of the prices.
  b. What percent of the data lies within one standard deviation of the mean?
  c. What percent of the data lies within two standard deviations of the mean?
  d. What percent of the data lies within three standard deviations of the mean?

FIGURE 9.50
Price (in dollars) for a quart of motor oil

Price per Quart	Number of Stores
0.99	2
1.09	5
1.19	10
1.29	13
1.39	9
1.49	3

13. To study the output of a machine that fills boxes with cereal, a quality control engineer weighed 150 boxes of Brand X cereal. The frequency distribution in Figure 9.51 summarizes his findings. Find the standard deviation of the weight of the boxes of cereal.

FIGURE 9.51
Amount of Brand X cereal per box

x = Weight (in ounces)	Number of Boxes
$15.3 \leq x < 15.6$	13
$15.6 \leq x < 15.9$	24
$15.9 \leq x < 16.2$	84
$16.2 \leq x < 16.5$	19
$16.5 \leq x < 16.8$	10

14. To study the efficiency of its new price-scanning equipment, a local supermarket monitored the amount of time its customers had to wait in line. The frequency distribution in Figure 9.52 summarizes the findings. Find the standard deviation of the amount of time spent in line.

FIGURE 9.52
Time spent waiting in a supermarket checkout line

x = Time (in minutes)	Number of Customers
$0 \leq x < 1$	79
$1 \leq x < 2$	58
$2 \leq x < 3$	64
$3 \leq x < 4$	40
$4 \leq x \leq 5$	35

15. The ages of over 52 million women who gave birth in the United States between June 1987 and June 1988 are given in Figure 9.53. Find the population standard deviation of the ages of these women.

FIGURE 9.53
Ages of women giving birth

Age	Number of Women
$18 \leq x < 25$	13,167,000
$25 \leq x < 30$	10,839,000
$30 \leq x < 35$	10,838,000
$35 \leq x < 40$	9,586,000
$40 \leq x < 45$	8,155,000

Source: U.S. Bureau of the Census

16. The age composition of the population of the United States in 1990 is given in Figure 9.54. Replace the interval "85 and over" with the interval $85 \leq x \leq 100$ and find the population standard deviation of the ages of all people in the United States.

FIGURE 9.54
Age composition of the population of the United States

Age	Number of People
$0 < x < 5$	18,354,000
$5 \leq x < 18$	45,250,000
$18 \leq x < 21$	11,727,000
$21 \leq x < 25$	15,011,000
$25 \leq x < 45$	80,755,000
$45 \leq x < 55$	25,223,000
$55 \leq x < 60$	10,532,000
$60 \leq x < 65$	10,616,000
$65 \leq x < 75$	18,107,000
$75 \leq x < 85$	10,055,000
85 and over	3,080,000

Source: U.S. Bureau of the Census

 Answer the following questions using complete sentences.

17. **a.** When studying the dispersion of a set of data, why are the deviations from the mean squared?
    b. What effect does squaring have on a deviation that is less than 1?
    c. What effect does squaring have on a deviation that is greater than 1?
    d. What effect does squaring have on the data's units?
    e. Why is it necessary to take a square root when calculating standard deviation?

18. Why do we use the midpoint of an interval when calculating the standard deviation of grouped data?

19. Explain why the data in Exercises 7, 8, and 11–14 are sample data, while the data in Exercises 9, 10, 15, and 16 are population data. Give a specific justification for each individual exercise.

Measures of Central Tendency and Dispersion on a Graphing Calculator

Calculating the Mean and the Standard Deviation: Ungrouped Data

In Examples 1, 2, and 3 of this section we found the mean, sample variance, and standard deviation of George's bowling scores. This work can be done quickly and easily on a graphing calculator.

On a TI-82/83:

- Enter the data in list L_1 as discussed in Section 9.1.
- Press $\boxed{\text{STAT}}$.
- Select "1-Var Stats" from the "CALC" menu.
- When "1-Var" appears on the screen, press $\boxed{\text{2nd}}$ $\boxed{L_1}$ $\boxed{\text{ENTER}}$.

On a TI-85:

- Enter the data as discussed in Section 9.1.
- Press $\boxed{\text{CALC}}$ (i.e., $\boxed{\text{2nd}}$ $\boxed{\text{M1}}$).
- The calculator will respond with the list names you selected previously. Press $\boxed{\text{ENTER}}$ twice to indicate that these two names are correct, or change them if they are incorrect.
- Press $\boxed{\text{1-VAR}}$ (i.e., $\boxed{\text{F1}}$).

On a TI-86:

- Enter the data as discussed in Section 9.1.
- Press $\boxed{\text{2nd}}$ $\boxed{\text{QUIT}}$, and put the calculator back into statistics mode by pressing $\boxed{\text{2nd}}$ $\boxed{\text{STAT}}$.
- Press $\boxed{\text{CALC}}$ and then $\boxed{\text{OneVa}}$ (i.e., $\boxed{\text{F1}}$ $\boxed{\text{F1}}$), and the screen will read "OneVar."
- Press $\boxed{\text{2nd}}$ $\boxed{\text{LIST}}$, and then $\boxed{\text{NAMES}}$ (i.e., $\boxed{\text{F3}}$).
- Press $\boxed{\text{xStat}}$ (i.e., $\boxed{\text{F2}}$), and the screen will read "OneVar xStat."
- Press $\boxed{\text{ENTER}}$.

The above steps will result in the first screen in Figure 9.55. This screen gives the mean, the sample standard deviation (S_x), the population standard deviation (σ_x) and the number of data points (n). The second green, which is not available on the TI-85, can be obtained by pressing the down arrow. It gives the minimum and maximum data points (minX and maxX respectively) as well as the median (Med).

FIGURE 9.55

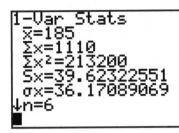

Calculating the Sample Variance

The above work does not yield the sample variance. To find it, follow these steps:

- Quit the statistics mode.
- Get S_x on the screen:

TI-82/83:	Press $\boxed{\text{VARS}}$, select "Statistics," and then select "S_x" from the "X/Y" menu.
TI-85:	Press $\boxed{\text{2nd}}\ \boxed{\text{VARS}}$, press $\boxed{\text{MORE}}$ until the "STAT" option appears, and select that option; then select "S_x."
TI-86:	Press $\boxed{\text{2nd}}\ \boxed{\text{STAT}}$, press $\boxed{\text{VARS}}$ (i.e., $\boxed{\text{F5}}$), and then $\boxed{S_x}$ (i.e., $\boxed{\text{F3}}$).

- Once S_x is on the screen, square it by pressing $\boxed{x^2}\ \boxed{\text{ENTER}}$.

The variance is 1570.

Calculating the Mean, the Variance, and the Standard Deviation: Grouped Data

To calculate the mean and standard deviation from a frequency distribution that utilizes grouped data, follow the steps above *except:*

On a TI-82/83:

- Enter the midpoints of the classes in list L_1.
- Enter the frequencies of those classes in list L_2. (With a TI-82, each frequency must be less than 100.)
- After "1-Var Stats" appears on the screen, press $\boxed{\text{2nd}}\ \boxed{L_1}\ \boxed{,}\ \boxed{\text{2nd}}\ \boxed{L_2}\ \boxed{\text{ENTER}}$.

On a TI-85:

- Enter the midpoints of the classes as x's.
- Enter the frequencies of those classes as the corresponding y's.

On a TI-86:

- Enter the midpoints of the classes in list xStat.
- Enter the frequencies of those classes in list fStat.
- After "OneVar" appears on the screen, press $\boxed{\text{2nd}}\ \boxed{\text{LIST}}$, and then $\boxed{\text{NAMES}}$ (i.e., $\boxed{\text{F3}}$).
- Press $\boxed{\text{xStat}}\ \boxed{,}\ \boxed{\text{fStat}}$, and the screen will read "OneVar xStat,fStat."
- Press $\boxed{\text{ENTER}}$.

EXERCISE

20. In a 1996 article, *Money* magazine (August 1996) claimed that "state regulation (of the insurance industry) is a joke." The article "ranks the states by the percentage of legislators with links to insurers who sit on committees focusing on insurance issues." Those rankings are:

Mississippi 40%	Louisiana 38%	Arkansas 37%
Virginia 35%	North Carolina 33%	Missouri 32%
Florida 31%	Alabama 26%	Georgia 25%
Texas 25%	West Virginia 24%	Indiana 21%
Ohio 18%	Minnesota 18%	Iowa 17%
North Dakota 17%	Pennsylvania 15%	Arizona 14%

Wyoming 14%	Illinois 14%	Kentucky 13%
Idaho 13%	Delaware 13%	Utah 12%
Kansas 12%	Wisconsin 10%	Hawaii 10%
New Mexico 10%	Rhode Island 9%	Maryland 9%
New Jersey 8%	New York 8%	South Carolina 8%
Washington 8%	Tennessee 8%	Maine 8%
Oklahoma 8%	New Hampshire 7%	Connecticut 5%

(Unlisted states were ranked 0%.)

Find the mean and population standard deviation of the data for all 50 states.

9.4

THE NORMAL DISTRIBUTION

In Section 9.1 we constructed a histogram for the weights of bags of corn chips filled by one specific machine; that histogram is shown in Figure 9.56. Notice that most of the data points are at the center of the distribution (most bags weighed close to 16 ounces), few data points are far from the center, and the histogram is nearly symmetric (the left and right sides are almost identical).

FIGURE 9.56
Weights of bags of corn chips

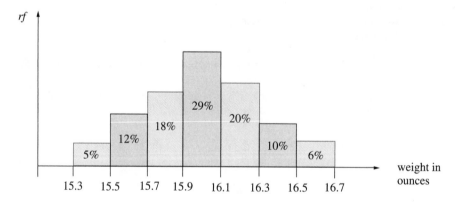

If twice as many groups had been used, the histogram would probably look like that in Figure 9.57. If even more groups had been used, the histogram would resemble the **bell-shaped curve** in Figure 9.58. Such a curve is said to describe a **normal distribution,** which is a symmetric distribution in which the mean μ is at the center, most data points are clus-

FIGURE 9.57
Weights of bags of corn
chips, with more groups

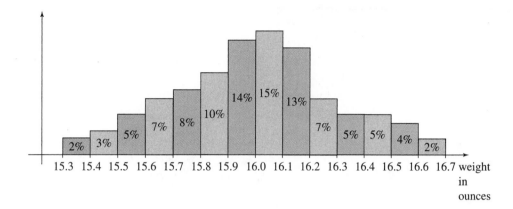

FIGURE 9.58
A normal distribution's bell-
shaped curve

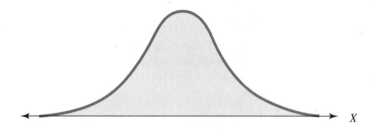

tered at the center, and few data points are far from the center. (This is an informal definition of a normal distribution. The normal definition involves the numbers e and π, as well
as the mean and standard deviation of the distribution.)

With a histogram, the relative frequency of any one group is given by the area of the
rectangle over that group's interval. Similarly, with a normal distribution, the relative frequency of any particular interval is given by the area under that distribution's bell-shaped
curve over that particular interval. Normal distributions are very common and have been
studied so thoroughly that relative frequencies can be found for *any* interval, as long as the
mean and standard deviation are known. This is not the case with frequency distributions,
where relative frequencies can be found only for intervals corresponding to the previously
selected groups. For example, in Figure 9.56 we can easily see that the relative frequency
of the interval from 15.9 to 16.1 is 29%; however, we cannot determine the relative frequency of the interval from 16.0 to 16.23.

Usually, the word *probability* is used in the context of a normally distributed quantity, while the phrase *relative frequency* is used in the context of a frequency distribution. This is done to emphasize the fact that with a frequency distribution, a group's
relative frequency is the direct result of counting the number of data points in that group
(i.e., finding the group's frequency), while with a normal distribution, an interval's
probability is computed on a theoretical level and is not the direct result of counting
anything.

Carl Friedrich Gauss

1777–1855

Dubbed the Prince of Mathematics, Carl Gauss is considered by many to be one of the greatest mathematicians of all time. At the age of 3, Gauss is said to have discovered an arithmetic error in his

father's bookkeeping. The child prodigy was encouraged by his teachers and excelled throughout his early schooling. When he was 14, Gauss was introduced to Ferdinand, the Duke of Brunswick. Impressed with the youth, the duke gave Gauss a yearly stipend and sponsored his education for many years.

In 1795 Gauss enrolled at Göttingen University, where he remained for three years. While at Göttingen, Gauss had complete academic freedom; he was not required to attend lectures, he had no required conferences with professors or tutors, and he did not take exams. Much of his time was spent studying

independently in the library. For reasons unknown to us, Gauss left the university in 1798 without a diploma. Instead, he sent his dissertation to the University of Helmstedt and in 1799 was awarded his degree without the usual oral examination.

In 1796 Gauss began his famous mathematical diary. Discovered 40 years after his death, the 146 sometimes cryptic entries exhibit the diverse range of topics that Gauss pondered and pioneered. The first entry was Gauss's discovery (at the age of 19) of a method for constructing a 17-sided polygon with a compass and a straightedge. Other entries include important results in

---

It is known that for *any* normally distributed quantity, slightly more than two-thirds of the data points (68.26%) lie within one standard deviation of the mean, 95.44% lie within two standard deviations of the mean, and virtually all (99.74%) lie within three standard deviations of the mean. This is shown in Figure 9.59, where μ is used to represent the distribution's mean, and σ is used to represent its standard deviation.

EXAMPLE 1 The heights of a large group of people are known to be normally distributed. Group members' mean height is $\mu = 66.5$ inches, and the standard deviation is $\sigma = 2.4$ inches. (Here, μ and σ are used rather than $\bar{x}$ and s, because the group of people is the population; it is not a sample used to approximate some larger population.) Find the intervals corresponding to one, two, and three standard deviations from the mean, and find the probability that a person in that group is in each of those intervals.

Solution The phrase "one standard deviation from the mean" refers to all heights within the interval $(\mu - \sigma, \mu + \sigma) = (66.5 - 2.4, 66.5 + 2.4) = (64.1, 68.9)$. For any normal distribution, 68.26% of the data lies within one standard deviation of the mean, so 68.26% of the group is between 64.1 and 68.9 inches tall. This means that the probability that a group member is between 64.1 and 68.9 inches tall is 68.26%; that is, $p(64.1 < x < 68.9) = 68.26\%$, where x measures group members' heights.

number theory, algebra, calculus and analysis, astronomy, electricity, and magnetism, the foundations of geometry, and probability.

At the dawn of the 19th century, Gauss began his lifelong study of astronomy. On January 1, 1801, the Italian astronomer Giuseppe Piazzi discovered Ceres, the first of the known planetoids (minor planets or asteroids). Piazzi and others observed Ceres for 41 days, until it was lost behind the sun. Because of his interest in the mathematics of astronomy, Gauss turned his attention to Ceres. Working with a minimum amount of data, he successfully calculated the orbit of Ceres. At the end of the year, the planetoid was rediscovered in exactly the spot that Gauss had predicted!

To obtain the orbit of Ceres, Gauss utilized his method of least squares, a technique for dealing with experimental error closely related to the line of best fit (or least-squares line) discussed in Section 1.3. Letting x represent the error between an experimentally obtained value and the true value it represents, Gauss's theory involved minimizing x^2—that is, obtaining the least square of the error. Theorizing that the probability of a small error was higher than that of a large error, Gauss subsequently developed the normal distribution, or bell-shaped curve, to explain the probabilities of the random errors. Because of his pioneering efforts, some mathematicians refer to the normal distribution as the Gaussian distribution.

In 1807, Gauss became director of the newly constructed observatory at Göttingen; he held the position until his death some 50 years later.

> **THEORIA**
> **MOTVS CORPORVM**
> **COELESTIVM**
>
> IN
>
> SECTIONIBVS CONICIS SOLEM AMBIENTIVM
>
> AVCTORE
>
> CAROLO FRIDERICO GAVSS
>
> HAMBVRGI SVMTIBVS FRID. PERTHES ET I. H. BESSER
> 1809.
>
> Venditur

Published in 1809, Gauss's *Theoria Motus Corporum Coelestium* (*Theory of the Motion of Heavenly Bodies*) contained rigorous methods of determining the orbits of planets and comets from observational data via the method of least squares. It is a landmark in the development of modern mathematical astronomy and statistics.

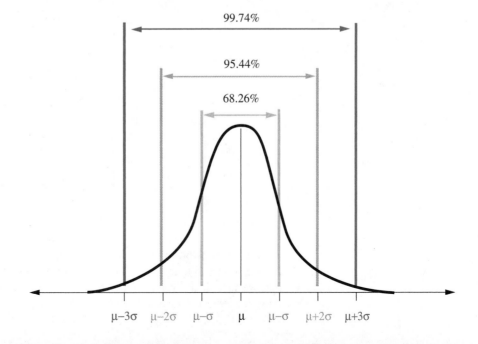

FIGURE 9.59
The spread of a normal distribution

Similarly, "two standard deviations from the mean" refers to all heights within the interval $(\mu - 2\sigma, \mu + 2\sigma) = (66.5 - 2 \cdot 2.4, 66.5 + 2 \cdot 2.4) = (61.7, 71.3)$. For any normal distribution, 95.44% of the data lies within two standard deviations of the mean, so $p(61.7 < x < 71.3) = 95.44\%$.

Finally, "three standard deviations from the mean" refers to all heights within the interval $(\mu - 3\sigma, \mu + 3\sigma) = (66.5 - 3 \cdot 2.4, 66.5 + 3 \cdot 2.4) = (59.3, 73.7)$. For any normal distribution, 99.74% of the data lies within three standard deviations of the mean, so $p(59.3 < x < 73.7) = 99.74\%$. ●

To find probabilities involving data that are normally distributed, we must find the area of the appropriate region under the bell-shaped curve. To find $p(a < x < b)$, the probability that x is between a and b, we must determine the area under the curve between a and b, as shown in Figure 9.60. To find $p(x > b)$, we must determine the area under the curve to the right of b. To find $p(x < a)$, we must determine the area under the curve to the left of a, as shown in Figure 9.61.

In Chapter 6, we found that the probability of the sample space is 1; for this reason, the total area under the curve is 1 (or 100%). The bell-shaped curve is symmetric, with the mean μ at the center, so the area on each side of μ is 0.5 (or 50%).

FIGURE 9.60

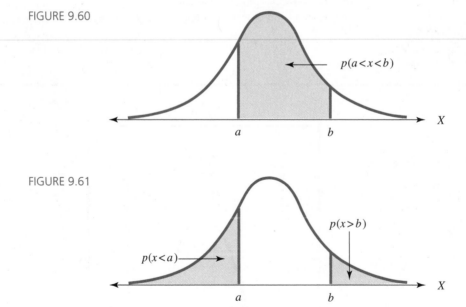

FIGURE 9.61

The Standard Normal Distribution

The **standard normal distribution** is the normal distribution whose mean is 0 and standard deviation is 1, as shown in Figure 9.62. The standard normal distribution is also called the **z-distribution;** we will always use the letter z to refer to the standard normal. By convention, we will use the letter x to refer to any other normal distribution.

FIGURE 9.62
The standard normal distribution (mean = 0; standard
deviation = 1)

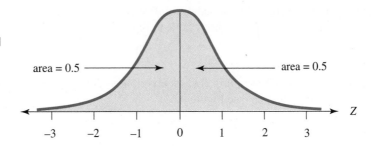

Tables have been developed for finding areas under the standard normal curve. A few statistical calculators will also give these areas. We will use the table in Appendix II to find $p(0 < z < c)$, the probability that z is between 0 and a positive number c, as shown in Figure 9.63. The table in Appendix II is known as the **body table** because it gives the probability of an interval located in the middle, or *body*, of the bell curve.

FIGURE 9.63

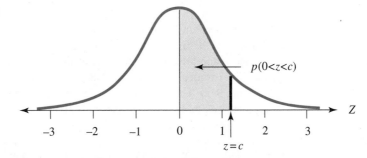

Area found by using the body table (Appendix II)

The tapered end of a bell curve is known as a *tail.* To find the probability of a tail—that is, to find $p(z > c)$ or $p(z < -c)$ where c is a positive real number—subtract the probability of the corresponding body from 0.5, as shown in Figure 9.64.

FIGURE 9.64

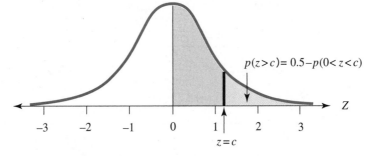

Area of a tail, found by subtracting the corresponding body area from 0.5

EXAMPLE 2 Find the following probabilities (that is, the areas), where z represents the standard normal distribution. Use the body table in Figure 9.66.

a. $p(0 < z < 1.25)$ b. $p(z > 1.25)$

Solution a. As a first step, it is always advisable to draw a picture of the z-curve and shade in the desired area (see Figure 9.65).

FIGURE 9.65

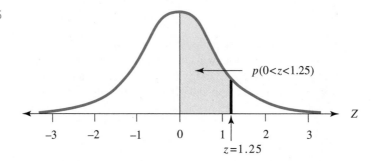

The z-numbers are located along the left edge and the top of the table. Locate the whole number and the first-decimal-place part of the number (1.2) along the left edge; then locate the second-decimal-place part of the number (0.05) along the top. The desired probability (or area) is in the row labeled "1.2" and the column labeled "0.05," as shown in Figure 9.66. Thus, $p(0 < z < 1.25) = 0.3944$. (Figure 9.66 is a portion of the more complete body table given in Appendix II.)

FIGURE 9.66

z	0.00	0.01	0.02	0.03	0.04	0.05	0.06	0.07	0.08	0.09
°										
°										
°										
1.1	0.3643	0.3665	0.3686	0.3708	0.3729	0.3749	0.3770	0.3790	0.3810	0.3830
1.2	0.3849	0.3869	0.3888	0.3907	0.3925	0.3944	0.3962	0.3980	0.3997	0.4015
1.3	0.4032	0.4049	0.4066	0.4082	0.4099	0.4115	0.4131	0.4147	0.4162	0.4177
°										
°										
°										

b. To find the area of a tail, we subtract the corresponding body area from 0.5, as shown in Figure 9.67. Therefore,

FIGURE 9.67

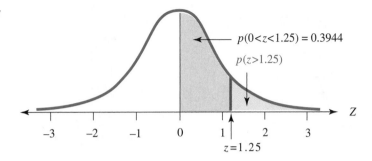

$$p(z > 1.25) = 0.5 - p(0 < z < 1.25)$$
$$= 0.5 - 0.3944$$
$$= 0.1056$$

The body table can also be used to find areas other than those given explicitly as $p(0 < z < c)$ and $p(z > c)$, where c is a positive number. By adding or subtracting two areas, we can find probabilities of the type $p(a < z < b)$, where a and b are positive or negative numbers, and probabilities of the type $p(z < c)$, where c is a positive or negative number.

EXAMPLE 3   Find the following probabilities (or areas), where z represents the standard normal distribution.

a. $p(0.75 < z < 1.25)$ **b.** $p(-0.75 < z < 1.25)$

Solution   **a.** Because the required region, shown in Figure 9.68, doesn't begin at $z = 0$, we cannot look up the desired area directly in the body table. Whenever z is between two nonzero numbers, we will take an indirect approach to finding the required area.

FIGURE 9.68

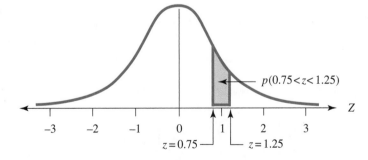

The total area under the curve from $z = 0$ to $z = 1.25$ can be divided into two portions: the area under the curve from 0 to 0.75 and the area under the curve from 0.75 to 1.25.

To find the area of the "strip" between $z = 0.75$ and $z = 1.25$, we *subtract* the area of the smaller body (from $z = 0$ to $z = 0.75$) from that of the larger body (from $z = 0$ to $z = 1.25$), as shown in Figure 9.69 on page 448.

FIGURE 9.69
Area of a strip

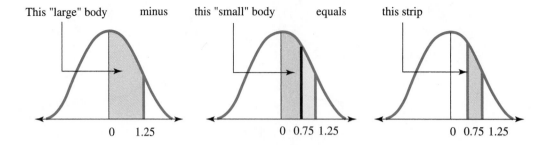

$$\text{area of strip} = \text{area of large body} - \text{area of small body}$$
$$p(0.75 < z < 1.25) = p(0 < z < 1.25) - p(0 < z < 0.75)$$
$$= 0.3944 - 0.2734 \qquad \text{From Appendix II}$$
$$= 0.1210$$

Therefore, $p(0.75 < z < 1.25) = 0.1210$. Hence, we could say that about 12% of the z-distribution lies between $z = 0.75$ and $z = 1.25$.

b. The required region, shown in Figure 9.70, can be divided into two regions: the area from $z = -0.75$ to $z = 0$ and the area from $z = 0$ to $z = 1.25$.

FIGURE 9.70

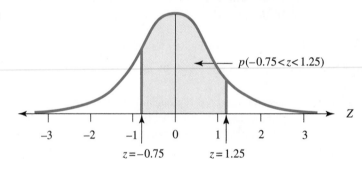

To find the total area of the region between $z = -0.75$ and $z = 1.25$, we *add* the area of the "left" body (from $z = -0.75$ to $z = 0$) to the area of the "right" body (from $z = 0$ to $z = 1.25$), as shown in Figure 9.71.

FIGURE 9.71

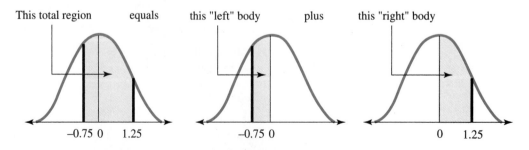

This example is different from our previous examples in that it contains a negative z-number. A glance at the tables reveals that negative numbers are not included! However, recall that normal distributions are symmetric. Therefore, the area of the body from $z = -0.75$ to $z = 0$ is the same as that from $z = 0$ to $z = 0.75$; that is, $p(-0.75 < z < 0) = p(0 < z < 0.75)$. Thus,

total area of region = area of left body + area of right body

$$p(-0.75 < z < 1.25) = p(-0.75 < z < 0) + p(0 < z < 1.25)$$
$$= p(0 < z < 0.75) + p(0 < z < 1.25)$$
$$= 0.2734 + 0.3944 \qquad \text{From Appendix II}$$
$$= 0.6678$$

Hence, we could say that about 67% of the z-distribution lies between $z = -0.75$ and $z = 1.25$. •

EXAMPLE 4   Find the following probabilities (the areas), where z represents the standard normal distribution.

    a. $p(z < 1.25)$ **b.** $p(z < -1.25)$

Solution   **a.** The required region is shown in Figure 9.72. Because 50% of the distribution lies to the left of 0, we can add 0.5 to the area of the body from $z = 0$ to $z = 1.25$:

FIGURE 9.72

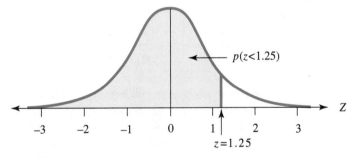

$$p(z < 1.25) = p(z < 0) + p(0 < z < 1.25)$$
$$= 0.5 + 0.3944$$
$$= 0.8944$$

Therefore, $p(z < 1.25) = 0.8944$. Hence, we could say that about 89% of the z-distribution lies to the left of $z = 1.25$.

    b. The required region is shown in Figure 9.73. Because a normal distribution is symmetric, the area of the left tail ($z < -1.25$) is the same as the area of the corresponding right tail ($z > 1.25$).

FIGURE 9.73

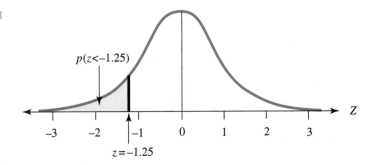

Therefore,

$$p(z < -1.25) = p(z > 1.25)$$
$$= 0.5 - p(0 < z < 1.25)$$
$$= 0.5 - 0.3944$$
$$= 0.1056$$

Hence, we could say that about 11% of the z-distribution lies to the left of $z = -1.25$.   •

Converting a Normal Distribution to the Standard Normal Distribution

U.S. residents measure temperatures in degrees Fahrenheit, even though the world standard is degrees centigrade. When traveling in another country, a U.S. resident would have to convert temperatures from centigrade to Fahrenheit in order to judge how hot or cold it is. A similar situation arises when we are working with a normal distribution. Suppose we know that a large set of data is normally distributed with a mean value of 68 and a standard deviation of 4. What percent of the data lies between 68 and 74? In other words, what is $p(68 < x < 74)$? To find this probability we must first convert the normal distribution to the standard normal distribution and then look up the appropriate z-numbers.

Notice that 74 is $74 - 68 = 6$ units away from the mean value of 68. Furthermore, this distance of 6 units is 1.5 standard deviations, because $6/4 = 1.5$ (or $1.5 \cdot 4 = 6$). Thus, $x = 74$ corresponds to $z = 1.5$ (each is 1.5 standard deviations to the right of the mean), and $x = 68$ corresponds to $z = 0$ (each is its own distribution's mean). And the question, "What is $p(68 < x < 74)$?" corresponds to the question, "What is $p(0 < z < 1.5)$?" According to Appendix II, the answer to either question is 0.4332.

We can think of a z-number as measuring the number of standard deviations from the mean. In the above paragraph we determined that $x = 74$ corresponds to $z = 1.5$ by subtracting the mean $\mu = 68$ from $x = 74$ and dividing the result by the standard deviation $\sigma = 4$. More generally, to find the z-number corresponding to a given x-number, first subtract the mean μ from the x-number and obtain $x - \mu$. Then divide the result by σ and obtain $\frac{x - \mu}{\sigma}$. Thus, $z = \frac{x - \mu}{\sigma}$.

Converting a Normal Distribution into the Standard Normal z
If a variable x is normally distributed, then every value of x has a corresponding z-number. The z-number that corresponds to a specific value of x is $$z = \frac{x - \mu}{\sigma}$$ where μ is the mean and σ is the standard deviation of the variable x.

EXAMPLE 5   The heights of members of a large group of people are normally distributed. Their mean height is 68 inches, and the standard deviation is 4 inches. What percentage of these people are the following heights?

a. taller than 73 inches
b. between 60 and 75 inches

Solution a. Let x represent the height of a randomly selected person. We need to find $p(x > 73)$, the area of a tail, as shown in Figure 9.74.

FIGURE 9.74

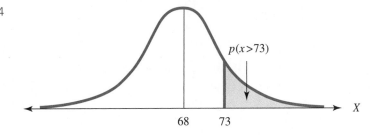

First, we must convert $x = 73$ to its corresponding z-number.

$$z = \frac{x - \mu}{\sigma}$$

$$= \frac{73 - 68}{4}$$

$$= 1.25$$

Therefore,

$$p(x > 73) = p(z > 1.25)$$
$$= 0.5 - p(0 < z < 1.25)$$
$$= 0.5 - 0.3944 \quad \text{From Appendix II}$$
$$= 0.1056$$

Approximately 11% of the people will be taller than 73 inches.

b. We need to find $p(60 < x < 75)$, the area of the central region shown in Figure 9.75. Notice that we will be adding the areas of the two bodies.

FIGURE 9.75

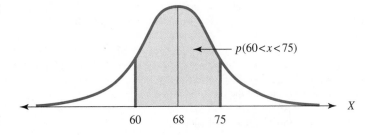

First, we convert $x = 60$ and $x = 75$ to their corresponding z-numbers:

$$p(60 < x < 75) = p\left(\frac{60 - 68}{4} < z < \frac{75 - 68}{4}\right)$$ Using the conversion formula

$$= p(-2.00 < z < 1.75)$$

$$= p(-2.00 < z < 0) + p(0 < z < 1.75)$$ Expressing the area as two bodies

$$= p(0 < z < 2.00) + p(0 < z < 1.75)$$ Using symmetry

$$= 0.4772 + 0.4599$$ From Appendix II

$$= 0.9371$$

Approximately 94% of the people will be between 60 and 75 inches tall.   •

9.4

EXERCISES

1. What percent of the standard normal z-distribution lies between the following values?
 a. $z = 0$ and $z = 1$ **b.** $z = -1$ and $z = 0$
 c. $z = -1$ and $z = 1$ NOTE: This interval represents one standard deviation of the mean.

2. What percent of the standard normal z-distribution lies between the following values?
 a. $z = 0$ and $z = 2$ **b.** $z = -2$ and $z = 0$
 c. $z = -2$ and $z = 2$ NOTE: This interval represents two standard deviations of the mean.

3. What percent of the standard normal z-distribution lies between the following values?
 a. $z = 0$ and $z = 3$ **b.** $z = -3$ and $z = 0$
 c. $z = -3$ and $z = 3$ NOTE:This interval represents three standard deviations of the mean.

4. What percent of the standard normal z-distribution lies between the following values?
 a. $z = 0$ and $z = 1.5$ **b.** $z = -1.5$ and $z = 0$
 c. $z = -1.5$ and $z = 1.5$ NOTE: This interval represents one and one-half standard deviations of the mean.

5. A population is normally distributed with mean 24.7 and standard deviation 2.3.
 a. Find the intervals representing one, two, and three standard deviations of the mean.
 b. What percentage of the data lies in each of the intervals in (a)?
 c. Draw a sketch of the bell curve, showing the given data as well as the solutions to parts (a) and (b).

6. A population is normally distributed with mean 18.9 and standard deviation 1.8.

a. Find the intervals representing one, two, and three standard deviations of the mean.
b. What percentage of the data lies in each of the intervals in (a)?
c. Draw a sketch of the bell curve, showing the given data as well as the solutions to parts (a) and (b).

7. Find the following probabilities.
 a. $p(0 < z < 1.62)$ **b.** $p(1.30 < z < 1.84)$
 c. $p(-0.37 < z < 1.59)$ **d.** $p(z < -1.91)$
 e. $p(-1.32 < z < -0.88)$ **f.** $p(z < 1.25)$

8. Find the following probabilities.
 a. $p(0 < z < 1.42)$ **b.** $p(1.03 < z < 1.66)$
 c. $p(-0.87 < z < 1.71)$ **d.** $p(z < -2.06)$
 e. $p(-2.31 < z < -1.18)$ **f.** $p(z < 1.52)$

9. Find c such that each of the following is true.
 a. $p(0 < z < c) = 0.1331$ **b.** $p(c < z < 0) = 0.4812$
 c. $p(-c < z < c) = 0.4648$ **d.** $p(z > c) = 0.6064$
 e. $p(z > c) = 0.0505$ **f.** $p(z < c) = 0.1003$

10. Find c such that each of the following is true.
 a. $p(0 < z < c) = 0.3686$ **b.** $p(c < z < 0) = 0.4706$
 c. $p(-c < z < c) = 0.2510$ **d.** $p(z > c) = 0.7054$
 e. $p(z > c) = 0.0351$ **f.** $p(z < c) = 0.2776$

11. A population is normally distributed with mean 36.8 and standard deviation 2.5. Find the following probabilities.
 a. $p(36.8 < x < 39.3)$ **b.** $p(34.2 < x < 38.7)$
 c. $p(x < 40.0)$ **d.** $p(32.3 < x < 41.3)$
 e. $p(x = 37.9)$ **f.** $p(x > 37.9)$

12. A population is normally distributed with mean 42.7 and standard deviation 4.7. Find the following probabilities.

a. $p(42.7 < x < 47.4)$ **b.** $p(40.9 < x < 44.1)$
c. $p(x < 50.0)$ **d.** $p(33.3 < x < 52.1)$
e. $p(x = 45.3)$ **f.** $p(x > 45.3)$

13. The mean weight of a box of cereal filled by a machine is 16.0 ounces, with a standard deviation of 0.3 ounce. If the weights of all the boxes filled by the machine are normally distributed, what percent of the boxes will weigh the following amounts?
    a. less than 15.5 ounces
    b. between 15.8 and 16.2 ounces

14. The amount of time required to assemble a component on a factory assembly line is normally distributed with a mean of 3.1 minutes and a standard deviation of 0.6 minute. Find the probability that a randomly selected employee will take the given amount of time to assemble the component.
    a. more than 4.0 minutes
    b. between 2.0 and 2.5 minutes

15. The time it takes an acrylic paint to dry is normally distributed. If the mean is 2 hours 36 minutes with a standard deviation of 24 minutes, find the probability that the drying time will be as follows.
    a. less than 2 hours 15 minutes
    b. between 2 and 3 hours
    HINT: Convert everything to minutes (or to hours).

16. The shrinkage in length of a certain brand of blue jeans is normally distributed with a mean of 1.1 inch and a standard deviation of 0.2 inch. What percent of this brand of jeans will shrink the following amounts?
    a. more than 1.5 inches
    b. between 1.0 and 1.25 inches

17. The mean volume of a carton of milk filled by a machine is 1.0 quart, with a standard deviation of 0.06 quart. If the volumes of all the cartons are normally distributed, what percent of the cartons will contain the following amounts?
    a. at least 0.9 quart
    b. at most 1.05 quarts

18. The amount of time between taking a pain reliever and getting relief is normally distributed with a mean of 23 minutes and a standard deviation of 4 minutes. Find the probability that the time between taking the medication and getting relief is as follows.
    a. at least 30 minutes
    b. at most 20 minutes

19. The results of a statewide exam for assessing the mathematics skills of realtors were normally distributed with a mean score of 72 and a standard deviation of 12. The realtors who scored in the top 10% are to receive a special certificate, while those in the bottom 20% will be required to attend a remedial workshop.

a. What score does a realtor need in order to receive a certificate?
b. What score will dictate that the realtor attend the workshop?

20. Professor Harde assumes that exam scores are normally distributed and wants to grade "on the curve." The mean score was 58, with a standard deviation of 16.
    a. If she wants 14% of the students to receive an A, find the minimum score to receive an A.
    b. If she wants 19% of the students to receive a B, find the minimum score to receive a B.

21. The time it takes an employee to package the components of a certain product is normally distributed with $\mu = 8.5$ minutes and $\sigma = 1.5$ minutes. In order to boost productivity, management has decided to give special training to the 34% of employees who took the greatest amount of time to package the components. Find the amount of time taken to package the components that will indicate that an employee should get special training.

22. The time it takes an employee to package the components of a certain product is normally distributed with $\mu = 8.5$ and $\sigma = 1.5$ minutes. As an incentive, management has decided to give a bonus to the 20% of employees who took the shortest amount of time to package the components. Find the amount of time taken to package the components that will indicate that an employee should get a bonus.

23. The probability $p(0 < z < 1)$ refers to the area under a bell-shaped curve over the interval $0 < z < 1$.
    a. Give a similar geometric description of the probability $p(z = 1)$.
    b. Use part (a) to determine $p(z = 1)$.
    c. Use part (b) to determine why $p(0 < z < 1) = p(0 < z \leq 1)$.
    d. How does $p(0 < z < 1)$ compare with $p(0 \leq z \leq 1)$?

✏️ Answer the following questions using complete sentences.

24. Compare and contrast the distinction made in this section between *relative frequency* and *probability* with the distinction made in Chapter 6.

25. What are the characteristics of a normal distribution?

26. Are all distributions of data normally distributed? Support your answer with an example.

27. Why is the total area under a bell curve equal to 1?

28. Why are there no negative z-numbers in the body table?

29. When converting an x-number to a z-number, what does a negative z-number tell you about the location of the x-number?

30. Is it logical to assume that the heights of all high school students in the United States are normally distributed? Explain.

31. Is it reasonable to assume that the ages of all high school students in the United States are normally distributed? Explain.

32. Who invented the normal distribution? What did he use it for?

9.5

BINOMIAL EXPERIMENTS

Some candidates for political office consider it desirable to be listed first on the ballot, reasoning that a voter who knows nothing about any of the candidates is inclined to choose the first name he or she sees. Frequently, a random drawing determines who gets the first listing so that the ballot's format does not give either political party a long-term advantage. Ostensibly using this method, the county clerk of Essex County, New Jersey, awarded the ballot's first listing to Democrats 40 out of 41 times. The consistency with which the Democrats won the drawing, plus the fact that the clerk was a Democrat, made some question the randomness of the drawing. (*Source:* "Essex Democrats Stay Lucky," *The New York Times,* Aug. 13, 1985).

The use of a random drawing to determine who is first on the ballot is an example of a **binomial experiment.**

Binomial Experiment Definition
A **binomial experiment** is an experiment that satisfies the following conditions.   • The outcome can be classified as either "success" or "failure."   • The experiment is repeated several times (each repetition is called a **trial,** or a **Bernoulli trial**).   • The trials' outcomes are independent, so that the probability of success is the same for each trial.

Each of these conditions is met in a random drawing. The outcomes of the random drawing can be classified as success or failure. If the top line is assigned to a Democrat, we can say that the experiment's outcome is "success," and if it is assigned to a Republican, we can say that the outcome is "failure." (When used in this context, "success" and "failure" are not meant to connote "good" and "bad"; they are merely traditional labels for the two types of outcomes in a binomial experiment.) Furthermore, the county clerk's experiment was repeated several times—41 times to be exact. And, finally, if the drawing was truly random, the experiment's outcome in any given year would not be affected by the outcome in a previous year, so the repetitions are independent.

Binomial Probabilities

To determine if the Essex County clerk's drawing was truly random, it would be helpful to know the probability that a random drawing would result in listing a Democrat first 40 out of 41 times. Before discussing that probability, we'll discuss the probability of a similar but simpler experiment.

EXAMPLE 1 A fair die is rolled three times in a row. Getting a 1 is a success, and getting any other outcome is a failure.

a. Why is this a binomial experiment?
b. Find the probability of having 1 success followed by 2 failures.
c. Find the probability of having 1 success and 2 failures, in any order.

Solution   a. We are told that the experiment is repeated, and the outcomes are already categorized as success and failure. In Section 6.5 we determined that the outcomes of repeatedly rolling a die are independent. Thus, the described experiment is a binomial experiment.

b. To find the number of elements in the sample space, use the fundamental principle of counting. Draw three boxes, one for each roll of the die. The first roll can be performed in any of 6 different ways, as can the second and third rolls. Thus, there are

$$\boxed{6} \cdot \boxed{6} \cdot \boxed{6}$$

elements in the sample space.

   To find the number of elements in the event, draw the same 3 boxes. The first roll, a success, could be performed in only 1 way. The second roll, a failure, could be performed in any of 5 ways, as could the third roll. Thus, there are

$$\boxed{1} \cdot \boxed{5} \cdot \boxed{5}$$
$$\underset{\text{success failure failure}}{\uparrow \qquad \uparrow \qquad \uparrow}$$

elements in the event.

   The probability of having 1 success followed by 2 failures is

$$\frac{n(E)}{n(S)} = \frac{1 \cdot 5 \cdot 5}{6 \cdot 6 \cdot 6} = \frac{25}{216} = 0.1157 \ldots \approx 0.12.$$

c. The one success can now occur at any of the three rolls. Regardless of when it occurs, the probability is the same, as shown in Figure 9.76.

FIGURE 9.76

Sequence of Trials	Probability of That Sequence
Success, failure, failure	$\dfrac{1 \cdot 5 \cdot 5}{6 \cdot 6 \cdot 6}$
Failure, success, failure	$\dfrac{5 \cdot 1 \cdot 5}{6 \cdot 6 \cdot 6}$
Failure, failure, success	$\dfrac{5 \cdot 5 \cdot 1}{6 \cdot 6 \cdot 6}$

Thus, the probability of having 1 success and 2 failures, in any order, is

$$\frac{1 \cdot 5 \cdot 5}{6 \cdot 6 \cdot 6} + \frac{5 \cdot 1 \cdot 5}{6 \cdot 6 \cdot 6} + \frac{5 \cdot 5 \cdot 1}{6 \cdot 6 \cdot 6} = 3 \cdot \frac{1 \cdot 5 \cdot 5}{6 \cdot 6 \cdot 6} = 0.3472 \ldots \approx 0.35$$

   In Example 1, success is "getting a 1," and the probability of success is $\frac{1}{6}$. Failure is "getting a 2, 3, 4, 5, or 6," and the probability of failure is $\frac{5}{6}$. There are 3 different sequences of trials that contain 1 success and 2 failures. If we rewrite the solution to Example 1(c) as

$$3 \cdot \frac{1 \cdot 5 \cdot 5}{6 \cdot 6 \cdot 6} = 3 \cdot \left(\frac{1}{6}\right)^1 \cdot \left(\frac{5}{6}\right)^2$$

then it's easy to see the role of these numbers.

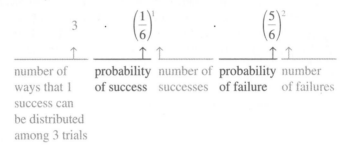

3	$\cdot$	$\left(\frac{1}{6}\right)^1$	$\cdot$	$\left(\frac{5}{6}\right)^2$	
↑	↑	↑	↑	↑	
number of ways that 1 success can be distributed among 3 trials	probability of success	number of successes	probability of failure	number of failures	

EXAMPLE 2   Find the probability that a truly random drawing would result in listing a Democrat first 40 out of 41 times.

Solution   We will use the "Which Counting Technique" flowchart from Section 5.4 to determine the number of ways that 40 successes can be distributed among 41 trials.

Q: What is being selected?
A: Forty slots to be labeled as "success," out of 41 slots in a row.
Q: Can the selected items (slots labeled "success") be repeated?
A: No, once a slot is labeled "success," it can't be reselected and relabeled.
Q: Is there more than one category (of slots)?
A: No, a slot is a slot.
Q: Does the order of selection (of slots) matter?
A: No, it doesn't matter when we label a particular slot as "success."

Thus, we use combinations, and the number of ways that 40 successes can be distributed among 41 trials is

$$_{41}C_{40}$$

If the drawing were truly random, then the probability of success and the probability of failure would each be $\frac{1}{2}$.

The probability of having 40 successes and 1 failure, in any order, is

$_{41}C_{40}$	$\cdot$	$\left(\frac{1}{2}\right)^{40}$	$\cdot$	$\left(\frac{1}{2}\right)^1$
↑	↑	↑	↑	↑
number of ways that 40 successes can be distributed among 41 trials	probability of success	number of successes	probability of failure	number of failures

$$_{41}C_{40} \cdot \left(\frac{1}{2}\right)^{40} \cdot \left(\frac{1}{2}\right)^{1} = \frac{41!}{1! \cdot 40!} \cdot \left(\frac{1}{2}\right)^{41}$$

$$= \frac{41}{2^{41}}$$

$$= 1.8644 \ldots \times 10^{-11}$$

$$\approx 0.00000000002$$

This is an extremely small probability. The New Jersey Supreme Court noted that there is almost no chance that the Democrats would have won a truly random drawing so consistently and recommended that a new procedure be used. •

Generalizing from Examples 1 and 2, we obtain the following formula.

Binomial Probability Formula

If p is the probability of success in a single trial of a binomial experiment and q is the probability of failure, then the probability of x successes in a sequence of n independent trials is

$$P(x \text{ successes}) = {}_{n}C_{x}p^{x}q^{n-x}$$

EXAMPLE 3 Notel makes computer chips. Its production process has a 1% defect rate; that is, 1% of its chips are defective. Furthermore, these defects are known to occur at random. Find the probability that a shipment of 2000 includes

a. no defective chips
b. one defective chip
c. less than two defective chips

Solution If we consider success to be "a defective chip," then the probability of success is $p = 1\% = 0.01$, and the probability of failure is $q = 1 - p = 0.99$. There are $n = 2000$ independent trials.

a. We are to find the probability of $x = 0$ successes. That probability is

$$P(x \text{ successes}) = {}_{n}C_{x} \quad p^{x} \quad q^{n-x}$$
$$P(x = 0) = ({}_{2000}C_{0}) \ (0.01)^{0} \ (0.99)^{(2000-0)}$$

$$= \frac{2000!}{2000! \cdot 0!} \cdot 1 \ \cdot \ 0.99^{2000}$$

$$= 1 \cdot 0.99^{2000}$$

$$= 1.8637 \ldots \times 10^{-9}$$

$$\approx 0.000000002$$

b. We are to find the probability of $x = 1$ success. That probability is

$$P(x = 1) = _{2000}C_1 \quad (0.01)^1 (0.99)^{(2000 - 1)}$$

$$= \frac{2000!}{1999! \cdot 1!} \cdot 0.01^1 \cdot 0.99^{1999}$$

$$= 2000 \cdot 0.01 \cdot 0.99^{1999}$$

$$= 3.7651 \ldots \times 10^{-8}$$

$$\approx 0.00000004$$

c. We are to find the probability of $x < 2$ successes. That probability is

$$P(x < 2) = P(x = 0) + P(x = 1)$$

$$= 1.8637 \ldots \times 10^{-9} + 3.7651 \ldots \times 10^{-8}$$

$$= 3.9515 \ldots \times 10^{-8}$$

$$\approx 0.00000004$$

Binomial Distributions and Expected Value

In Example 3, we found that it is extremely unlikely that Notel's shipment of 2000 chips would include no defective chips, 1 defective chip, or less than 2 defective chips. How many defective chips should Notel expect to have in that shipment? Common sense implies that if 1% of Notel's product is defective, then Notel should expect 1% of a shipment of 2000 chips to be defective. In order to see if our intuition is correct, we need to compute the expected value of the number of defective chips.

We'll start by discussing the expected value of a similar but simpler experiment. As always, calculating an expected value involves creating a probability distribution. A probability distribution for a binomial experiment is called a **binomial distribution.**

EXAMPLE 4　A fair coin is tossed four times.

a. Why is this a binomial experiment?
b. Create a probability distribution for the experiment, and use the probability distribution to find the most likely number of heads.
c. Use the probability distribution to find the expected number of heads.
d. Create a histogram for the experiment.

Solution　a. We can categorize heads as success and tails as failure. The experiment is repeated. The outcomes of these repetitions are independent, because

p(heads on the 2nd toss | heads on the 1st toss)

$= p$(heads on the 2nd toss).

Thus, the described experiment is a binomial experiment.

b. The probability of success is

$$p = p(\text{heads}) = \frac{1}{2}$$

Similarly, the probability of failure is $q = \frac{1}{2}$. The probability of no heads in four tosses is

$$P(x = 0) = (_4C_0)(0.5)^0(0.5)^4$$
$$= \frac{4!}{4! \cdot 0!} \cdot 1 \cdot 0.5^4$$
$$= 1 \cdot 0.5^4$$
$$= 0.0625$$

Similar calculations yield the probability distribution in Figure 9.77.

FIGURE 9.77

Number of Heads	Probability
$x = 0$	$p(x = 0) = (_4C_0)(0.5)^0(0.5)^4 = 0.0625$
$x = 1$	$p(x = 1) = (_4C_1)(0.5)^1(0.5)^3 = 0.25$
$x = 2$	$p(x = 2) = (_4C_2)(0.5)^2(0.5)^2 = 0.375$
$x = 3$	$p(x = 3) = (_4C_3)(0.5)^3(0.5)^1 = 0.25$
$x = 4$	$p(x = 4) = (_4C_4)(0.5)^4(0.5)^0 = 0.0625$

The most likely number of heads is $x = 2$, because its probability is higher than all of the other outcomes' probabilities.

c. To find the expected number of heads, multiply the value of each outcome of the experiment by its probability and add the results. Therefore,

   expected number of heads
   $$= (0)(0.0625) + (1)(0.25) + (2)(0.375) + (3)(0.25) + (4)(0.0625)$$
   $$= 2$$

d. The histogram is shown in Figure 9.78. The first bar corresponds to $x = 0$; its height is 0.0625, since $p(x = 0) = 0.0625$, and its base goes from $x = -0.5$ to $x = 0.5$. The second bar corresponds to $x = 1$; its height is $p(x = 1) = 0.25$, and its base goes from $x = 0.5$ to $x = 1.5$.

FIGURE 9.78

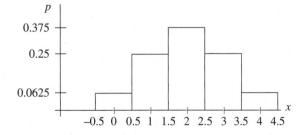

Our intuition tells us that if the probability of heads is $\frac{1}{2}$, then we should expect half of our four tosses to result in heads. The expected value found in Example 4 is in agreement with our intuition; there we found that the expected number of heads is 2. This always happens; the expected number of successes in a binomial experiment is the product of the number of trials n and the probability of success p.

Expected Number of Successes in a Binomial Experiment

The expected number of successes in a binomial experiment with n trials and a probability of success p is np.

EXAMPLE 5 Find the expected number of defective chips in Notel's shipment of 2000 chips.

Solution We could calculate this by creating a probability distribution for the experiment and then using the probability distribution to calculate the expected value, as we did in Example 5. This would be quite laborious, since it would entail computing $p(x = 0)$ through $p(x = 2000)$. Instead, we will use the above formula.

$$\text{expected number of defects} = np$$
$$= 2000 \cdot 0.01$$
$$= 20$$

Notel should expect to find 20 defective chips in a shipment of 2000. Anything is possible, though—there could be 0 defects, or even 2000. ●

9.5

EXERCISES

1. Find the probability that a truly random drawing would result in listing a Democrat first *at least* 40 out of 41 times.

2. Find the probability that a truly random drawing would result in listing a Democrat first *at least* 39 out of 41 times.

3. A certain binomial experiment is repeated 12 times. The probability of success is 0.2. Find the probability of 3 successes.

4. A certain binomial experiment is repeated 15 times. The probability of success is 0.35. Find the probability of 7 successes.

5. The expression
$$_{18}C_{10}\,0.8^{10}0.2^{8}$$
refers to the probability of a certain binomial experiment.
  a. How many trials are there?
  b. How many successes are there?
  c. How many failures are there?
  d. What is the probability of success?
  e. What is the probability of failure?

6. The expression
$$_{32}C_{15}\,0.66^{15}0.34^{17}$$
refers to the probability of a certain binomial experiment.
  a. How many trials are there?
  b. How many successes are there?
  c. How many failures are there?
  d. What is the probability of success?
  e. What is the probability of failure?

7. A fair die is rolled 4 times in a row. Getting a 1 or 2 is a success, and getting any other outcome is a failure.
  a. Why is this a binomial experiment?
  b. What is the probability of success?
  c. What is the probability of failure?
  d. List all of the outcomes that involve 2 successes and 2 failures, using the letters s and f [one of these outcomes is (s,s,f,f)]. Count the number of outcomes on your list.
  e. Use combinations to find the number of outcomes that involve 2 successes and 2 failures. Compare this number to the solution of (d).

f. Find the probability of having 2 successes and 2 failures, in any order.

8. A fair die is rolled 5 times in a row. Getting a 1 or 2 is a success, and getting any other outcome is a failure.
   a. Why is this a binomial experiment?
   b. What is the probability of success?
   c. What is the probability of failure?
   d. List all of the outcomes that involve 2 successes and 3 failures, using the letter *s* and *f* [one of these outcomes is (*s,s,f,f,f*)]. Count the number of outcomes on your list.
   e. Use combinations to find the number of outcomes that involve 2 successes and 3 failures. Compare this number to the solution of (d).
   f. Find the probability of having 2 successes and 3 failures, in any order.

9. A fair coin is flipped 3 times.
   a. Find the probability of getting no heads.
   b. Find the probability of getting 1 head.
   c. Find the probability of getting 2 heads.
   d. Find the probability of getting 3 heads.
   e. Create a binomial distribution for this experiment.
   f. Create a histogram for this experiment.

10. A fair coin is flipped 2 times.
   a. Find the probability of getting no heads.
   b. Find the probability of getting 1 head.
   c. Find the probability of getting 2 heads.
   d. Create a binomial distribution for this experiment.
   e. Create a histogram for this experiment.

In Exercises 11–14, two experiments are described, one of which is binomial. Determine which experiment is binomial and which isn't, and explain why.

11. Flip a coin 7 times, and observe how many times heads comes up. Flip a coin once, and observe whether heads comes up.

12. Successively deal 10 cards from a full deck, and observe how many cards are hearts. Successively roll a single die 10 times, and observe how many rolls resulted in a 3.

13. Successively roll a single die 10 times, and observe how many rolls resulted in a 1 or 2, how many resulted in a 3 or 4, and how many resulted in a 5 or 6. Successively roll a single die 10 times, and observe how many rolls resulted in an even number and how many resulted in an odd number.

14. Install four identical nondefective wheel bearings on a bus, and measure how long it takes for each to fail as a result of wearing out. Install four identical nondefective wheel bearings on a bus, and measure how long it takes for each to fail as a result of the explosion of a bomb on the bus.

15. The Everglo Light Bulb Company attempts to control quality by randomly choosing 10 bulbs from its assembly line for testing. If any (i.e., one or more) of those bulbs are defective, production is temporarily stopped while the machines are inspected. Find the probability that production is stopped if the defect rate is actually:
   a. 1%   **b.** 10%

16. The Everglo Light Bulb Company attempts to control quality by randomly choosing 25 bulbs from its assembly line for testing. If two or more of those bulbs are defective, production is temporarily stopped while the machines are inspected. Find the probability that production is stopped if the defect rate is actually:
   a. 1%   **b.** 10%

17. The SUDS diagnostic test is commonly used to detect the presence of AIDS/HIV. Of those who test negative, only 0.0004% actually have the disease. However, of those who test positive, only 49% actually have the disease, due to the rarity of the disease and the relatively higher probability of a false postive. (See Section 6.8, Exercise 2 for more information.)
   a. If an AIDS testing agency informs 1000 people that they tested negative for AIDS/HIV, how many of those people should the agency expect to actually have AIDS/HIV?
   b. What is the probability that none of the 1000 people actually have AIDS/HIV?
   c. If an AIDS testing agency informs 1000 people that they tested positive for AIDS/HIV, how many of those people should the agency expect to actually have AIDS/HIV?

18. A common test for extrasensory perception (ESP) involves a deck of 25 cards, with 5 different shapes (circle, cross, wave, star, and square) each appearing on 5 cards. (See the illustration on page 462.) The cards are placed in a random order, and the experimenter and the subject are positioned so that the experimenter can see the cards but the subject can see neither the cards nor the experimenter. One at a time, the experimenter looks at each of the top 10 cards and concentrates on the image on the card while the subject guesses the shape.
   a. What is the probability that the subject makes a successful guess on a single card, if that subject does not have ESP?
   b. How many correct guesses should the experimenter expect the subject to make, if that subject does not have ESP?
   c. What is the probability that a subject who does not have ESP makes the number of correct guesses found in part (b)?
   d. What is the probability that a subject who does not have ESP makes no more than the number of correct guesses found in part (b)?

e. What is the probability that a subject who does not have ESP makes more than the number of correct guesses found in part (b)?

19. A public college is hiring a new mathematics professor. According to data supplied by the state, 34% of all qualified, available mathematicians who reside in the state are members of minority groups. Two hundred twelve people apply for the job, 152 of whom are found to be qualified and available. Ten are selected for the final interview. If these 10 are chosen without regard to race, and if minority applicants tend to be as well prepared as majority applicants, find the probability that:

a. none of the 10 are minorities
b. fewer than 2 are minorities
c. 2 or more are minorities

20. There is a simple equation that relates p and q. Determine what that equation is, and use probability rules to justify it.

21. In Example 3, it was stated that $P(x < 2) = P(x = 0) + P(x = 1)$. What probability rule is being applied here? Justify the application of this rule.

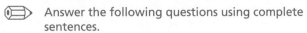 Answer the following questions using complete sentences.

22. Discuss why an experiment's trials' outcomes must be independent in order to apply the binomial probability formula. Consider the solution to Example 1 if, instead of rolling a die three times, three cards were dealt from a deck of six cards.

23. In Example 3 regarding Notel's chips, it was stated that "these defects are known to occur at random." Discuss whether this does or does not matter. Consider the case if it were known that most of the defects occur in chips made on Mondays.

24. In Example 4, the most likely number of heads was found to be 2, and the expected number of heads was also found to be 2. Discuss whether the most likely outcome and the expected value are the same for all experiments. Consider the case where the expected value is not an actual outcome.

25. Exercises 9 and 10 asked you to use probabilities to create a histogram.
a. What is the difference between such a histogram and one that shows only frequencies?
b. What is the difference between such a histogram and one that shows only relative frequencies?
c. Is a probability histogram more similar to a frequency histogram or to a relative frequency histogram? Why?

26. In Example 4(d), why does the histogram's first bar go from $x = -0.5$ to $x = 0.5$? Why does the second bar go from $x = 0.5$ to $x = 1.5$?

BINOMIAL PROBABILITIES ON A GRAPHING CALCULATOR

The binomial probability formula states that in a binomial experiment with probability of success p and probability of failure q, the probability of x successes in n trials is

$$P(x \text{ successes}) = {}_nC_x \, p^x q^{n-x}$$

Thus, in an experiment with 10 trials and probability of success 0.3, the probability of x successes is

$$P(x \text{ successes}) = {}_{10}C_x \, 0.3^x 0.7^{10-x}$$

If this function is stored as Y_1, the function can easily be evaluated for any value of x in its domain using the methods described in Section 1.1.

EXERCISES

27. What is the domain of the above function?

28. If a husband and wife each have blood type AB, then the probability that their child has blood type A is $\frac{1}{4}$.

  a. If the family has five children, create a probability distribution for the number of children with blood type A.

  b. Create a histogram for the number of children with blood type A.

  c. What is the most likely number of children that will have blood type A?

  d. What is the expected number of children that will have blood type A?

29. If a husband and wife each have blood type AB, then the probability that their child has blood type AB is $\frac{1}{2}$.

  a. If the family has five children, create a probability distribution for the number of children with blood type AB.

  b. Create a histogram for the number of children with blood type AB.

  c. What is the most likely number of children that will have blood type AB?

  d. What is the expected number of children that will have blood type AB?

9.6

THE NORMAL APPROXIMATION TO THE BINOMIAL DISTRIBUTION

Binomial Distributions and Histograms

In Example 5 of Section 9.5, we found that Notel should expect to find 20 defective chips in a shipment of 2000 chips. A shipment with 20 defective chips is acceptable to Notel, but a shipment with 25 or more defective chips would allow the purchaser of the chips to impose a fine on Notel. Notel executives are concerned about the probability of being fined. This probability is

$$p(x \geq 25) = p(x = 25) + p(x = 26) + \cdots + p(x = 1999) + p(x = 2000)$$

Computing $p(x \geq 25)$ would be quite toilsome. However, the normal distribution provides us with an acceptable way of approximating this probability. Before discussing that problem, we'll discuss a simpler but similar problem.

EXAMPLE 1   Figure 9.79 shows the binomial distribution for tossing a fair coin four times, from Example 4 of Section 9.5. Use this distribution to find the probability that the experiment results in between one and three heads (inclusive).

FIGURE 9.79

Number of Heads	Probability
$x = 0$	$p(x = 0) = (_4C_0)\,(0.5)^0(0.5)^4 = 0.0625$
$x = 1$	$p(x = 1) = (_4C_1)\,(0.5)^1(0.5)^3 = 0.25$
$x = 2$	$p(x = 2) = (_4C_2)\,(0.5)^2(0.5)^2 = 0.375$
$x = 3$	$p(x = 3) = (_4C_3)\,(0.5)^3(0.5)^1 = 0.25$
$x = 4$	$p(x = 4) = (_4C_4)\,(0.5)^4(0.5)^0 = 0.0625$

Solution

$$p(1 \le x \le 3) = p(x = 1) + p(x = 2) + p(x = 3)$$
$$= 0.25 + 0.375 + 0.25 = 0.875$$

The histogram for the binomial distribution in Example 1 is shown in Figure 9.80. The height of the second bar is $p(x = 1) = 0.25$. However, each of the bars has a base of width 1, so the area of each bar is the same as its height (area = base · height = 1 · height = height). Thus, the probability calculated in Example 1 is the sum of the areas of the histogram's second, third, and fourth bars.

FIGURE 9.80

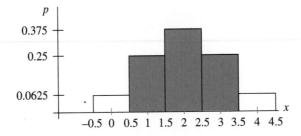

The Normal Approximation to the Binomial Distribution

Figure 9.81 shows a normal curve superimposed over the histogram of Figure 9.80. As the graph suggests, the shaded area under the normal curve is approximately equal to the sum of the areas of the histogram's second, third, and fourth bars. We can find that area easily using the body table for the standard normal distribution, as long as we know the binomial distribution's mean and standard deviation.

FIGURE 9.81

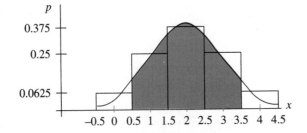

In Section 9.5 we found that the expected value, or the mean, of a binomial distribution with n trials, a probability of success p, and a probability of failure q is $\mu = np$. It can be shown that the standard deviation is $\sigma = \sqrt{npq}$.

Mean and Standard Deviation of a Binomial Distribution

The mean and standard deviation of a binomial distribution that consists of n trials with probability of success p and probability of failure q are, respectively,

$$\mu = np \quad \text{and} \quad \sigma = \sqrt{npq}$$

EXAMPLE 2   Use the normal distribution to approximate the probability found in Example 1.

Solution   We need to calculate the area under the normal curve that corresponds to the histogram's second, third, and fourth bars. By inspecting the graph, we can tell that the area is of a region that runs from $x = 0.5$ to $x = 3.5$. Thus, we need $p(0.5 < x < 3.5)$.

The distribution's mean is

$$\mu = np = 4 \cdot \frac{1}{2} = 2$$

and its standard deviation is

$$\sigma = \sqrt{npq} = \sqrt{4\left(\frac{1}{2}\right)\left(\frac{1}{2}\right)} = \sqrt{1} = 1$$

The probability is

$$p(0.5 < x < 3.5) = p\left(\frac{0.5 - 2}{1} < z < \frac{3.5 - 2}{1}\right)$$

$$= p(-1.5 < z < 1.5)$$
$$= 2 \cdot p(0 < z < 1.5) \qquad \text{Using symmetry}$$
$$= 2 \cdot 0.4332 \qquad \text{From the standard normal table}$$
$$= 0.8664 \qquad\qquad\qquad\qquad\qquad\qquad •$$

Notice that the solution of Example 2 is close to the solution of Example 1. There were only a few computations in Example 1, so using the normal distribution to approximate the binomial distribution did not involve a significant time savings. It will, however, in Example 3.

EXAMPLE 3   A shipment of 2000 chips that includes 25 or more defective chips would allow the purchaser of the chips to impose a fine on Notel. Find the probability that Notel is fined.

Solution   The probability $p(x \geq 25)$ is the sum of the areas of the Notel histogram's bars from $x = 25$ through $x = 2000$. The first of these bars runs from $x = 24.5$ to $x = 25.5$, so we're after the area of the bars to the right of $x = 24.5$. This is approximately equal to the area under the normal curve, to the right of $x = 24.5$. Thus, we will use the normal distribution to calculate $p(x > 24.5)$.

The distribution's mean is

$$\mu = np = 2000 \cdot 0.01 = 20$$

and its standard deviation is

$$\sigma = \sqrt{npq} = \sqrt{2000\,(0.01)\,(0.99)} = \sqrt{19.8}$$

The probability is

$$p(x > 24.5) = p\left(z > \frac{24.5 - 20}{\sqrt{19.8}}\right)$$

$$= p(z > 1.0112\ldots)$$

$$\approx p(z > 1.01)$$

$$= 0.5 - p(0 < z < 1.01)$$

$$= 0.5 - 0.3438 \qquad \text{From the standard normal table}$$

$$= 0.1562$$

This means that, in the long run, approximately 16% of these shipments will have 25 or more defective chips. Notel needs to weigh the cost of decreasing the number of defective chips against their negative effect on Notel's business. ●

EXAMPLE 4 Notel has decided to decrease the number of fines it must pay by changing the cutoff point from 25 defects in a shipment of 2000 to some higher number. What cutoff point would make the probability of paying a fine 0.1 or less?

Solution The question asks us to find a number c such that $p(x > c) = 0.1$. Thus, the area of the right tail is 0.1, and the area of the right body is 0.4. The right body area in the tables closest to 0.4 is 0.3997, and this area corresponds to $z = 1.28$. Therefore, the number c that we are seeking lies 1.28 standard deviations above the mean. All that remains is to convert $z = 1.28$ to a value of x.

$$z = \frac{x - \mu}{\sigma}$$

$$1.28 = \frac{x - 20}{\sqrt{19.8}}$$

$$1.28 \cdot \sqrt{19.8} = x - 20$$

$$x = 1.28 \cdot \sqrt{19.8} + 20 = 25.6956\ldots$$

This means that a cutoff point of $25.6956\ldots$ would make the probability of paying a fine 0.1. The number of defects must be a whole number, and a cutoff point of 25 generates more fines, so the new cutoff point should be 26. Thus, if Notel paid a fine for shipments that include 26 or more defects, then less than 10% of the shipments will incur fines. ●

9.6

EXERCISES

1. A certain binomial experiment is repeated 1000 times. The probability of success is 0.2. Use the normal distribution to approximate the probability of 230 or more successes.

2. A certain binomial experiment is repeated 200 times. The probability of success is 0.3. Use the normal distribution to approximate the probability of 75 or fewer successes.

3. A certain binomial experiment is repeated 420 times. The probability of success is 0.4. Use the normal distribution to approximate the probability of between 175 and 195 successes, inclusive.

4. A certain binomial experiment is repeated 1000 times. The probability of success is 0.7. Use the normal distribution to approximate the probability of between 660 and 740 successes, inclusive.

5. The SUDS diagnostic test is commonly used to detect the presence of AIDS/HIV. Of those who test negative, only 0.0004% actually have the disease. However, of those who test positive, only 49% actually have the disease, due to the rarity of the disease and the relatively higher probability of a false positive. If an AIDS testing agency informs 1000 people that they tested positive for AIDS/HIV, find the probability that ten or more don't have the disease.

6. Airlines generally overbook their planes, assuming that some ticket holders will be "no-shows"—that is, neither make their ticketed flight nor cancel their reservation. Over the years, Trans Global Air has found that 7% of their ticket holders are no-shows. Their Boeing 747's hold 418 passengers.
   a. How many tickets should Trans Global sell for a given flight if their goal is to fill all seats, on the average?
   b. If Trans Global sells the number of tickets in part (a), what is the probability that no ticketed passenger will have to give up his or her seat?
   c. If Trans Global sells the number of tickets in part (a), what is the probability that at least one ticketed passenger will have to give up his or her seat?
   d. If Trans Global sells the number of tickets in part (a), what is the probability that the flight will have one or more empty seats?

7. A common test for extrasensory perception (ESP) involves a deck of 25 cards, with 5 different shapes (circle, cross, wave, star, and square) each appearing on 5 cards. The cards are placed in a random order, and the experimenter and the subject are positioned so that the experimenter can see the cards but the subject can see neither the cards nor the experimenter. One hundred times, the experimenter looks at a card and concentrates on the image on the card, while the subject guesses the shape.
   a. The experimenter does not consider a subject to have ESP unless the subject makes so many correct guesses that the probability of doing that well or better by random guessing is at most 0.1. How many correct guesses must a subject make in order to be considered to have ESP?
   b. How many correct guesses must a subject make if the standard is 0.01 rather than 0.1?
   c. How many correct guesses must a subject make if the standard is 0 rather than 0.1?

8. Proposition Q calls for an increase in Douglas County's property taxes. Suppose that 55% of the voters actually support this proposition.
   a. If a newspaper interviews five registered voters, what is the probability that a majority (i.e., three or more) opposes the proposition?
   b. If 400 people are polled, what is the probability that a majority opposes the proposition?
   c. What do the solutions to (a) and (b) imply about polls correctly reporting a population's majority opinion?

9. A public college is hiring a new mathematics professor. According to data supplied by the state, 34% of all qualified, available mathematicians who reside in the state are members of minorities. Two hundred twelve people apply for the job, 152 of whom are found to be qualified and available. If these 152 are chosen without regard to race, and if minority applicants tend to be just as well prepared as majority applicants, find the probability that:
   a. none of them are minorities
   b. 45 or fewer are minorities
   c. One quarter or more are minorities

10. In Figure 9.81, the area under the normal curve, from $x = 0.5$ to $x = 1.5$, is very close to the area of the histogram's second bar. How can the former region be cut up and redistributed so that these two areas appear to be more similar than they do in Figure 9.81?

11. Figure 9.82 shows histograms for two different binomial distributions. Each histogram has a normal curve superimposed over it.

FIGURE 9.82

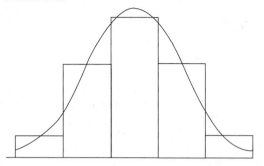 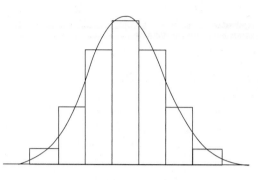

a. What is n in the histogram on the left? What is n in the histogram on the right? How can you tell what n is by looking at the histogram?

b. In which case does the normal distribution more closely approximate the binomial distribution? Why?

CHAPTER 9 REVIEW

Terms

bell-shaped curve
binomial distribution
binomial experiment
body table
data point
descriptive statistics
deviation, or deviation
    from the mean
extreme value, or outlier

frequency
frequency distribution
group boundaries
histogram
inferential statistics
mean
measures of central
    tendency

measures of dispersion
median
mode
normal distribution
population
relative frequency
relative frequency density
sample

standard deviation
standard normal
    distribution, or
    z-distribution
statistics
summation notation
trial, or Bernoulli trial
variance

Review Exercises

1. In order to study the composition of families in Winslow, Arizona, 40 randomly selected married couples were surveyed to determine the number of children in each family. The following results were obtained:

3 1 0 4 1 3 2 2 0 2 0 2 2 1
4 3 1 1 3 4 2 1 3 0 1 0 2 5
1 2 3 0 0 1 2 3 1 2 0 2

  a. Organize the given data by creating a frequency distribution.
  b. Find the mean number of children per family.
  c. Find the median number of children per family.
  d. Find the mode number of children per family.

  e. Find the standard deviation of the number of children per family.

2. The frequency distribution in Figure 9.83 lists the number of hours per day that a randomly selected sample of teenagers spent watching television. Where possible, determine what percent of the teenagers spent the following number of hours watching television.

  a. less than 4 hours
  b. not less than 6 hours
  c. at least 2 hours
  d. less than 2 hours
  e. at least 4 hours but less than 8 hours
  f. more than 3.5 hours

FIGURE 9.83

x = Hours per Day	Frequency
$0 \le x < 2$	23
$2 \le x < 4$	45
$4 \le x < 6$	53
$6 \le x < 8$	31
$8 \le x \le 10$	17

3. In order to study the efficiency of its new oil-changing system, a local service station monitored the amount of time it took to change the oil in customers' cars. The frequency distribution in Figure 9.84 summarizes the findings.
   a. Find the mean amount of time to change the oil in a car.
   b. Find the standard deviation of the amount of time to change the oil in a car.
   c. Construct a histogram to represent the data.

FIGURE 9.84

x = Time (in minutes)	Number of Customers
$3 \le x < 6$	18
$6 \le x < 9$	42
$9 \le x < 12$	64
$12 \le x < 15$	35
$15 \le x \le 18$	12

4. If your scores on the first four exams (in this class) are 74, 65, 85, and 76, what score do you need on the next exam for your overall mean to be at least 80?

5. The mean salary of 12 men is $37,000, and the mean salary of 8 women is $28,000. Find the mean salary of all 20 people.

6. Timo and Henke golfed five times during their vacation. Their scores are given in Figure 9.85.
   a. Find the mean score of each golfer. Who has the lowest mean?
   b. Find the sample standard deviation of each golfer's scores.
   c. Who is the more consistent golfer? Why?

FIGURE 9.85

Timo	103	99	107	93	92
Henke	101	92	83	96	111

7. Suzanne surveyed the prices for a quart of a certain brand of motor oil. The sample data, in dollars per quart, are summarized in Figure 9.86.
   a. Find the mean and standard deviation of the prices.
   b. What percentage of the data lies within one standard deviation of the mean?
   c. What percentage of the data lies within two standard deviations of the mean?
   d. What percentage of the data lies within three standard deviations of the mean?

FIGURE 9.86

Price per Quart	Number of Stores
0.99	2
1.09	3
1.19	7
1.29	10
1.39	14
1.49	4

8. What percentage of the standard normal z-distribution lies in the following intervals?
   a. between $z = 0$ and $z = 1.75$
   b. between $z = -1.75$ and $z = 0$
   c. between $z = -1.75$ and $z = 1.75$

9. All incoming freshmen at a major university are given a diagnostic mathematics exam. The scores are normally distributed with a mean of 420 and a standard deviation of 45. If the student scores less than a certain score, he or she will have to take a review course. Find the cutoff score at which 34% of the students would have to take the review course.

10. The Wechsler Adult Intelligence Scale (WAIS) is one of the most common IQ tests for adults. The distribution of scores on the WAIS is approximately normal with mean 100 and standard deviation 15.
    a. What is the probability that an adult has a WAIS score of 100 or higher? Explain how that probability can be found without performing any calculations.

b. What is the probability that an adult has a WAIS score of 125 or higher?

c. Find the WAIS score that marks the most intelligent 10% of the population.

11. A certain standardized test consists of 20 multiple-choice questions. Each question is accompanied by five possible answers, only one of which is correct. One test taker's strategy is to choose each answer at random, without even reading the questions.

  a. How many correct answers should this test taker expect to get?

  b. What is the probability that he or she would get a perfect score?

c. What is the probability that he or she would get a score of 90%?

12. A certain standardized test consists of 20 multiple-choice questions. Each question is accompanied by five possible answers, only one of which is correct. One test taker's strategy is to use common sense to eliminate three answers and to choose at random from the remaining two answers.

  a. How many correct answers should this test taker expect to get?

  b. What is the probability that he or she would get a perfect score?

  c. What is the probability that he or she would get a score of 90%?

FINANCE 10

Webster's Dictionary defines **finance** as "the system that includes the circulation of money, the granting of credit, the making of investments, and the provision of banking facilities." In this chapter, we will investigate the mathematics of two aspects of finance: loans and nonspeculative investments—such as savings accounts, money-market accounts, certificates of deposit, and annuities. (Speculative investments are investments that involve some risk, such as stocks and real estate.)

The mathematics of finance is an important topic for consumers and businesspeople alike. Most people borrow money, usually to finance a college education, a car, or a home or to cover credit card charges. Most businesses borrow money, usually to purchase equipment or inventory, to pay operating expenses, or to pay taxes. Most people save money, through a savings account at a bank, a money-market account, a savings bond, or some form of annuity. Most businesses save money, usually through a savings account, a money-market account, commercial paper (a loan from one business to another), or a bankers' acceptance or repurchase agreement (a loan from a business to a bank).

Loans and investments are very similar financial transactions; both involve the flow of money from one party to another, the return of the money to its source, and the payment of a fee to the source for the use of the money. When you

471

make a deposit in your savings account, you view the transaction as an investment, but the bank views it as a loan; you are lending the bank your money, which they will lend to another customer, perhaps to buy a house. When you borrow money to buy a car, you view that transaction as a loan, but the bank views it as an investment; the bank is investing its money in you in order to make a profit.

In this chapter we explore some of the different forms of loans and investments, along with the various ways in which they are calculated, so that you can make the best choice when you have to make a financial decision. We investigate the different types of automobile loans available and discuss how home loans work. We also show how you can accumulate an incredibly large amount of money by saving only $50 a month.

10.1

SIMPLE INTEREST

When a lender lends money to a borrower, the lender usually expects to be repaid more than was lent. The original amount of money lent is called the **principal,** or **present value;** if some of principal has been paid back, then the portion that remains unpaid is called the **outstanding principal,** or **balance.** The total amount of money the lender is paid back is called the **future value;** this includes the original amount lent and the lender's profit, or **interest.** How much interest will be paid depends on the **interest rate** (usually expressed as a percent per year); the **term,** or the length of time before the debt is repaid; and how the interest is calculated.

Many short-term loans and some other forms of investment are calculated as simple interest. **Simple interest** means that the amount of interest is computed as a percent-per-year of the principal.

Simple Interest Formula

The *simple interest* I on a principal P at an annual rate of interest r for t years is

$I = Prt$

The future value FV is the total of the principal and the interest; therefore,

$$FV = P + I$$
$$= P + Prt$$
$$= P(1 + rt)$$

Simple Interest Future Value Formula

The future value FV of a principal P at an annual rate of interest r for t years is

$FV = P(1 + rt)$

One of the most common uses of simple interest is for a short-term (i.e., a year or less) simple interest loan that requires a single lump sum payment at the end of the term. Businesses routinely obtain these loans to purchase equipment or inventory, to pay operating expenses, or to pay taxes.

EXAMPLE 1   South Face, a mountaineering and camping store, borrowed $340,000 at 5.1% for 120 days to purchase inventory for the Christmas season. Find the interest on the loan.

Solution   We are given $P = 340,000$, $r = 5.1\% = 0.051$, and $t = 120$ days $= 120$ days $\cdot \frac{1 \text{ year}}{365 \text{ days}} = \frac{120}{365}$ years. Using the simple interest formula, we have

$$I = Prt$$

$$= 340,000 \cdot 0.051 \cdot \frac{120}{365}$$

$$= 5700.8219 \ldots$$

$$\approx \$5,700.82$$

This means that South Face has agreed to pay the lender $340,000 plus $5,700.82 interest at the end of 120 days.   •

There is an important distinction between the variables *FV, I, P, r,* and *t*: *FV, I,* and *P* measure amounts of money, whereas *r* and *t* do not. For example, consider the interest rate *r* and the interest *I*. Frequently, people confuse these two. However, the interest rate *r* is a percentage, and the interest *I* is an amount of money; in Example 1, the interest rate is 5.1% and the interest is $5700.82. To emphasize this distinction, we will always use capital letters for variables that measure amounts of money, and small letters for other variables. Hopefully, this notation will help you avoid substituting 5.1% = 0.051 for *I* when it should be substituted for *r*.

In Example 1, we naturally used 365 days per year; but some institutions traditionally count a year as 360 days and a month as 30 days (especially if that tradition works in their favor). This is a holdover from the days before calculators and computers—the numbers were simply easier to work with. Also, we used normal round-off rules to round 5700.8219 . . . to 5,700.82; some institutions round off some interest calculations in their favor. In this book, we will count a year as 365 days and use normal round-off rules (unless stated otherwise).

The issue of how a financial institution rounds off its interest calculations might seem unimportant; after all, we're talking about a difference of a fraction of a penny. However, consider one classic form of computer crime: the round-down fraud, performed on a computer system that processes a large number of accounts. Frequently, such systems use the normal round-off rules in their calculations and keep track of the difference between the theoretical account balance if no rounding off is done and the actual account balance with rounding off. Whenever that difference reaches or exceeds 1 cent, the extra penny is deposited in (or withdrawn from) the account. A fraudulent computer programmer can write the program so that the extra penny is deposited in *his or her* account. This fraud is difficult to detect because the accounts appear to be balanced. While each individual gain is small, the total gain can be quite large if a large number of accounts is processed on a regular basis.

EXAMPLE 2 To pay its taxes, Espree Clothing borrowed $185,000 at 7.3% for four months. Find the future value of the loan.

Solution We are given $P = 185{,}000$, $r = 7.3\% = 0.073$, and $t = 4 \text{ months} = 4 \text{ months} \cdot \frac{1 \text{ year}}{12 \text{ months}} = \frac{4}{12}$ years. Using the simple interest future value formula, we have

$$FV = P(1 + rt)$$

$$= 185{,}000\left(1 + .073 \cdot \frac{4}{12}\right)$$

$$= 189{,}501.6666\ldots$$

$$\approx \$189{,}501.67$$

This means that Espree has agreed to pay the lender $189,501.67 at the end of four months.

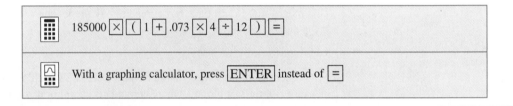

185000 ×ᵇ (1 + .073 × 4 ÷ 12) =

With a graphing calculator, press ENTER instead of =

✔ Notice that the answer is somewhat higher than the original $185,000, as it should be. •

Technically, Example 2 should specify which months are involved, because months vary from 28 to 31 days in length. The four-month time span from January through April does not have the same number of days as the four-month time span from March through June and thus would generate a slightly different amount of interest. The calculation done in Example 2 gives a very good approximation of the future value after *any* four-month time span, but it is *only* an approximation.

A written contract signed by the lender and the borrower is called a **loan agreement,** or a **note.** The **maturity value** of the note (or just the **value** of the note) refers to the note's future value. Thus, the value of the note in Example 2 was $189,501.67.

EXAMPLE 3 Find the amount of money that must be invested now at a $5\frac{3}{4}\%$ interest rate so that it will be worth $1000 in 2 years.

Solution We are asked to find the present value, or principal P, that will generate a future value of $1000.

We know that $FV = 1000$, $r = 5\frac{3}{4}\% = 0.0575$, and $t = 2$. Using the future value formula, we have

$$FV = P(1 + rt)$$
$$1000 = P(1 + 0.0575 \cdot 2)$$
$$P = \frac{1000}{1 + 0.0575 \cdot 2} \qquad \text{Dividing}$$
$$= 896.86098 \ldots$$
$$\approx \$896.86$$

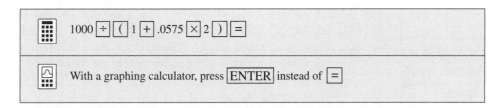

1000 ÷ (1 + .0575 × 2) =

With a graphing calculator, press ENTER instead of =

✔ Notice that the answer is somewhat smaller than $1000, as it should be.   •

Add-on Interest

Purchases made at car dealers, appliance stores, and furniture stores can be financed through the store itself. Frequently, this type of loan involves **add-on interest,** which consists of a simple interest charge on the loan amount, distributed equally over each payment.

EXAMPLE 4 Flo and Eddie Turkel purchase $1300 worth of kitchen appliances at Link's Appliance Mart. They put $200 down and agree to pay the balance at a 10% add-on rate for two years. Find their monthly payment.

Solution We are given P = loan amount = $1300 - 200 = 1100$, $r = 10\% = 0.10$, and $t = 2$ years. The total amount due is the future value:

$$FV = P(1 + rt)$$
$$= 1100\,(1 + 0.10 \cdot 2)$$
$$= 1320$$

The total amount due is spread out over 24 monthly payments; therefore, monthly payment $= \frac{1320}{24} = \$55$.

Simple Discount Loans

So far, the loans discussed in this section have been **simple interest loans,** where the loan amount is a specifically agreed-upon amount, the interest is a percent of the loan amount, and the loan amount plus interest must be paid at maturity. In the case of a **simple discount loan,** however, the *amount to be repaid* is a specifically agreed-upon amount, the interest is a percent *of the amount to be repaid,* and the loan amount is the amount to be paid *less* interest. The interest that is subtracted is called the **discount.** The money that the borrower receives after the interest has been subtracted is called the **proceeds.**

The discount is a percent per year of the future value, so if we rewrite the simple interest formula, we get:

Simple Discount Formula

The simple discount D on a future value FV at an annual rate of interest r for T years is

$$D = FV \cdot r \cdot t$$

EXAMPLE 5   Ramblin Productions has a contract specifying that it will be paid $60,000 in one year as payment for its work on a TV commercial. Ramblin needs money now to pay bills, so it obtains a simple discount loan of $60,000 for one year at a 13% discount rate. Find the discount and the proceeds.

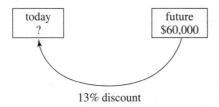

Solution   We are given $FV = 60,000$, $r = 13\% = 0.13$, $t = 1$ year. Using the simple discount formula, we have

$$D = FV \cdot r \cdot t$$
$$= 60,000 \cdot 0.13 \cdot 1$$
$$= 7800$$

The discount is $7800.

The proceeds are the money that the borrower receives after the interest has been subtracted, and Ramblin receives $60,000 − $7800 = $52,200. The proceeds are $52,200.

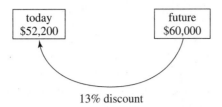

13% discount

In Example 5, Ramblin is charged a simple discount rate of 13%. This is not to be confused with a simple *interest* rate of 13%. Recall that simple interest means that the amount of interest is computed as a percent per year *of the principal,* while simple discount means that the amount of interest is computed as a percent per year *of the future value.*

EXAMPLE 6 Find the simple interest rate of the loan in Example 5.

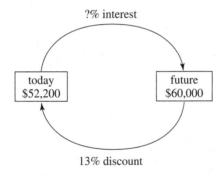

Solution To do this, we must recompute Ramblin's loan as a simple interest loan. We have found that $P = 52{,}200$, $I = 7800$, and $t = 1$ year. Using the simple interest formula, we have

$$I = P \cdot r \cdot t$$
$$7800 = 52{,}200 \cdot r \cdot 1$$
$$\frac{7800}{52{,}200} = \frac{52{,}200 \cdot r \cdot 1}{52{,}200}$$
$$r = \frac{7800}{52{,}200}$$
$$= 0.149425\ldots$$
$$\approx 0.1494 = 14.94\%$$

By tradition, the rate is rounded to the nearest hundredth of one percent, so the simple interest rate is 14.94%. This means that the dollar amounts of Ramblin's 13% simple discount loan would be the same as those of a 14.94% simple interest loan. If Ramblin is not aware of the simple interest rate of this loan, it is probably paying a lot more interest than it realized. •

Discounting a Note

Discounting a note is a procedure in which a note due at some point in the future can be converted to cash by its holder at an earlier date. The holder of the note sells the note for the future value of that note less a discount.

EXAMPLE 7   Shmata Shirt Co. bought collar stays from the Permerect Collar Stay Co. for $1500. Schmata Shirt financed this purchase with a 90-day note from Permerect at 8% simple interest. Thirty days later, Permerect needed cash, so it discounted Shmata's note to the Bag-o-Bucks Factor Co. at a 9% discount rate. Find the discount and the proceeds.

Solution   First, determine the future value of the note. We are given that $P = 1500$, $r = 8\% = 0.08$, and $t = 90$ days $= 90$ days $\cdot \frac{1\ \text{year}}{365\ \text{days}} = \frac{90}{365}$ years. Using the simple interest future value formula, we have

$$FV = P(1 + rt)$$

$$= 1500\left(1 + 0.08 \cdot \frac{90}{365}\right)$$

$$= 1529.5890\ldots$$

$$\approx \$1529.59$$

Next, determine the discount. We are given $r = 9\% = 0.09$, $t = 60$ days (why 60, not 90?) $= 60$ days $\cdot \frac{1\ \text{year}}{365\ \text{days}} = \frac{60}{365}$ years, and we have found that $FV = 1529.59$. Using the simple discount formula, we have

$$D = \quad FV \quad \cdot \quad r \quad \cdot \quad t$$

$$D = 1529.59 \cdot 0.09 \cdot \frac{60}{365}$$

$$= 22.6295\ldots$$

$$\approx \$22.63$$

The discount is $22.63. The proceeds are then $1529.59 − $22.63 = $1506.96. This means that Permerect received $1506.96 from Bag-o-Bucks 30 days after the sale was made, rather than the $1500 it might have received at the point of sale. It also means that Permerect received $1506.96 from Bag-o-Bucks 60 days before the note was due, rather than the $1529.59 it would have received from Shmata Shirt if it had waited those 60 days. •

10.1

EXERCISES

In Exercises 1–6, find the simple interest of the given loan amounts:

1. loan amount $2000 at 8% for 3 years

2. loan amount $35,037 at 6% for 2 years

3. loan amount $420 at $6\frac{3}{4}$% for 9 months

4. loan amount $8950 at $9\frac{1}{2}$% for 10 months

5. loan amount $1410 at $12\frac{1}{4}$% for 325 days

6. loan amount $5682 at $11\frac{3}{4}$% for 278 days

In Exercises 7–10, find the future value of the given present values:

7. present value of $3670 deposited at $2\frac{3}{4}$% for 7 years

8. present value of $4719 deposited at 14.1% for 11 years

9. present value of $12,430 deposited at $5\frac{7}{8}$% for 2 years and 3 months

10. present value of $172.39 deposited at 6% for 3 years and 7 months

In Exercises 11–16, find the maturity value of the given loan amounts:

11. $1400 borrowed at $7\frac{1}{8}$% for 9 months.

12. $3250 borrowed at $8\frac{1}{2}$% for 1 year and 1 month.

13. $5900 borrowed at $14\frac{1}{2}$% for 112 days.

14. $2720 borrowed at $12\frac{3}{4}$% for 275 days.

15. $16,500 borrowed at $11\frac{7}{8}$% from April 1 to July 10 of the same year.

16. $2234 borrowed at $12\frac{1}{8}$% from March 10 to December 20 of the same year.

In Exercises 17–22, find the present value of the given future value.

17. future value $8600 at $9\frac{1}{2}$% simple interest for 3 years

18. future value $420 at $5\frac{1}{2}$% simple interest for 2 years

19. future value $1112 at $3\frac{5}{8}$% simple interest for 1 year and 11 months

20. future value $5750 at $4\frac{7}{8}$% simple interest for 2 years and 2 months

21. future value $1311 at $6\frac{1}{2}$% simple interest for 317 days

22. future value $4200 at $6\frac{3}{4}$% simple interest for 509 days

23. How much must be deposited now at 6% interest so that in 1 year and 8 months an account will contain $3000?

24. How much must be deposited now at $5\frac{7}{8}$% interest so that in 2 years and 7 months an account will contain $1900?

25. Fred Murtz just received his income tax refund of $1312.82, and he needs $1615 to buy a new stereo system. If his money can earn $6\frac{7}{8}$% interest, for how long must he invest his tax refund?

26. Sam Spade inherited $7000. He wishes to buy a used car, but the type of car he wants to buy typically sells for around $8000. If his money can earn $6\frac{1}{2}$% interest, for how long must he invest his money?

27. Alice Cohen buys a two-year-old Honda from a car dealer for $9000. She puts $500 down and finances the rest through the dealer at 13% add-on interest. If she agrees to make 36 monthly payments, find the size of each payment.

28. Sven Lundgren buys a three-year-old Chevrolet Celebrity from a car dealer for $4600. He puts $300 down and finances the rest through the dealer at 12.5% add-on interest. If he agrees to make 24 monthly payments, find the size of each payment.

29. Ray and Teresa Martinez buy a bedroom set at Fowler's Furniture for $3700. They put $500 down and finance the rest through the store at 9.8% add-on interest. If they agree to make 36 monthly payments, find the size of each payment.

30. Helen and Dick Davis buy a refrigerator at Appliance Barn for $1200. They put $100 down and finance the rest through the store at 11.6% add-on interest. If they agree to make 36 monthly payments, find the size of each payment.

In Exercises 31–34, find

   a. the discount
   b. the proceeds
   c. the simple interest rate of the given simple discount loan

31. future value $1200 at 11% simple discount rate for 2 years

32. future value $3000 at $12\frac{1}{2}$% simple discount rate for 3 years

33. future value $1750 at $11\frac{7}{8}$% simple discount rate for 1 year and 6 months

34. future value $3210 at $12\frac{3}{8}$% simple discount rate for 1 year and 11 months

35. The Country Oak Cabinet Company bought wood from the Douglass Lumber for $29,500. Country Oak financed this purchase with a 60-day note from Douglass Lumber at $7\frac{1}{2}$%

interest. Five days later, Douglass discounted Country Oak's note to the Bag-o-Bucks Finance Co. at a 9% discount rate. Find and interpret:

a. the discount

b. the proceeds

36. The Square Wheel Bicycle Store bought bicycles from Eurobike Distributors for $153,200. Square Wheel financed this purchase with a 150-day note from Eurobike at $4\frac{1}{2}\%$ interest. Thirty days later, Eurobike discounted Square Wheel's note to the Friendly Factor Co. at a 8.5% discount rate. Find and interpret:

a. the discount

b. the proceeds

37. Donovan and Pam Hamilton bought a house from Edward Gurney for $162,500. Typically, a purchaser will make a down payment to the seller of 10% to 20% of the purchase price and will borrow the rest from their bank. However, the Hamiltons did not have sufficient savings for a 10% down payment, and Mr. Gurney was a motivated seller. In lieu of a 10% down payment, Mr. Gurney accepted a 5% down payment at the time of the sale and a promissory note from the Hamiltons for an additional 5%, due in four years. The note required the Hamiltons to make monthly interest payments to Mr. Gurney at 10% interest until the note expired. The Hamiltons obtained a loan from their bank for the remaining 90% of the purchase price. The bank in turn paid the sellers the remaining 90% of the purchase price, less a sales commission (of 6% of the purchase price) paid to the seller's and the buyers' real estate agents.

a. Find the Hamiltons' down payment.

b. Find the amount that the Hamiltons borrowed from their bank.

c. Find the amount that the Hamiltons borrowed from Mr. Gurney.

d. Find the Hamiltons' monthly interest payment to Mr. Gurney.

e. Find Mr. Gurney's total income from all aspects of the down payment.

f. Find Mr. Gurney's income from the Hamiltons' bank.

g. Find Mr. Gurney's total income from all aspects of the sale.

38. George and Peggy Fulwider bought a house from Sally Sinclair for $233,500. Typically, a purchaser will make a down payment to the seller of 10% to 20% of the purchase price and will borrow the rest from their bank. However, the Fulwiders did not have sufficient savings for a 10% down payment, and Ms. Sinclair was a motivated seller. In lieu of a 10% down payment, Ms. Sinclair accepted a 5% down payment at the time of the sale and a promissory note from the Fulwiders for an additional 5%, due in four years. The note required the Fulwiders to make monthly interest payments to Ms. Sinclair at 10% interest until the note expired. The Fulwiders obtained a loan from their bank for the remaining 90% of the purchase price. The bank in turn paid the seller the remaining 90% of the purchase price, less a sales commission (of 6% of the purchase price) paid to the seller's and the buyers' real estate agents.

a. Find the Fulwiders' down payment.

b. Find the amount that the Fulwiders borrowed from their bank.

c. Find the amount that the Fulwiders borrowed from Ms. Sinclair.

d. Find the Fulwiders' monthly interest payment to Ms. Sinclair.

e. Find Ms. Sinclair's total income from all aspects of the down payment.

f. Find Ms. Sinclair's income from the Fulwiders' bank.

g. Find Ms. Sinclair's total income from all aspects of the sale.

39. The Clintons bought a house from the Bushes for $389,400. In lieu of a 20% down payment, the Bushes accepted a 10% down payment at the time of the sale and a promissory note from the Clintons for an additional 10%, due in four years. The Clintons also agreed to make monthly interest payments to the Bushes at 11% interest until the note expires. The Clintons obtained a loan from their bank for the remaining 80% of the purchase price. The bank in turn paid the sellers the remaining 80% of the purchase price, less a sales commission (of 6% of the purchase price) paid to the sellers' and the buyers' real estate agents.

a. Find the Clintons' down payment.

b. Find the amount that the Clintons borrowed from their bank.

c. Find the amount that the Clintons borrowed from the Bushes.

d. Find the Clintons' monthly interest payment to the Bushes.

e. Find the Bushes' total income from all aspects of the down payment.

f. Find the Bushes' income from the Clintons' bank.

g. Find the Bushes' total income from all aspects of the sale.

40. Sam Needham bought a house from Sheri Silva for $238,300. In lieu of a 20% down payment, Ms. Silva accepted a 10% down payment at the time of the sale and a promissory note from Mr. Needham for an additional 10%, due in four years. Mr. Needham also agreed to make monthly interest payments to Ms. Silva at 9% interest until the note expires. Mr. Needham obtained a loan from his bank for the remaining 80% of the purchase price. The bank in turn paid Ms. Silva the remaining 80% of the purchase price, less a sales commission (of 6% of the purchase price) paid to the seller's and the buyer's real estate agents.

a. Find Mr. Needham's down payment.

b. Find the amount that Mr. Needham borrowed from his bank.

c. Find the amount that Mr. Needham borrowed from Ms. Silva.

d. Find Mr. Needham's monthly interest payment to Ms. Silva.

e. Find Ms. Silva's total income from all aspects of the down payment.

f. Find Ms. Silva's income from Mr. Needham's bank.

g. Find Ms. Silva's total income from all aspects of the sale.

41. Develop a formula for the proceeds P of a loan that has a maturity value M at a simple discount rate r for t years.
HINT: Pattern your work after the development of the simple interest future value formula in this section.

42. Use the formula developed in Exercise 41 to find the proceeds in Exercise 31.

43. Use the formula developed in Exercise 41 to find the proceeds in Exercise 32.

44. Use the formula developed in Exercise 41 to find the proceeds in Exercise 33.

45. Use the formula developed in Exercise 41 to find the proceeds in Exercise 34.

 Answer the following questions using complete sentences.

46. Could Exercises 1–22 all be done with the simple interest formula? How? Could Exercises 1–22 all be done with the simple interest future value formula? How? Why do we have both formulas?

47. Which is always higher, future value or principal? Why?

10.2

COMPOUND INTEREST

Many forms of investment (including savings accounts) earn **compound interest,** in which case interest is periodically paid on the existing account balance, which includes both the original principal and previous interest payments. This form of interest results in significantly higher earnings over a long period of time. The **compounding period** (usually annually, semiannually, quarterly, monthly, or daily) refers to the frequency at which interest is computed and deposited. The effects of the compounding period, the interest rate, and the time interval can be much more dramatic with compound interest than with simple interest. It is important that you understand these effects in order to make wise financial decisions.

EXAMPLE 1 $1000 is deposited in an account where it earns 8% interest compounded quarterly. Find the account balance (future value) after six months, using the simple interest future value formula to compute the balance at the end of each compounding period.

Solution Because interest is compounded quarterly in this case, interest is computed and deposited every quarter of a year.
At the end of the first quarter, $P = 1000$, $r = 8\% = 0.08$, and $t =$ one quarter or $\frac{1}{4}$ year.

$$FV = P(1 + rt)$$

$$= 1000\left(1 + 0.08 \cdot \frac{1}{4}\right)$$

$$= 1000(1 + 0.02)$$

$$= \$1020$$

At the end of the second quarter, $P = 1020$ (the new principal), $r = 0.08$, and $t = \frac{1}{4}$ (the length of time, in years, that the new principal has been in the account).

$$FV = P(1 + rt)$$

$$= 1020\left(1 + 0.08 \cdot \frac{1}{4}\right)$$

$$= 1020(1 + 0.02)$$

$$= \$1040.40$$

At the end of six months, the account balance is $1040.40.

This process would become tedious if we were computing the balance after 20 years. Therefore, compound interest problems are usually solved with their own formula.

Notice that for each quarter's calculation in Example 1 we multiplied the annual rate of 8% (0.08) by the time 1 quarter ($\frac{1}{4}$ year) and got 2% (0.02). This 2% is the **quarterly rate** (or more generally, the **periodic rate**). A periodic rate is any rate that is prorated in this manner from an annual rate.

If i is the periodic interest rate, then the future value (the account balance in the case of an investment or the maturity value in the case of a loan) at the end of the first period is

$$FV = P(1 + i)$$

Because this is the account balance at the beginning of the second period, it becomes the new principal (or present value). Substituting the new $P(1 + i)$ for P, we have an account balance at the end of the second period of

$$FV = P(1 + i) \cdot (1 + i)$$

$$= P(1 + i)^2$$

$P(1 + i)^2$ therefore becomes the account balance at the beginning of the third period, and the future value at the end of the third period is:

$$FV = [P(1 + i)^2] \cdot (1 + i) \qquad \text{Substituting } P(1 + i)^2 \text{ for } P$$

$$= P(1 + i)^3$$

We can generalize this procedure by replacing the exponent with n to represent any number of periods. Thus we get the compound interest formula.

Compound Interest Formula

At the end of n periods, the future value FV of an initial principal P subject to compound interest at a periodic interest rate i for n periods is

$$FV = P(1 + i)^n$$

Notice that we have maintained our variables tradition—i and n do not measure amounts of money, so they are not capital letters. We now have three interest-related variables:

- r, the annual interest rate (not an amount of money)
- i, the periodic interest rate (not an amount of money)
- I, the interest itself (an amount of money)

EXAMPLE 2   Recompute Example 1 with the compound interest formula.

Solution   Compounding quarterly, we have $P = 1000$, $i = \frac{1}{4}$ of $8\% = 2\% = 0.02$, and $n = \frac{1}{2}$ year = $\frac{1}{2}$ year $\cdot \frac{4 \text{ quarters}}{1 \text{ year}} = 2$ quarters (i and n are *periodic* figures).

$$FV = P(1 + i)^n$$
$$= 1000 (1 + 0.02)^2$$
$$= \$1040.40$$

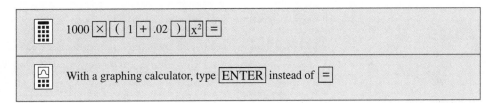

This is a slightly easier calculation than that done in Example 1.   •

Comparing the future value derived from compound interest in Example 2 with the future value derived from simple interest, we find little difference. Computing future value at simple interest, we have $P = 1000$, $r = 8\% = 0.08$, and $t = 6$ months $= \frac{1}{2}$ year (r and t are *annual* figures).

$$FV = P(1 + rt)$$
$$= 1000\left(1 + 0.08 \cdot \frac{1}{2}\right)$$
$$= \$1040.00$$

In this instance, compound interest gives us an extra 40 cents. As we shall see, however, the difference between simple interest and compound interest isn't always so insignificant.

EXAMPLE 3   In 1777, Jacob DeHaven, a wealthy Pennsylvania merchant, responded to a desperate plea from George Washington when it looked as if the Revolutionary War was about to be lost. He lent the Continental Congress \$450,000 to rescue Washington's troops at Valley Forge. When the war was over, Mr. DeHaven unsuccessfully tried to collect what was owed him. He died penniless in 1812. In 1990, his descendants sued the U.S. government for repayment of the original amount plus interest at the then-prevailing rate of 6%. (*Source: New York Times*, May 27, 1990.) How much did the government owe his descendents on the 1990 anniversary of the loan, if the interest is:

a. compounded monthly
b. simple interest

Solution   **a.** To find the future value at monthly compounded interest, we have $P = 450,000$ and $i = \frac{1}{12}$th of $6\% = \frac{0.06}{12}$. Thus

$$n = 213 \text{ years} = 213 \text{ years} \cdot \frac{12 \text{ months}}{1 \text{ year}} = 2556 \text{ months} \quad (i \text{ and } n \text{ are } periodic \text{ figures})$$

$$FV = P(1 + i)^n$$

$$= 450,000\left(1 + \frac{0.06}{12}\right)^{2556}$$

$$= 1.547627234\ldots \times 10^{11}$$

$$\approx \$154,762,723,400$$

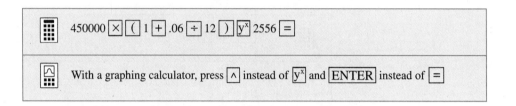

Notice how much easier this method is than the method of Example 1. (We would have to make 2556 calculations!)

b. To find the future value at simple interest, we have $P = 450,000$, $r = 6\% = 0.06$, and $t = 213$ years (r and t are *annual* figures).

$$FV = P(1 + rt)$$

$$= 450,000(1 + 0.06 \cdot 213)$$

$$= \$6,201,000$$

In Example 3, the future value with compound interest is almost 25,000 times the future value with simple interest. Over longer periods of time, compound interest is immensely more profitable to the investor than simple interest, because compound interest gives interest on interest. Similarly, compounding more frequently (daily rather than quarterly, for example) is more profitable to the investor.

The effects of the size of the time interval and the compounding period can be seen in Figures 10.1–10.3.

FIGURE 10.1

	Future Value of $1000 at 10% Interest	
	In 1 year	In 10 years
Simple Interest	$1100.00	$2000.00
Compounded Annually	$1100.00	$2593.74
Compounded Quarterly	$1103.81	$2685.06
Compounded Monthly	$1104.71	$2707.04
Compounded Daily	$1105.16	$2717.91

FIGURE 10.2
Future value of $1000
invested at 10% interest in
1 year

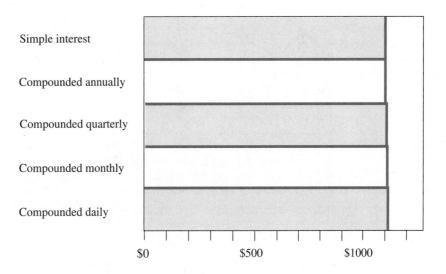

Simple interest

Compounded annually

Compounded quarterly

Compounded monthly

Compounded daily

$0 $500 $1000

FIGURE 10.3
Future value of $1000
invested at 10% interest in
10 years

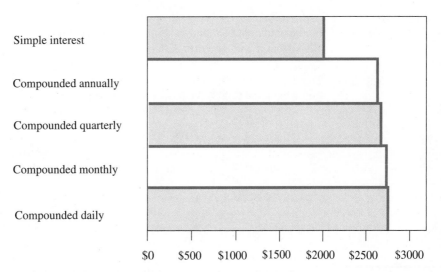

Simple interest

Compounded annually

Compounded quarterly

Compounded monthly

Compounded daily

$0 $500 $1000 $1500 $2000 $2500 $3000

EXAMPLE 4   **a.** Find the future value of $2500 deposited in an account where it earns 10.3% interest compounded daily for 15 years.
b. Find the interest earned.

Solution   **a.** Compounding daily, we have $P = 2500$, $i = \frac{1}{365}$ of $10.3\% = \frac{0.103}{365}$, and $n = 15$ years $=$ 15 years $\cdot \frac{365 \text{ days}}{1 \text{ year}} = 5475$ days

$$FV = P(1 + i)^n$$

$$= 2500\left(1 + \frac{.103}{365}\right)^{5475}$$

$$= 11{,}717.37496\ldots$$

$$\approx \$11{,}717.37$$

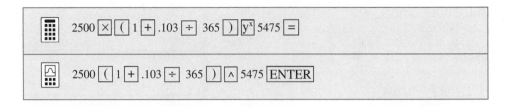

 WARNING: If you compute $\frac{.103}{365}$ separately, you will get a long decimal. Do not round off that decimal, because the resulting answer could be inaccurate. Doing the calculation all at once (as shown above) avoids this difficulty.

b. The principal is $2500, and the total of principal and interest is $11,717.37; therefore, the interest is $11,717.37 − $2500 = $9217.37.

EXAMPLE 5   Find the amount of money that must be invested now at $7\frac{3}{4}\%$ interest compounded annually so that it will be worth $2000 in 3 years.

Solution   The question actually asks us to find the present value, or principal P, that will generate a future value of $2000. We have $FV = 2000$, $i = 7\frac{3}{4}\% = .0775$, and $n = 3$.

$$FV = P(1 + i)^n$$
$$2000 = P(1 + .0775)^3$$
$$P = \frac{2000}{1.0775^3}$$
$$= 1598.7411 \ldots$$
$$\approx \$1598.74$$

Annual Yield

Which investment is the more profitable—one that pays 5.8% compounded daily or one that pays 5.9% compounded quarterly? It's difficult to tell; certainly, 5.9% is a better rate than 5.8%, but compounding daily is better than compounding quarterly. The two rates cannot be directly compared because of their different compounding frequencies. The way to tell which is the better investment is to find the annual yield of each.

The **annual yield** (or just **yield**) of a compound interest deposit is the *simple interest rate* that has the same future value as the compound rate would have in one year. The annual yields of two different investments can be compared because they are both simple interest rates. Annual yield provides the consumer with a uniform basis for comparison. The annual yield should be slightly higher than the compound rate, because compound interest is slightly more profitable than simple interest over a short period of time. The compound rate is sometimes called the **nominal rate** to distinguish it from the yield (here, nominal means *named* or *stated*).

Most banks advertise their
yields as well as their rates.

THE GW MAXIMUM YIELD CD 6-Month Term, $2,500 Deposit Tier	
9.27% Yield	**8.90%** Rate
THE GW MAXIMUM YIELD CD 1 Year Term, $2,500 Deposit Tier	
9.21% Yield	**8.85%** Rate

Interest compounded monthly.

Because the annual yield is the simple interest rate that has the same future value as the compound rate in one year, the formula for the future value at the simple interest rate will equal the formula for the future value at the compound rate (with r being the annual yield).

$$FV \text{ (compound interest)} = FV \text{ (simple interest)}$$
$$P(1 + i)^n = P(1 + rt)$$

EXAMPLE 6   Find the annual yield of $2500 deposited in an account where it earns 10.3% interest compounded daily for 15 years.

Solution

Compound interest

$P = 2500$

$i = \dfrac{1}{365}$ of $10.3\% = \dfrac{.103}{365}$

$n = 1$ year $= 365$ days

(*i* and *n* are *periodic* figures)

Simple interest

$P = 2500$

$r =$ unknown annual yield

$t = 1$ year

(*r* and *t* are *annual* figures)

$$P(1 + i)^n = P(1 + rt)$$

$$2500 \left(1 + \frac{.103}{365}\right)^{365} = 2500\,(1 + r \cdot 1)$$

$$\left(1 + \frac{.103}{365}\right)^{365} = 1 + r \qquad \text{Dividing by 2500}$$

$$r = \left(1 + \frac{.103}{365}\right)^{365} - 1$$

$$= .10847530\ldots \approx 10.85\%$$

✔ Notice that the annual yield is slightly higher than the compound rate, as it should be.

By tradition, we round this to the nearest hundredth of a percent, so the annual yield is 10.85%. This means that, in one year's time, 10.3% compounded daily has the same effect as 10.85% simple interest. For any period of time longer than a year, 10.3% compounded daily will yield *more* interest than would 10.85% simple interest, because compound interest gives interest on interest. ●

Notice that in Example 6, the principal of $2500 canceled out. If the principal were two dollars or two million dollars, it would still cancel out, and the annual yield of 10.3% compounded daily would still be 10.85%. The principal doesn't matter in computing the annual yield. Also notice that the 15 years did not enter into the calculation; annual yield is *always* based on a *one-year* period, so that we can compare various investment terms to a standard.

EXAMPLE 7   Find the annual yield corresponding to a nominal rate of 8.4% compounded monthly.

Solution   We are told neither the principal nor the time, but (as discussed above) these variables don't affect the annual yield. Compounding monthly, we have $i = \frac{1}{12}$ of $8.4\% = \frac{0.084}{12}$, $n = 1$ year $= 12$ months, and $t = 1$ year.

$$FV \text{ (compounded monthly)} = FV \text{ (simple interest)}$$

$$P(1 + i)^n = P(1 + rt)$$

$$(1 + i)^n = 1 + rt \qquad \text{Dividing by } P$$

$$\left(1 + \frac{0.084}{12}\right)^{12} = 1 + r \cdot 1$$

$$r = \left(1 + \frac{0.084}{12}\right)^{12} - 1$$

$$r = 0.08731066 \dots$$

$$\approx 8.73\%$$

> ✔ Notice that the annual yield is slightly higher than the compound rate, as it should be. •

In Example 7, we found that 8.4% compounded monthly generates an annual yield of 8.73%. This means that 8.4% compounded monthly has the same effect as does 8.73% simple interest, in one year's time. Furthermore, as Figure 10.4 indicates, 8.4% compounded monthly has the same effect as does 8.73% compounded annually *for any time period.*

FIGURE 10.4

For a Principal of $1000	After 1 Year	After 10 Years
FV at 8.4% compounded monthly	$1087.31	$2309.60
FV at 8.73% simple interest	$1087.30	$1873.00
FV at 8.73% compounded annually	$1087.30	$2309.37

Notice that all three rates have the same future value after one year (the 1¢ difference is due to rounding off the annual yield to 8.73%). However, after ten years, the simple interest has fallen way behind, while the 8.4% compounded monthly and 8.73% compounded annually remain the same (except for the round-off error). This always happens. The annual yield is the simple interest rate that has the same future value the compound rate would have in one year; it is also the annually compounded rate that has the same future value the nominal rate would have after any amount of time.

An annual yield formula does exist, but the annual yield can be calculated efficiently without it, as shown above. The formula is developed in the exercises.

10.2

EXERCISES

In Exercises 1–6, find the periodic rate that corresponds to the given compound rate, if the rate is compounded (a) quarterly, (b) monthly, (c) daily, (d) biweekly (every two weeks), and (e) semi-monthly (twice a month). Do not *round off the periodic rate.*

1. 12% **2.** 6% **3.** 3.1%

4. 6.8% **5.** 9.7% **6.** 10.1%

In Exercises 7–10, find the number of periods that corresponds to the given time span, if a period is (a) a quarter of a year, (b) a month, and (c) a day. (Ignore leap years.)

7. $8\frac{1}{2}$ years **8.** $9\frac{3}{4}$ years

9. 30 years **10.** 45 years

In Exercises 11–16, find and interpret the future value of the given amount.

11. $3000 at 6% compounded annually for 15 years

12. $7300 at 7% compounded annually for 13 years

13. $5200 at $6\frac{3}{4}$% compounded quarterly for $8\frac{1}{2}$ years

14. $36,820 at $7\frac{7}{8}$% compounded quarterly for 4 years

15. $1960 at $4\frac{1}{8}$% compounded daily for 17 years (ignore leap years)

16. $12,350 at 6% compounded daily for $10\frac{1}{2}$ years (ignore leap years)

In Exercises 17–22, find and interpret the annual yield corresponding to the given nominal rate.

17. 8% compounded monthly

18. $5\frac{1}{2}$% compounded quarterly

19. $4\frac{1}{4}$% compounded daily

20. $12\frac{5}{8}$% compounded daily

21. 10% compounded (a) quarterly, (b) monthly, and (c) daily

22. $12\frac{1}{2}$% compounded (a) quarterly, (b) monthly, and (c) daily

In Exercises 23–26, find and interpret the present value that will generate the given future value.

23. $1000 at 8% compounded annually for 7 years

24. $9280 at $9\frac{3}{4}$% compounded monthly for 2 years 3 months

25. $3758 at $11\frac{7}{8}$% compounded monthly for 17 years 7 months

26. $4459 at $10\frac{3}{4}$% compounded quarterly for 4 years

For Exercises 27–31, note the following information: A certificate of deposit (or CD) is an agreement between a bank and a saver in which the bank guarantees an interest rate and the saver commits to leaving his or her deposit in the account for an agreed-upon period of time.

27. First National Bank offers two-year CDs at 9.12% compounded daily, and Citywide Savings offers two-year CDs at 9.13% compounded quarterly. Compute the annual yield for each institution and determine which is more advantageous for the consumer.

28. National Trust Savings offers five-year CDs at 8.25% compounded daily, and Bank of the Future offers five-year CDs at 8.28% compounded annually. Compute the annual yield for each institution, and determine which is more advantageous for the consumer.

29. Verify the annual yield for the five-year CD quoted in the bank sign in Figure 10.5.

FIGURE 10.5

30. Verify the yield for the one-year CDs quoted in the savings bank advertisement below.

Worrying what to do with your money? Bury it in Jean Lafitte Savings Bank's 1-year "High Seas" CDs! You'll rest in peace.

Rate: 8.7% compounded monthly
Yield: 10.16%

31. Verify the yield for the two-year CDs quoted in the savings bank advertisement on page 491.

It's not just a great deal –
it's highway robbery!
2-year CDs from
Cole Younger Savings Bank:

| 9.3% interest, compounded daily |
| 10.74% yield |

32. Verify the yield for six-month money market accounts quoted in the Great Western bank advertisement on page 487.

33. When Jason Levy was born, his grandparents deposited $3000 into a special account for Jason's college education. The account earned $6\frac{1}{2}\%$ interest compounded daily.
  a. How much will be in the account when Jason is 18?
  b. If, on becoming 18, Jason arranged for the monthly interest to be sent to him, how much would he receive each 30-day month?

34. When Alan Cooper was born, his grandparents deposited $5000 into a special account for Alan's college education. The account earned $7\frac{1}{4}\%$ interest compounded daily.
  a. How much will be in the account when Alan is 18?
  b. If, on becoming 18, Alan arranged for the monthly interest to be sent to him, how much would he receive each 30-day month?

For Exercises 35–38, note the following information: An Individual Retirement Account *(or* IRA*) is an account in which the saver does not pay income tax on the amount deposited but is not allowed to withdraw the money until retirement. (The saver pays income tax at that point, but his or her tax bracket is much lower then.)*

35. At age 27, Lauren Johnson deposited $1000 into an IRA, where it earns $7\frac{7}{8}\%$ compounded monthly. What will it be worth when she retires at 65?

36. At age 36, Dick Shoemaker deposited $2000 into an IRA, where it earns $8\frac{1}{8}\%$ compounded semiannually. What will it be worth when he retires at 65?

37. Marlene Silva wishes to have an IRA that will be worth $100,000 when she retires at age 65.
  a. How much must she deposit at age 35 at $8\frac{3}{8}\%$ compounded daily?
  b. If at age 65 she arranges for the monthly interest to be sent to her, how much would she receive each 30-day month?

38. David Murtha wishes to have an IRA that will be worth $150,000 when he retires at age 65.

  a. How much must he deposit at age 26 at $6\frac{1}{8}\%$ compounded quarterly?
  b. If, at age 65, he arranges for the monthly interest to be sent to him, how much will he receive each 30-day month?

39. In December 1996, Bank of the West offered six-month CDs at 5.0% interest compounded monthly.
  a. Find the CD's annual yield.
  b. How much would a $1000 CD be worth at maturity?
  c. How much interest would you earn?
  d. What percent of the original $1000 is this interest?
  e. The answer to part (d) is not the same as that of part (a). Why?
  f. The answer to part (d) is close to, but not exactly half, that of part (a). Why?

40. In December 1996, Bank of the West offered six-month CDs at 5.0% interest compounded monthly, and one-year CDs at 5.20% interest compounded monthly. Maria Ruiz bought a six-month $2000 CD, even though she knew she wouldn't need the money for at least a year, because it was predicted that interest rates would rise.
  a. Find the future value of Maria's CD.
  b. Six months later Maria's CD came to term, and in the intervening time the interest rates had risen. She reinvests the principal and interest from her first CD in a second six-month CD that pays 5.31% interest compounded monthly. Find the future value of Maria's second CD.
  c. Would Maria have been better off if she bought a one-year CD in December 1996?
  d. If Maria's second CD pays 5.46% interest compounded monthly, rather than 5.31%, would she be better off with the two six-month CDs or the one-year CD?

41. Develop a formula for the annual yield of a compound interest rate.
  HINT: Follow the procedure given in Example 7, but use the letters *i* and *n* in the place of numbers.

In Exercises 42–46, use the formula found in Exercise 41 to compute the annual yield corresponding to the given nominal rate.

42. $9\frac{1}{2}\%$ compounded monthly

43. $7\frac{1}{4}\%$ compounded quarterly

44. $12\frac{3}{8}\%$ compounded daily

45. $15\frac{5}{8}\%$ compounded (a) semiannually, (b) quarterly, (c) monthly, (d) daily, (e) biweekly, and (f) semimonthly.

46. $10\frac{1}{2}\%$ compounded (a) semiannually, (b) quarterly, (c) monthly, (d) daily, (e) biweekly, and (f) semimonthly.

Answer the following questions using complete sentences.

47. Why is there no work involved in finding the annual yield of a given simple interest rate?

48. Why is there no work involved in finding the annual yield of a given compound interest rate, when that rate is compounded annually?

49. Which should be higher, the annual yield of a given rate compounded quarterly or compounded monthly? Explain why, *without* performing any calculations or referring to any formulas.

50. Why should the annual yield of a given compound interest rate be higher than the compound rate? Why should it be only slightly higher? Explain why, *without* performing any calculations or referring to any formulas.

51. Explain the difference between simple interest and compound interest.

52. The simple interest rate corresponding to a given simple discount rate is somewhat similar in concept to the annual yield of a given compound interest rate. Compare and contrast these two concepts.

53. *Money Magazine* and other financial publications regularly list the top-paying money-market funds, the top-paying bond funds, and the top-paying CDs and their yields. Why do they list yields rather than interest rates and compounding periods?

54. Equal amounts are invested in two different accounts. One account pays simple interest, and the other pays compound interest, at the same rate. When will the future values of the two accounts be the same?

DOUBLING TIME ON A GRAPHING CALCULATOR

Simple interest is a very straightforward concept. If an account earns 5% simple interest, then 5% of the principal is paid for each year that the principal is in the account. In one year the account earns 5% interest, in two years it earns 10% interest, in three years it earns 15% interest, and so on.

It is not nearly so easy to get an intuitive grasp of compound interest. If an account earns 5% interest compounded daily, then it does not earn 5% interest in one year, and it does not earn 10% interest in two years.

Annual yield is one way of gaining an intuitive grasp of compound interest. If an account earns 5% interest compounded daily, then it will earn 5.13% interest in one year (because the annual yield is 5.13%). But it does not earn 10.26% (twice 5.13%) interest in two years.

Doubling time is another way of gaining an intuitive grasp of compound interest. **Doubling time** is the amount of time it takes for an account to double in value; that is, it's the amount of time it takes for the future value to become twice the principal. To find the doubling time for an account that earns 5% interest compounded daily, substitute $2P$ for the future value and solve the resulting equation.

$$FV = P(1 + i)^n \qquad \text{Compound interest future value formula}$$

$$2P = P\left(1 + \frac{0.05}{365}\right)^n \qquad \text{Substituting}$$

$$2 = \left(1 + \frac{0.05}{365}\right)^n \qquad \text{Dividing by } P$$

How do you solve this for n? One method involves logarithms; you may have learned that method in Intermediate Algebra. Another method involves the graphing calculator.

EXAMPLE 8 Use a graphing calculator to find the doubling time for an account that earns 5% interest compounded daily.

Solution   Step 1   *Use the calculator to graph two different equations:*

$$y = 2 \quad \text{and} \quad y = \left(1 + \frac{0.05}{365}\right)^x$$

Follow the procedure discussed in Section 1.0. Fill in the "Y=" screen as shown in Figure 10.6. If you use the "ZoomStandard" command, you will get the graph shown in Figure 10.6.

FIGURE 10.6

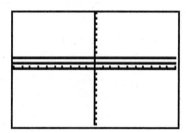

Step 2   *Adjust the calculator's screen so that it shows the point at which these two equations intersect.* To do this, we must determine appropriate values of xmin, xmax, ymin, and ymax.

In the equation $y = (1 + \frac{0.05}{365})^x$, x measures time and y measures money. Neither can be negative, so set xmin and ymin to 0.

We can get a rough idea of how big x should be by finding the doubling time for simple interest (a much easier calculation):

$FV = P(1 + rt)$	Simple interest future value formula
$2P = P(1 + 0.05t)$	Substituting $2P$ for FV
$2 = 1 + 0.05t$	Dividing by P
$1 = 0.05t$	
$t = \dfrac{1}{0.05} = 20$	

The doubling time for 5% *simple* interest is 20 years. Simple interest isn't as productive as compound interest, so the doubling time for 5% *compound* interest will be less than 20 years. In the equation $y = (1 + \frac{0.05}{365})^x$, x measures time in days, and 20 years $= 20 \cdot 365 = 7300$ days, so set xmax to 7300.

The value of ymax should correspond to xmax $= 7300$:

$$y = \left(1 + \frac{0.05}{365}\right)^{7300}$$

$$= 2.71809 \ldots$$

$$\approx 2.7$$

Set ymax to 2.7.

Step 3   *Press the* GRAPH *button to obtain the screen shown in Figure 10.7. Notice that* we can now see the point of intersection.

FIGURE 10.7

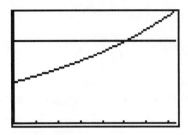

Step 4 *Use the calculator to find the point of intersection.* Follow the procedure discussed in Section 1.2 and summarized below.

TI-82/83:	Select the "intersect" option from the "CALC" menu
TI-85/86:	Select "MATH" for the "GRAPH" menu, and then "ISECT" from the "GRAPH MATH" menu

Figure 10.8 shows the result of this work.

FIGURE 10.8

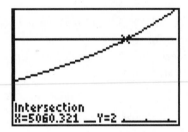

The intersection is at (5060.321, 2). This means that it takes 5060.321 days (or about $\frac{5060.321}{365} = 13.86389315\ldots \approx 13.9$ years) for money invested at 5% interest compounded daily to double.

✔ The solution can be checked by substituting 5060.321 for x in the equation

$$2 = \left(1 + \frac{0.05}{365}\right)^x$$

The right side is

$$\left(1 + \frac{0.05}{365}\right)^x = \left(1 + \frac{0.05}{365}\right)^{5060.321}$$

$$= 2.000000005$$

This solution is mathematically (almost) correct, as shown in the above box. Practically speaking, however, the ".321" part of the solution doesn't make sense. If interest is compounded daily, then at the end of each day, your account is credited with that day's interest. After 5060 days, your account would contain slightly less than twice the original principal. After 5060.321 days, your account balance would not have changed, since interest won't be credited until the end of the day. After 5061 days, your account would contain slightly more than twice the original principal. The doubling time is 5061 days $\approx$ 13.9 years.

●

EXERCISES

55. The solution in Example 8 didn't quite check; we found that the y-value at the point of intersection is 2.000000005, not 2. Why is there a discrepancy?

56. If $1000 is deposited into an account that earns 5% interest compounded daily, the doubling time is approximately 5061 days.
   a. Find the amount in the account after 5061 days.
   b. Find the amount in the account after 2 · 5061 days.
   c. Find the amount in the account after 3 · 5061 days.
   d. Find the amount in the account after 4 · 5061 days.
   e. What conclusion can you make?

57. Complete parts (a)–(e). (Give the number of periods and the number of years, rounded to the nearest hundredth.)
   a. Find the doubling time corresponding to 5% interest compounded annually.
   b. Find the doubling time corresponding to 5% interest compounded monthly.
   c. Find the doubling time corresponding to 5% interest compounded quarterly.
   d. Find the doubling time corresponding to 5% interest compounded daily.
   e. Discuss the effect of the compounding period on doubling time.

58. Complete parts (a)–(d). (Give the number of years, rounded to the nearest hundredth.)

   a. Find the doubling time corresponding to 6% interest compounded annually.
   b. Find the doubling time corresponding to 7% interest compounded annually.
   c. Find the doubling time corresponding to 10% interest compounded annually.
   d. Discuss the effect of the interest rate on doubling time.

59. If you invest $10,000 at 8.125% interest compounded daily, how long will it take for you to accumulate $15,000? How long will it take for you to accumulate $100,000? (Give the number of periods and the number of years, rounded to the nearest hundredth.)

60. If you invest $15,000 at $9\frac{3}{8}$% interest compounded daily, how long will it take for you to accumulate $25,000? How long will it take for you to accumulate $100,000? (Give the number of periods and the number of years, rounded to the nearest hundredth.)

61. If you invest $20,000 at $6\frac{1}{4}$% interest compounded daily, how long will it take for you to accumulate $30,000? How long will it take for you to accumulate $100,000? (Give the number of periods and the number of years, rounded to the nearest hundredth.)

10.3

ANNUITIES

Many people have long-term financial goals and limited means with which to accomplish them. Your goal may be to save $3000 over the next four years for your college education, to save $10,000 over the next 10 years for the down payment on a home, to save $30,000 over the next 18 years to finance your new baby's college education, or to save $300,000 over the next 40 years for your retirement. It seems incredible, but each of these goals can be achieved by saving only $50 a month (if interest rates are favorable)! All you need to do is start an annuity.

An **annuity** is simply a sequence of equal, regular payments into an account where each payment receives compound interest. Because most annuities involve relatively small periodic payments, they're affordable to the average person. Over longer periods of time, the payments themselves start to amount to a significant sum, but it's really the power of compound interest that makes annuities so amazing. If you pay $50 a month into an annuity for the next 40 years, then your total payment is

$$\frac{\$50}{\text{month}} \cdot \frac{12 \text{ months}}{\text{year}} \cdot 40 \text{ years} = \$24{,}000$$

However, if the annuity pays 10% interest compounded monthly, after 40 years the account will contain over $316,000!

A *Christmas Club* is an annuity that is set up to save for Christmas shopping. A Christmas Club participant makes regular equal deposits, and the deposits and the resulting interest are released to the participant in December when the money is needed. Christmas Clubs are different from other annuities in that they span a short amount of time—a year at most—and thus earn only a small amount of interest. (People set them up to be sure they're putting money aside rather than to generate interest.) Our first few examples will deal with Christmas Clubs because their short time span makes it possible to see how an annuity actually works.

Calculating Short-Term Annuities

EXAMPLE 1 On August 12, Patty Leitner joined a Christmas Club through her bank. For the next three months, she would deposit $200 at the beginning of each month. The money would earn $8\frac{3}{4}\%$ interest compounded monthly, and on December 1 she could withdraw her money for shopping. Use the compound interest formula to find the future value of the account.

Solution We are given $P = 200$, and $r = \frac{1}{12}$ of $8\frac{3}{4}\% = \frac{0.0875}{12}$.

First, calculate the future value of the first payment (made on September 1), using $n = 3$ (it will receive interest during September, October, and November).

$$FV = P(1 + i)^n$$

$$= 200\left(1 + \frac{0.0875}{12}\right)^3$$

$$= 204.40697 \ldots$$

$$\approx \$204.41$$

Next, calculate the future value of the second payment (made on October 1), using $n = 2$ (it will receive interest during October and November).

$$FV = P(1 + i)^n$$

$$= 200\left(1 + \frac{0.0875}{12}\right)^2$$

$$= 202.9273 \ldots$$

$$\approx \$202.93$$

To calculate the future value of the third payment (made on November 1), use $n = 1$ (it will receive interest during November).

$$FV = P(1 + i)^n$$

$$= 200\left(1 + \frac{0.0875}{12}\right)^1$$

$$= 201.45833\ldots$$

$$\approx \$201.46$$

The payment schedule and interest earned are illustrated in Figure 10.9.

FIGURE 10.9

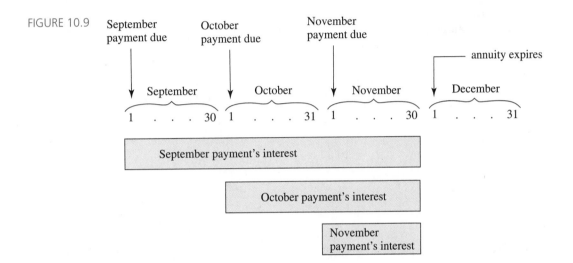

The future value of Patty's annuity is then the sum of the future values of each payment:

$$FV \approx \$204.41 + \$202.93 + \$201.46$$

$$= \$608.80$$

Patty's deposits will total $600.00; therefore, she will earn $8.80 interest on her deposits.

The **payment period** of an annuity is the time between payments: in Example 1, the payment period was one month. The *term* is the time from the beginning of the first payment period to the end of the last; the term of Patty's Christmas Club was three months. When an annuity has *expired* (that is, when its term is over), the entire account or any portion of it may be withdrawn. Naturally, any portion not withdrawn would continue to receive interest.

Most annuities are **simple,** that is, their compounding period is the same as their payment period (for example, if payments are made monthly, then interest is compounded monthly). In this book, we will work only with simple annuities. An **annuity due** is one for which each payment is due at the beginning of its time period; Patty's annuity in Example 1 was an annuity due because the payments were due at the *beginning* of each month. An **ordinary annuity** is an annuity for which each payment is due at the end of its time period; as the name implies, this form of annuity is more typical. As we will see in the next example, the difference is one of accounting.

EXAMPLE 2   Dan Bach also joined a Christmas Club through his bank. His was just like Patty's, except his payments were due at the end of each month, and his first payment was due September 30. Use the compound interest formula to find the future value of the account.

Solution   This is an *ordinary* annuity because payments are due at the *end* of each month; interest is compounded monthly. From Example 1, we know that $P = 200$, $i = \frac{1}{12}$ of $8\frac{3}{4}\% = \frac{0.0875}{12}$.

To calculate the future value of the first payment (made on September 30), use $n = 2$ (this payment will receive interest during October and November).

$$FV = P(1 + i)^n$$

$$= 200\left(1 + \frac{0.0875}{12}\right)^2$$

$$= 202.9273\ldots$$

$$\approx \$202.93$$

To calculate the future value of the second payment (made on October 31), use $n = 1$ (this payment will receive interest during November).

$$FV = P(1 + i)^n$$

$$= 200\left(1 + \frac{0.0875}{12}\right)^1$$

$$= 201.45833\ldots$$

$$\approx \$201.46$$

To calculate the future value of the third payment (made on November 30), note that no interest is earned, because the payment is due November 30 and the annuity expires December 1. Therefore,

$$FV = \$200$$

Dan's payment schedule and interest payments are illustrated in Figure 10.10 on page 499. The future value of Dan's annuity is the sum of the future values of each payment:

$$FV \approx \$200 + \$201.46 + \$202.93$$

$$= \$604.39$$

Dan earned $4.39 interest on his deposits.

FIGURE 10.10

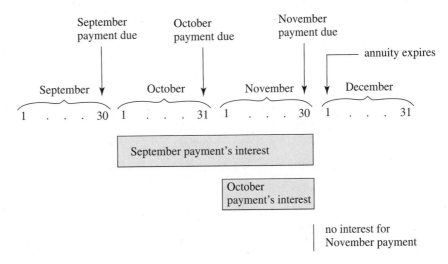

In Examples 1 and 2, why did Patty earn more interest than Dan? The reason is that each of her payments was made a month earlier and therefore received an extra month's interest. In fact, we could find the future value of Patty's account by giving Dan's future value one more month's interest:

$$604.39 \cdot \left(1 + \frac{0.0875}{12}\right)^1 = 608.80$$

Dan's FV · extra interest = Patty's FV

The difference between an ordinary annuity and an annuity due is strictly an accounting difference, because any ordinary annuity in effect will become an annuity due if you leave all funds in the account for one extra period.

Calculating Long-Term Annuities

The procedure followed in Examples 1 and 2 reflects what actually happens with annuities, and it works fine for a small number of payments. However, most annuities are long term. If we were to calculate the future value of an annuity with a term of 30 years of monthly payments, the above procedure would be very laborious! Long-term annuities should be calculated using a new formula.

For an *ordinary* annuity with payment *pymt*, periodic rate *i*, and a term of *n* payments, the first payment receives interest for $n - 1$ periods (payment is made at the end of the first period, so it received no interest for that period). Its future value is then

$$FV(\text{first pymt}) = pymt(1 + i)^{n-1}$$

The last payment receives no interest (under the annuity), because it is due at the end of the last period and it expires the next day. Its future value is simple

$$FV(\text{last pymt}) = pymt$$

The next-to-last payment receives one period's interest, so its future value is

$$FV(\text{next-to-last pymt}) = pymt(1 + i)^1$$

The future value of the annuity is the sum of all of these future values of individual payments:

$$FV = pymt + pymt(1 + i)^1 + pymt(1 + i)^2 + \cdots + pymt(1 + i)^{n-1}$$

In order to get a shortcut formula from all this, we can first multiply each side of this equation by $(1 + i)$ and then subtract the original equation from the result. This leads to a lot of cancelling.

$$FV(1+i) = pymt(1 + i) + pymt(1 + i)^2 + \cdots + pymt(1 + i)^{n-1} + pymt(1 + i)^n$$

$$\text{minus: } FV = pymt + pymt(1 + i)^1 + pymt(1 + i)^2 + \cdots + pymt(1 + i)^{n-1}$$

$$\text{equals: } FV(1 + i) - FV = pymt(1 + i)^n - pymt \qquad \text{Subtracting}$$

$$FV(1 + i - 1) = pymt[(1 + i)^n - 1] \qquad \text{Factoring}$$

$$FV(i) = pymt[(1 + i)^n - 1]$$

$$FV = pymt \frac{(1 + i)^n - 1}{i} \qquad \text{Dividing by } i$$

This is the future value of the ordinary annuity.

Ordinary Annuity Formula

The future value FV of an ordinary annuity with payment size $pymt$, period rate i, and a term of n payments is

$$FV(\text{ordinary}) = pymt \frac{(1 + i)^n - 1}{i}$$

Notice that in the ordinary annuity formula, the amount $pymt$ is multiplied by the fraction $\frac{(1 + i)^n - 1}{i}$. Some sources refer to this fraction as $s_{\overline{n}|i}$, and refer to the payment ($pymt$) as "rent." We will not use this notation and terminology.

As we saw in Examples 1 and 2, the difference between an ordinary annuity and an annuity due is that with an annuity due each payment receives one more period's interest (because each payment is made at the beginning of the period). Thus, the future value of an annuity due is the future value of an ordinary annuity plus one more period's interest.

$$FV \text{ (of annuity due)} = FV \text{ (of ordinary annuity)} \cdot (1 + i)$$

$$= pymt \frac{(1 + i)^n - 1}{i} (1 + i)$$

Annuity Due Formula

The future value FV of an annuity due with payment size $pymt$, periodic rate i, and a term of n payments is

$$FV(\text{due}) = FV(\text{ordinary}) \cdot (1 + i)$$

$$= pymt \frac{(1 + i)^n - 1}{i} (1 + i)$$

**Tax-Deferred
Annuities**

A **tax-deferred annuity (TDA)** is an annuity that is set up to save for retirement. Money is automatically deducted from the participant's paychecks until retirement, and the federal (and perhaps state) tax deduction is computed *after* the annuity payment has been deducted, resulting in significant tax savings. In some cases, the employer also makes a regular contribution to the annuity.

The following example involves a long-term annuity. Usually, the interest rate of a long-term annuity varies somewhat from year to year. In this case, calculations must be viewed as predictions, not guarantees.

EXAMPLE 3

Jim Moran just got a new job, and he immediately set up a tax-deferred annuity to save for retirement. He arranged to have $200 taken out of each of his monthly checks, which will earn $8\frac{3}{4}\%$ interest. Due to the tax-deferring effect of the TDA, his take-home pay went down by only $115. Jim just had his 30th birthday and his ordinary annuity will come to term when he is 65. Find the future value of the annuity.

Solution

This is an ordinary annuity, with $pymt = 200$, $i = \frac{1}{12}$ of $8\frac{3}{4}\% = \frac{0.0875}{12}$, and $n = 35$ years $= 35$ years $\cdot \frac{12 \text{ months}}{1 \text{ year}} = 420$ monthly payments.

$$FV = pymt \frac{(1 + i)^n - 1}{i}$$

$$= 200 \frac{\left(1 + \dfrac{0.0875}{12}\right)^{420} - 1}{\dfrac{0.0875}{12}}$$

$$\approx \$552,539.96$$

Because $\frac{0.0875}{12}$ occurs twice in the calculation, compute it first and put it into your calculator's memory. After storing the periodic rate, type

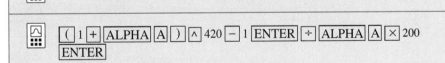

How much of the future value of the annuity in Example 3 is interest? Jim made 420 payments of $200 each, so he paid a total of $420 \cdot \$200 = \$84,000$. The interest is then $\$552,539.96 - \$84,000 = \$468,539.96$. The interest is almost 6 times as large as the total of Jim's payments! The magnitude of the earnings illustrates the amazing power of annuities and the effect of compound interest over a long period of time.

EXAMPLE 4

Find the future value of the annuity in Example 3 if Jim's payments were made at the beginning of each period.

Solution

$$FV(\text{due}) = FV(\text{ordinary}) \cdot (1 + i)$$

$$= 552{,}539.96 \cdot \left(1 + \frac{0.0875}{12}\right) \qquad \text{From Example 3}$$

$$= \$556{,}568.897 \ldots$$

$$\approx \$556{,}568.90$$

Sinking Funds

A **sinking fund** is an annuity in which the future value is a specific amount of money that will be used for a certain purpose, such as a child's education or the down payment on a home.

EXAMPLE 5 Jeff and Pat Schoppert wish to set up a sinking fund to save for their new baby's college education. How much would they have to have deducted from their biweekly paycheck in order to have $30,000 in 18 years, at $9\frac{1}{4}\%$ interest? Assume the account is an ordinary annuity.

Solution This is an ordinary annuity, with $i = \frac{1}{26}$ of $9\frac{1}{4}\% = \frac{0.0925}{26}$, $n = 18$ years $= 18$ years $\cdot \frac{26 \text{ periods}}{1 \text{ year}} = 468$ periods, and $FV = 30{,}000$.

$$FV = pymt \, \frac{(1 + i)^n - 1}{i}$$

$$30{,}000 = pymt \, \frac{\left(1 + \dfrac{0.0925}{26}\right)^{468} - 1}{\dfrac{0.0925}{26}}$$

We want to find the value of *pymt*. To do so, we must divide 30,000 by the fraction on the right side of the equation. Because the fraction is so complicated, it's best to first calculate the fraction and then multiply its reciprocal by 30,000.

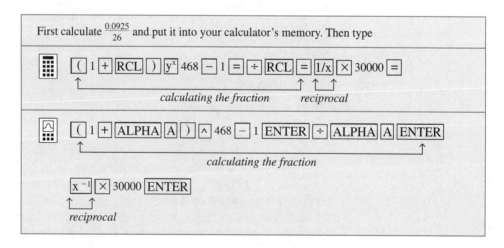

This gives *pymt* = 24.995038 . . .; Jeff and Pat would need to have only $25.00 taken out of each of their biweekly paychecks in order to save $30,000 in 18 years. Notice that, in order to have exactly $30,000 saved, they would need to deduct exactly $24.995048 . . . from their paychecks. They will actually have a few dollars more than $30,000, because the paycheck deduction was rounded up. •

EXAMPLE 6   If in Example 5 Jeff and Pat Schoppert have exactly $25.00 deducted from each paycheck, find the following:

a. the actual future value of their annuity
b. the portion that is their contribution
c. the total interest

Solution   a. This is an ordinary annuity, with $pymt = 25.00$, $i = \frac{1}{26}$ of $9\frac{1}{4}\% = \frac{0.0925}{26}$, and $n = 18$ years $= 18$ years $\cdot \frac{26 \text{ periods}}{1 \text{ year}} = 468$ periods.

$$FV = pymt \frac{(1 + i)^n - 1}{i}$$

$$FV = 25 \frac{\left(1 + \dfrac{0.0925}{26}\right)^{468} - 1}{\dfrac{0.0925}{26}}$$

$$= 30005.955 \ldots$$

$$\approx \$30,005.96$$

b. Their contribution is 468 payments of $25 each $= 468 \cdot \$25 = \$11,700$.
c. The interest is then $\$30,005.96 - \$11,700.00 = \$18,305.96$. •

The interest is more than one and a half times as large as the total of the Schoppert's payments! While this isn't quite as incredible as Jim Moran's interest, which was almost six times as large as the total of his payments (see Examples 3 and 4), it is still a very attractive investment.

Present Value of an Annuity

The **present value of an annuity** refers to the lump sum that can be deposited at the beginning of the annuity's term, at the same interest rate and with the same compounding period, that would yield the same amount as the annuity. This value can help the saver understand his or her options; it refers to an alternative way of saving the same amount of money in the same time. It is called the present value because it refers to the single action that the saver can take *in the present* (i.e., at the beginning of the annuity's term) that would have the same effect as would the annuity.

EXAMPLE 7   What lump sum deposit would Pat and Jeff Schoppert from Example 6 have to make (at the same interest rate) in order to have as much money in 18 years as their annuity would give them?

Solution   We are asked to find the present value of their annuity. Because we have already found the future value, what remains is a compound interest problem. Interest is compounded biweekly; from Example 6, we know that $i = \frac{1}{26}$ of $9\frac{1}{4}\% = \frac{0.0925}{26}$, $n = 468$, and $FV = 30005.956$.

$$FV = P(1 + i)^n$$

$$30005.955 = P\left(1 + \frac{0.0925}{26}\right)^{468}$$

$$P = \frac{30005.955}{\left(1 + \frac{0.0925}{26}\right)^{468}}$$

$$= 5693.6451\ldots$$

$$\approx \$5693.65$$

This means that the Schopperts would have to deposit $5693.65 at their baby's birth in order to save as much money as the annuity would yield. The Schopperts chose an annuity over a lump sum deposit because they couldn't afford to tie up $5700 for 18 years, but they could afford to deduct $25 out of each paycheck. •

EXAMPLE 8   Find the present value of an ordinary annuity which has $200 monthly payments for 25 years, where the account receives $10\frac{1}{2}\%$ interest.

Solution   We could find the future value of the annuity and then find the lump sum deposit whose future value matches it, as we did in the previous example. However, it's simpler to do the calculation all at once. The key is to realize that the future value of the lump sum must equal the future value of the annuity:

$$\text{future value of lump sum} = \text{future value of annuity}$$

$$P(1 + i)^n = pymt\,\frac{(1 + i)^n - 1}{i}$$

Lump Sum Compounding Monthly

$$P = ?$$

$$i = \frac{1}{12} \text{ of } 10\frac{1}{2}\% = \frac{0.105}{12}$$

$$n = 25 \text{ years}$$

$$= 25 \text{ years} \cdot \frac{12 \text{ months}}{1 \text{ year}}$$

$$= 300 \text{ months}$$

Annuity (Ordinary Annuity)

$$pymt = 200$$

$$i = \frac{0.105}{12}$$

$$n = 300 \text{ months}$$

Note that i and n are the same with the annuity and the lump sum.

$$P(1 + i)^n = pymt\frac{(1 + i)^n - 1}{i}$$

$$P\left(1 + \frac{.105}{12}\right)^{300} = 200\frac{\left(1 + \frac{.105}{12}\right)^{300} - 1}{\frac{.105}{12}}$$

First, calculate the right side, as with any annuity calculation. Then divide by $(1 + \frac{.105}{12})^{300}$ to find P.

$$P = 21182.363 \ldots$$
$$P \approx \$21,182.36$$

This means that one would have to make a lump sum deposit of more than $21,000 in order to have as much money after 25 years as with monthly $200 annuity payments. •

Present Value of Annuity Formula

FV (lump sum) $= FV$ (annuity)

$$P(1 + i)^n = pymt \, \frac{(1 + i)^n - 1}{i}$$

The present value is the lump sum P.

There is a special formula for finding the present value of an ordinary annuity, but computations can be done quite efficiently without it. The formula will be developed in the exercises.

10.3

EXERCISES

In Exercises 1–4, find the future value of the given annuity.

1. Ordinary annuity, $120 monthly payment, $5\frac{3}{4}\%$ interest, 1 year

2. Ordinary annuity, $175 monthly payment, $6\frac{1}{8}\%$ interest, 11 years

3. Annuity due, $100 monthly payment, $5\frac{7}{8}\%$ interest, 4 years

4. Annuity due, $150 monthly payment, $6\frac{1}{4}\%$ interest, 13 years

5. On February 8, Bert Sarkis joined a Christmas Club. His bank will automatically deduct $75 from his checking account at the end of each month and deposit it into his Christmas Club account, where it will earn 7% interest. The account comes to term on December 1. Find:
   a. the future value of the account
   b. Bert's total contribution to the account
   c. the total interest

6. On March 19, Rachael Westlake joined a Christmas Club. Her bank will automatically deduct $110 from her checking account at the end of each month and deposit in it into her Christmas Club account, where it will earn $6\frac{7}{8}\%$ interest. The account comes to term on December 1. Find:
   a. the future value of the account
   b. Rachel's total contribution to the account
   c. the total interest

7. On February 23, Ginny Deus joined a Christmas Club. Her bank will automatically deduct $150 from her checking account at the beginning of each month and deposit it into her Christmas Club account, where it will earn $7\frac{1}{4}\%$ interest. The account comes to term on December 1. Find:
   a. the future value of the account
   b. Ginny's total contribution to the account
   c. the total interest

8. On January 19, Lynn Knight joined a Christmas Club. Her bank will automatically deduct $100 from her checking account at the beginning of each month and deposit it into her Christmas Club account, where it will earn 6% interest. The account comes to term on December 1. Find:
   a. the future value of the account
   b. Lynn's total contribution to the account
   c. the total interest

9. Pat Gilbert recently set up a TDA to save for her retirement. She arranged to have $175 taken out of each of her monthly checks; it will earn $10\frac{1}{2}$% interest. She just had her 39th birthday, and her ordinary annuity comes to term when she is 65. Find:
   a. the future value of the account
   b. Pat's total contribution to the account
   c. the total interest

10. Dick Eckel recently set up a TDA to save for his retirement. He arranged to have $110 taken out of each of his biweekly checks; it will earn $9\frac{7}{8}$% interest. He just had his 29th birthday, and his ordinary annuity comes to term when he is 65. Find:
    a. the future value of the account
    b. Dick's total contribution to the account
    c. the total interest

11. Sam Whitney recently set up a TDA to save for his retirement. He arranged to have $290 taken out of each of his monthly checks; it will earn 11% interest. He just had his 45th birthday, and his ordinary annuity comes to term when he is 65. Find:
    a. the future value of the account
    b. Sam's total contribution to the account
    c. the total interest

12. Art Dull recently set up a TDA to save for his retirement. He arranged to have $50 taken out of each of his biweekly checks; it will earn $9\frac{1}{8}$% interest. He just had his 30th birthday, and his ordinary annuity comes to term when he is 65. Find:
    a. the future value of the account
    b. Art's total contribution to the account
    c. the total interest

In Exercises 13–18, find and interpret the present value of the given annuity.

13. The annuity in Exercise 1
14. The annuity in Exercise 2
15. The annuity in Exercise 5

16. The annuity in Exercise 6
17. The annuity in Exercise 9
18. The annuity in Exercise 10

In Exercises 19–24, find the monthly payment that will yield the given future value.

19. $100,000 at $9\frac{1}{4}$% interest for 30 years; ordinary annuity.
20. $45,000 at $8\frac{7}{8}$% interest for 20 years; ordinary annuity.
21. $250,000 at $10\frac{1}{2}$% interest for 40 years; ordinary annuity.
22. $183,000 at $8\frac{1}{4}$% interest for 25 years; ordinary annuity.
23. $250,000 at $10\frac{1}{2}$% interest for 40 years; annuity due.
24. $183,000 at $8\frac{1}{4}$% interest for 25 years; annuity due.

25. Mr. and Mrs. Gonzales set up a TDA in order to save for their retirement. They agreed to have $100 deducted from each of Mrs. Gonzales's biweekly paychecks, which will earn $8\frac{1}{8}$% interest.
    a. Find the future value of their ordinary annuity if it comes to term after they retire in $35\frac{1}{2}$ years.
    b. After retiring, the Gonzaleses convert their annuity to a savings account, which earns 6.1% interest compounded monthly. At the end of each month, they withdraw $650 for living expenses. Complete the following chart for their post-retirement account:

Month Number	Account Balance at Beginning of the Month	Interest for the Month	Withdrawal	Account Balance at End of the Month
1				
2				
3				
4				
5				

26. Mr. and Mrs. Jackson set up a TDA in order to save for their retirement. They agreed to have $125 deducted from each of Mrs. Jackson's biweekly paychecks, which will earn $7\frac{5}{8}$% interest.
    a. Find the future value of their ordinary annuity if it comes to term after they retire in $32\frac{1}{2}$ years.

b. After retiring, the Jacksons convert their annuity to a savings account, which earns 6.3% interest compounded monthly. At the end of each month, they withdraw $700 for living expenses. Complete the following chart for their post-retirement account:

Month Number	Account Balance at Beginning of the Month	Interest for the Month	Withdrawal	Account Balance at End of the Month
1				
2				
3				
4				
5				

27. Jeanne and Harold Kimura want to set up a TDA that will generate sufficient interest at maturity to meet their living expenses, which they project to be $950 per month.

   a. Find the amount needed at maturity to generate $950 per month interest if they can get $6\frac{1}{2}$% interest compounded monthly.

   b. Find the monthly payment they would have to put into an ordinary annuity to obtain the future value found in part (a) if their money earns $8\frac{1}{4}$% and the term is 30 years.

28. Susan and Bill Stamp want to set up a TDA which will generate sufficient interest at maturity to meet their living expenses, which they project to be $1200 per month.

   a. Find the amount needed at maturity to generate $1200 per month interest if they can get $7\frac{1}{4}$% interest compounded monthly.

   b. Find the monthly payment they would have to make into an ordinary annuity to obtain the future value found in part (a) if their money earns $9\frac{3}{4}$% and the term is 25 years.

29. Toni Torres wants to save $1200 in the next two years to use as a down payment on a new car. If her bank offers her 9% interest, what monthly payment would she need to make into an ordinary annuity in order to reach her goal?

30. Fred and Melissa Furth's daughter Sally will be a freshman in college in six years. To help cover their extra expenses, the Furths decide to set up a sinking fund of $12,000. If the account pays 7.2% interest and they wish to make quarterly payments, find the size of each payment.

31. Anne Geyer buys some land in Utah. She agrees to pay the seller a lump sum of $65,000 in five years. Until then, she will make monthly simple interest payments to the seller at 11% interest.

   a. Find the amount of each interest payment.

   b. Anne sets up a sinking fund to save the $65,000. Find the size of her semiannual payments if her payments are due at the end of every six-month period and her money earns $8\frac{3}{8}$% interest.

   c. Prepare a table showing the amount in the sinking fund after each deposit.

32. Chrissy Fields buys some land in Oregon. She agrees to pay the seller a lump sum of $120,000 in six years. Until then, she will make monthly simple interest payments to the seller at 12% interest.

   a. Find the amount of each interest payment.

   b. Chrissy sets up a sinking fund to save the $120,000. Find the size of her semiannual payments if her money earns $10\frac{3}{4}$% interest.

   c. Prepare a table showing the amount in the sinking fund after each deposit.

33. Develop a formula for the present value of an ordinary annuity by solving the present value of annuity formula for *P* and simplifying.

34. Use the formula developed in Exercise 33 to find the present value of the annuity in Exercise 2.

35. Use the formula developed in Exercise 33 to find the present value of the annuity in Exercise 1.

36. Use the formula developed in Exercise 33 to find the present value of the annuity in Exercise 6.

37. Use the formula developed in Exercise 33 to find the present value of the annuity in Exercise 5.

 Answer the following questions using complete sentences.

38. Write a paragraph or two in which you explain (in your own words) the difference between an ordinary annuity and an annuity due.

39. Compare and contrast an annuity with a lump-sum investment that receives compound interest. What are the relative advantages and disadvantages of each?

ANNUITIES ON A GRAPHING CALCULATOR

EXERCISES

40. a. Use the method discussed in Section 10.2 on doubling time to find how long it takes for an annuity to have a balance of $500,000, if the ordinary annuity requires monthly $200 payments that earn 5% interest. Notice that payments are made on a monthly basis, so the answer must be a whole number of months.

   b. State what values you used for xmin, xmax, xscl, ymin, ymax, and yscl. Explain how you arrived at those values.

   c. Draw a freehand sketch of the graph obtained on your calculator.

41. Analyze the effect of the interest rate on annuities by redoing Exercise 40 with interest rates of 6%, 8%, and 10%. Discuss the impact of the increased rate.

42. Analyze the effect of the payment period on annuities by redoing Exercise 40 with twice-monthly payments of $100 each that earn 5% interest. Discuss the impact of the altered period.

10.4

AMORTIZED LOANS

An **amortized loan** is a loan for which the loan amount, plus interest, is paid off in a series of regular equal payments. In Section 10.1, we looked at one type of amortized loan—the add-on interest loan. A second type of amortized loan is the simple interest amortized loan. There is an important difference between these two types of loans that any potential borrower should be aware of: *The payments are smaller with a simple interest amortized loan than with an add-on interest loan* (assuming, naturally, that the loan amounts, interest rates, and number of payments are the same).

A **simple interest amortized loan** is really a type of annuity; specifically, it is an ordinary annuity whose future value is the loan amount plus compound interest. This raises a puzzling question. How can a simple interest amortized loan be defined as an annuity whose future value is the loan amount plus *compound* interest and still be called a *simple* interest loan? We will see.

The following formula for a simple interest amortized loan is based on the definition above, which requires that the future value of the ordinary annuity equal the future value of the loan amount.

Simple Interest Amortized Loan Formula

future value of annuity = future value of loan amount

$$pymt\, \frac{(1 + i)^n - 1}{i} = P(1 + i)^n$$

where *pymt* is the loan payment, *i* is the periodic interest rate, *n* is the number of periods, and *P* is the present value or loan amount.

Algebraically, the above formula could be used to determine any one of the four unknowns (*pymt, i, n,* and *P*) if the other three are known. In Section 10.3, we used it to find the present value *P* of an annuity. In this section, we use it to find the size of the payment of a simple interest amortized loan.

EXAMPLE 1 Heidi Ochikubo buys a car for $13,518.77. She makes a $1000 down payment and finances the balance through a four-year simple interest amortized loan from her bank. She is charged 12% interest. Find the following.

a. the size of her monthly payment
b. the total interest for the loan

Solution **a.** We are given $P = 13,518.77 - 1000.00 = 12,518.77$, $i = \frac{1}{12}$ of $12\% = 0.01$, and $n = 4$ years $= 4$ years $\cdot \frac{12 \text{ months}}{\text{year}} = 48$ months.

future value of annuity = future value of loan amount

$$pymt\, \frac{(1 + i)^n - 1}{i} = P(1 + i)^n$$

$$pymt\, \frac{(1 + 0.01)^{48} - 1}{0.01} = 12{,}518.77(1 + 0.01)^{48}$$

To find *pymt,* we need to divide the right side by the fraction on the left side or, equivalently, multiply by its reciprocal. First, find the fraction on the left side, as with any annuity calculation; then multiply the right side by the fraction's reciprocal.

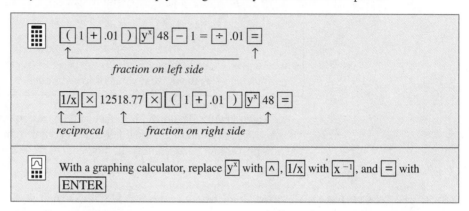

We get *pymt* = 329.667 . . . ≈ $329.67.

> ✔ Heidi borrowed $12,518.77, and each payment includes principal and interest; therefore, the payment must be larger than 12,518.77/48 = $260.81.

b. Heidi has agreed to make 48 payments of $329.67 each, for a total of 48 · $329.67 = $15,824.16. Of this, $12,518.77 is principal, so the balance of $15,824.16 − $12,518.77 = $3305.39 is interest. ●

Amortization Schedules

Each payment of an amortized loan includes principal and interest. A *simple interest amortized loan* is so named because the interest portion of each payment is simple interest on the outstanding principal. An **amortization schedule** is a list of several periods of payments showing the principal and interest portions of those payments and the outstanding principal (or **balance**) after each payment is made.

Most lending agencies provide the borrower with an amortization schedule on an annual basis. The data on the amortization schedule are important to the borrower for two reasons. First, the borrower needs to know the total interest paid, for tax purposes. (Interest paid on a home loan is usually deductible from the borrower's income tax, and interest paid on a loan by a business is usually deductible.) Second, the borrower needs the data if he or she is considering paying off the loan early. Such prepayment could save money, because an advance payment would be all principal and would not include any interest; however, the lending institution might charge a **prepayment penalty** that would absorb some or all of the interest savings.

EXAMPLE 2 Prepare an amortization schedule for the first two months of Heidi Ochikubo's simple interest amortized loan in Example 1.

Solution From Example 1, we have loan amount = $12,518.77, paid in 48 monthly payments of $329.67.

For any simple interest loan, the interest portion of each payment is simple interest on the outstanding principal, so we use the simple interest formula $I = Prt$ to compute the interest. Recall that r and t are annual figures.

For each payment, $r = 12\% = 0.12$ and $t = 1$ month $= \frac{1}{12}$ year.

For payment number 1, $P = \$12,518.77$ (the amount borrowed).

$$I = Prt$$

$$= 12,518.77 \cdot 0.12 \cdot \frac{1}{12}$$

$$= 125.1877$$

$$\approx \$125.19$$

The principal portion of payment number 1 is

$\$329.67 - \$125.19 = \$204.48$ Payment minus interest portion

The outstanding principal or balance is

$\$12,518.77 - \$204.48 = \$12,314.29$ Previous principal minus principal portion

For payment number 2, $P = \$12,314.29$ (the outstanding principal).

$$I = Prt$$

$$= 12,314.29 \cdot 0.12 \cdot \frac{1}{12}$$

$$= 123.1429$$

$$\approx \$123.14$$

The principal portion of payment number 2 is

$\$329.67 - \$123.14 = \$206.53$

The balance is

$\$12,314.29 - \$206.53 = \$12,107.76$

The chart in Figure 10.11 is the amortization schedule for the first two months of Heidi's loan. Notice that the principal portion increased and the interest portion decreased after the first payment. This pattern continues throughout the life of the loan, and the final payment is mostly principal. After each payment, the amount due is somewhat smaller, so the interest on the balance is somewhat smaller as well.

FIGURE 10.11

Payment Number	Principal Portion	Interest Portion	Total Payment	Balance
0	—	—	—	$12,518.77
1	$204.48	$125.19	$329.67	$12,314.29
2	$206.53	$123.14	$329.67	$12,107.76

EXAMPLE 3 Comp-U-Rent needs to borrow $60,000 in order to increase its inventory of rental computers. The company is confident that the expanded inventory will generate sufficient extra income to allow it to pay off the loan in a short amount of time, so it wishes to borrow the money for only three months. First National Bank offered a simple interest amortized loan at $8\frac{3}{4}\%$ interest.

a. Find what the monthly payment would be with First National Bank.
b. Prepare an amortization schedule for the loan.

Solution **a.** First National Bank offered a simple interest amortized loan; we have $P = \$60,000$, $i = \frac{1}{12}$ of $8\frac{3}{4}\% = \frac{0.0875}{12}$, and $n = 3$ months.

future value of annuity = future value of loan amount

$$pymt \frac{(1 + i)^n - 1}{i} = P(1 + i)^n$$

$$pymt \frac{\left(1 + \dfrac{0.0875}{12}\right)^3 - 1}{\dfrac{0.0875}{12}} = 60,000\left(1 + \frac{0.0875}{12}\right)^3$$

 First compute the fraction on the left side. Then find *pymt* by taking the reciprocal of that fraction and multiplying it by the right side.

We get

$$pymt = 20,292.376$$
$$\approx \$20,292.38$$

✔ Comp-U-Rent borrowed $60,000, and each payment includes principal and interest; therefore, the payment must be larger than $60,000/3 = $20,000.

b. For each payment, $r = 8\frac{3}{4}\% = 0.0875$ and $t = 1$ month $= \frac{1}{12}$ year. For payment number 1, $P = \$60,000$ (the amount borrowed).

$I = Prt$

$= 60,000 \cdot 0.0875 \cdot \dfrac{1}{12}$

$= \$437.50$

The principal portion of payment number 1 is

$\$20,292.38 - \$437.50 = \$19,854.88$ Payment minus interest portion

The balance is

$$\$60,000.00 - \$19,854.88 = \$40,145.12 \qquad \text{Previous principal minus}$$
principal portion

For payment number 2, $P = \$40,145.12$.

$$I = Prt$$
$$= 40,145.12 \cdot 0.0875 \cdot \frac{1}{12}$$
$$= 292.72483 \ldots$$
$$\approx \$292.72$$

The principal portion of payment number 2 is

$$\$20,292.38 - \$292.72 = \$19,999.66$$

The balance is

$$\$40,145,12 - \$19,999.66 = \$20,145.46$$

For payment number 3, $P = \$20,145.46$.

$$I = Prt$$
$$= 20,145.46 \cdot 0.0875 \cdot \frac{1}{12}$$
$$= 146.89398 \ldots$$
$$\approx \$146.89$$

The principal portion of payment number 3 is

$$\$20,292.38 - \$146.89 = \$20,145.49$$

The balance is

$$\$20,145.46 - \$20,145.49 = -\$0.03$$

Negative three cents can't be right. After the final payment is made, the balance *must be* $0.00. The discrepancy arises from the fact that we rounded off the payment size from $20,292.376 to $20,292.38. If there were some way the borrower could make a monthly payment that is not rounded off, the calculation above would have yielded an amount due of $0.00. To repay the exact amount owed, we must compute the last payment differently.

The principal portion of payment number 3 *must* be $20,145.46, because that is the balance, and this payment is the only chance to pay it. The payment must also include $146.89 interest, as calculated above. Thus, the last payment is the sum of the principal due and the interest on that principal:

$$\$20,145.46 + \$146.89 = \$20,292.35$$

The balance is then $20,145.46 − $20,145.46 = $0.00, as it should be.

The amortization schedule is given in Figure 10.12, and the schedule steps are outlined in Figure 10.13.

FIGURE 10.12

Payment Number	Principal Portion	Interest Portion	Total Payment	Balance
0	—	—	—	$60,000.00
1	$19,854.88	$437.50	$20,292.38	$40,145.12
2	$19,999.66	$292.72	$20,292.38	$20,145.46
3	$20,145.46	$146.89	$20,292.35	$0.00

Amortization Schedule Steps

1. Find interest on balance—use the simple interest formula.
2. Principal portion is payment minus interest portion.
3. New balance is previous balance minus principal portion.
   For the last period:
4. Principal portion is the previous balance.
5. Total payment is sum of principal portion and interest portion.

FIGURE 10.13
Amortization schedule steps

Payment Number	Principal Portion	Interest Portion	Total Payment	Balance
0	—	—	—	loan amount
first through next-to-last	total payment minus interest portion	simple interest on previous balance; use $I = Prt$	use Simple Interest Amortized Loan Formula	previous balance minus this payment's principal portion
last	previous amount due	simple interest on previous balance; use $I = Prt$	principal portion plus interest portion	$0.00

In preparing an amortization schedule, the new balance is the previous balance minus the principal portion. This means that only the principal portion of a payment goes toward paying off the loan; the interest portion is the lender's fee for the use of its money. In Example 3, the interest portion of Comp-U-Rent's first loan payment was $437.50, and the principal portion was $19,854.88, so almost all of that payment went toward paying off the loan. However, if Comp-U-Rent's loan was for 30 years rather than 3 months, the interest portion of the first payment would remain at $437.50, but the principal portion would be $34.52, and very little of that payment would go toward paying off the loan.

It is extremely depressing for first-time home purchasers to discover how little of their initial payment actually goes toward paying off the loan. The homeowner has a few alternatives: paying off the loan early, finding a new loan with a better interest rate, or "creatively altering" the traditional structure of a simple interest amortized loan. Popular examples of the third alternative include altering the payment period from monthly to bi-weekly, and altering the loan's duration from 30 years to 20 or 15 years. These alternatives are discussed in the exercises.

Another example of altering the structure of the loan is making one extra payment each year. This extra payment is all principal, and it creates some very complex accounting that is beyond the scope of this book; the option is a viable alternative only if the loan does not have a prepayment penalty. An informed source in the banking world has told us that if the borrower makes one extra payment each year for the life of a 30-year loan, then the loan will be paid off approximately 10 years early!

Finding an Unpaid Balance

Prepaying a loan has some real advantages. Any additional amount paid beyond the required monthly principal and interest must be classified by the lender as payment toward the principal. The size of the monthly payment is unaffected by a prepayment, but the interest portion of future monthly payments will be decreased because less principal is owed. Furthermore, the loan will be paid off early for the same reason.

Prepaying also has some disadvantages. Obviously, the borrower must have sufficient savings to afford prepayments. Using part of one's savings for prepayment means that that money is no longer earning interest. Also, one's income tax would go up as a result of prepayment if the interest portion of each payment is deductible.

If a borrower is thinking about paying off his or her loan early, several things should be considered. Some lenders charge a prepayment penalty—an extra fee assessed if the loan is prepaid. The lender can do this only if the loan agreement allows it; when you sign a loan agreement, you should determine whether a prepayment penalty is included. If a borrower is considering prepaying a loan, he or she should weigh the advantage of not making the monthly payments against the interest he or she could earn by investing the money that will be used to prepay. This involves knowing the unpaid balance. If an amortization schedule has not already been prepared, the borrower can easily compute the unpaid balance by subtracting the current value of the annuity from the current value of the loan.

Unpaid Balance Formula

unpaid balance = current value of loan amount − current value of annuity

$$= P(1 + i)^n - pymt \frac{(1 + i)^n - 1}{i}$$

where *pymt* is the loan payment, i is the periodic interest rate, n is the number of periods *from the beginning of the loan to the present,* and P is the loan amount.

A common error is using this formula is to let n equal the number of periods in the entire life of the loan, rather than the number of periods from the beginning of the loan until the time of prepayment.

EXAMPLE 4   Ten years ago, Rob and Shelly Golumb bought a home for $140,000. They paid the sellers a 20% down payment and obtained a simple interest amortized loan for the balance from their bank at $10\frac{3}{4}\%$ for 30 years.

Rob and Shelly Golumb won $100,000 on a TV game show!

Recently, they won $100,000 on a TV game show. They are considering paying off their home loan with that money, and they want to know how much of their winnings would be left.

a. Find the size of the Golumbs' monthly payment.
b. Find the unpaid balance of their loan.
c. Find what portion of their winnings would be left after prepaying their loan.
d. Find the amount of interest they would save by prepaying.

Solution **a.** *Finding their monthly payment:* We are given that down payment = 20% of $140,000 = $28,000, P = loan amount = $140,000 − $28,000 = $112,000, $i = \frac{1}{12}$ of $10\frac{3}{4}\% = \frac{0.1075}{12}$, and n = 30 years = 30 years $\cdot \frac{12 \text{ months}}{1 \text{ year}}$ = 360 months.

future value of annuity = future value of loan amount

$$pymt \frac{(1 + i)^n - 1}{i} = P(1 + i)^n$$

$$pymt \frac{\left(1 + \dfrac{0.1075}{12}\right)^{360} - 1}{\dfrac{0.1075}{12}} = 112{,}000\left(1 + \dfrac{0.1075}{12}\right)^{360}$$

Computing the fraction on the left side and multiplying its reciprocal by the right side, we get

$$pymt = 1{,}045.4991 \ldots \approx \$1045.50$$

b. *Finding the unpaid balance:* The Golumbs have made payments on their loan for 10 years = 120 months, so n = 120 (not 360!).

unpaid balance = current value of loan amount − current value of annuity

$$= P(1 + i)^n - pymt \frac{(1 + i)^n - 1}{i}$$

$$= 112{,}000\left(1 + \frac{0.1075}{12}\right)^{120} - 1045.50 \frac{\left(1 + \dfrac{0.1075}{12}\right)^{120} - 1}{\dfrac{0.1075}{12}}$$

$$= 102{,}981.4236\ldots \approx \$102{,}981.42$$

First, compute $\frac{0.1075}{12}$ and put it in memory. Then type

112000 ✕ ((1 + RCL)) yˣ 120 − 1045.50 ✕
(((1 + RCL)) yˣ 120 − 1) ÷ RCL =

112000 ✕ ((1 + ALPHA A)) ^ 120 − 1045.50 ✕
(((1 + ALPHA A)) ^ 120 − 1) ÷ ALPHA A ENTER

This means that after 10 years of payments on a loan of $112,000, the Golumbs still owe $102,981.42!

c. *Finding the portion of their winnings that would be left.* The Golumbs won $100,000, so they could not afford to prepay the loan unless they used more than their game show winnings.

d. *Finding the amount of interest they would save by prepaying:* If they prepaid, they would not have to make their remaining 240 monthly payments of $1045.50 each, a total of

240 · $1045.50 = $250,920.00

Of this amount, $102,981.42 is principal, and the interest they would save by prepaying is

$250,920.00 − $102,981.42 = $147,938.58

They would have to weigh this savings and the extra money they would have from not making monthly payments against the interest they would receive from investing all or part of their winnings.   •

10.4

EXERCISES

In the following exercises, all loans are simple interest amortized loans with monthly payments, unless labeled otherwise.

In Exercises 1–6, find (a) the monthly payment and (b) the total interest for the given simple interest amortized loan.

1. $5000 at $9\frac{1}{2}\%$ for 4 years
2. $8200 at $10\frac{1}{4}\%$ for 6 years
3. $10,000 at $6\frac{1}{8}\%$ for 5 years
4. $20,000 at $7\frac{3}{8}\%$ for $5\frac{1}{2}$ years

5. $155,000 at $9\frac{1}{2}\%$ for 30 years

6. $289,000 at $10\frac{3}{4}\%$ for 35 years

7. Wade Ellis buys a new car for $16,113.82. He puts 10% down and obtains a simple interest amortized loan for the balance at $11\frac{1}{2}\%$ interest for 4 years.
   a. Find his monthly payment.
   b. Find the total interest.
   c. Prepare an amortization schedule for the first 2 months of the loan.

8. Guy de Primo buys a new car for $9837.91. He puts 10% down and obtains a simple interest amortized loan for the balance at $8\frac{7}{8}\%$ for 4 years.
   a. Find his monthly payment.
   b. Find the total interest.
   c. Prepare an amortization schedule for the first 2 months of the loan.

9. Chris Burditt bought a house for $212,500. He put 20% down and obtained a simple interest amortized loan for the balance at $10\frac{7}{8}\%$ for 30 years.
   a. Find his monthly payment.
   b. Find the total interest.
   c. Prepare an amortization schedule for the first 2 months of the loan.
   d. Most lenders will approve a home loan only if the total of all the borrower's monthly payments, including the home loan payment, is no more than 38% of the borrower's monthly income. How much must Mr. Burditt make in order to qualify for the loan?

10. Shirley Trembley bought a house for $187,600. She put 20% down and obtained a simple interest amortized loan for the balance at $6\frac{3}{8}\%$ for 30 years.
    a. Find her monthly payment.
    b. Find the total interest.
    c. Prepare an amortization schedule for the first 2 months of the loan.
    d. Most lenders will approve a home loan only if the total of all the borrower's monthly payments, including the home loan payment, is no more than 38% of the borrower's monthly income. How much must Ms. Trembley make in order to qualify for the loan?

11. Ray and Helen Lee bought a house for $189,500. They made a 10% down payment, borrowed 80% from their bank for 30 years at 11.5%, and convinced the seller to take a second mortgage for the remaining 10%. That 10% is due in full in 5 years (this is called a *balloon payment*), and the Lees agreed to make monthly interest-only payments to the seller at 12% simple interest in the interim.
    a. Find the Lees' down payment.
    b. Find the amount that the Lees borrowed from their bank.

c. Find the amount that the Lees borrowed from the seller.
d. Find the Lees' monthly payment to the bank.
e. Find the Lees' monthly interest payment to the seller.
f. If the Lees save for their balloon payment with a sinking fund, find the size of the necessary monthly payments into that fund if their money earns 6% interest.
g. Find the Lees' total monthly payment for the first 5 years.
h. Find the Lees' total monthly payment for the last 25 years.

12. Jack and Laurie Worthington bought a house for $163,700. They made a 10% down payment, borrowed 80% from their bank for 30 years at 12%, and convinced the seller to take a second mortgage for the remaining 10%. That 10% is due in full in 5 years, and the Worthingtons agreed to make monthly interest-only payments to the seller at 12% simple interest in the interim.
    a. Find the Worthingtons' down payment.
    b. Find the amount that the Worthingtons borrowed from their bank.
    c. Find the amount that the Worthingtons borrowed from the seller.
    d. Find the Worthingtons' monthly payment to the bank.
    e. Find the Worthingtons' monthly interest payment to the seller.
    f. If the Worthingtons save for their balloon payment with a sinking fund, find the size of the necessary monthly payments into that fund if their money earns 7% interest.
    g. Find the Worthingtons' total monthly payment for the first 5 years.
    h. Find the Worthingtons' total monthly payment for the last 25 years.

13. Dennis Lamenti wants to buy a new car that costs $15,829.32. He has two possible loans in mind. One loan is through the car dealer; it is a 4-year add-on interest loan at $7\frac{3}{4}\%$ and requires a down payment of $1000. The second loan is through his bank; it is a 4-year simple interest amortized loan at $8\frac{7}{8}\%$ and requires a 10% down payment.
    a. Find the monthly payment for each loan.

b. Find the total interest paid for each loan.

c. Which loan should Dennis choose? Why?

14. Barry Wood wants to buy a used car that costs $4000. He has two possible loans in mind. One loan is through the car dealer; it is a 3-year add-on interest loan at 6% and requires a down payment of $300. The second loan is through his credit union; it is a 3-year simple interest amortized loan at 9.5% and requires a 10% down payment.

a. Find the monthly payment for each loan.

b. Find the total interest paid for each loan.

c. Which loan should Barry choose? Why?

15. Investigate the effect of the term on simple interest amortized automobile loans by finding the monthly payment and the total interest for a loan of $11,000 at $9\frac{7}{8}\%$ interest for the following terms.

a. 3 years **b.** 4 years **c.** 5 years

16. Investigate the effect of the interest rate on simple interest amortized automobile loans by finding the monthly payment and the total interest for a 4-year loan of $12,000 at the following interest rates.

a. $8\frac{1}{2}\%$ **b.** $8\frac{3}{4}\%$ **c.** 9% **d.** 10%

17. Investigate the effect of the interest rate on home loans by finding the monthly payment and the total interest for a 30-year loan of $100,000 at the following interest rates.

a. 6% **b.** 7% **c.** 8%
d. 9% **e.** 10% **f.** 11%

18. Some lenders are now offering 15-year home loans. Investigate the effect of the term on home loans by finding the monthly payment and total interest for a loan of $100,000 at 10% for the following terms.

a. 30 years **b.** 15 years

19. Verify (a) the monthly payments and (b) the interest savings in the savings and loan advertisement shown at right.

20. The home loan in Exercise 19 presented two options. The 30-year option required a smaller monthly payment. A consumer who chooses the 30-year option could take the savings in the monthly payment generated by that option and invest that savings in an annuity. At the end of 15 years, the annuity might be large enough to pay off the 30-year loan. Determine whether this is a wise plan, if the annuity's interest rate is 7.875%. (Disregard the tax ramifications of this approach.)

21. Some lenders are now offering loans with biweekly payments rather than monthly payments. Investigate the effect of this option on home loans by finding the payment and total interest on a 30-year loan of $100,000 at 10% interest if payments are made (a) monthly and (b) biweekly.

22. Pool-N-Patio World needs to borrow $75,000 to increase its inventory for the upcoming summer season. The owner is

confident that he will sell most if not all of the new inventory during the summer, so he wishes to borrow the money for only 4 months. His bank has offered him a simple interest amortized loan at $7\frac{3}{4}\%$ interest.

a. Find the size of the monthly bank payment.

b. Prepare an amortization schedule for all 4 months of the loan.

23. Slopes R Us needs to borrow $120,000 to increase its inventory of ski equipment for the upcoming season. The owner is confident that she will sell most if not all of the new inventory during the winter, so she wishes to borrow the money for only 5 months. Her bank has offered her a simple interest amortized loan at $8\frac{7}{8}\%$ interest.

a. Find the size of the monthly bank payment.

b. Prepare an amortization schedule for all 5 months of the loan.

24. The owner of Video Extravaganza is opening a second store and needs to borrow $93,000. Her success with the first store has made her confident that she will be able to pay off her loan quickly, so she wishes to borrow the money for only 4 months. Her bank has offered her a simple interest amortized loan at $9\frac{1}{8}\%$ interest.

a. Find the size of the monthly bank payment.

b. Prepare an amortization schedule for all 4 months of the loan.

25. The Green Growery Nursery needs to borrow $48,000 to increase its inventory for the upcoming summer season. The owner is confident that he will sell most if not all of the new plants during the summer, so he wishes to borrow the money for only 4 months. His bank has offered him a simple interest amortized loan at $9\frac{1}{4}\%$ interest.

a. Find the size of the monthly bank payment.

b. Prepare an amortization schedule for all 4 months of the loan.

In Exercises 26 and 27, you are asked to prepare an amortization schedule for an add-on interest loan. This schedule should have the same data as does an amortization schedule for a simple interest amortized loan, but the computational procedure is different. Use the information in Section 10.1 on add-on interest loans to help you determine this procedure.

26. Prepare an amortization schedule for the first two months of each of Barry Wood's two possible loans in Exercise 14. By comparing the schedules, what advantages can you see in one loan over the other?

27. Prepare an amortization schedule for the first three months of each of Dennis Lamenti's two possible loans in Exercise 13. By comparing the schedules, what advantages can you see in one loan over the other?

28. This is an exercise in buying a car. It involves choosing a car and selecting the car's financing. Write a paper describing all of the following points.

a. You may not be in a position to buy a car now. If so, fantasize about your future. What job do you have? How long have you had that job? What is your salary? If you're married, does your spouse work? Do you have a family? What needs will your car fulfill? Make your fantasy realistic. Briefly describe what has happened between the present and your future fantasy. (If you are in a position to buy a car now, discuss these points on a more realistic level.)

b. Go shopping for a car. Look at new cars, used cars, or both. Read newspaper and magazine articles about your choices (see, for example, *Consumer Reports, Motor Trend,* and *Road and Track*). Discuss in detail the car you selected and why you did so. How will your selection fulfill your (projected) needs? What do the newspaper and magazines say about your selection? Why did you select a new/used car?

c. Go shopping for financing. Many banks, savings and loans, and credit unions have information on car loans available on request. Get all of the information you need from at least two lenders, but do not bother the staff unnecessarily. You may be able to find the necessary information in the newspaper. Perform all appropriate computations yourself—do not have the lenders tell you the payment size. Summarize the appropriate data (including the down payment, payment size, interest rate, duration of loan, type of loan, and loan fees) in your paper, and discuss which loan you would choose. Explain how you would be able to afford your purchase.

29. Wade Ellis buys a new car for $16,113.82. He puts 10% down and obtains a simple interest amortized loan for the balance at $11\frac{1}{2}\%$ interest for 4 years. Three years and 2 months later, he sells his car. Find the unpaid balance on his loan.

30. Guy de Primo buys a new car for $9837.91. He puts 10% down and obtains a simple interest amortized loan for the balance at $10\frac{7}{8}\%$ interest for 4 years. Two years and 6 months later, he sells his car. Find the unpaid balance on his loan.

31. Gary Kersting buys a house for $212,500. He puts 20% down and obtains a simple interest amortized loan for the balance at $10\frac{7}{8}\%$ interest for 30 years. Eight years and 2 months later, he sells his house. Find the unpaid balance on his loan.

32. Shirley Trembley buys a house for $187,600. She puts 20% down and obtains a simple interest amortized loan for the balance at $11\frac{3}{8}\%$ interest for 30 years. Ten years and 6 months later, she sells her house. Find the unpaid balance on her loan.

33. Harry and Natalie Wolf have a 3-year-old loan with which they purchased their house; their interest rate is $13\frac{3}{8}\%$. Since they obtained this loan, interest rates have dropped, and they can now get a loan for $8\frac{7}{8}\%$ through their credit union. Because of this, the Wolfs are considering refinancing their home. Each loan is a 30-year simple interest amortized loan, and neither has a prepayment penalty. The existing loan is for $152,850, and the new loan would be for the current amount due on the old loan.

a. Find their monthly payment with the existing loan.

b. Find the loan amount for their new loan.

c. Find the monthly payment with their new loan.

d. Find the total interest they will pay if they do *not* get a new loan.

e. Find the total interest they will pay if they *do* get the new loan.

f. Should the Wolfs refinance their home? Why?

34. Russ and Roz Rosow have a 10-year-old loan with which they purchased their house; their interest rate is $10\frac{5}{8}\%$. Since they obtained this loan, interest rates have dropped, and they can now get a loan for $9\frac{1}{4}\%$ through their credit union. Because of this, the Rosows are considering refinancing their home. Each loan is a 30-year simple interest amortized loan, and neither has a prepayment penalty. The existing loan is for $112,000, and the new loan would be for the current amount due on the old loan.

a. Find the monthly payment with their existing loan.
b. Find the loan amount for their new loan.
c. Find the monthly payment with their new loan.
d. Find the total interest they will pay if they do *not* get a new loan.
e. Find the total interest they will pay if they *do* get the new loan.
f. Should the Rosows refinance their home? Why?

35. Michael and Lynn Sullivan have a 10-year-old loan for $187,900 with which they purchased their home. They just sold their highly profitable import-export business and are considering paying off their home loan. Their loan is a 30-year simple interest amortized loan at $10\frac{1}{2}\%$ and has no prepayment penalty.

a. Find the size of their monthly payment.
b. Find the unpaid balance of the loan.
c. Find the amount of interest they would save by prepaying.
d. The Sullivans decided that if they paid off their loan, they would deposit the equivalent of half their monthly payment into an annuity. If the ordinary annuity pays 9% interest, find its future value after 20 years.
e. The Sullivans decided that if they did not pay off their loan, they would deposit an amount equivalent to their unpaid balance into an account that pays $9\frac{3}{4}\%$ interest compounded monthly. Find the future value of this account after 20 years.
f. Should the Sullivans prepay their loan? Why?

36. Charlie and Ellen Wilson have a 25-year-old loan for $47,000 with which they bought their home. The Wilsons have retired and are living on a fixed income, so they are contemplating paying off their home loan. Their loan is a 30-year simple interest amortized loan at $4\frac{1}{2}\%$ and has no prepayment penalty. They also have savings of $73,000, which they have invested in a certificate of deposit currently paying $8\frac{1}{4}\%$ simple interest. Should they pay off their home loan? Why?

For Exercises 37–39, note the following information. An ad-justable rate mortgage (or ARM) is, as the name implies, a mortgage for which the interest rate is allowed to change; as a result, the payment changes too. At first, an ARM costs less than a fixed rate mortgage—its initial interest rate is usually two or three percentage points lower. As time goes by, the rate is adjusted; as a result, it may or may not continue to hold this advantage.

37. Trustworthy Savings offers a 30-year adjustable rate mortgage with an initial rate of 5.375%. The rate and the required payment are adjusted annually. Future rates are set at 2.875 percentage points above the 11th District Federal Home Loan Bank's cost of funds. Currently, that cost of funds is 4.839%. The loan's rate is not allowed to rise more than 2 percentage points in any one adjustment, nor is it allowed to rise above 11.875%. Trustworthy Savings also offers a 30-year fixed rate mortgage with an interest rate of 7.5%.

a. Find the monthly payment for the fixed rate mortgage on a loan amount of $100,000.
b. Find the monthly payment for the ARM's first year on a loan amount of $100,000.
c. How much would the borrower save in the mortgage's first year by choosing the adjustable rather than the fixed rate mortgage?
d. Find the unpaid balance at the end of the ARM's first year.
e. Find the interest rate and the value of *n* for the ARM's second year, if the 11th District Federal Home Loan Bank's cost of funds does not change during the loan's first year.
f. Find the monthly payment for the ARM's second year, if the 11th District Federal Home Loan Bank's cost of funds does not change during the loan's first year.
g. How much would the borrower save in the mortgage's first two years by choosing the adjustable rather than the fixed rate mortgage, if the 11th District Federal Home Loan Bank's cost of funds does not change during the loan's first year?
h. Discuss the advantages and disadvantages of an adjustable rate mortgage.

38. American Dream Savings Bank offers a 30-year adjustable rate mortgage with an initial rate of 4.25%. The rate and the required payment are adjusted annually. Future rates are set at 3 percentage points above the one-year treasury bill rate. Currently, the one-year treasury bill rate is 5.42%. The loan's rate is not allowed to rise more than 2 percentage points in any one adjustment, nor is it allowed to rise above 10.25%. American Dream Savings Bank also offers a fixed rate mortgage with an interest rate of 7.5%.

a. Find the monthly payment for the fixed rate mortgage on a loan amount of $100,000.
b. Find the monthly payment for the ARM's first year on a loan amount of $100,000.
c. How much would the borrower save in the mortgage's first year by choosing the adjustable rather than the fixed rate mortgage?
d. Find the unpaid balance at the end of the ARM's first year.
e. Find the interest rate and the value of *n* for the ARM's second year if the treasury bill rate does not change during the loan's first year.

f. Find the monthly payment for the ARM's second year, if the treasury bill rate does not change during the loan's first year.

g. How much would the borrower save in the mortgage's first two years by choosing the adjustable rather than the fixed rate mortgage, if the treasury bill rate does not change during the loan's first year?

h. Discuss the advantages and disadvantages of an adjustable rate mortgage.

39. Bank Two offers a 30-year adjustable rate mortgage with an initial rate of 3.95%. This initial rate is in effect for the first six months of the loan, after which it is adjusted on a monthly basis. The monthly payment is adjusted annually. Future rates are set at 2.45 percentage points above the 11th District Federal Home Loan Bank's cost of funds. Currently, that cost of funds is 4.839%.

a. Find the monthly payment for the ARM's first year on a loan amount of $100,000.

b. Find the unpaid balance at the end of the ARM's first six months.

c. Find the interest portion of the seventh payment if the cost of funds does not change.

d. Usually, the interest portion is smaller than the monthly payment, and the difference is subtracted from the unpaid balance. However, the interest portion found in part (c) is *larger* than the monthly payment found in part (a), and the difference is *added* to the unpaid balance found in part (b). Why would this difference be added to the unpaid balance? What effect will this have on the loan?

e. The situation described in part (d) is called *negative amortization.* Why?

f. What is there about the structure of Bank Two's loan that allows negative amortization?

 Answer the following questions using complete sentences.

40. Why are the computations for the last period of an amortization schedule different from those for all preceding periods?

41. What is there about the structure of an add-on interest loan that makes its payments larger than that of a simple interest amortized loan?

AMORTIZATION SCHEDULES ON A COMPUTER

Interest paid on a home loan is deductible from the borrower's income taxes, and interest paid on a loan by a business is usually deductible. A borrower with either of these types of loans needs to know the total interest paid on the loan during the last year. The way to determine the total interest paid during a given year is to prepare an amortization schedule for that year. Typically, the lender provides the borrower with an amortization schedule; however, it's not uncommon for this schedule to arrive after taxes are due. In this case the borrower must either do the calculation personally or pay his or her taxes without the benefit of the mortgage deduction and then file an amended set of tax forms after the amortization schedule has arrived.

Computing a year's amortization schedule is rather tedious, and neither a scientific calculator nor a graphing calculator will relieve you of the tedium. The best tool for the job is a computer, combined either with the Amortrix computer program available with this book or with a computerized spreadsheet. Each is discussed below.

Amortization Schedules and Amortrix

The Amortrix computer program available with this book will enable you to compute an amortization schedule for any time period. Amortrix is available for both Macintosh and Windows-based computers.

When you start Amortrix, a main menu appears. Use the mouse to click on the "Amortization Schedule" option. Once you're into the "Amortization Schedule" part of the program, type the loan amount (P), and press "return." Follow the instructions that appear on the screen to enter the annual interest rate, the payment period, and the number of payment periods (n). The area labeled "Expressionist" is a calculator; you may use it to compute the number of payment periods—just use the mouse to press Expressionist's buttons.

If you need to change any of the above information, either use the mouse to click on the appropriate box or type the letter in brackets in that box's label (for example, type "A" for the box labeled "Loan [A]mount").

Once all the necessary information is entered, use the mouse to click on the box labeled "[C]reate Table" (or just type "C"). This causes the schedule for the first 12 payments to appear; to see the schedule for other payments, use the arrow buttons in the lower-right corner of the screen. An Amortrix amortization schedule is shown in Figure 10.14. You can print a copy of a portion of the table by clicking on the box labeled "[P]rint Table."

`FIGURE 10.14

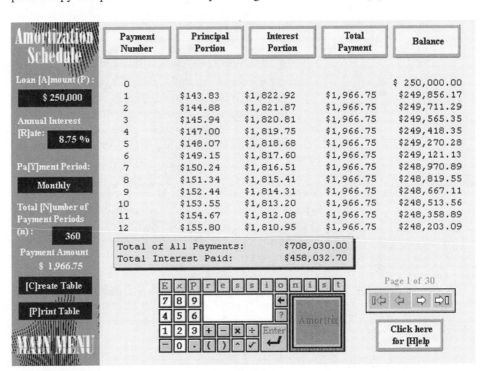

Payment Number	Principal Portion	Interest Portion	Total Payment	Balance
0				$ 250,000.00
1	$143.83	$1,822.92	$1,966.75	$249,856.17
2	$144.88	$1,821.87	$1,966.75	$249,711.29
3	$145.94	$1,820.81	$1,966.75	$249,565.35
4	$147.00	$1,819.75	$1,966.75	$249,418.35
5	$148.07	$1,818.68	$1,966.75	$249,270.28
6	$149.15	$1,817.60	$1,966.75	$249,121.13
7	$150.24	$1,816.51	$1,966.75	$248,970.89
8	$151.34	$1,815.41	$1,966.75	$248,819.55
9	$152.44	$1,814.31	$1,966.75	$248,667.11
10	$153.55	$1,813.20	$1,966.75	$248,513.56
11	$154.67	$1,812.08	$1,966.75	$248,358.89
12	$155.80	$1,810.95	$1,966.75	$248,203.09

Loan [A]mount (P): $ 250,000
Annual Interest [R]ate: 8.75 %
Pa[Y]ment Period: Monthly
Total [N]umber of Payment Periods (n): 360
Payment Amount $ 1,966.75
[C]reate Table
[P]rint Table
MAIN MENU

Total of All Payments: $708,030.00
Total Interest Paid: $458,032.70

Expressionist
Amortrix
Page 1 of 30
Click here for [H]elp

The program will *not* correctly compute the last loan payment; it will compute the last payment in the same way it computes all other payments, rather than in the way shown earlier in this section. You will have to correct this last payment if you use the computer to prepare an amortization schedule for a time period that includes the last payment.

Amortization Schedules and Computerized Spreadsheets

A **spreadsheet** is a large piece of paper marked off in rows and columns. Accountants use spreadsheets to organize numerical data and perform computations. A **computerized spreadsheet,** such as Microsoft Excel or Lotus 1-2-3, is a computer program that mimics the appearance of a paper spreadsheet. It frees the user from performing any computations; instead, it allows the user to merely give instructions on how to perform those computations.

When you start a computerized spreadsheet, you see something that looks like a table waiting to be filled in. The rows are labeled with numbers and the columns with letters, as shown in Figure 10.15.

FIGURE 10.15

	A	B	C	D	E
1					
2					
3					
4					
5					

The individual boxes are called **cells.** The cell in column A, row 1 is called cell A1; the cell below it is called cell A2 because it is in column A, row 2.

A computerized spreadsheet is an ideal tool to use in creating an amortization schedule. We will illustrate this process by preparing the amortization schedule for a four-month, $175,000 loan at 7.75% interest; such a loan requires a monthly payment of $44,458.65.

Step 1 *Label the columns.* Use the mouse and/or the arrow buttons to move to cell A1, type in "Payment Number," and press "Return" or "Enter." Move to cell B1, type in "Principal Portion," and press "Return" or "Enter." Type the remaining amortization schedule labels in cells C1, D1, and E1. The columns' widths can be adjusted—your instructor will tell you how.

Step 2 *Fill in row 2 with payment 0 information.* Move to cell A2, type in "0," and press "Return" or "Enter." Move to cell E2, type in "175000," and Press "Return" or "Enter." After completing this step your spreadsheet should look like that in Figure 10.16.

FIGURE 10.16

	A	B	C	D	E
1	Payment Number	Principal Portion	Interest Portion	Total Payment	Balance
2	0				175000
3					
4					
5					
6					

Step 3 *Fill in row 3 with instructions on how to compute payment 1 information.*

- Move to cell D3 and type in "44458.65," the total payment.
- Column A should eventually contain the numbers 0, 1, 2, 3, and 4. Since each of these numbers is 1 more than the previous number, the instructions in cell A3 should say to add 1 to the contents in cell A2. To do this, type "= A2 + 1" or "+ A2 + 1," and press "Return" or "Enter." Notice that these computational instructions begin with either the "=" symbol or the "+" symbol. Computational instructions are always preceded with one of these symbols. (Your software may accept only one of these symbols.)
- Cell C3 should contain instructions on computing interest on the previous balance, using the simple interest formula $I = Prt$.

$$I = P \cdot r \cdot t$$

$$= \text{(the previous balance)} \cdot (0.0775) \cdot \left(\frac{1}{12}\right)$$

$$= E2 * \frac{.0775}{12} \qquad \text{Computers use * for multiplication}$$

Move to cell C3 and type in either "= E2 * .0775 / 12" or "+ E2 * .0775 / 12", and press "Return" or "Enter."

- Cell B3 should contain instructions on computing the payment's principal portion.

$$\text{principal portion} = \text{payment} - \text{interest portion}$$
$$= D3 - C3$$

Move to cell B3 and type in either "= D3 − C3" or "+ D3 − C3."

- Cell E3 should contain instructions on computing the new balance.

$$\text{new balance} = \text{previous balance} - \text{principal portion}$$
$$= E2 - B3$$

Move to cell E3 and type in either "= E2 − B3" or "+ E2 − B3."

After completing this step, your spreadsheet should look like that in Figure 10.17.

FIGURE 10.17

	A	B	C	D	E
1	Payment Number	Principal Portion	Interest Portion	Total Payment	Balance
2	0				175000
3	1	43328.44167	1130.20833	44458.65	131671.5583
4					
5					
6					

Notice that the cells in row 3 show more decimal places than is appropriate. We will take care of this later.

Step 4 *Fill in rows 4 and 5 with instructions on how to compute payment 2 and payment 3 information.* Payment 2 and payment 3 computations are just like payment 1 computations, so all we have to do is copy the payment 1 instructions from row 3 and paste them into rows 4 and 5. *To copy the payment 1 instructions from row 3:*

- Use your mouse to move to cell A3.
- Press the mouse's button, and move the mouse to the right until cells B3, C3, D3, and E3 are highlighted, and release the button. Be careful that you do not highlight any other cells.
- Use your mouse to move to the word "Edit" at the very top of the screen.
- Press the mouse's button and move the mouse down until the word "Copy" is highlighted, and release the button.

To paste the payment 1 instructions into rows 4 and 5, do the following.

- Use your mouse to move to cell A4.
- Press the mouse's button, and move the mouse to the right and down until cells A4, B4, C4, D4, E4, A5, B5, C5, D5, and E5 are highlighted.
- Use your mouse to move to the word "Edit" at the very top of the screen.
- Press the mouse's button and move the mouse down until the word "Paste" is highlighted, and release the button.

After completing this step, your spreadsheet should look like that in Figure 10.18.

FIGURE 10.18

	A	B	C	D	E
1	Payment Number	Principal Portion	Interest Portion	Total Payment	Balance
2	0				175000
3	1	43328.44167	1130.20833	44458.65	131671.5583
4	2	43608.27119	850.378814	44458.65	88063.28715
5	3	43889.90794	568.742063	44458.65	44173.37921
6					

Step 5 *Fill in row 6 with instructions on how to compute information on the last payment.* Except for the principal portion and the total payment, the last payment computations are just like previous payment computations, so we can do more copying and pasting.

- Copy the instructions from cell A5 and paste them into cell A6.
- Copy the instructions from cell C5 and paste them into cell C6.

- Copy the instructions from cell E5 and paste them into cell E6.
- Cell B6 should contain the principal portion of the last payment, which is equal to the previous balance. Move to cell B6 and type in either "= E5" or "+ E5," and press "Return" or "Enter."
- Cell D6 should contain instructions on computing the last total payment.

    total payment = principal portion + interest portion
    $$= B6 + C6$$

Move to cell D6 and type in either "= B6 + C6" or "+ B6 + C6."

After completing this step, your spreadsheet should look like that in Figure 10.19.

FIGURE 10.19

	A	B	C	D	E
1	Payment Number	Principal Portion	Interest Portion	Total Payment	Balance
2	0				175000
3	1	43328.44167	1130.20833	44458.65	131671.5583
4	2	43608.27119	850.378814	44458.65	88063.28715
5	3	43889.90794	568.742063	44458.65	44173.37921
6	4	44173.37921	285.286407	44458.665617	0

Step 6 *Format the contents of columns B, C, D, and E in currency style.* Your instructor will tell you how. After this formatting, your spreadsheet should look like that in Figure 10.20.

FIGURE 10.20

	A	B	C	D	E
1	Payment Number	Principal Portion	Interest Portion	Total Payment	Balance
2	0				$175,000.00
3	1	$43,328.44	$1,130.21	$44,458.65	$131,671.56
4	2	$43,608.27	$850.38	$44,458.65	$88,063.29
5	3	$43,889.91	$568.74	$44,458.65	$44,173.38
6	4	$44,173.38	$285.29	$44,458.67	$0

Important Note: Most amortization schedules involve more than four payments. In this case, the Step 4 instructions apply to all payments other than the first and the last.

Finding the Total Interest Paid

Once an amortization schedule is prepared, it is a simple matter to find the total interest paid—just find the sum of the interest portions of the appropriate payments. If you are using Amortrix, compute this sum with your own calculator or Expressionist. If you are using a computerized spreadsheet, you can use the spreadsheet to compute the sum. In the example illustrated in Figure 10.19, the total interest paid is the sum of cells C3 through C6. To find this quantity, go to cell C7 (or any other empty cell) and type "=sum(C3:C6)" or "@sum(C3:C6)" or "@sum(C3..C6)" and press "Return" or "Enter."

EXERCISES

42. Use a computer to prepare an amortization schedule for a simple interest amortized loan of $100,000 at 7.5% interest for 1 year.

In Exercises 43–50:

   a. Use a computer to prepare an amortization schedule for the given loan's first year.
   b. Find the amount that could be deducted from the borrower's taxable income (i.e, find the total interest paid) in the loan's first year if the first payment was made in January. (Interest on a car loan is only deductible in some circumstances. For the purpose of these exercises assume that this interest is deductible.)
   c. Find the unpaid balance at the beginning of the loan's last year.
   d. Use a computer to prepare an amortization schedule for the given loan's last year. If you are using a computerized spreadsheet, this will entail using the solution to part (c) in place of the loan amount and starting with payment numbers more appropriate than 0 and 1.

   e. Find the amount that could be deducted from the borrower's taxes (i.e., the total interest paid) in the loan's last year.
   (Answers to parts (b), (c), and (e) are given in the back of the book. Your answers may vary from those given by a few cents.)

43. The loan in Exercise 7

44. The loan in Exercise 8

45. The loan in Exercise 9

46. The loan in Exercise 10

47. A 30-year simple interest amortized home loan for $350,000 at $14\frac{1}{2}\%$ interest

48. A 5-year simple interest amortized car loan for $29,500 at $13\frac{1}{4}\%$ interest

49. A 5-year simple interest amortized car loan for $10,120 at $7\frac{7}{8}\%$ interest

50. A 30-year simple interest amortized home loan for $109,000 at $9\frac{1}{2}\%$ interest

10.5

ANNUAL PERCENTAGE RATE ON A GRAPHING CALCULATOR

A *simple interest loan* is any loan for which the interest portion of each payment is simple interest on the outstanding principal. A *simple interest amortized loan* fulfills this requirement; in fact, we compute an amortization schedule for a simple interest amortized loan by finding the simple interest on the outstanding principal.

EXAMPLE 1   Chris Dant paid $18,327 for a new car. Her dealer offered her a five-year add-on interest loan at 8.5% interest.

    a. Find the size of her monthly payment.

    b. Find the principal portion and the interest portion of each payment.

    c. Determine if this loan is a simple interest loan.

Solution   **a.** We are given $P = 18{,}327$, $r = 0.085$, and $t = 5$ years. The total interest charge is simple interest on the loan amount:

$$I = Prt$$
$$= 18{,}327 \cdot 0.085 \cdot 5$$
$$= 7788.975$$
$$\approx \$7788.98.$$

The total of principal and interest is then:

$$P + I = 18{,}327 + 7788.98 = \$26{,}115.98$$

This amount is equally distributed over 60 monthly payments, so the monthly payment is

$$\frac{26{,}115.98}{60} = 435.26633 \ldots \approx \$435.27$$

  b. The total amount due consists of $18,327 in principal and $7788.98 in interest. The total amount due is equally distributed over 60 monthly payments, so the principal and interest are also equally distributed over 60 monthly payments. The principal portion of each payment is then

$$\frac{18{,}327}{60} = \$305.45$$

and the interest portion is

$$\frac{7788.98}{60} = 129.81633 \ldots \approx \$129.82.$$

  c. The loan is a simple interest loan if the interest portion of each payment is simple interest on the outstanding principal. For the first payment, the outstanding principal is $18,327; simple interest on this amount is

$$I = Prt$$
$$= 18{,}327 \cdot 0.085 \cdot \frac{1}{12}$$
$$= 129.81625 \approx \$129.82$$

In part (b) we found that the interest portion of each payment is also $129.82; thus, for the first payment, the interest portion is simple interest on the outstanding principal. For the second payment, the outstanding principal is

$$18{,}327 - 305.45 = 18{,}021.55$$

Simple interest on this amount is

Historical Note

Truth in Lending Act

The Truth in Lending Act was signed by President Lyndon Johnson in 1968. Its original intent was to promote credit shopping by requiring lenders to use uniform language and calculations and to make full disclosure of credit charges, so that consumers could shop for the most favorable credit terms. The Truth in Lending Act is interpreted by the Federal Reserve Board's Regulation Z.

During the 1970s, Congress amended the act many times, and corresponding changes were made in Regulation Z. These changes resulted in a huge increase in the length and complexity of the law. Its scope now goes beyond disclosure to grant significant legal rights to the borrower.

$$I = Prt$$

$$= 18{,}021.55 \cdot 0.085 \cdot \frac{1}{12}$$

$$= 127.65264\ldots \approx \$127.65$$

However, the interest portion of each payment is $129.82; thus, for the second payment, the interest portion is *more than* simple interest on the outstanding principal. The loan is *not* a simple interest loan; it requires higher interest payments than would a simple interest loan at the same rate.

•

Whenever a loan is not a simple interest loan, the Truth in Lending Act requires the lender to disclose the annual percentage rate to the borrower. The **annual percentage rate** (or A.P.R.) of an add-on interest loan is the simple interest rate that makes the dollar amounts the same if the loan is recomputed as a simple interest amortized loan.

EXAMPLE 2   Find the A.P.R. of the add-on interest loan in Example 1.

Solution   Substitute the dollar amounts into the simple interest amortized loan formula and solve for the interest rate i.

$$pymt \, \frac{(1 + i)^n - 1}{i} = P(1 + i)^n$$

$$435.27 \, \frac{(1 + i)^{60} - 1}{i} = 18{,}327(1 + i)^{60}$$

Solving this equation without the aid of a graphing calculator would be difficult, if not impossible. To solve it with a graphing calculator:

- Enter $435.27 \dfrac{(1 + x)^{60} - 1}{x}$ for Y_1.
- Enter $18,327(1 + x)^{60}$ for Y_2.
- Graph the two functions.
- Find the point at which the two functions intersect, as discussed in Section 1.2 and summarized below.

TI-82/83:	Select the "intersect" option from the "CALC" menu
TI-85/86:	Select "MATH" for the "GRAPH" menu, and then "ISECT" from the "GRAPH MATH" menu

The graph is shown in Figure 10.21.

FIGURE 10.21

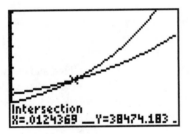

```
Intersection
X=.0124369 __Y=38474.183 _
```

The intersection is at $i = 0.0124369$. This is a monthly rate; the corresponding annual rate is $12 \cdot 0.0124369 = 0.1492428 \approx 14.92\%$. The 8.5% add-on interest loan in Example 1 has an A.P.R. of 14.92%. This means that the add-on interest loan requires the same monthly payment that a 14.92% simple interest amortized loan would have.

The Truth in Lending Act, which requires a bank to divulge its loan's A.P.R., allows a tolerance of one-eighth of 1% ($=0.125\%$) in the claimed A.P.R. Thus, the dealership would be legally correct if it stated that the A.P.R. was between $14.92428\% - 0.125\% = 14.79928\%$ and $14.92428\% + 0.125\% = 15.04928\%$.

Finance Charges

Sometimes **finance charges** other than the interest portion of the monthly payment are associated with a loan; these charges must be paid when the loan agreement is signed. For example, a **point** is a finance charge that is equal to 1% of the loan amount; a **credit report fee** pays for a report on the borrower's credit history, including any late or missing payments and the size of all outstanding debts; an **appraisal fee** pays for the determination of the current market value of the property (the auto, boat, or home) to be purchased with the loan. The Truth in Lending Act requires the lender to inform the borrower of the total finance charge, which includes the interest, the points, and some of the fees (see Figure 10.22 on page 532). Most of the finance charges must be paid before the loan is awarded, so in essence the borrower must pay money now in order to get more money later. The law says that this means the lender is not really borrowing as much as he or she thinks. According to the law, the actual amount loaned is the loan amount minus all points and those fees included in the finance charge, and the A.P.R. is the rate that reconciles the payment and this actual loan amount.

FIGURE 10.22
Sample portion of a federal
truth-in-lending disclosure
statement (loan amount =
$120,765.90)

Costs Included in the Prepaid Finance Charge		Other Costs Not Included in the Finance Charge:	
2 points	$2415.32	appraisal fee	$ 70.00
prorated interest	$1090.10	credit report	$ 60.00
prepaid mortgage insurance	$ 434.50	closing fee	$ 670.00
loan fee	$1242.00	title insurance	$ 202.50
document preparation fee	$ 80.00	recording fee	$ 20.00
tax service fee	$ 22.50	notary fee	$ 20.00
processing fee	$ 42.75	tax and insurance escrow	$ 631.30
subtotal	$5327.17	subtotal	$1673.80

EXAMPLE 3 Glen and Tanya Hansen bought a home for $140,000. They paid the sellers a 20% down payment and obtained a simple interest amortized loan for the balance from their bank at $10\frac{3}{4}\%$ for 30 years. The bank in turn paid the sellers the remaining 80% of the purchase price, less a 6% sales commission paid to the sellers' and the buyers' real estate agents. (The transaction is illustrated in Figure 10.23.) The bank charged the Hansens two points, plus fees totaling $3247.60; of these fees, $1012.00 were included in the finance charge.

FIGURE 10.23

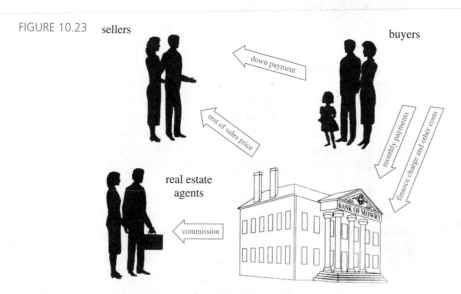

a. Find the size of the Hansens' monthly payment.
b. Find the total interest paid.
c. Compute the total finance charge.
d. Find the A.P.R.

Solution   **a.** *Finding their monthly payment:* We are given down payment = 20% of $140,000 = $28,000, P = loan amount = $140,000 − $28,000 = $112,000, $i = \frac{1}{12}$ of $10\frac{3}{4}\% = \frac{0.1075}{12}$, and n = 30 years = 30 years · (12 months)/(1 year) = 360 months.

$$\text{future value of annuity} = \text{future value of loan amount}$$

$$pymt\, \frac{(1 + i)^n - 1}{i} = P(1 + i)^n$$

$$pymt\, \frac{\left(1 + \dfrac{0.1075}{12}\right)^{360} - 1}{\dfrac{0.1075}{12}} = 112{,}000\left(1 + \frac{0.1075}{12}\right)^{360}$$

Computing the fraction on the left side and multiplying its reciprocal by the right side, we get

$$pymt = 1045.4991 \approx \$1045.50$$

b. *Finding the total interest paid:* The total interest paid is the total amount paid minus the amount borrowed. The Hansens agreed to make 360 monthly payments of $1045.50 each, for a total of 360 · $1045.50 = $376,380.00. Of this, $112,000 is principal; therefore, the total interest is

$$376{,}380 - 112{,}000 = \$264{,}380$$

c. *Computing their total finance charge:*

$$\begin{aligned}
2 \text{ points} = 2\% \text{ of } \$112{,}000 = \$\ \ \ 2240 \\
\text{included fees} = \$\ \ \ 1012 \\
\text{total interest paid} = \underline{\$264{,}380} \qquad \text{From part (b)} \\
\text{total finance charge} = \$267{,}632
\end{aligned}$$

d. *Finding the A.P.R.:* The A.P.R. is the simple interest rate that makes the dollar amounts the same if the loan is recomputed using the legal loan amount (loan amount less points and fees) in place of the actual loan amount. The legal loan amount is

$$P = \$112{,}000 - \$3252 = \$108{,}748$$

Substitute the dollar amounts into the simple interest amortized loan formula and solve for the interest rate i.

$$pymt\, \frac{(1 + i)^n - 1}{i} = P(1 + i)^n$$

$$1045.50\, \frac{(1 + i)^{360} - 1}{i} = 108{,}748(1 + i)^{360}$$

To solve this with a graphing calculator:

- enter $1045.50\, \dfrac{(1 + x)^{360} - 1}{x}$ for Y_1
- enter $108{,}748(1 + x)^{360}$ for Y_2

- graph the two functions
- find the point at which the two functions intersect

The graph is shown in Figure 10.24.

FIGURE 10.24

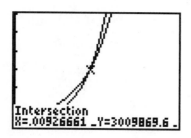

The intersection is at $i = 0.00926661$. This is a monthly rate; the corresponding annual rate is $12 \cdot 0.00926661 = 0.11119932 \approx 11.12\%$.

The $10\frac{3}{4}\%$ simple interest amortized loan has an A.P.R. of 11.12%. This means that the $10\frac{3}{4}\%$ loan requires the same monthly payment that a 11.12% loan with no points or fees would have.

•

If you need to obtain a loan, it is not necessarily true that the lender with the lowest interest rate will give you the least expensive loan. One lender may charge more points or higher fees than does another lender. One lender may offer an add-on interest loan, while another lender offers a simple interest amortized loan. These differences can have a significant impact on the cost of a loan. If two lenders offer loans at the same interest rate but differ in any of these ways, that difference will be reflected in the A.P.R. "Lowest interest rate" does *not* mean "least expensive loan," but "lowest A.P.R." *does* mean "least expensive loan."

A.P.R. Steps

1. *Compute the payment.*
   a. Use the methods of Section 10.1 with an add-on interest loan.
   b. Use the methods of Section 10.4 with a simple interest amortized loan.
2. *Use the simple interest amortized loan formula to compute the A.P.R.*
   a. Substitute the payment from Step 1 for *pymt*.
   b. If there are points and fees, substitute the legal loan amount (legal loan amount = loan amount less points and those fees included in the finance charge) for *P*.
   c. Use your graphing calculator to find the periodic interest rate.
   d. Convert the periodic interest rate to an annual rate. This annual rate is the A.P.R.
3. *Determine if the claimed A.P.R. is legally correct.* If the lender claims a certain A.P.R., The Truth in Lending Act allows a tolerance of one-eighth of 1% (= 0.125%) in the claimed A.P.R.

10.5

EXERCISES

1. Wade Ellis buys a new car for $16,113.82. He puts 10% down and obtains a simple interest amortized loan for the balance at $11\frac{1}{2}\%$ interest for 4 years. If loan fees included in the finance charge total $814.14, find the A.P.R.

2. Guy de Primo buys a new car for $9837.91. He puts 10% down and obtains a simple interest amortized loan for the balance at $10\frac{7}{8}\%$ interest for 4 years. If loan fees included in the finance charge total $633.87, find the A.P.R.

3. Chris Burditt bought a house for $212,500. He put 20% down and obtained a simple interest amortized loan for the balance at $10\frac{7}{8}\%$ interest for 30 years. If Chris paid 2 points and $4728.60 in fees, $1318.10 of which are included in the finance charge, find the A.P.R.

4. Shirley Trembley bought a house for $187,600. She put 20% down and obtained a simple interest amortized loan for the balance at $11\frac{3}{8}\%$ for 30 years. If Shirley paid 2 points and $3427.00 in fees, $1102.70 of which are included in the finance charge, find the A.P.R.

5. Jennifer Tonda wants to buy a used car that costs $4600. The used car dealer has offered her a 4-year add-on interest loan that requires no down payment at 8% annual interest, with an A.P.R. of $14\frac{1}{4}\%$.
   a. Find the monthly payment.
   b. Verify the A.P.R.

6. Melody Shepherd wants to buy a used car that costs $5300. The used car dealer has offered her a 4-year add-on interest loan that requires a $200 down payment at 7% annual interest, with an A.P.R. of 10%.
   a. Find the monthly payment.
   b. Verify the A.P.R.

7. Anne Scanlan is buying a used car that costs $10,340. The used car dealer has offered her a 5-year add-on interest loan at 9.5% interest, with an A.P.R. of 9.9%. The loan requires a 10% down payment.
   a. Find the monthly payment.
   b. Verify the A.P.R.

8. Stan Loll bought a used car for $9800. The used car dealer offered him a 4-year add-on interest loan at 7.8% interest, with an A.P.R. of 8.0%. The loan requires a 10% down payment.
   a. Find the monthly payment.
   b. Verify the A.P.R.

9. Susan Chin is shopping for a car loan. Her savings and loan offers her a simple interest amortized loan for 4 years at 9% interest. Her bank offers her a simple interest amortized loan for 4 years at 9.1% interest. Which is the less expensive loan?

10. Stephen Tamchin is shopping for a car loan. His credit union offers him a simple interest amortized loan for 4 years at 7.1% interest. His bank offers him a simple interest amortized loan for 4 years at 7.3% interest. Which is the less expensive loan?

11. Ruben Lopez is shopping for a home loan. Really Friendly Savings and Loan offers him a 30-year simple interest amortized loan at 9.2% interest, with an A.P.R. of 9.87%. The Solid and Dependable Bank offers him a 30-year simple interest amortized loan at 9.3% interest, with an A.P.R. of 9.80%. Which loan would have the lower payments? Which loan would be the least expensive, taking into consideration monthly payments, points, and fees? Justify your answers.

12. Keith Moon is shopping for a home loan. Sincerity Savings offers him a 30-year simple interest amortized loan at 8.7% interest, with an A.P.R. of 9.12%. Pinstripe National Bank offers him a 30-year simple interest amortized loan at 8.9% interest, with an A.P.R. of 8.9%. Which loan would have the lower payments? Which loan would be the least expensive, taking into consideration monthly payments, points, and fees? Justify your answers.

13. This is an exercise in buying a home. It involves choosing a home and selecting the home's financing. Write a paper describing all of the following points.
   a. You may not be in a position to buy a home now. If so, fantasize about your future. What job do you have? How long have you had that job? What is your salary? If you're married, does your spouse work? Do you have a family? What needs will your home fulfill? Make your future fantasy realistic. Briefly describe what has happened between the present and your future fantasy. (If you are in a position to buy a home now, discuss these points on a more realistic level.)
   b. Go shopping for a home. Look at houses, condominiums, or both. Look at new homes, used homes, or both. (Used homes can easily be visited by going to an "open house," where the owners are gone and the real estate agent allows interested parties to inspect the home. Open houses are probably listed in your local newspaper.) Read appropriate newspaper and magazine articles (for example, in the real estate section

of your local newspaper). Discuss in detail the home you selected and why you did so. How will your selection fulfill your (projected) needs? Why did you select a house/condominium? Why did you select a new/used home? Explain your choice of location, house size, and features of the home.

c. Go shopping for financing. Many banks, savings and loans, and credit unions have information on home loans available on request. Get all of the information you need from at least two lenders, but do not bother the staff unnecessarily. You may be able to find the necessary information in the newspaper. Perform all appropriate computations yourself—do not have lenders tell you the payment size. Summarize the appropriate data in your paper and discuss which loan you would choose. Include in your discussion the down payment, the presence or absence of a prepayment penalty, the duration of the loan, the interest rate, the A.P.R., the payment size, and other terms of the loan.

d. Also discuss the real estate taxes (your instructor will provide you with information on the local tax rate) and the effect of your purchase on your income taxes (interest paid on a home loan is deductible from the borrower's income taxes).

e. Most lenders will approve a home loan only if the total of all the borrower's monthly payments, including the home loan payment, real estate taxes, credit card payments, and car loan payments, is no more than 38% of the borrower's monthly income. Discuss your ability to qualify for the loan.

Answer the following questions using complete sentences.

14. If the A.P.R. of a simple interest amortized home loan is equal to the loan's interest rate, does the loan require any fees? Does it require any points?

15. Compare and contrast the annual percentage rate of a loan with the annual yield of a compound interest rate.

CHAPTER 10 REVIEW

Terms

add-on interest loan	compound interest	loan agreement, or note	proceeds
adjustable rate mortgage	compounding period	maturity value, or value	simple interest
amortization schedule	computerized spreadsheet	nominal rate	simple interest amortized
amortized loan	discount	outstanding principal, or	loan
annual percentage rate of	discount loan	balance	simple interest loan
an add-on interest loan	discounting a note	payment period	sinking fund
annual percentage rate of	doubling time	periodic interest rate	spreadsheet
a simple interest	finance	point	tax-deferred annuity
amortized loan	finance charges	prepayment penalty	(TDA)
annuity	future value	present value	term
annuity due versus	interest	present value of an annuity	yield, or annual yield
ordinary annuity	interest rate	principal	

Formulas

Simple Interest: $I = Prt$	Ordinary Annuity: $FV(\text{ordinary}) = pymt \dfrac{(1 + i)^n - 1}{i}$
Simple Interest Future Value: $FV = P(1 + rt)$	
	Annuity Due: $FV(\text{due}) = FV(\text{ordinary}) \cdot (1 + i)$
Simple Discount: $D = FV \cdot r \cdot t$	
	Present Value of Annuity: $FV(\text{lump sum}) = FV(\text{annuity})$ (Find the lump sum.)
Compound Interest: $FV = P(1 + i)^n$	
Annual Yield: $FV(\text{compound interest}) = FV(\text{simple interest})$ (Find the simple interest rate.)	Simple Interest Amortized Loan: $FV(\text{ordinary annuity}) = FV(\text{compound interest})$ (Find *pymt*.)

Review Exercises

1. Find the interest earned by a deposit of $8140 at $9\frac{3}{4}\%$ simple interest for 11 years.

2. Find the interest earned by a deposit of $8140 at $9\frac{3}{4}\%$ interest compounded monthly for 11 years.

3. Find the maturity value of a loan of $3550 borrowed at $12\frac{1}{2}\%$ simple interest for 1 year and 2 months.

4. Find the future value of $3550 deposited at $12\frac{1}{2}\%$ interest compounded monthly for 1 year and 2 months.

5. Lynn Knight inherited $7000. She wants to buy a used car, but the type of car she wants typically sells for around $8000. If her money can earn $7\frac{1}{2}\%$ simple interest, how long must she invest her money?

6. George and Martha Simpson bought a house from Sue Sanchez for $205,500. In lieu of a 20% down payment, Ms. Sanchez accepted 5% down at the time of the sale and a promissory note from the Simpsons for the remaining 15%, due in 8 years. The Simpsons also agreed to make monthly interest payments to Ms. Sanchez at 12% simple interest until the note expires.
   a. Find the Simpsons' monthly interest-only payment to Ms. Sanchez.
   b. Find Ms. Sanchez's total income from all aspects of the down payment.
   c. Find Ms. Sanchez's total income from all aspects of the sale of the house, including the down payment.

7. Extremely Trustworthy Savings offers 5-year CDs at 7.63% compounded annually, and Bank of the South offers 5-year CDs at 7.59% compounded daily. Compute the annual yield for each institution and determine which offering is more advantageous for the consumer.

8. Tien Ren Chiang wants to have an IRA that will be worth $250,000 when he retires at age 65.
   a. How much must he deposit at age 25 at $10\frac{1}{8}\%$ compounded quarterly?
   b. If at age 65 he arranged for the monthly interest to be sent to him, how much would he receive each month? (Assume that he will continue to receive $10\frac{1}{8}\%$ interest, compounded monthly.)

9. Find the future value of an ordinary annuity with monthly payments of $200 that earns $6\frac{1}{8}\%$ interest, after 11 years.

10. Find the present value of the annuity in Exercise 9.

11. Find the future value of an annuity due with monthly payments of $200 that earns $6\frac{1}{8}\%$ interest, after 11 years.

12. Matt and Leslie Silva want to set up a TDA that will generate sufficient interest on maturity to meet their living expenses, which they project to be $1300 per month.
   a. Find the amount needed at maturity to generate $1300 per month interest if they can get $8\frac{1}{4}\%$ interest compounded monthly.

b. Find the monthly payment they would have to make into an ordinary annuity to obtain the future value found in part (a) if their money earns $9\frac{3}{4}\%$ and the term is 30 years.

13. Ben Suico buys a new car for $13,487.31. He puts 10% down and obtains a simple interest amortized loan for the rest at $10\frac{7}{8}\%$ interest for 5 years.
   a. Find his monthly payment.
   b. Find the total interest.
   c. Prepare an amortization schedule for the first 2 months of the loan.

d. If loan fees included in the finance charge total $633.87, verify the lender's statement that the A.P.R. is 11.4%.
e. Mr. Suico decides to sell his car 2 years and 6 months after he bought it. Find the unpaid balance on his loan.

14. Scott Frei wants to buy a used car that costs $6200. The used car dealer has offered him a 4-year add-on interest loan that requires a $200 down payment at 9.9% annual interest with an A.P.R. of 10%.
   a. Find the monthly payment.
   b. Verify the A.P.R.

USING A GRAPHING CALCULATOR

The following discussion was written specifically for Texas Instruments graphing calculators; however, it frequently applies to other brands as well. Read the following discussion with your calculator close at hand. When a calculation is discussed, do that calculation on your calculator.

The Enter Button

A graphing calculator will never perform a calculation until the ENTER button is pressed. To add 3 and 2, type

$$3 \; \boxed{+} \; 2 \; \boxed{\text{ENTER}}$$

and the display will read "5." To square 4, type

$$4 \; \boxed{x^2} \; \boxed{\text{ENTER}}$$

and the display will read "16." If the ENTER button isn't pressed, the calculation will not be performed.

The 2nd and Alpha Buttons

Most calculator buttons have more than one label and more than one use; you select from these uses with the 2nd and ALPHA buttons. For example, one button is labeled "x^2" on the button itself, "$\sqrt{}$" above the button, and either "I" or "K" above and to the right of the button. If it is used without the 2nd or ALPHA buttons, it will square a number. Typing

$$4 \; \boxed{x^2} \; \boxed{\text{ENTER}}$$

makes the display read "16," since $4^2 = 16$. If it is used with the 2nd button, it will take the square root of a number. Typing

$$\boxed{\text{2nd}} \; \boxed{\sqrt{}} \; 4 \; \boxed{\text{ENTER}}$$

makes the display read "2," since $\sqrt{4} = 2$. If it is used with the ALPHA button, it will print the letter I (or K).

Notice that to square 4 you press the $\boxed{x^2}$ button *after* the 4, but to take the square root of 4 you press the $\boxed{\sqrt{}}$ *before* the 4. This is because Texas Instruments graphing calculators are designed so that the way you type something is as similar as possible to the way it is written algebraically. When you write 4^2, you write the 4 first and then the squared symbol; thus, on your graphing calculator you press the 4 first and then the $\boxed{x^2}$ button. When you write $\sqrt{4}$, you write the square root symbol first and then the 4; thus, on your graphing calculator you press the $\boxed{\sqrt{}}$ button first and then the 4.

Frequently, the two operations that share a button are operations that "undo" each other. For example, typing

$$3 \ \boxed{x^2} \ \boxed{\text{ENTER}}$$

makes the display read "9," since $3^2 = 9$, and typing

$$\boxed{\text{2nd}} \ \boxed{\sqrt{}} \ 9 \ \boxed{\text{ENTER}}$$

makes the display read "3," since $\sqrt{9} = 3$. This is done as a memory device; it is easier to find the various operations on the keyboard if the two operations that share a button also share a relationship.

Two operations that *always* undo each other are called **inverses.** The x^2 and $\sqrt{x}$ operations are not inverses because $(-3)^2 = 9$, but $\sqrt{9} \neq -3$. However, there is an inverse-type relationship between the x^2 and $\sqrt{x}$ operations—they undo each other sometimes. When two operations share a button, they are inverses or they share an inverse-type relation.

Correcting Typing Errors

If you've made a typing error *and you haven't yet pressed* $\boxed{\text{ENTER}}$, you can correct that error with the $\boxed{\leftarrow}$ button. For example, if you typed "5 × 2 + 7" and then realized you wanted "5 × 3 + 7" you can replace the incorrect 2 with a 3 by pressing the $\boxed{\leftarrow}$ button until the 2 is flashing and then pressing "3."

If you realize that you've made a typing error *after* you pressed $\boxed{\text{ENTER}}$, just press $\boxed{\text{2nd}} \ \boxed{\text{ENTRY}}$ to reproduce the previously entered line. Then correct the error with the $\boxed{\leftarrow}$ button as described above.

The $\boxed{\text{INS}}$ button allows you to insert a character. For example, if you typed "5 × 27" and you meant to type "5 × 217", press the $\boxed{\leftarrow}$ button until the 7 is flashing and then insert a 1 by typing

$$\boxed{\text{2nd}} \ \boxed{\text{INS}} \ 1$$

The $\boxed{\text{DEL}}$ button allows you to delete a character. For example, if you typed "5 × 217" and you meant to type "5 × 27", press the $\boxed{\leftarrow}$ button until the 1 is flashing and then press $\boxed{\text{DEL}}$.

If you haven't yet pressed $\boxed{\text{ENTER}}$, the $\boxed{\text{CLEAR}}$ button allows you to erase an entire line. If you have pressed $\boxed{\text{ENTER}}$, the $\boxed{\text{CLEAR}}$ button clears everything off of the screen.

The Subtraction Symbol and the Negative Symbol

If you read "3 − −2" aloud, you could say "three subtract negative two" or "three minus minus two" and you would be understood. The expression is understandable even if the distinction between the negative symbol and the subtraction symbol is not made clear. With a calculator, however, this distinction is crucial. The subtraction button is labelled "−" and the negative button is labelled "(−)."

EXAMPLE 1   Calculate $3 - -2$, both (a) by hand and (b) with a calculator.

Solution   a. $3 - -2 = 3 + 2 = 5$

b. Type

3 $\boxed{-}$ $\boxed{(-)}$ 2 $\boxed{\text{ENTER}}$

and the display will read "5."
If you had typed

3 $\boxed{-}$ $\boxed{-}$ 2 $\boxed{\text{ENTER}}$ or   3 $\boxed{(-)}$ $\boxed{-}$ 2 $\boxed{\text{ENTER}}$

the calculator would have responded with an error message.

●

The Multiplication Button

In algebra, we do not use "×" for multiplication. Instead, we use "x" as a variable, and we use "·" for multiplication. However, Texas Instruments graphing calculators use "×" as the label on the multiplication button, and "*" for multiplication on the display screen. (The "variable x" button is labeled either "X,T,θ" "X,T,θ,n" or "x-VAR.") This is one of the few instances where you don't type things in the same way that you write them algebraically.

Order of Operations and Use of Parentheses

Graphing calculators are programmed so that they follow the order of operations. That is, they perform calculations in the following order:

1. Parentheses-enclosed work
2. Exponents
3. Multiplication and
   Division, from left to right
4. Addition and
   Subtraction, from left to right

You can remember this order by remembering the mnemonic "PEMDAS."

EXAMPLE 2   Calculate $2(3 + 4)$, both (a) by hand and (b) with a calculator.

Solution   a. $2(3 + 4) = 2(7)$ Parentheses-enclosed work comes first
   $= 14$ Multiplication follows parentheses-enclosed work

b. Type

2 $\boxed{\times}$ $\boxed{(}$ 3 $\boxed{+}$ 4 $\boxed{)}$ $\boxed{\text{ENTER}}$

and the display will read "14."

●

In the instructions to Example 2, notice that we wrote "$2(3 + 4)$" rather than "$2 \cdot (3 + 4)$"; in this case it's not necessary to write the multiplication symbol. Similarly, it's not necessary to type the multiplication symbol. Example 2(b) could be computed by typing

$$2 \;\boxed{(}\; 3 \;\boxed{+}\; 4 \;\boxed{)}\; \boxed{\text{ENTER}}$$

EXAMPLE 3 Calculate $2 \cdot 3^3$, both (a) by hand and (b) with a calculator.

Solution **a.** $2 \cdot 3^3 = 2 \cdot 27$ Exponents come before multiplication
$= 54$

b. To do this we must use the exponent button, which is labeled "$\wedge$." Type

$$2 \;\boxed{\times}\; 3 \;\boxed{\wedge}\; 3 \;\boxed{\text{ENTER}}$$

and the display will read "54." •

EXAMPLE 4 Calculate $(2 \cdot 3)^3$, both (a) by hand and (b) with a calculator.

Solution **a.** $(2 \cdot 3)^3 = 6^3$ Parentheses-enclosed work comes first
$= 216$

b. Type

$$\boxed{(}\; 2 \;\boxed{\times}\; 3 \;\boxed{)}\; \boxed{\wedge}\; 3 \;\boxed{\text{ENTER}}$$

and the display will read "216." •

In Example 3 the exponent applies only to the 3 because the order of operations dictates that exponents come before multiplication. In Example 4 the exponent applies to the $(2 \cdot 3)$ because the order of operations dictates that parentheses-enclosed work comes before exponents. In each example, the way you type the problem matches the way it is written algebraically because the calculator is programmed to follow the order of operations. Sometimes, however, the way you type a problem doesn't match the way it is written algebraically.

EXAMPLE 5 Calculate $\frac{2}{3 \cdot 4}$ with a calculator.

Solution **WRONG**   It is incorrect to type

$$2 \;\boxed{\div}\; 3 \;\boxed{\times}\; 4 \;\boxed{\text{ENTER}}$$

even though that matches the way the problem is written algebraically. According to the order of operations, multiplication and division are done from left to right, so the above typing is algebraically equivalent to

$$= \frac{2}{3} \cdot 4 \qquad \text{First dividing then multiplying, since division is on the left and} \\ \text{multiplication is on the right}$$

$$= \frac{2}{3} \cdot \frac{4}{1} = \frac{2 \cdot 4}{3}$$

which is not what we want. The difficulty is that the large fraction bar in the expression $\frac{2}{3 \cdot 4}$ groups the "$3 \cdot 4$" together in the denominator; in the above typing, nothing groups the "$3 \cdot 4$" together, and only the 3 ends up in the denominator.

RIGHT   The calculator needs parentheses inserted in the following manner:

$$\frac{2}{(3 \cdot 4)}$$

Thus, it is correct to type

2 ÷ (3 × 4) ENTER

This makes the display read 0.166666667, the correct answer.

ALSO RIGHT   It is correct to type

2 ÷ 3 ÷ 4 ENTER

According to the order of operations, multiplication and division are done from left to right, so the above typing is algebraically equivalent to

$$\frac{2}{3} \div 4 \qquad \text{Doing the left-hand division first}$$

$$= \frac{2}{3} \cdot \frac{1}{4} \qquad \text{Inverting and multiplying}$$

$$= \frac{2}{3 \cdot 4}$$

which is what we want. *When you're calculating something that involves only multiplication and division and you don't use parentheses, the* × *button places a factor in the numerator and the* ÷ *button places a factor in the denominator.*   •

EXAMPLE 6   Calculate $\frac{2}{3/4}$ with a calculator.

Solution   **WRONG**   It is incorrect to type

2 ÷ 3 ÷ 4 ENTER

even though that matches the way the problem is written algebraically. As discussed in Example 5, this typing is algebraically equivalent to

$$\frac{2}{3 \cdot 4}$$

which is not what we want.

RIGHT   The calculator needs parentheses inserted in the following manner:

$$\frac{2}{(3 \div 4)}$$

Thus, it is correct to type

2 ÷ (3 ÷ 4) ENTER

since, according to the order of operations, parentheses-enclosed work is done first. This makes the display read 2.66666667, the correct answer. •

EXAMPLE 7   Calculate $\frac{2+3}{4}$ with a calculator.

Solution   **WRONG**   It is incorrect to type

$$2 \boxed{+} 3 \boxed{÷} 4 \boxed{\text{ENTER}}$$

even though that matches the way the problem is written algebraically. According to the order of operations, division is done before addition, so this typing is algebraically equivalent to

$$2 + \frac{3}{4}$$

which is not what we want. The large fraction bar in the expression $\frac{2+3}{4}$ groups the "2 + 3" together in the numerator; in the above typing, nothing groups the "2 + 3" together, and only the 3 ends up in the numerator.

RIGHT   The calculator needs parentheses inserted in the following manner:

$$\frac{(2+3)}{4}$$

Thus it is correct to type

$$\boxed{(} 2 \boxed{+} 3 \boxed{)} \boxed{÷} 4 \boxed{\text{ENTER}}$$

This makes the display read 1.25, the correct answer.

ALSO RIGHT   It is correct to type

$$2 \boxed{+} 3 \boxed{\text{ENTER}} \boxed{÷} 4 \boxed{\text{ENTER}}$$

The first $\boxed{\text{ENTER}}$ makes the calculator perform all prior calculations before continuing. This too makes the display read 1.25, the correct answer. •

Memory

The **memory** is a place to store a number for later use, without having to write it down. Graphing calculators have a memory for each letter of the alphabet; that is, you can store one number in memory A, a second number in memory B, and so on. Pressing

TI-85/86: 5 $\boxed{\text{STO}}$ $\boxed{\text{A}}$ $\boxed{\text{ENTER}}$	
TI-82/83: 5 $\boxed{\text{STO}}$ $\boxed{\text{ALPHA}}$ $\boxed{\text{A}}$ $\boxed{\text{ENTER}}$	

will store 5 in memory A. Similar keystrokes will store in memory B. Pressing $\boxed{\text{ALPHA}}$ $\boxed{\text{A}}$ will recall what has been stored in memory A.

EXAMPLE 8   Calculate

$$\frac{3 + \dfrac{5 + 7}{2}}{4}$$

(a) by hand, and (b) by first calculating $\frac{5 + 7}{2}$ and storing the result.

Solution   **a.** $\dfrac{3 + \dfrac{5 + 7}{2}}{4} = \dfrac{3 + \dfrac{12}{2}}{4} = \dfrac{3 + 6}{4} = \dfrac{9}{4} = 2.25$

b. First calculate $\frac{5 + 7}{2}$ and store the result in memory A. The large fraction bar in this expression groups the "5 + 7" together in the numerator; in our typing, we must group the "5 + 7" together with parentheses. The calculator needs parentheses inserted in the following manner:

$$\frac{(5 + 7)}{4}$$

Type

TI-85/86:	$\boxed{(}$ 5 $\boxed{+}$ 7 $\boxed{)}$ $\boxed{\div}$ 2 $\boxed{\text{STO}\rightarrow}$ $\boxed{\text{A}}$ $\boxed{\text{ENTER}}$
TI-82/83:	$\boxed{(}$ 5 $\boxed{+}$ 7 $\boxed{)}$ $\boxed{\div}$ 2 $\boxed{\text{STO}\rightarrow}$ $\boxed{\text{ALPHA}}$ $\boxed{\text{A}}$ $\boxed{\text{ENTER}}$

What remains is to compute $\frac{3 + A}{4}$. Again the large fraction bar groups the "3 + A" together, so the calculator needs parentheses inserted in the following manner:

$$\frac{(3 + A)}{4}$$

Type

$\boxed{(}$ 3 $\boxed{+}$ $\boxed{\text{ALPHA}}$ $\boxed{\text{A}}$ $\boxed{)}$ $\boxed{\div}$ 4 $\boxed{\text{ENTER}}$

and the display will read "2.25."   ●

EXAMPLE 9   Calculate

$$\frac{3 + \dfrac{5 + 7}{2}}{4}$$

using one line of instructions, and without using the memory.

Solution   There are two large fraction bars, one grouping the "5 + 7" together and one grouping the "3 + $\frac{5 + 7}{2}$" together. In our typing, we must group each of these together with parentheses.

The calculator needs parentheses inserted in the following manner:

$$\frac{\left(3 + \dfrac{(5 + 7)}{2}\right)}{4}$$

Type

[(] 3 [+] [(] 5 [+] 7 [)] [÷] 2 [)] [÷] 4 [ENTER]

and the display will read "2.25."

EXAMPLE 10   **a.** Use the quadratic formula and your calculator's memory to solve

$$2.3x^2 + 4.9x + 1.5 = 0$$

b. Use your calculator's memory to check your answers.

Solution   **a.** According to the quadratic formula, if $ax^2 + bx + c = 0$ then

$$x = \frac{-b \pm \sqrt{b^2 - 4ac}}{2a}$$

For our problem, $a = 2.3$, $b = 4.9$, and $c = 1.5$. This gives

$$x = \frac{-4.9 \pm \sqrt{4.9^2 - 4 \cdot 2.3 \cdot 1.5}}{2 \cdot 2.3}$$

One way to do this calculation is to calculate the radical, store it, and then calculate the two fractions.

Step 1   *Calculate the radical.* To do this, type

TI-82:	[2nd] [√‾] [(] 4.9 [x²] [−] 4 [×] 2.3 [×] 1.5 [)] [STO→] [ALPHA] [A] [ENTER]
TI-85/86:	[2nd] [√‾] [(] 4.9 [x²] [−] 4 [×] 2.3 [×] 1.5 [)] [STO→] [A] [ENTER]
TI-83:	[2nd] [√‾] 4.9 [x²] [−] 4 [×] 2.3 [×] 1.5 [)] [STO→] [ALPHA] [A] [ENTER]

This makes the display read "3.195309062" (possibly with more decimal places) and stores the number in the memory A. Notice the use of parentheses.

Step 2   *Calculate the first fraction.* The first fraction is

$$\frac{-4.9 + \sqrt{4.9^2 - 4 \cdot 2.3 \cdot 1.5}}{2 \cdot 2.3}$$

However, the radical has already been calculated and stored in memory A, so this is equivalent to

$$\frac{-4.9 + A}{2 \cdot 2.3}$$

Type

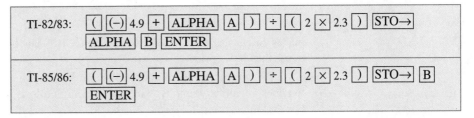

This makes the display read "−.3705849866" (possibly with more decimal places) and stores the number in memory B.

Step 3 *Calculate the second fraction.* The second fraction is

$$\frac{-4.9 - \sqrt{4.9^2 - 4 \cdot 2.3 \cdot 1.5}}{2 \cdot 2.3} = \frac{-4.9 - A}{2 \cdot 2.3}$$

Memories A and B are already in use, so we will store this in memory C. Instead of retyping the line in Step 2 with the "+" changed to a "−" and the "B" to a "C," press ⟦2nd⟧ ⟦ENTRY⟧ to reproduce that line, and use the ⟦←⟧ button to make these changes. Press ⟦ENTER⟧ and the display will read −1.759849796 (possibly with more decimal places) and that number will be stored in memory C. The solutions to $2.3x^2 + 4.9x + 1.5 = 0$ are $x = -0.370584966$ and $x = -1.759849796$. (The TI-85/86 shows 2 more decimal places.)

Step 4 *Check your solutions.* These two solutions are stored in memories B and C; they can be checked by seeing if they satisfy the equation $2.3x^2 + 4.9x + 1.5 = 0$. To check the first solution, type

2.3 ⟦×⟧ ⟦ALPHA⟧ ⟦B⟧ ⟦x²⟧ ⟦+⟧ 4.9 ⟦×⟧ ⟦ALPHA⟧ ⟦B⟧ ⟦+⟧ 1.5 ⟦ENTER⟧

and the display should read either "0" or a number very close to 0.

Scientific Notation

Typing

4000000 ⟦×⟧ 8000000 ⟦ENTER⟧

makes the display read "3.2E13" rather than "32000000000000." This is because the calculator does not have enough room on its display for "32000000000000." When the display shows "3.2E13" read it as "3.2×10^{13}," which is written in scientific notation. Literally, "3.2×10^{13}" means "multiply 3.2 by 10, thirteen times," but as a shortcut you can interpret it as "move the decimal point in the '3.2' thirteen places to the right."

Typing

.0000005 ⟦×⟧ .0000007 ⟦ENTER⟧

makes the display read "3.5E-13" rather than "0.00000000000035," because the calculator does not have enough room on its display for "0.00000000000035." Read "3.5E-13" as "3.5×10^{-13}." Literally, this means "divide 3.5 by 10 thirteen times," but as a short-cut you can interpret it as "move the decimal point in the '3.5' thirteen places to the left."

You can type a number in scientific notation by using the button labeled "EE" (which stands for "Enter Exponent"). For example, typing

5.2 EE 8 ENTER

makes the display read "520000000." (If the "EE" label is above the button you will need to use the 2nd button.)

Be careful that you don't confuse the EE button with the ∧ button. The EE button does *not* allow you to type in an exponent; it allows you to type in scientific notation. For example, typing

3 EE 4 ENTER

makes the display read "30000," since $3 \times 10^4 = 30,000$. Typing

3 ∧ 4 ENTER

makes the display read "81," since $3^4 = 81$.

EXERCISES

In Exercises 1–32, use your calculator to perform the given calculation. The correct answer is given in brackets. In your homework, write down what you type to get that answer.

1. $-3 - -5$ [2]

2. $-6 - 3$ [−9]

3. $4 - -9$ [13]

4. $-6 - -8$ [2]

5. $-3 - (-5 - -8)$ [−6]

6. $-(-4 - 3) - (-6 - -2)$ [11]

7. $-8 \cdot -3 \cdot -2$ [−48]

8. $-8 \cdot -3 - 2$ [22]

9. $(-3)(-8) - (-9)(-2)$ [6]

10. $2(3 - 5)$ [−4]

11. $2 \cdot 3 - 5$ [1]

12. $4 \cdot 11^2$ [484]

13. $(4 \cdot 11)^2$ [1936]

14. $4 \cdot (-11)^2$ [484]

15. $4 \cdot (-3)^3$ [−108]

16. $(4 \cdot -3)^3$ [−1728]

17. $\dfrac{3 + 2}{7}$ [0.7142857]

18. $\dfrac{3 \cdot 2}{7}$ [0.8571429]

19. $\dfrac{3}{2 \cdot 7}$ [0.2142857]

20. $\dfrac{3 \cdot 2}{7 \cdot 5}$ [0.1714286]

21. $\dfrac{3 + 2}{7 \cdot 5}$ [0.1428571]

22. $\dfrac{3 \cdot -2}{7 + 5}$ [−0.5]

23. $\dfrac{3}{7/2}$ [0.8571429]

24. $\dfrac{3/7}{2}$ [0.2142857]

25. 1.8^2 [3.24]

26. $\sqrt{1.8}$ [1.3416408]

27. $47{,}000{,}000^2$ [2.209×10^{15}]

28. $\sqrt{0.0000000000027}$ [1.643168×10^{-6}]

29. $(-3.92)^7$ [14,223.368737]

30. $(5.72 \times 10^{19})^4$ [1.070494×10^{79}]

31. $(3.76 \times 10^{-12})^{-5}$ [1.330641×10^{57}]

32. $(3.76 \times 10^{-12}) - 5$ [−5]

In Exercises 33–36, perform the given calculation (a) by hand, (b) with a calculator using the memory, and (c) with a calculator using one line of instruction and without using the memory. In your homework for parts (b) and (c), write down what you type. Answers are not given in the back of the book.

33. $\dfrac{\dfrac{9-12}{5}+7}{2}$

34. $\dfrac{\dfrac{4-11}{6}+8}{7}$

35. $\dfrac{\dfrac{7+9}{5}+\dfrac{8-14}{3}}{3}$

36. $\dfrac{\dfrac{4-16}{5}+\dfrac{7-22}{2}}{5}$

In Exercises 37 and 38, use your calculator to solve the given equation for x. Check your two answers, as shown in Example 10. In your homework, write down what you type to get the answers, and what you type to check the answers. Answers are not given in the back of the book.

37. $4.2x^2 + 8.3x + 1.1 = 0$

38. $5.7x^2 + 12.3x - 8.1 = 0$

39. Discuss the use of parentheses in Example 10(a) Step 1. Why are they necessary? What would happen if they were omitted?

40. Discuss the use of parentheses in Example 10(a) Step 2. Why are they necessary? What would happen if they were omitted?

41. a. Calculate
$$\dfrac{\dfrac{5+7.1}{3}+\dfrac{2-7.1}{5}}{7}$$
using one line of instruction and without using the memory. In your homework, write down what you type as well as the solution.

b. Use the ⎡2nd⎤ ⎡ENTRY⎤ feature to calculate
$$\dfrac{\dfrac{5+7.2}{3}+\dfrac{2-7.2}{5}}{7}$$

c. Use the ⎡2nd⎤ ⎡ENTRY⎤ feature to calculate
$$\dfrac{\dfrac{5+9.3}{3}+\dfrac{2-4.9}{5}}{7}$$

42. a. Calculate
$$\dfrac{3+\dfrac{5+\dfrac{6-8.3}{2}}{3}}{9}$$
using one line of instruction and without using the memory. In your homework, write down what you type as well as the solution.

b. Use the ⎡2nd⎤ ⎡ENTRY⎤ feature to calculate
$$\dfrac{3+\dfrac{5-\dfrac{6-8.3}{2}}{3}}{9}$$

c. Use the ⎡2nd⎤ ⎡ENTRY⎤ feature to calculate
$$\dfrac{3-\dfrac{5+\dfrac{6-8.3}{2}}{3}}{9}$$

43. a. What is the result of typing "8.1 ⎡EE⎤ 4?"
b. What is the result of typing "8.1 ⎡EE⎤ 12?"
c. Why do the instructions in part (b) yield an answer in scientific notation, while the instructions in part (a) yield an answer that's not in scientific notation?
d. By using the ⎡MODE⎤ button, your calculator can be reset so that all answers will appear in scientific notation. Describe how this can be done.

BODY TABLE FOR THE STANDARD NORMAL DISTRIBUTION

$p(0 < z < c)$

z	0.00	0.01	0.02	0.03	0.04	0.05	0.06	0.07	0.08	0.09
0.0	0.0000	0.0040	0.0080	0.0120	0.0160	0.0199	0.0239	0.0279	0.0319	0.0359
0.1	0.0398	0.0438	0.0478	0.0517	0.0557	0.0596	0.0636	0.0675	0.0714	0.0753
0.2	0.0793	0.0832	0.0871	0.0910	0.0948	0.0987	0.1026	0.1064	0.1103	0.1141
0.3	0.1179	0.1217	0.1255	0.1293	0.1331	0.1368	0.1406	0.1443	0.1480	0.1517
0.4	0.1554	0.1591	0.1628	0.1664	0.1700	0.1736	0.1772	0.1808	0.1844	0.1879
0.5	0.1915	0.1950	0.1985	0.2019	0.2054	0.2088	0.2123	0.2157	0.2190	0.2224
0.6	0.2257	0.2291	0.2324	0.2357	0.2389	0.2422	0.2454	0.2486	0.2517	0.2549
0.7	0.2580	0.2611	0.2642	0.2673	0.2704	0.2734	0.2764	0.2794	0.2823	0.2852
0.8	0.2881	0.2910	0.2939	0.2967	0.2995	0.3023	0.3051	0.3078	0.3106	0.3133
0.9	0.3159	0.3186	0.3212	0.3238	0.3264	0.3289	0.3315	0.3340	0.3365	0.3389
1.0	0.3413	0.3438	0.3461	0.3485	0.3508	0.3531	0.3554	0.3577	0.3599	0.3621
1.1	0.3643	0.3665	0.3686	0.3708	0.3729	0.3749	0.3770	0.3790	0.3810	0.3830
1.2	0.3849	0.3869	0.3888	0.3907	0.3925	0.3944	0.3962	0.3980	0.3997	0.4015
1.3	0.4032	0.4049	0.4066	0.4082	0.4099	0.4115	0.4131	0.4147	0.4162	0.4177
1.4	0.4192	0.4207	0.4222	0.4236	0.4251	0.4265	0.4279	0.4292	0.4306	0.4319
1.5	0.4332	0.4345	0.4357	0.4370	0.4382	0.4394	0.4406	0.4418	0.4429	0.4441
1.6	0.4452	0.4463	0.4474	0.4484	0.4495	0.4505	0.4515	0.4525	0.4535	0.4545
1.7	0.4554	0.4564	0.4573	0.4582	0.4591	0.4599	0.4608	0.4616	0.4625	0.4633
1.8	0.4641	0.4649	0.4656	0.4664	0.4671	0.4678	0.4686	0.4692	0.4699	0.4706
1.9	0.4713	0.4719	0.4726	0.4732	0.4738	0.4744	0.4750	0.4756	0.4761	0.4767
2.0	0.4772	0.4778	0.4783	0.4788	0.4793	0.4798	0.4803	0.4808	0.4812	0.4817
2.1	0.4821	0.4826	0.4830	0.4834	0.4838	0.4842	0.4846	0.4850	0.4854	0.4857
2.2	0.4861	0.4864	0.4868	0.4871	0.4875	0.4878	0.4881	0.4884	0.4887	0.4890
2.3	0.4893	0.4896	0.4898	0.4901	0.4904	0.4906	0.4909	0.4911	0.4913	0.4916
2.4	0.4918	0.4920	0.4922	0.4925	0.4927	0.4929	0.4931	0.4932	0.4934	0.4936
2.5	0.4938	0.4940	0.4941	0.4943	0.4945	0.4946	0.4948	0.4949	0.4951	0.4952
2.6	0.4953	0.4955	0.4956	0.4957	0.4959	0.4960	0.4961	0.4962	0.4963	0.4964
2.7	0.4965	0.4966	0.4967	0.4968	0.4969	0.4970	0.4971	0.4972	0.4973	0.4974
2.8	0.4974	0.4975	0.4976	0.4977	0.4977	0.4978	0.4979	0.4979	0.4980	0.4981
2.9	0.4981	0.4982	0.4982	0.4983	0.4984	0.4984	0.4985	0.4985	0.4986	0.4986
3.0	0.4987	0.4987	0.4987	0.4988	0.4988	0.4989	0.4989	0.4989	0.4990	0.4990

ANSWERS TO SELECTED EXERCISES

CHAPTER 1

SECTION 1.0

1. a. rise = 4; run = 2; $m = 2$
    b. rise = -4; run = -2; $m = 2$
    d.

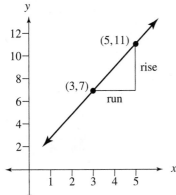

c. and d.

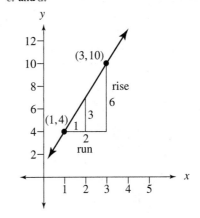

3. a.

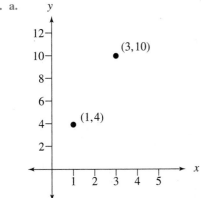

    b. m = 3

5. a.

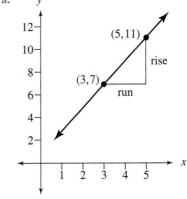

    b. $m = 2$
    c. $m = 2$
    d. Yes

7. a.

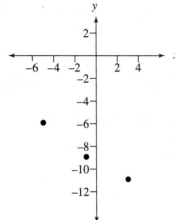

b. $m = -\frac{3}{4}$

c. $m = -\frac{1}{2}$

d. No

9. a.

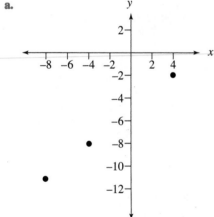

b. $m = \frac{3}{4}$

c. $m = \frac{3}{4}$

d. Yes

11. a. $m = -4$

c.

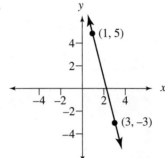

13. a. $m = \frac{7}{8}$

c.

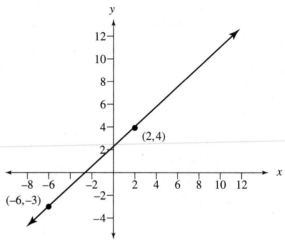

15. a. (1.30, 42,000); (1.40, 41,000)

b. $m = -10,000$

d. 40,000

e. 41,500

27. a. $n = -10,000p + 55,000$

b. 40,500

c.

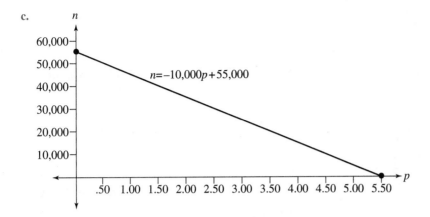

$n = -10,000p + 55,000$

29. **a.** (1000, 900) (800, 780)
    b. $m = \frac{3}{5} = 0.60$
    c. $n = 0.6c + 300$
    d.

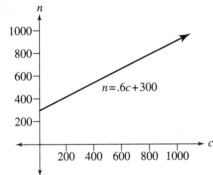

$n = .6c + 300$

e. intercept = $300

31. **a.** (1970, 74.7); (1990, 78.8)
    b. $m = 0.205$
    c. $e = .205t - 329.15$
    d.

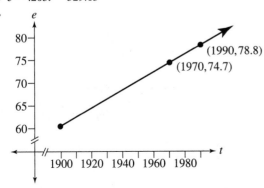

(1990, 78.8)
(1970, 74.7)

e. $t = 2020$
f. $e = 80.85$

33. **a.** $y = -3x + 9$
    b. slope = -3; intercept = 9
    c.

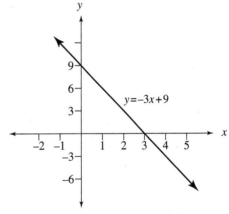

$y = -3x + 9$

35. **a.** $y = \dfrac{-5}{7}x + 5$
    b. slope = $-\frac{5}{7}$; intercept = 5

c.

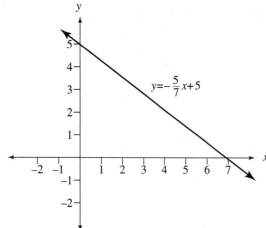

37. a.

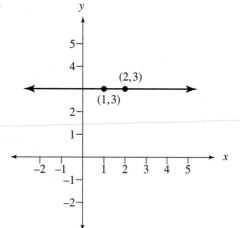

b. $m = 0$

c. $y = 3$

d. possible answers: (3, 3), (4, 3), (5, 3)

e. y is the same in both ordered pairs, so $y = 3$.

39. a. $m = 0$ b. $y = 7$

41. a. undefined   b. $x = -3$

43. a. undefined   b. $x = 2$

45. a–j Different drawings are possible.

SECTION 1.1

1. a. $P(1000) = 310$ (million)

b. 310 million is the world population in the year 1000.

3. a. $g(-2) = 4$

b. The square of -2 is 4.

5. a. $f(3) = 2$

b. The point with x coordinate 3 on the graph has y coordinate 2.

7. a. p(Nixon) = Republican

b. The political party of President Nixon is the Republican Party.

9. a. yes   b. yes

11. no; (Democrat, Clinton) or (Democrat, Carter)

13. a. yes

b. no; (statistics, Mr. Kersling) or (statistics, Ms. Landre)

15. a. yes

b. x and y can be any real number.

17. no; (16, 4) or (16, -4)

19. a. yes

b. $\{x \mid -4 \le x \le 2\}$ $\{y \mid -2 \le y \le 3\}$

21. a. no; (1, 2.5) or (1, -2)

23. a. yes

b. $\{x \mid -4 \le x \le 3\}$; $\{y \mid -3 \le y \le 4\}$

25. a. C   b. D   c. B   d. A

27. $\{y \mid y \ge -1\}$

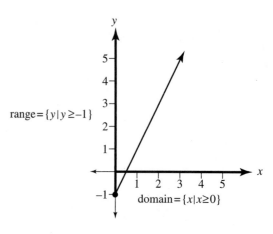

29. $\{y \mid 4 \le y \le 24\}$

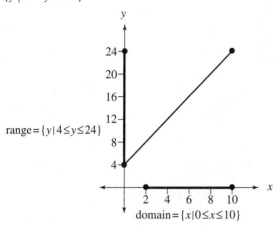

range=$\{y|4\le y\le24\}$

domain=$\{x|0\le x\le10\}$

31. a. -6 **b.** -6 **c.** 6

33. a. 16 **b.** -16 **c.** -16

35. a. $12x + 5$ **b.** $12x + 20$ **c.** $12x - 2$

37. a. $6x - 2$ **b.** $6x - 1$

39. b. $N(p) = -10,000p + 55,000$

41. b. $C(n) = \frac{3}{5}n + 300$

43. $C(N(p)) = -6000p + 33,300$

45. a. $(0, 14.7), (33, 29.4)$
  b. $m = 0.445$ lbs/sq in./ft
  c. $p = 0.445d + 14.7$
  d. $f(d) = 0.445d + 14.7$
  e. $\{d \mid d \ge 0\}; \{p \mid p \ge 14.7\}$
  f. $41.4; 2.818$
  g. $59.2; 4.03$
  h. $5601.27; 381.04$

i.

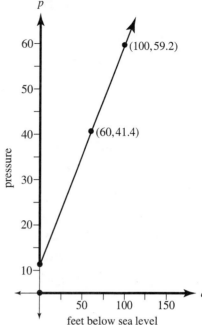

47. a. $(1, 32)$
  b. $(0, 0); m = 32$
  c. $s = 32t$
  d. $S(t) = 32t$
  e. 320 ft/sec
  f. 64 ft/sec

49. a. $T = 0.03t + 15$
  c. $t = 2103$

SECTION 1.2

1. a. 750 shirts
  b. $R(x) = 9.5x; \$7125$
  c. $\$3425$
  d. $\$3700$

3. a. $\$1325$
  b. $\$4950$
  c. $P(x) = 5x - 50$

5. a. Variable costs include metal tubing, bearing, wages of workers, and workman's compensation. Fixed costs include managers' salaries, and fire insurance.
  b. $C(x) = 57.19x + 125,418; \{x \mid x \ge 0\}; \{c \mid c \ge \$125,418\}$
  $R(x) = 118.53x; \{x \mid x \ge 0\}; (r \mid r \ge 0\}$
  $P(x) = 61.34x - 125,418; \{x \mid x \ge 0\}; \{p \mid p \ge -\$125,418\}$

d. Costs = $271,595.64; Revenue = $302,962.68; Profit = $31,367.04

e. Costs = $213,147.46; Revenue = $181,825.02; Profit = −31,322.44

f.

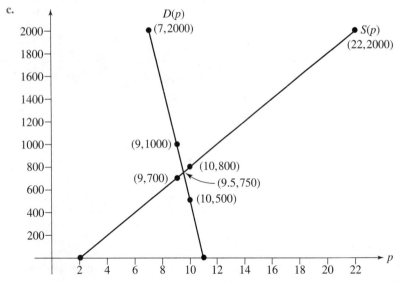

g. $118.53; $57.19; $61.34

7. a. 700; 1000   b. 800; 500

c.

9. a. (5, 408.24,), (10, 220.32)

b. $d(p) = -37.548p + 596.16$

c. $\{p \mid 0 \le p \le 15.9\}$; $\{d \mid 0 \le d \le 596.2\}$

d.

11. a. $S(p) = 5p + 100$

b. $\{p \mid p \ge 0\}$; $\{s \mid s \ge 100\}$

c. 12¢ higher

d. $s \ge 160$ million gallons per month

e.

13. a. $V(t) = -1928.57t + 13,500$
  b. \$11,571.43
  c. \$9642.86
  d. \$1928.57
  e. \$1928.57

15. a. $R(x) = 45x$
  b. $P(x) = -0.0667x^2 + 34.91x - 2542.88$
  c. $m(x) = -0.1334x + 34.8433$
  d. $-\$1.31$; decrease production to 261,000 units
  e. \$24.17; increase production to 261,000 units

17.

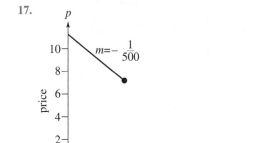

19.

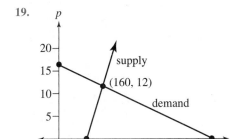

SECTION 1.3

1. e. $m = -1.125$
  f. relative error at $x = 1$: 0.05

5. a. relative errors: $x = 1, 0.0820$; $x = 5, 0.0487$; $x = 10$, 0.0138
  b. $r = 0.998$
  c. Yes; because the points are close to the line, the relative errors are small and the correlation coefficient is near 1.

7. a. 2 and 3
  b. 4 and 5
  c. y does not predictably increase or decrease as x increases.

9. a. (11, 255.4), (23, 250.4), (33, 247.6), (42, 244.6), (45, 241.4), (54, 239.4), (64, 234.1), (67, 231.1), (75, 229.4), (80, 228.8), (85, 226.3)
  b. slope $= -0.403$; intercept $= 260.2$; $r = -0.995$
  c. $\hat{y} = -0.403x + 260.2$
  d. 219.8 seconds
  e. 2024

11. a. and **c.**

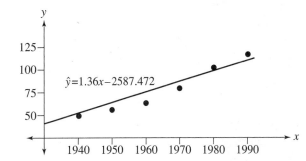

  a. yes
  b. $y = 1.36x - 2587.472$
  d. 132.53 (million)
  e. 1984
  f. $r = 0.980$

13. a. and **c.**

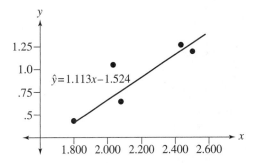

  a. yes
  b. $\hat{y} = 1.113x - 1.524$
  d. 1.537 million
  e. 3.840 million
  f. $r = 0.934$

15. a. (75, 10.92), (60, 25.44), (50, 40.44), (40, 41.52), (25, 45.00)
  b. slope $= -0.6985$; intercept $= 67.588$; $r = -0.931$
  c. $\hat{d} = -0.6985p + 67.588$
  d. $R(p) = -0.6985p^2 + 67.588p$
  f.

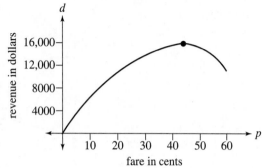

  g. 48 cents; revenue $= \$16,348.80$
  h. 50 cents; they would want a fare that was easy to collect.

CHAPTER 1 REVIEW

1. a. $m = -8$
  b.

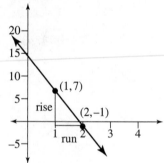

2. a. $m = 0$

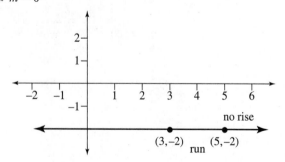

3. a. undefined

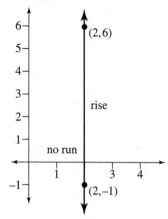

4. a. $m = \frac{7}{4}$
  b.

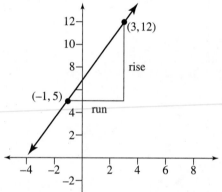

9. a. $y = \frac{x}{3} - \frac{3}{2}$
  b. $m = \frac{1}{3}$; y-intercept $= -\frac{3}{2}$
  c.

10. a. $y = -2x + 6$

b. $m = -2$; y-intercept $= 6$

c.

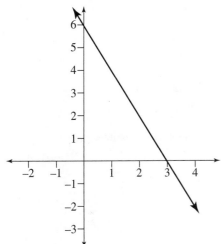

11. a. yes

b. $E(x) =$ enrollment in class x

12. a. no

b. Two classes may have the same enrollment.

13. a. yes

b. $\{x \mid x$ is any real number$\}$
$\{y \mid y$ is any real number$\}$

14. a. yes

b. $\{x \mid x$ is any real number$\}$; $\{y \mid y \geq 0\}$

15. no; $(16, 4)$, $(16, -4)$

16. a. function

b. $\{x \mid -4 \leq x \leq 1\}$; $\{y \mid -4 \leq y \leq 3\}$

17. a. not a function

18. $\{y \mid -7 \leq y \leq 23\}$

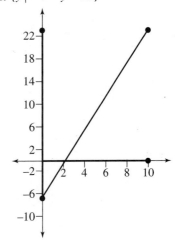

19. a. -5 **b.** -7

20. a. $4x^2 - 4x + 1$

b. $4x^2 - 1$

c. $2x^2 - 1$

21. a. $2x^2 + 1$ **b.** $4x^2 + 4x + 1$

22. a. $(212, 100)$, $(32, 0)$

b. $m = \frac{5}{9}$; a change of $5°$ C corresponds to one of $9°$ F.

c. $C = \frac{5}{9} F - \frac{160}{9}$

d. $C = f(F) = \frac{5}{9} F - \frac{160}{9}$

e. $-\frac{160}{9}$ or $-17.78°$ F

f. $37°$ C

g. $F = \frac{9}{5}C + 32$

h. $77°$ C

23. a. Fixed costs: wages for managers. Variable costs: down, cotton fabric and thread, wages for assembly workers, gas and electric, machine repair.

b. $C(x) = 38x + 111,412$; $\{x \mid x \geq 0\}$ $\{y \mid y \geq 111,412\}$
$R(x) = 349x$; $\{x \mid x \geq 0\}$ $\{y \mid y \geq 0\}$
$P(x) = 311x - 111,412$; $\{x \mid x \geq 0\}$ $\{y \mid y \geq -111,412\}$

d. $126,384$; $137,506$; $11,122$

e. $123,648$; $112,378$; $-11,270$

f.

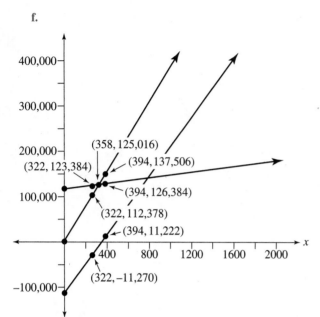

g. $349; $38

25. a. $V(t) = -1028.57t + 7200$
    b. $4114.29
    c. $1028.57
    d. $1028.57

26. a. yes
    b. $\hat{y} = -0.08638x + 11.8636$
    d. 8,408,400 cows
    e. in 2003
    f. -0.95; a strong linear trend is exhibited

CHAPTER 2

SECTION 2.0

1. yes

3. no

5. no

7. one solution

9. infinite number of solutions

11. one solution

13. yes

15. no

17. no

29. $(4, -2, 1)$

31. $(2, 5, 7)$

33. Such a system could have no solutions, one solution, or an infinite number of solutions. If the three equations were unique, the system could have no solutions or one solution.

SECTION 2.1

1. a. 3×2 b. none

3. a. 2×1 b. column

5. a. 1×2 b. row

7. a. 3×3 b. square

9. a. 3×1 b. column

11. 22

13. 41

15. 3

17. -11

19. -3

21. $\begin{bmatrix} 2 & 7 & 11 \\ 3 & -2 & 15 \end{bmatrix}$

23. $\begin{bmatrix} 2 & -1 & 5 \\ 3 & -1 & 41 \end{bmatrix}$

25. $\begin{bmatrix} 2 & 3 & -7 & 53 \\ 5 & -2 & 12 & 19 \\ 1 & 1 & 1 & 55 \end{bmatrix}$

27. $\begin{bmatrix} 5 & 0 & 1 & 2 \\ 3 & -1 & 0 & 15 \\ 2 & 2 & 2 & 53 \end{bmatrix}$

29. $\begin{bmatrix} 1 & 3 & 5 \\ 0 & 7 & 14 \end{bmatrix}$

31. $\begin{bmatrix} 1 & 0 & \frac{39}{5} \\ 0 & 1 & \frac{7}{5} \end{bmatrix}$

33. $\begin{bmatrix} 1 & 1 & 2 & 6 \\ 0 & -3 & 0 & -9 \\ 0 & 1 & -11 & -4 \end{bmatrix}$

35. $(9, 2)$

37. $(0.5281, -0.6205, 0)$

SECTION 2.2

1. 9 River models and 2 Ocean models

3. 10 oz. liver and 1 cup of Velveeta surprise

5. 10 barrels Green Beauty, $13\frac{1}{3}$ barrels No Moors Man, and 2 barrels Purismal

7. no solution

9. infinite number of solutions; $(-3 - 5z, 8 + 7z, z)$; some solutions include: $(-3, 8, 0), (-8, 15, 1), (-13, 22, 2)$

15. **a.** $x_1 + x_2 = 2700$
$\quad\quad x_2 + x_3 = 2300$
$\quad\quad x_3 + x_4 = 2600$
b. $(3000 - x_4, x_4 - 300, 2600 - x_4, x_4)$
c. one meter on A St. between 1st Ave. and 2nd Ave.
d. $x_1 = 1100$ cars/hr, $x_2 = 1600$ cars/hr, $x_3 = 700$ cars/hr

17. no solution

19. $(4, 0, -7)$

SECTION 2.3

1.

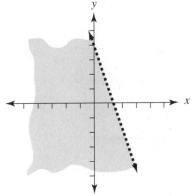

3.

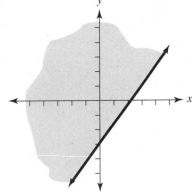

5. Note: Graphing calculators can't graph vertical lines.

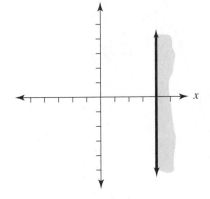

7.

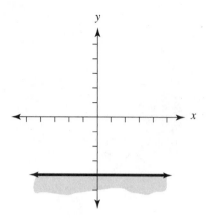

13. a.

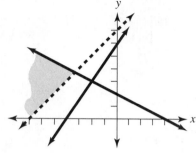

b. unbounded

c. $\left(-\dfrac{10}{3}, \dfrac{11}{3}\right)$

9. a.

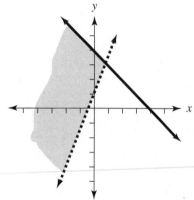

b. unbounded

c. $(1, 3)$

15. a.

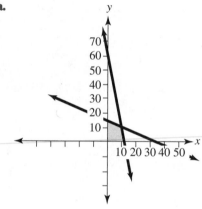

b. bounded

c. $(0, 0), (0, 14), (12, 0), (10, 10)$

11. a.

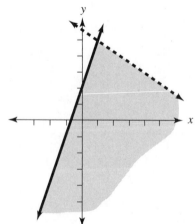

b. unbounded

c. $(1, 5)$

17. a.

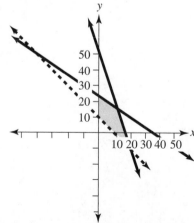

b. bounded

c. $(0, 10)$, $(10, 0)$, $(17\frac{1}{7}, 0)$ $(0, 23\frac{2}{11})$, $(12, 15)$

19. a.

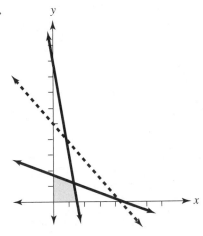

b. bounded

c. $(0, 0)$, $(0, 1.70)$, $(1\frac{1}{2}, 0)$, $(1.3, 1.2)$

21. a.

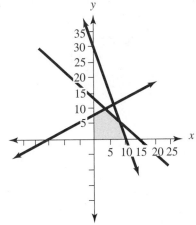

b. bounded

c. $(0, 0)$, $(10, 0)$, $(0, 8)$, $(8, 6)$, $(4, 10)$

SECTION 2.4

1. x = number of shrubs, y = number of trees; $1x + 3y \leq 100$

3. x = number of hardbacks, y = number of paperbacks; $4.50x + 1.25y \geq 5000$

5. x = number of refrigerators, y = number of dishwashers; $63x + 41y \leq 1650$

7. no floor lamps and 45 table lamps, for a profit of $1485

9. 60 pounds of Morning Blend and 120 pounds of South American Blend each day for a profit of $480

11. 15,000 loaves to Shopgood and 20,000 loaves to Rollie's each week for shipping costs of $3000

13. Global has many choices, including the following:
15 Orvilles and 30 Wilburs
18 Orvilles and 28 Wilburs
21 Orvilles and 26 Wilburs
Each of these generates a cost of $720,000.

15. a. 20 large and 30 small, for a profit of $9500
b. 40 large and no small, for a profit of $10,000
c. 40

17. 0 Mexican; 0 Colombian

19. 150 minutes; $0

29. a. and **b.** no maximum   **c.** $(6, 0)$ **d.** 12

31. 3 weeks at Detroit and 6 weeks at Los Angeles, for a minimum cost of $3,780,000

CHAPTER 2 REVIEW

1. one solution; the lines intersect

2. no solutions

3. infinite number of solutions: the lines are parallel

4. $(3, -1)$

5. infinite number of solutions; all ordered pairs of the form $(x, -2x + 4)$; some particular solutions: $(0, 4)$ $(1, 2)$ $(2, 0)$

6. no solutions

7. $(\frac{53}{28}, -\frac{75}{28}, -\frac{3}{2})$

8. none of these; 3×2

9. square; 3×3

10. column matrix; 3×1

11. row; 1×2

15. 5 Dog Yums and 4 Arf O Vite

16.

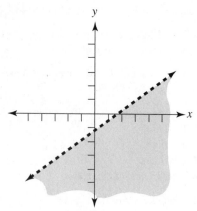

17.

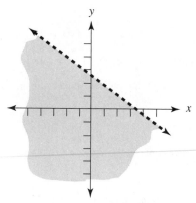

18.

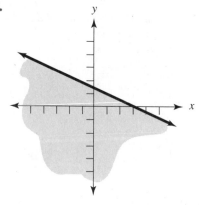

19.

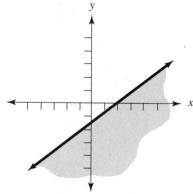

20. $\left(\frac{3}{2}, \frac{1}{2}\right)$

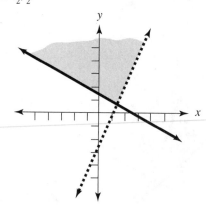

21. $\left(\frac{44}{17}, -\frac{15}{17}\right)$

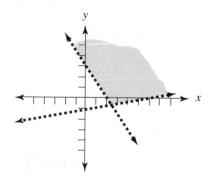

22. $(-3, -23)$

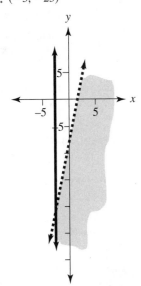

23. $(4, -\frac{3}{2})$

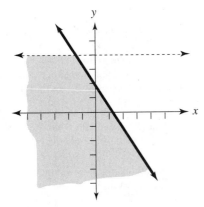

24. no solution

25. $(3, 0), (4, 0)$

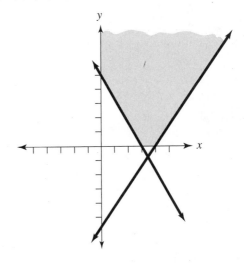

26. 30 of each assembly for a maximum income of $16,500

27. 750 Sunnys and 375 Iwas for a maximum profit of $240,000

CHAPTER 3

SECTION 3.1

1. $3x_1 + 2x_2 + s_1 = 5$

3. $3.41x_1 + 9.20x_2 + 6.16x_3 + s_1 = 45.22$

5. a. $6.99x_1 + 3.15x_2 + 1.98x_3 \le 42.15$
   b. $6.99x_1 + 3.15x_2 + 1.98x_3 + s_1 = 42.15$
   c. x_1 = pounds of meat, x_2 = pounds of cheese, x_3 = loaves of bread, s_1 = unused money

7. a. $5x_1 + 4x_2 \le 480$
   b. $5x_1 + 4x_2 + s_1 = 480$

c. x_1 = feet of pipe, x_2 = number of elbows, s_1 = unused time in minutes

9. a. $24x_1 + 36x_2 + 56x_3 + 72x_4 \le 50,000$
   b. $24x_1 + 36x_2 + 56x_3 + 72x_4 + s_1 = 50,000$
   c. x_1 = number of twin beds, x_2 = number of double beds, x_3 = number of queen-size beds, x_4 = number of king-size beds, s_1 = unused space in cubic feet

11. $(x_1, x_2, s_1, s_2) = (0, 19, 0, 12), z = 22$

13. $(x_1, x_2, s_1, s_2, s_3) = (0, 0, 7.8, 9.3, 0.5), z = 9.6$

15. $(x_1, x_2, s_1, s_2, s_3, s_4) = (25, 73, 0, 32, 46, 0)$, $z = 63$

17.
$$\begin{bmatrix} 3 & 4 & 1 & 0 & 0 & 40 \\ 4 & 7 & 0 & 1 & 0 & 50 \\ -2 & -4 & 0 & 0 & 1 & 0 \end{bmatrix}$$
$(x_1, x_2, s_1, s_2) = (0, 0, 40, 50)$, $z = 0$

19.
$$\begin{bmatrix} 112 & -3 & 1 & 0 & 0 & 0 & 370 \\ 1 & 1 & 0 & 1 & 0 & 0 & 70 \\ 47 & 19 & 0 & 0 & 1 & 0 & 512 \\ -12.10 & -43.86 & 0 & 0 & 0 & 1 & 0 \end{bmatrix}$$
$(x_1, x_2, s_1, s_2, s_3) = (0, 0, 370, 70, 512)$, $z = 0$

21.
$$\begin{bmatrix} 5 & 3 & 9 & 1 & 0 & 0 & 0 & 10 \\ 12 & 34 & 100 & 0 & 1 & 0 & 0 & 10 \\ 52 & 7 & 12 & 0 & 0 & 1 & 0 & 10 \\ -4 & -7 & -9 & 0 & 0 & 0 & 1 & 0 \end{bmatrix}$$
$(x_1, x_2, x_3, s_1, s_2, s_3) = (0, 0, 0, 10, 10, 10)$, $z = 0$

23.
$$\begin{bmatrix} 50 & 30 & 1 & 0 & 0 & 2400 \\ 0.90 & 1.50 & 0 & 1 & 0 & 190 \\ -1.2 & -4 & 0 & 0 & 1 & 0 \end{bmatrix}$$
$(x_1, x_2, s_1, s_2) = (0, 0, 2400, 190)$, $z = 0$

25.
$$\begin{bmatrix} \frac{1}{2} & \frac{1}{4} & 1 & 0 & 0 & 200 \\ \frac{1}{2} & \frac{3}{4} & 0 & 1 & 0 & 330 \\ -3.5 & -4 & 0 & 0 & 1 & 0 \end{bmatrix}$$
$(x_1, x_2, s_1, s_2) = (0, 0, 200, 330)$, $z = 0$

SECTION 3.2

1. pivot on 19

3. pivot on 19

5. a. pivot on 4

b.
$$\begin{bmatrix} 0 & \frac{7}{4} & 1 & \frac{1}{4} & 0 & \frac{5}{2} \\ 1 & \frac{1}{4} & 0 & \frac{1}{4} & 0 & \frac{1}{2} \\ 0 & -\frac{5}{2} & 0 & \frac{3}{2} & 1 & 3 \end{bmatrix}$$

c. $(x_1, x_2, s_1, s_2) = (\frac{1}{2}, 0, 2\frac{1}{2}, 0)$, $z = 3$

7. a. pivot on 1 (row 1, column 1)

b.
$$\begin{bmatrix} 1 & 1 & 5 & 0 & 0 & 3 \\ 0 & -3 & -14 & 1 & 0 & 3 \\ 0 & 6 & 28 & 0 & 1 & 24 \end{bmatrix}$$

c. $(x_1, x_2, s_1, s_2) = (3, 0, 0, 3)$, $z = 24$

9. Each day, make no bread and 80 cakes, for a profit of $320. This uses all available time and leaves $70 unspent.

11. Each week, make 15 sofas and 40 chairs, for a profit of $21,750. This leaves no extra time in either shop.

13. Prepare 20,000 liters of House White, no Premium White, and 5,000 liters of Sauvignon Blanc, for a profit of $30,000. This uses all the grapes.

15. Prepare 300 pounds of Smooth Sipper and no Kona Blend, for a profit of $900. This leaves 100 pounds each of Colombian and Arabian beans.

23. $(x_1, x_2, x_3, x_4, s_1, s_2, s_3) = (0, 3.21, 0, 0, 2.16, 71.30, 0)$; $z = 2745.87$

25. $(x_1, x_2, x_3, x_4, s_1, s_2, s_3) = (0, 0, 1.88, 1.84, 0, 0)$; $z = 275.49$

27. Make 9.6 coats and no vests, resulting in a profit of $1680. There will be 2.8 square feet of unused leather each day and no unused hours on the sewing machine.

29. Each week make 342.86 floor lamps, 0 table lamps, and 171.42 desk lamps, resulting in a profit of $25,714.10 and leaving 80 unused hours in the wood shop, 0 unused hours in the metal shop, 2.86 unused hours in the electrical shop, and 0 unused hours in testing.

31. Order 127 Packard Rear Projection televisions, resulting in a profit of $91,200, 767 cubic feet of unused warehouse space, and no unused money. (The following answer can also be justified: Order 126 televisions, resulting in a profit of $90,720, 890 cubic feet of unused space, and $300 unused money.)

33. Make 27,632 liters of House Wine, no Premium Red, 7895 liters of Zinfandel, and 11,000 liters of Cabernet, resulting in a profit of $75,632.55. There will be 18,947 pounds of pinot noir grapes left, but all other grapes will be used.

SECTION 3.3

1. $(x_1, x_2, s_1, s_2, s_3) = (5, 0, 35, 20, 0)$; $w = 10$

3. $(x_1, x_2, s_1, s_2, s_3, s_4) = (44.29, 20, 0, 5.71, 34.29, 0)$, $z = 251.06$

5. $(x_1, x_2, s_1, s_2, s_3, s_4) = (48.33, 0, 120, 0, 28.33, 38)$, $z = 251.33$

7. $(x_1, x_2, s_1, s_2, s_3) = (8, 2, 97, 0, 0)$, $w = 22$

9. Produce 480,000 pounds of whole sprouts, 90,000 pounds of sprout pieces, and 120,000 pounds of sprouts salsa, which result in production costs of $1,117,500. No excess product is produced.

11. a. City Electronics should order 110 Packards and 140 Bells, resulting in a profit of $58,400. This will result in 840 cubic feet of unused space in their warehouse, no extra money, and no surplus televisions.

    b. City Electronics should order 160 Packards and 100 Bells, resulting in a profit of $58,000. This will result in 240 cubic feet of unused space in their warehouse, no extra money, and no surplus televisions.

15. Sunny should ship 180 Boomman stereos from the Los Angeles warehouse to the Los Angeles store. They should ship 95 stereos from the Los Angeles warehouse and 25 from the Portland warehouse to the store in San Francisco. Total shipping costs will be $1016.75. There will be no surplus stereos sent to either store. The Los Angeles warehouse will have 0 stereos left, and the Portland warehouse will have 185 stereos left.

17. South reservoir should send 20 million gallons of water to Crockett and none to Valona. North reservoir should send 10 million gallons of water to Crockett and 10 million gallons to Valona, resulting in a total cost of $350,000. This results in no unused water at the South reservoir and 20 million gallons of unused water at the North reservoir, with no surplus water in Crockett or Valona.

SECTION 3.4

1. a. 14 chairs and 4 sofas should be produced to earn the maximum profit of $2800.

    b. One extra hour of carpentry will increase profit by $6.67.

    c. One extra hour of finishing will increase profit by $120.

    d. One extra hour of upholstery will not increase profit.

    e. They should hire an additional finishing worker and pay up to $120 per hour.

    f. They should retrain some upholstery workers as carpenters or finishers, as they are not using all the upholstery workers.

3. a. New constraint: $6x_1 + 5x_2 \leq 40 + x$.

b.
$$\begin{bmatrix} 6 & 5 & 1 & 0 & 0 & 40 + x \\ 200 & 100 & 0 & 1 & 0 & 1000 \\ -240 & -160 & 0 & 0 & 1 & 0 \end{bmatrix}$$

c.
$$\begin{bmatrix} 0 & 1 & .5 & -0.015 & 0 & 5 + \frac{x}{2} \\ 1 & 0 & -0.25 & 0.0125 & 0 & 2.5 - \frac{x}{4} \\ 0 & 0 & 20 & 0.6 & 1 & 1400 + 20x \end{bmatrix}$$

d. $z = \$1440$; $x_1 = 2$; $x_2 = 6$

e. $z = \$1460$; $x_1 = 1.75$; $x_2 = 6.5$

f. $z = \$1680$; $x_1 = 3$; $x_2 = 6$

5. a. An increase in time for labor decreases the number of coffee tables made.

    b. There is a decrease of $\frac{1}{4}$ coffee tables per hour increase in time for labor.

    c. a maximum of 10 such decreases

    d. a maximum of 10 hours increase in time for labor

7. a. a maximum of 12 additional hours for carpentry

    b. a maximum of 2 additional hours for finishing

    c. You can increase hours for upholstery as much as you like; profit will not be affected.

    d. If carpentry hours are increased to 108 and finishing hours are increased to 20, then the maximum profit is $3120, when 16 chairs and 4 sofas are made.

9. a. One extra hour of labor will increase production of cakes by 2, and profit by $8.00.

    b. up to 23.33 more hours

    c. One extra dollar for materials will not increase profit.

    d. No matter how much more is available for materials, profit will not increase unless labor is increased.

11. a. One extra lb of French colombard grapes will increase profit by $0.75.

    b. up to 15,000 lbs

    c. One extra lb of savignon blanc grapes will increase profit by $1.75.

    d. up to 35,000 lbs

SECTION 3.5

1. Minimize $w = 12y_1 + 48y_2$
subject to
$$8y_1 + 7y_2 \geq 5$$
$$4y_1 + 11y_2 \geq 3$$
$$9y_1 + 913y_2 \geq 4$$

3. Maximize $z = 84x_1 + 68x_2 + 19x_3$
subject to
$$4x_1 + 15x_2 + 2x_3 \leq 36$$
$$7x_1 + 21x_2 + 17x_3 \leq 16$$

5. a. $(y_1, y_2) = (0, 9.7)$; $w = 11,200$; minimize w

    b. $(x_1, x_2) = (0, 0.33)$; $z = 11,200$; maximize z

7. a. $(x_1, x_2) = (19.2, 18.7)$; $z = 12.8$; maximize z

    b. $(y_1, y_2) = (0.91, 0.58)$; $w = 12.8$; minimize w

9. $(y_1, y_2) = (8, 2)$; $w = 22$

11. The winery should produce 8000 cases of the premium label and 22,000 cases of the economy label to achieve the minimum production cost of $1,368,000.

13. The cafeteria should serve 4.15 oz of chili macs per meal and 239.9 oz of potatoes per meal to achieve the minimum cost of $5.42 per meal.

15. The company should produce 480,000 lbs of canned whole Brussels sprouts, 90,000 lbs of canned Brussels sprouts pieces, and 120,000 lbs of canned Brussels sprouts salsa to achieve the minimum production cost of $1,117,500.

CHAPTER 3 REVIEW

1. a. 30 of each   **b.** $16,500

   c. 150; for each additional tweeter, income increases $150.

   d. 50; for each additional mid-range speaker, income increases $50.

   e. 0; you are not using all of the current woofer supply, so adding additional woofers to the supply will not increase income.

2. Make 24 model 110s, 16 model 220s, and 50 model 330s, with no tweeters, mid-range speakers, or woofers left over. This gives a maximum income of $26,780.

3. a. 750 Sunnys and 357 Iwas

   b. $240,000

   c. For every additional cubic foot of warehouse space made available, profit increases by $20.

4. a. To achieve the minimum cost of $0.75 per day, feed 0.667 oz of Chinchilla Vanilla and 1.33 oz of Science Chinchilla. The slack in the protein constraint is 0, meaning this diet provides the minimum amount of protein. The slack in the carbohydrate constraint is 0.33, meaning this diet provides 0.33 units of carbohydrate above the minimum. The slack in the vitamins constraint is 0, meaning this diet provides the minimum amount of vitamins.

   b. 0.667 oz of Chinchilla Vanilla and 1.33 oz of Science Chinchilla.

CHAPTER 4

SECTION 4.0

1. a. $\begin{bmatrix} 8 & 5 \\ 2 & 4 \end{bmatrix}$ **b.** $\begin{bmatrix} 8 & 5 \\ 2 & 4 \end{bmatrix}$ **c.** yes, since $M + N = N + M$

3. a. $\begin{bmatrix} 14 & 5 \\ 4 & 8 \end{bmatrix}$ **b.** $\begin{bmatrix} 14 & 5 \\ 4 & 8 \end{bmatrix}$

   c. yes, since $(M + N) + P = M + (N + P)$

5. a. $\begin{bmatrix} 8 & 0 \\ -3 & 10 \end{bmatrix}$ **b.** $\begin{bmatrix} 23 & 0 \\ -7 & 22 \end{bmatrix}$

   c. $\begin{bmatrix} 7 & 0 \\ -5 & 16 \end{bmatrix}$ **d.** $\begin{bmatrix} 7 & 0 \\ -22 & 16 \end{bmatrix}$

7. a. $[1 \quad 16]$ **b.** does not exist   **c.** does not exist

9. a. does not exist   **b.** $\begin{bmatrix} 20 & 2 \\ 22 & 12 \\ 54 & 4 \end{bmatrix}$ **c.** does not exist

11. a. does not exist   **b.** $\begin{bmatrix} 73 \\ -35 \end{bmatrix}$ **c.** does not exist

13. a. $[34 \quad 20]$ **b.** $[34 \quad 20]$

   c. yes, since $F(B + G) = FB + FG$

15. sale price: $84; regular price: $113

17. Blondies: $3.45; SliceMan's: $3.70

19. a. and b. sale price: $54; regular price: $72

   c. $2A = A + A$

21. Jose: 3.07; Sylvie: 3.50; Eloise: 3.00

25. $\begin{bmatrix} 3 & -2 \\ 4 & 0 \end{bmatrix}$

27. does not exist

29. $\begin{bmatrix} 19 & 7 & 34 \\ 74 & 0 & -11 \\ 13 & -2 & 44 \end{bmatrix}$

31. $x = \frac{1}{6}; y = 1; z = 12; w = 10$

49. a. $\begin{bmatrix} 0 & 0 & 1 & 0 & 0 \\ 1 & 0 & 0 & 1 & 0 \\ 0 & 1 & 0 & 0 & 1 \\ 0 & 0 & 1 & 0 & 0 \\ 1 & 0 & 0 & 1 & 0 \end{bmatrix}$

   b. $\begin{bmatrix} 0 & 0 & 1 & 0 & 0 \\ 1 & 0 & 0 & 1 & 0 \\ 0 & 1 & 0 & 0 & 1 \\ 0 & 0 & 1 & 0 & 0 \\ 1 & 0 & 0 & 1 & 0 \end{bmatrix}$

   c. $\begin{bmatrix} 0 & 1 & 1 & 0 & 0 \\ 1 & 0 & 1 & 1 & 0 \\ 1 & 1 & 0 & 1 & 1 \\ 0 & 0 & 1 & 0 & 1 \\ 1 & 0 & 1 & 1 & 0 \end{bmatrix}$

   d. The zeroes represent no connection, while the ones represent a connection either direct or through one other city.

e. $\begin{bmatrix} 1 & 2 & 1 & 1 & 1 \\ 1 & 1 & 3 & 1 & 1 \\ 3 & 1 & 2 & 3 & 1 \\ 1 & 1 & 1 & 1 & 2 \\ 1 & 1 & 3 & 1 & 1 \end{bmatrix}$

Each entry represents the number of ways to get from a starting point to a destination.

f. They should add a flight from L.A. to Chicago, as the current shortest flight has 4 legs. They should also add a flight from San Francisco to Atlanta, as the shortest flight has 4 legs.

SECTION 4.1

1. $AB = BA = \begin{bmatrix} 1 & 0 \\ 0 & 1 \end{bmatrix}$

3. $AB = BA = \begin{bmatrix} 1 & 0 & 0 \\ 0 & 1 & 0 \\ 0 & 0 & 1 \end{bmatrix}$

13. a. $4x - y = 11; 2x - 2y = -1$

15. $(4, -1)$

17. $(-4, -3)$

19. $(12, 1)$

21. $(1, 3, -5)$

23. 540 river models; 120 oceangoing

25. 1 Velveeta Surprise and 10 ounces of liver

29. a. $\begin{bmatrix} \dfrac{+d}{ad-bc} & \dfrac{-b}{ad-bc} \\ \dfrac{-c}{ad-bc} & \dfrac{+a}{ad-bc} \end{bmatrix}$

b. $\begin{bmatrix} 1 & 0 \\ 0 & 1 \end{bmatrix}$ c. $\begin{bmatrix} 1 & 0 \\ 0 & 1 \end{bmatrix}$

SECTION 4.2

1. $\begin{bmatrix} \dfrac{3}{52} & \dfrac{15}{52} & -\dfrac{1}{52} \\ \dfrac{19}{260} & -\dfrac{61}{260} & \dfrac{11}{260} \\ -\dfrac{1}{10} & -\dfrac{1}{10} & \dfrac{1}{10} \end{bmatrix}$

3. $\begin{bmatrix} \dfrac{416}{5415} & -\dfrac{307}{5415} & -\dfrac{5}{1083} & \dfrac{26}{1083} \\ \dfrac{66}{361} & -\dfrac{87}{722} & -\dfrac{30}{361} & -\dfrac{49}{722} \\ \dfrac{29}{722} & \dfrac{11}{1444} & -\dfrac{23}{361} & -\dfrac{27}{1444} \\ -\dfrac{88}{1805} & \dfrac{477}{3610} & \dfrac{8}{361} & -\dfrac{11}{722} \end{bmatrix}$

5. a. $(\frac{25}{52}, \frac{89}{260}, \frac{9}{10})$

b. $(3\frac{1}{2}, -\frac{9}{10}, -\frac{4}{10})$

7. a. $(\frac{382}{1805}, -\frac{9307}{722}, -\frac{10,035}{1444}, \frac{12,289}{3610})$

b. $(\frac{15,068}{5415}, \frac{7757}{361}, \frac{3408}{361}, \frac{1811}{1805})$

9. $(41.5, 55, 57.5)$

SECTION 4.3

1. 20,112 corn; 3471 fish

3. 12,373 corn; 2697 fish

5. a. $P = \begin{bmatrix} 3279 \\ 4754 \end{bmatrix}$

b. The entries in the T matrix are the amounts of coal and steel needed. The entries in the D matrix are the amounts demanded.

7. a. $P = \begin{bmatrix} 12,367 \\ 10,888 \\ 11,834 \end{bmatrix}$

b. The entries in the T matrix are the amounts of automobiles, steel, and plastic needed. The entries in the D matrix are the amounts demanded.

11. agriculture: 667.7 million; manufacturing: 910.8 million; energy: 776.3 million

CHAPTER 4 REVIEW

1. $\begin{bmatrix} 71 & -11 \\ 10 & -32 \end{bmatrix}$

2. $\begin{bmatrix} -15 & 26 \\ 52 & 54 \end{bmatrix}$

3. $\begin{bmatrix} 20 & 5 \\ 16 & 4 \end{bmatrix}$

4. $[24]$

5. $\begin{bmatrix} 53 \\ 26 \end{bmatrix}$

6. not possible

7. not possible

8. $\begin{bmatrix} 22 & 40 \\ 22 & -7 \end{bmatrix}$

9. -47

10. $\begin{bmatrix} 59/47 & 9/47 \\ -22/47 & -40/47 \end{bmatrix}$

11. yes; $A + (B + C) = (A + B) + C$ if A, B, and C are the same dimensions.

12. a. no: $A - B = B - A$
  b. no: $A - (B - C) = (A - B) - C$

13. a. no: $AB \neq BA$
  b. yes, assuming the dimensions are correct: $(AB)C = A(BC)$
  c. yes, assuming the dimensions are correct: $A(B + C) = AB + AC$

14. Al: 3.57; Tipper: 2.86

17. a.
$$\begin{bmatrix} 0 & 0 & 0 & 0 & 1 \\ 0 & 0 & 0 & 1 & 0 \\ 0 & 0 & 0 & 1 & 0 \\ 0 & 1 & 1 & 0 & 1 \\ 1 & 0 & 0 & 1 & 0 \end{bmatrix} \begin{matrix} H \\ L \\ P \\ S \\ T \end{matrix}$$

  b.
$$\begin{bmatrix} 1 & 0 & 0 & 1 & 0 \\ 0 & 1 & 1 & 0 & 1 \\ 0 & 1 & 1 & 0 & 1 \\ 1 & 0 & 0 & 3 & 0 \\ 0 & 1 & 1 & 0 & 2 \end{bmatrix}$$
  Each entry indicates the number of routes between the cities having 2 legs.

  c.
$$\begin{bmatrix} 1 & 0 & 0 & 1 & 1 \\ 0 & 1 & 1 & 1 & 1 \\ 0 & 1 & 1 & 1 & 1 \\ 1 & 1 & 1 & 3 & 1 \\ 1 & 1 & 1 & 1 & 2 \end{bmatrix}$$
  Each entry indicates the number of routes between the cities having 1 or 2 legs.

  d.
$$\begin{bmatrix} 1 & 1 & 1 & 1 & 3 \\ 1 & 1 & 1 & 4 & 1 \\ 1 & 1 & 1 & 4 & 1 \\ 1 & 4 & 4 & 3 & 5 \\ 3 & 1 & 1 & 5 & 2 \end{bmatrix}$$
  Each entry indicates the number of routes between the cities having 1, 2, or 3 legs.

  e. Add Hong Kong to Los Angeles and Hong Kong to Portland because the shortest routes between these cities have 3 legs.

18.
$$\begin{bmatrix} \frac{2}{5} & -\frac{67}{15} & -\frac{14}{15} \\ -\frac{1}{5} & \frac{17}{15} & \frac{4}{5} \\ \frac{1}{5} & -\frac{26}{15} & -\frac{7}{15} \end{bmatrix}$$

21. DO WELL ON YOUR EXAM

23. $P = (I - T)^{-1}D$, where P is the production level needed to meet a demand D, given a technology matrix T; I is the identity matrix

CHAPTER 5

SECTION 5.1

1. a. well-defined
  b. not well-defined
  c. well-defined
  d. not well-defined

3. proper: $\varnothing$, {Lennon}, {McCartney}
  improper: {Lennon, McCartney}

5. proper: $\varnothing$, {yes}, {no}, {undecided}, {yes, no}, {yes, undecided}, {no, undecided}
  improper: {yes, no, undecided}

7. {Friday}

9. {Monday, Tuesday, Wednesday, Thursday}

11. {Friday, Saturday, Sunday}

13.

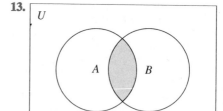

15.

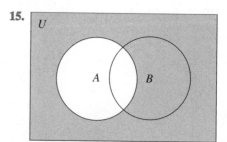

17.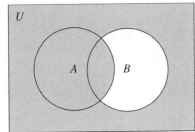

19.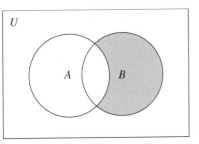

21.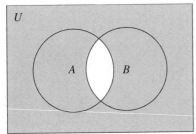

23. a. 21

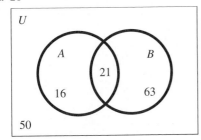

b. 0

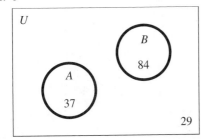

25. a.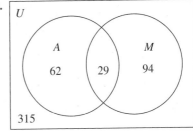

b. 37%

27. a.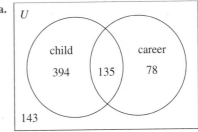

b. 18%

29. 42

31. 43

33. 8

35. 5

37. 16

39. 32

41. 6

43. 8

45. 0

47. a. {1, 2, 3} = A
 b. {1, 2, 3, 4, 5, 6} = B
 c. $E \subset F$
 d. $E \subset F$

49. a. ∅, {a}; 2
 b. ∅, {a}, {b}, {a, b}; 4
 c. ∅, {a}, {b}, {c}, {a, b}, {a, c}, {b, c}, {a, b, c}; 8
 d. ∅, {a}, {b}, {c}, {d}, {a, b}, {a, c}, {a, d}, {b, c}, {b, d}, {c, d}, {a, b, c}, {a, b, d}, {a, c, d}, {b, c, d}, {a, b, c, d}; 16
 e. yes; $2^{n(A)}$
 f. 64

61. (a)

63. (d)

SECTION 5.2

1. a. 143   **b.** 16   **c.** 49   **d.** 57

3. a. 408   **b.** 1343   **c.** 664   **d.** 149

5. a. 106   **b.** 448   **c.** 265   **d.** 159

7. a. $x + y - z$
  b. $x - z$
  c. $y - z$
  d. $w - (x + y - z) = w - x - y + z$

9. a. 12%   **b.** 13%

11. a. 5%   **b.** 21%

13. a. 0%   **b.** 54%

15. a. 44.0%   **b.** 12.8%

17. 16

19. $\{0, 4, 5\}$

21. $\{1, 2, 3, 6, 7, 8, 9\}$

23.
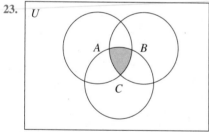

25.
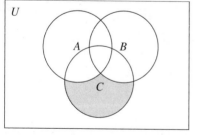

27.
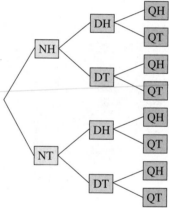

33. (d)

35. (d)

37. (b)

39. (e)

SECTION 5.3

1. a. 8   **b.**

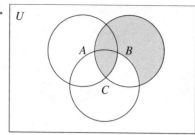

3. a. 12 **b.**

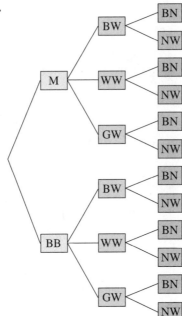

5. 24

7. 720

9. 2,646

11. 216

13. 1 billion

15. 10^{10} or 10 billion

17. 128

19. 540,000

21. 24

23. 3,628,800

25. $2.432902008 \times 10^{18}$

27. 17,280

29. a. 30 **b.** 360

31. 56

33. 70

35. 3,321

37. $1.046139494 \times 10^{13}$

39. 120

41. 35

43. 1

47. (b)

49. (a)

51. (a)

SECTION 5.4

1. a. 210 **b.** 35

3. a. 120 **b.** 1

5. a. 14 **b.** 14

7. a. 970,200 **b.** 161,700

9. a. $x!$ **b.** x

11. a. $x^2 - x$ **b.** $\dfrac{x^2 - x}{2}$

13. a. 6 **b.** $\{a, b\}, \{a, c\}, \{b, a\}, \{b, c\}, \{c, a\}, \{c, b\}$

15. a. 6 **b.** $\{a, b\}, \{a, c\}, \{a, d\}, \{b, c\}, \{b, d\}, \{c, d\}$

17. a. 479,001,600 **b.** 1

19. 24

21. 91

23. 2,184

25. a. 420 **b.** 1,001 **c.** 406

27. 2,598,960

29. a. 4512 **b.** 58,656 (including a full house as having 3 of a kind)

31. 1,098,240

33. 22,957,480

35. 376,992

37. $\frac{5}{36}$

41. a. 4th row **b.** nth row **c.** no, the third
d. yes **e.** nth row, $(r + 1)$th number

43. (c)

45. (c)

47. (e)

CHAPTER 5 REVIEW

2. a. {1, 3, 5, 7, 9}   **b.** {0, 2, 4, 6, 8}
c. {0, 1, 2, 3, 4, 5, 6, 7, 8, 9}   **d.** ∅

3. a. $A \cup B$ = {Maria, Nobuko, Leroy, Mickey, Kelly, Rachel, Deanna}
b. $A \cap B$ = {Leroy, Mickey}

4. proper: ∅, {Dallas}, {Chicago}, {Tampa}, {Dallas, Chicago}, {Dallas, Tampa}, {Chicago, Tampa}
improper: {Dallas, Chicago, Tampa}

5. a. 18
b.

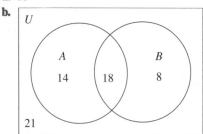

6. a. 1670   **b.** 733   **c.** 346   **d.** 330

7. 29%

8. a. {b, f}   **b.** {a, c, d, e, g, h, i}

9. a. 12
b.

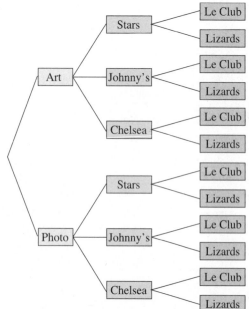

10. 180

11. 1,080,000

12. a. 3,628,800   **b.** 1
c. 531,360   **d.** 888,030

13. a. 165   **b.** 990

14. a. 32,760   **b.** 1365   **c.** 54,486,432,000

15. a. 660   **b.** 1540   **c.** 880

16. 45

17. 720

18. 133,784,560

19. 177,100,560

20. 5,245,786

22. a. row 7, number 4   **b.** row 7, number 5
c. same   **d.** nth row, $(r + 1)$th number

23. a. 3   **b.** 3   **c.** 1
d. 1   **e.** 8   **f.** $2^{n(S)} = 2^3 = 8$

24. $n(A) = 5$, $n(B) = 5$; equivalent; Roman numeral $\leftrightarrow$ number word

25. $n(C) = 450$, $n(D) = 450$; equivalent; $2n + 1 \leftrightarrow 2n$

26. $n(E) = 1$, $n(F) = 2$; not equivalent

27. a. $n \leftrightarrow n^2$ **b.** 29   **c.** $\sqrt{x}$
d. 20,736   **e.** n^2

28. a. 594   **b.** 264   **c.** 103   **d.** $\aleph_0$

CHAPTER 6

SECTION 6.1

1. picking one jellybean from the jar

3. $\frac{12}{35}$

5. $\frac{18}{35}$

7. $\frac{5}{7}$

9. 0

11. 12:23

13. 18:17

15. a. 13   **b.** $\frac{1}{4}$

17. a. 12   **b.** $\frac{3}{13}$

19. a. 4   **b.** $\frac{1}{13}$

21. picking one card from the well-shuffled deck

23. a. $\frac{1}{2}$ **b.** 1:1

25. a. $\frac{1}{13}$ **b.** 1:12

27. a. $\frac{1}{52}$ **b.** 1:51

29. a. $\frac{3}{13}$ **b.** 3:10

31. a. $\frac{40}{52} = \frac{10}{13}$ **b.** 40:12 = 10:3

33. a. $\frac{12}{52} = \frac{3}{13}$ **b.** 12:40 = 3:10

35. 1:4

37. $a: b - a$

39. $\frac{4}{11}$

41. a. $\frac{1}{2}$ **b.** 1:1

43. a. {(b, b), (b, g), (g, b), (g, g)}   **b.** {(b g), (g, b)}
  c. {(b, g), (g, b), (g, g)}   **d.** {(g, g)}
  e. $\frac{1}{2}$ **f.** $\frac{3}{4}$ **g.** $\frac{1}{4}$ **h.** 1:1   **i.** 3:1   **j.** 1:3

45. a. {(b, b, b), (b, b, g), (b, g, b), (g, b, b), (b, g, g), (g, b, g), (g, g, b), (g, g, g)}
  b. {(b, g, g), (g, b, g), (g, g, b)}
  c. {(b, g, g), (g, b, g), (g, g, b), (g, g, g)}
  d. {(g, g, g,)}   **e.** $\frac{3}{8}$ **f.** $\frac{1}{2}$ **g.** $\frac{1}{8}$ **h.** 3:5   **i.** 1:1
  j. 1:7

47. a. $\frac{1}{4}$ **b.** $\frac{1}{2}$ **c.** $\frac{1}{4}$ **d.** equally likely

49. different

51. a. Parkview District   **b.** 0.3517

SECTION 6.2

1. a.

Outcome	Probability
1	0.1
2	0.2
3	0.3
4	0.4

  b. 0.9

3. a.

Outcome	Probability
black	$\frac{1}{4}$
white	$\frac{1}{4}$
blue	$\frac{1}{2}$

  b. $\frac{3}{4}$

5. a.

Outcome	Probability
red	$\frac{1}{2}$
pink	$\frac{1}{2}$

  b. $\frac{1}{2}$ **c.** $\frac{1}{2}$ **d.** 0

7. a.

Outcome	Probability
carrier	$\frac{1}{2}$
no cystic fibrosis gene	$\frac{1}{2}$

  b. 0   **c.** $\frac{1}{2}$ **d.** $\frac{1}{2}$ **e.** 1

9. a.

Outcome	Probability
has sickle cell	$\frac{1}{4}$
has trait	$\frac{1}{2}$
disease free	$\frac{1}{4}$

  b. $\frac{1}{4}$ **c.** $\frac{1}{2}$ **d.** $\frac{1}{4}$

11. a.

Outcome	Probability
carrier	$\frac{1}{2}$
no Tay-Sachs gene	$\frac{1}{2}$

b. 0   **c.** $\frac{1}{2}$ **d.** 1

13. a.

Outcome	Probability
Hh	$\frac{1}{2}$
hh	$\frac{1}{2}$

b. $\frac{1}{2}$ **c.** $\frac{1}{2}$ **d.** $\frac{1}{2}$

SECTION 6.3

1. not mutually exclusive

3. mutually exclusive

5. not mutually exclusive

7. mutually exclusive

9. mutually exclusive

11. a. $\frac{1}{26}$ **b.** $\frac{7}{13}$ **c.** $\frac{25}{26}$

13. a. $\frac{1}{52}$ **b.** $\frac{4}{13}$ **c.** $\frac{51}{52}$

15. a. $\frac{2}{13}$ **b.** $\frac{5}{13}$ **c.** 0   **d.** $\frac{7}{13}$

17. a. $\frac{9}{13}$ **b.** $\frac{8}{13}$ **c.** $\frac{4}{13}$ **d.** 1

19. $\frac{12}{13}$

21. $\frac{10}{13}$

23. $\frac{11}{13}$

25. $\frac{9}{13}$

27. 9:5

29. 2:5, 5:2

31. *b:a*

33. 12:1

35. 10:3

37. 10:3

39. a. $\frac{71}{175}$ **b.** $\frac{104}{175}$

41. a. $\frac{151}{700}$ **b.** $\frac{97}{140}$

43. a. $\frac{8}{25}$ **b.** $\frac{17}{25}$

45. a. $\frac{9}{25}$ **b.** $\frac{16}{25}$

47. a. {2, 3, 4, 5, 6, 7, 8, 9, 10, 11, 12}

b.

Outcome	Probability
2	$\frac{1}{36}$
3	$\frac{2}{36}$
4	$\frac{3}{36}$
5	$\frac{4}{36}$
6	$\frac{5}{36}$
7	$\frac{6}{36}$
8	$\frac{5}{36}$
9	$\frac{4}{36}$
10	$\frac{3}{36}$
11	$\frac{2}{36}$
12	$\frac{1}{36}$

c. $\frac{5}{36}$ **d.** $\frac{10}{36} = \frac{5}{18}$ **e.** $\frac{26}{36} = \frac{13}{18}$

49. a. $\frac{2}{9}$ **b.** $\frac{7}{18}$

51. a. $\frac{1}{6}$ **b.** $\frac{3}{4}$

53. a. $\frac{1}{6}$ **b.** $\frac{1}{2}$

55. a. 0.35   **b.** 0.25   **c.** 0.15   **d.** 0.6

57. a. 0.20   **b.** 0.95   **c.** 0.8

59. a. 0.3517   **b.** 0.4673   **c.** 0.1655   **d.** 0.6526

SECTION 6.4

1. 0.71

3. a. $\frac{1}{13,983,816}$ **b.** $\frac{1}{22,957,480}$ **c.** $\frac{1}{18,009,460}$
d. about 64% more likely

5. 0.00003

7. a. 0.000003080   **b.** 0.000462062

9. a. 0.000001737   **b.** 0.000295263

11. $\frac{5}{26}$ is easiest; $\frac{6}{54}$ is hardest

13.

Outcome	Probability
8 winning spots	0.000004
7 winning spots	0.000160
6 winning spots	0.002367
5 winning spots	0.018303
4 winning spots	0.081504
fewer than 4 winning spots	0.897662

15. a. 0.0005   **b.** 0.00198   **c.** 0.00002   **d.** 0.00197

17. 0.05

19. 0.48

21. 0.64

23. 0.36

25. a. 0.08894   **b.** 0.4221   **c.** 0.4889

SECTION 6.5

1. 1.75

3. $10.05

7. bank

9. 0.59 or more

11. a. 0   **b.** $\frac{1}{16}$ **c.** $\frac{3}{8}$

13. −$0.28

15. −$0.55

17. at least $1200

19. backyard

SECTION 6.6

1. a. 0.23   **b.** 0.53   **c.** 0.14   **d.** 0.32   **e.** 0.08   **f.** 0.08

3. 0.20

5. 0.34

7. a.

Outcome	Probability
1	0.1
2	0.2
3	0.3
4	0.4

b. 0.1   **c.** 0.25

9. a. $\frac{3}{12} = \frac{1}{4}$ **b.** $\frac{2}{12} = \frac{1}{6}$ **c.** $\frac{1/6}{1/4} = \frac{2}{3}$

11. a. $\frac{1}{4}$ **b.** $\frac{4}{17}$ **c.** $\frac{1}{17}$

13. a. $\frac{1}{4}$ **b.** $\frac{13}{51}$ **c.** $\frac{13}{204}$

15. a. $p(B \mid A)$ **b.** $p(A')$ **c.** $p(C \mid A')$

17. a. $\frac{1}{6}$ **b.** $\frac{1}{3}$ **c.** 0   **d.** 1

19. a. $\frac{5}{36}$ **b.** $\frac{5}{18}$ **c.** 0   **d.** 1

21. a. $\frac{1}{12}$ **b.** $\frac{3}{10}$ **c.** 1

23. E_2 is most likely; E_3 is least likely.

25. 0.46

27. 0.0005

29. 0.0020

31. 0.14

33. 0.20

35. 0.07

37. 0.00646

39. 0.01328

41. 92%

43. 39%

45. 0.05

47. a. 20–40   **b.** 0.4733
c. yes; you are given that the renter lives in the Parkview district.

49. a. 0.41   **b.** 0.44   **c.** 0.35   **d.** yes
e. appears to be a bias against men   **f.** ?

51. a. 0.00187   **b.** 0.00180   **c.** 0.00543
d. 0.00227   **e.** 0.00160   **f.** 0.00340   **g.** ?

53. a. 0.86　　　**b.** 0.34　　　**c.** 0.66　　　**d.** $N' \mid W$

SECTION 6.7

1. a. dependent　　　**b.** not mutually exclusive

3. a. dependent　　　**b.** mutually exclusive

5. a. independent　　　**b.** not mutually exclusive

7. a. dependent　　　**b.** mutually exclusive

9. a. dependent　　　**b.** mutually exclusive

11. no

13. dependent

15. dependent

17. a. 0.000001　　　**b.** 4

19. a. $\frac{1}{6}$　**b.** $\frac{5}{6}$　**c.** 0.4823　**d.** 0.5177　**e.** $0.0355

SECTION 6.8

1. a. 0.04721　　**b.** 0.99999　　**c.** 0.95279　　**d.** 0.00001
　e. (c) is a false positive and (d) is a false negative
　f. (a) and (c)

5. a. 54.5%　　**b.** 45.5%　　**c.** 39.7%

7. a. 18.6%　　**b.** 56.1%　　**c.** 25.3%

11. a. 0.5　　**b.** 0.8947

13. a. 19.9%　　**b.** 65.4%　　**c.** 64.3%

CHAPTER 6 REVIEW

1. a. $\frac{1}{2}$; 1:1　　**b.** $\frac{1}{13}$; 1:12　　**c.** $\frac{1}{4}$; 1:3
　d. $\frac{1}{52}$; 1:51　　**e.** $\frac{4}{13}$; 4:9　　**f.** $\frac{12}{13}$, 12:1

2. a. Three coins are tossed.
　b. {(h, h, h), (h, h, t), (h, t, h), (t, h, h), (h, t, t), (t, h, t),
　　(t, t, h), (t, t, t)}
　c. {(h, t, t), (t, h, t), (t, t, h)}
　d. {(h, t, t), (t, h, t), (t, t, h), (t, t, t)}　**e.** $\frac{3}{8}$; 3:5　**f.** $\frac{1}{2}$; 1:1

3. a. $\frac{1}{6}$　**b.** $\frac{1}{18}$　**c.** $\frac{7}{18}$　**d.** $\frac{1}{6}$　**e.** $\frac{11}{18}$　**f.** $\frac{7}{18}$

4. a. 0.013　　**b.** 0.138　　**c.** 0.151　　　**d.** 0.000008

5. a. 0.005　　**b.** 0.069　　**c.** 0.074

6. a. 0.14　　　　　　**b.** no

7. a. $\frac{1}{2}$　　　　**b.** $\frac{1}{2}$

8. a. $\frac{1}{4}$　　　　**b.** $\frac{1}{2}$　　　　**c.** $\frac{1}{4}$

9. a. $\frac{1}{4}$　　　　**b.** $\frac{1}{2}$　　　　**c.** $\frac{1}{4}$

10. a. $\frac{1}{2}$　　　　**b.** $\frac{1}{2}$

11. a. 0　　　　　**b.** $\frac{1}{2}$　　　　**c.** $\frac{1}{2}$

12. a. 0.56; 0.39　　　　　　**b.** 0.40; 0.55
　c. 0.67; 0.33　　　　　　**d.** 0.50; 0.50
　e. urban area; urban area　　**f.** O'Neill; Bell
　g. no　　　　　　　　　　**h.** O'Neill

13. a. dependent; not mutually exclusive
　b. independent; not mutually exclusive
　c. dependent; mutually exclusive
　d. dependent; not mutually exclusive
　e. independent; not mutually exclusive

14. a. 42.9%　　　　**b.** 57.1%

15. a.

Outcome	Probability
9 spots	0.0000007243
8 spots	0.00003259
7 spots	0.000592
6 spots	0.005719
5 spots	0.032601
0–4 spots	0.961055

　b. −$0.249

16. The craps bet has a higher expected value.

CHAPTER 7

SECTION 7.1

1. a. p(current purchase is KickKola) = 0.14;
　　p(current purchase is not KickKola) = 0.86
　b. $P = [0.14 \quad 0.86]$

3. a. p(currently own) = 0.32; p(currently rent) = 0.68
　b. $P = [0.32 \quad 0.68]$

5. a. p(Silver's) = 0.48; p(Fitness Lab) = 0.37;
　　p(ThinNFit) = 0.15

b. $P = [0.48 \quad 0.37 \quad 0.15]$

7. a. p(next purchase is KickKola | current purchase is not Kick Kola) = 0.12; p(next purchase is KickKola | current purchase is KickKola) = 0.63; p(next purchase is not KickKola | current purchase is not KickKola) = 0.88; p(next purchase is not KickKola | current purchase is KickKola) = 0.37

b. $\begin{bmatrix} 0.63 & 0.37 \\ 0.12 & 0.88 \end{bmatrix}$

9. a. p(buy next residence | currently rent) = 0.12; p(rent next residence | currently own) = 0.03: p(rent next residence | currently rent) = 0.88; p(buy next residence | currently own) = 0.97

b. $\begin{bmatrix} 0.97 & 0.03 \\ 0.12 & 0.88 \end{bmatrix}$

11. a. p(Silver's next | currently Silver's) = 0.71; p(Fitness Lab next | currently Silver's) = 0.12; p(ThinNFit next | currently Silver's) = 0.17; p(Silver's next | currently Fitness Lab) = 0.32; p(Fitness Lab next | currently Fitness Lab) = 0.34; p(ThinNFit next | currently Fitness Lab) = 0.34; p(Silver's next | currently ThinNFit) = 0.02; p(Fitness Lab next | currently ThinNFit) = 0.02; p(ThinNFit next | currently ThinNFit) = 0.96

b. $\begin{bmatrix} 0.71 & 0.12 & 0.17 \\ 0.32 & 0.34 & 0.34 \\ 0.02 & 0.02 & 0.96 \end{bmatrix}$

13. a. 19%   **b.** 22%

15. a. 46%; 19%; 35%   **b.** 39%; 13%; 48%
   c. 33%; 10%; 57%

17. 27%

19. a. $P = [0.313 \quad 0.462 \quad 0.225]$

b. $\begin{bmatrix} 0.927 & 0.064 & 0.009 \\ 0.022 & 0.970 & 0.008 \\ 0.013 & 0.019 & 0.968 \end{bmatrix}$

   c. 30.3%; 47.2%; 22.4%
   d. 29.4%; 48.2%; 22.4%

21. a. $P = [0.296 \quad 0.091 \quad 0.613]$

b. $T = \begin{bmatrix} 0.879 & 0.040 & 0.081 \\ 0.236 & 0.764 & 0 \\ 0 & 0 & 1 \end{bmatrix}$

   c. 394 million acres; 113 million acres; 889 million acres
   d. 373 million acres; 102 million acres; 921 million acres

SECTION 7.2

1. $[\frac{2}{11} \quad \frac{9}{11}]$

3. $[0.2115 \quad 0.5577 \quad 0.2308]$

5. 24%

7. 10.8%; 4.5%; 84.6%

9. 80%; 20%; current trends continue, and residents' moving plans are realized

11. a. $\begin{bmatrix} 0.927 & 0.064 & 0.009 \\ 0.022 & 0.970 & 0.008 \\ 0.013 & 0.019 & 0.968 \end{bmatrix}$

   b. 21.2%: 58.3%: 20.5%

13. Toyonda: 17.8%; Nissota: 17.7%, Reo: 16.7%; Henry J: 25.5%; DeSoto: 19.4%; Hugo: 2.8%

SECTION 7.3

1. a. $T = \begin{bmatrix} 0.08 & 0.85 & 0 & 0 & 0 & 0.07 \\ 0 & 0.07 & 0.87 & 0 & 0 & 0.06 \\ 0 & 0 & 0.06 & 0.89 & 0 & 0.05 \\ 0 & 0 & 0 & 0.05 & 0.91 & 0.04 \\ 0 & 0 & 0 & 0 & 1 & 0 \\ 0 & 0 & 0 & 0 & 0 & 1 \end{bmatrix}$

   b. T is absorbing, since it is not possible to leave the states G and Q.
   c. 1.0864 years   **d.** 1.0870 years
   e. 3.8614 years   **f.** 2.0604 years
   g. between 75% and 78.1%
   h. 0.7839   **i.** 0.9069

3. a. 6.33 years   **b.** 12.35 years   **c.** 100%   **d.** 100%

5. a. $T = \begin{bmatrix} 0 & 0.3 & 0 & 0 & 0.7 & 0 \\ 0 & 0 & 0.4 & 0 & 0.6 & 0 \\ 0 & 0 & 0 & 0.5 & 0.5 & 0 \\ 0 & 0 & 0 & 0 & 0.4 & 0.6 \\ 0 & 0 & 0 & 0 & 1 & 0 \\ 0 & 0 & 0 & 0 & 0 & 1 \end{bmatrix} \begin{matrix} \text{New} \\ 1 \\ 2 \\ 3 \\ \text{Paid} \\ \text{Bad} \end{matrix}$

   b. T is absorbing, since it is not possible to leave the states paid up and bad debt.
   c. 0.964
   d. 0.12
   e. 1.48 months
   f. $4500; $54,000

7. a. 85%   **b.** 12.75%   **c.** 2.25%

d.

Years as Freshman	Probability
1	0.85
2	0.1275
3	0.0225
expected value = 1.1725	

CHAPTER 7 REVIEW

1. **a.** 21.5%   **b.** 25.9%   **c.** 29.6%

2. **a.** If current trends continue, the percentages will be 22.1%, 19.6%, and 58.3%. Thus, more condominiums and townhouses will be needed.
   b. If current trends continue, the percentages will be 17.6%, 23.9%, and 58.5%. Thus, significantly fewer apartments will be needed, and significantly more condominiums and townhouses will be needed.
   c. emigration and immigration

3. **a.** 8.125 years   **b.** 9.5 years

4. 0.47

CHAPTER 8

SECTION 8.1

1. The game is strictly determined. The row player should choose row 1, and the column player should choose column 1. The game value is 2, and the game favors the row player.

3. The game is not strictly determined. Possible moves: row player plays row 1, column player plays column 2, row player plays row 2, and column player plays column 1

5. The game is not strictly determined. Possible moves: row player plays row 1, column player plays column 2, row player plays row 2, and column player plays column 1.

7. The game is not strictly determined. Possible moves: row player plays row 1, column player plays column 3, row player plays row 2, column player plays column 1.

9. The game is strictly determined. The row player plays row 2 and the column player plays column 1. The game is fair.

11. $x \le 0$

13. **a.** it is a zero-sum game   **b.** it is strictly determined
    c. Runzgood should always lower prices and Petrolia should not change prices. The game's value is +1%. It favors Runzgood.

15. **a.** constant sum game   **b.** strictly determined
    c. pure strategy; each should choose River City, and each will get 50% of the business.

17. **a.** zero sum game   **b.** not strictly determined
    c. possible moves: Brown emphasizes North, Jorden emphasizes South, Brown splits money equally, Jorden emphasizes North

SECTION 8.2

1. **a.** not strictly determined; no saddlepoint
   b. no; if even responds with a pure strategy, odd will lose $0.50 per game on average.
   c. E(pure strategy) = $[1 - 2p_1, 2p_1 - 1]$

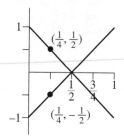

d.

e. 0
f. 1 finger $\frac{1}{2}$ of the time and 2 fingers $\frac{1}{2}$ of the time
g. yes

3. **a.** not strictly determined
   c. The row player should play row one $\frac{10}{19}$ of the time and row two $\frac{9}{19}$ of the time. The column player could not do better with a randomized strategy. The expected value of the game is $\frac{1}{19}$.

5. a. strictly determined

   b. The row player should play row 2 and the column player should play column 2. The game value is 3, and the game favors the row player.

7. a. not strictly determined

   c. The row player should play row one $\frac{1}{9}$ of the time and row two $\frac{8}{9}$ of the time. The column player could not do better with a randomized strategy. The expected value of the game is $-\frac{2}{3}$.

9. a. strictly determined

   b. The row player should play row 1 and the column player should play column 2. The game value is 4, and the game favors the row player.

11. Brown's optimal strategy is to play row 1 two out of three times and row 2 one out of three times. Jorden cannot do better with a randomized strategy. Brown will gain voters. The game value is 0.67.

13. a. wet year is better for rice; normal year is better for wheat

    b. $\frac{5}{7}$ wheat; $\frac{2}{7}$ rice

    c. wet: $714,000 profit; normal: $10,000,000 profit; dry: $714,000 profit

15. Both Marcos and Jean should be honest.

SECTION 8.3

1. The game is not strictly determined. The row player should play row 1 $\frac{10}{19}$ of the time and row 2 $\frac{9}{19}$ of the time.

3. The game is strictly determined.

5. The game is not strictly determined; $p_1 = 0.57$ $p_2 = 0.43$

7. The game is not strictly determined; $p_1 = \frac{1}{2}$ $p_2 = \frac{1}{2}$ $p_3 = 0$

9. Brown plays row 1 two out of three times and row 2 one out of three times

11. The game is not strictly determined. The odd player should play 1 with probability $\frac{1}{4}$, 2 with probability $\frac{1}{2}$, and 3 with probability $\frac{1}{4}$.

13. Each company should choose River City.

CHAPTER 8 REVIEW

1. a. The game is strictly determined. Both Discs-R-Us and 500 Channels should choose Elmwood. The game value is 50%, and the game does not favor either player. The pure strategies will result in each company getting 50% of the business.

2. a. The game is not strictly determined.

   b. Discs-R-Us should choose Gotham with probability $\frac{3}{10}$ and Elmwood with probability $\frac{7}{10}$. The game value is 55.8%, and the column player could not do better with a randomized strategy.

3. a. The game is strictly determined. Pure strategies: both Bloomsworthy and Mercy's have a sale; the game favors Bloomsworthy and the game value is 4%. Bloomsworthy will gain 4% of Mercy's customers.

CHAPTER 9

SECTION 9.1

1.

Number of Children	Tally	Frequency	Relative Frequency
0	⦀⦀ ⦀⦀⦀	8	0.2 = 20%
1	⦀⦀ ⦀⦀ ⦀⦀⦀	13	0.325 = 32.5%
2	⦀⦀ ⦀⦀⦀⦀	9	0.225 = 22.5%
3	⦀⦀ ⦀	6	0.15 = 15%
4	⦀⦀⦀	3	0.075 = 7.5%
5	⦀	1	0.025 = 2.5%
		$n = 40$	1.000 = 100%

3. a.

Speed (in mph)	Tally	Frequency	Relative Frequency
$51 \leq x < 56$	⦀⦀	2	0.05 = 5%
$56 \leq x < 61$	⦀⦀⦀⦀	4	0.1 = 10%
$61 \leq x < 66$	⦀⦀ ⦀⦀	7	0.175 = 17.5%
$66 \leq x < 71$	⦀⦀ ⦀⦀	10	0.25 = 25%
$71 \leq x < 76$	⦀⦀ ⦀⦀⦀⦀	9	0.225 = 22.5%
$76 \leq x < 81$	⦀⦀ ⦀⦀⦀	8	0.2 = 20%
		$n = 40$	1.000 = 100%

b.

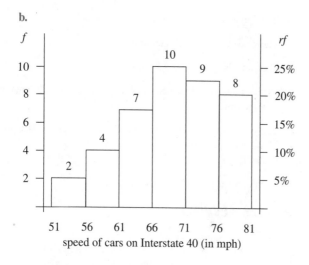

speed of cars on Interstate 40 (in mph)

7. a.

x = Hourly Wage	Frequency	Relative Frequency
$4.00 \leq x < 5.50$	21	0.138 or 13.8%
$5.50 \leq x < 7.00$	35	0.230 or 23.0%
$7.00 \leq x < 8.50$	42	0.276 or 27.6%
$8.50 \leq x < 10.00$	27	0.178 or 17.8%
$10.00 \leq x < 11.50$	18	0.118 or 11.8%
$11.50 \leq x < 13.00$	9	0.059 or 5.9%
Total	152	1.000 or 100%

5. a.

Time in Minutes	Frequency	Relative Frequency
$0 \leq x < 10$	101	0.144 or 14.4%
$10 \leq x < 20$	237	0.339 or 33.9%
$20 \leq x < 30$	169	0.241 or 24.1%
$30 \leq x < 40$	79	0.113 or 11.3%
$40 \leq x < 50$	51	0.073 or 7.3%
$50 \leq x < 60$	63	0.090 or 9.0%
Total	700	1.000 or 100%

b.

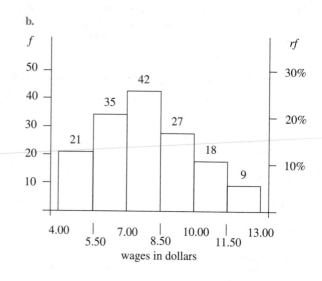

wages in dollars

b.

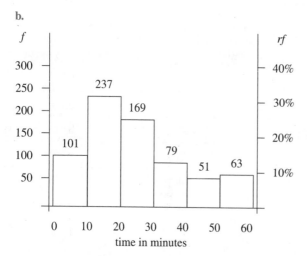

time in minutes

9. a.

Age	Frequency	Relative Frequency
$18 \leq x < 25$	13,167,000	.250 or 25%
$25 \leq x < 30$	10,839,000	.206 or 21%
$30 \leq x < 35$	10,838,000	.206 or 21%
$35 \leq x < 40$	9,586,000	.182 or 18%
$40 \leq x < 45$	8,155,000	.155 or 16%
Total	52,585,000	1.000 or 100%

b.

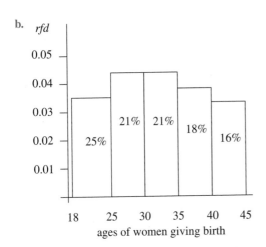

ages of women giving birth

11. a.

Age of Males	Frequency	Relative Frequency
$14 \leq x < 18$	111,000	0.019 or 2%
$18 \leq x < 20$	1,380,000	0.234 or 23%
$20 \leq x < 22$	1,158,000	0.197 or 20%
$22 \leq x < 25$	928,000	0.157 or 16%
$25 \leq x < 30$	847,000	0.144 or 14%
$30 \leq x < 35$	591,000	0.100 or 10%
$35 \leq x \leq 100$	878,000	0.149 or 15%
Total	5,893,000	1.000 or 100%

b.

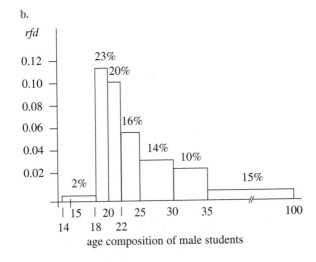

age composition of male students

13. a. 28.5%   **b.** 32%   **c.** not possible
d. 97.5%   **e.** 50%   **f.** 52.5%

19. f

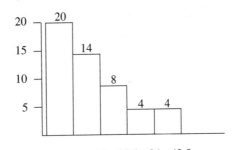

percentage of legislators with links
to insurers in the 50 states

SECTION 9.2

1. mean = 12.5; median = 11.5; mode = 9

3. mean = 1.45; median = 1.5; mode = 1.7

5. a. mean = 11; median = 10.5; mode = 9
  b. mean = 25.5; median = 10.5; mode = 9

7. a. mean = 7; median = 7; mode = none
  b. mean = 107; median = 107; mode = none

9. mean = 43.9; median = 46; mode = 46

11. mean = 7.6; median = 8; mode = 6 and 9

13. 16.028 oz

15. 96

17. 51.4 mph

19. a. $2,184.10   **b.** $4,535.04

21. a. 35.2   **b.** 35.9

23. a. 23.4   **b.** 33.5

SECTION 9.3

1. a. 16.4; 4.0   **b.** 16.4; 4.0

3. a. 0   **b.** 0

5. a. 22; 7.5   **b.** 1100; 374.2

7. a. Joey, 168; Dee Dee, 167; Joey has the higher mean.
  b. Joey, 30.9; Dee Dee, 18.2

c. Dee Dee is more consistent because her standard deviation is lower.

9. 3.21; 1.87

11. a. 8; 1.4   b. 71%   c. 94%   d. 100%

13. 0.285

15. 7.3

SECTION 9.4

1. a. 34.13%   b. 34.13%   c. 68.26%

3. a. 49.87%   b. 49.87%   c. 99.74%

5. a. one [22.4, 27.0]; two [20.1, 29.3]; three [17.8, 31.6]
   b. 68.26%; 95.44%; 99.74%
   c.

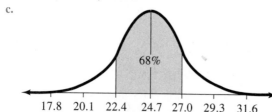

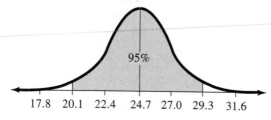

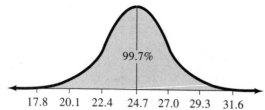

7. a. 0.4474   b. 0.0639   c. 0.5884
   d. 0.0281   e. 0.0960   f. 0.8944

9. a. 0.34   b. −2.08   c. 0.62
   d. −0.27   e. 1.64   f. −1.28

11. a. 0.3413   b. 0.6272   c. 0.8997
    d. 0.9282   e. 0   f. 0.3300

13. a. 4.75%   b. 49.72%

15. a. 0.1894   b. 0.7745

17. a. 95.25%   b. 79.67%

19. a. 87   b. 62

21. 9.1 min

23. a. The area of the line $Z = 1$ is 0.   b. 0

SECTION 9.5

1. 0.00000000002

3. 0.2362

5. a. 18   b. 10   c. 8   d. 0.8   e. 0.2

7. a. There are two possible outcomes: 1 or 2, or not 1 or 2.
   b. $\frac{1}{3}$ c. $\frac{2}{3}$
   d. $(s, s, f, f), (s, f, s, f), (s, f, f, s), (f, f, s, s), (f, s, s, f),$
      (f, s, f, s) 6 outcomes
   e. 6   f. $\frac{8}{27}$

9. a. $\frac{1}{8}$ b. $\frac{3}{8}$ c. $\frac{3}{8}$ d. $\frac{1}{8}$
   e.

Outcome	Probability
0	$\frac{1}{8}$
1	$\frac{3}{8}$
2	$\frac{3}{8}$
3	$\frac{1}{8}$

   f.

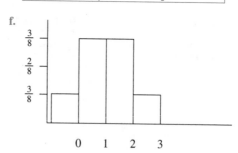

11. One flip is not a binomial experiment

13. Observing 1 or 2, 3 or 4, or 5 or 6 is not a binomial experiment

15. a. 0.0956   b. 0.651

17. a. 0.004   b. 0.996   c. 490

19. a. 0.0157   b. 0.0965   c. 0.9035

27. {0, 1, 2, 3, 4, 5, 6, 7, 8, 9, 10}

29. a.

Children with AB Type	Probability
0	$\frac{1}{32}$
1	$\frac{5}{32}$
2	$\frac{10}{32}$
3	$\frac{10}{32}$
4	$\frac{5}{32}$
5	$\frac{1}{32}$

b.

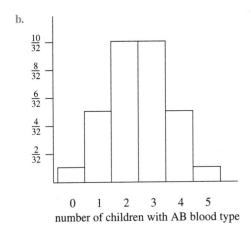

c. 2 or 3   d. 2.5

SECTION 9.6

1. 0.0099

3. 0.2546

5. 1

7. a. 25   b. 29   c. 32 or more

9. a. 3.7×10^{-28}, or approximately 0
   b. 0.1446   c. 0.9925

11. a. 4; 6   b. right

CHAPTER 9 REVIEW

1. a.

Number of Children	Tally	Frequency	Relative Frequency
0	ЖЖ III	8	0.2 = 20%
1	ЖЖ ЖЖ	10	0.25 = 25%
2	ЖЖ ЖЖ I	11	0.275 = 27.5%
3	ЖЖ II	7	0.175 = 17.5%
4	III	3	0.075 = 7.5%
5	I	1	0.025 = 2.5%
		$n = 40$	1.000 = 100%

b. 1.75   c. 2   d. 2   e. 1.3

2. a. 40%   b. 28%   c. 86%   d. 14%   e. 50%
   f. cannot determine

3. a. 10.17   b. 3.21
   c. f

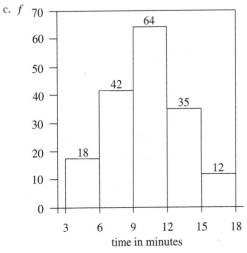

4. 100

5. $33,400

6. a. Timo, 98.8; Henke, 96.6; Henke has the lower mean.
   b. Timo, 6.4; Henke, 10.4
   c. Timo is more consistent because his standard deviation is lower.

7. a. $1.298; $0.131 b. 78% c. 95% d. 100%

8. a. 45.99% b. 45.99% c. 91.98%

9. 401

10. a. 0.5 b. 0.0475 c. 119.2 or higher

11. a. 4 b. $1.05 \times 10^{-14} \approx 0$ c. $3.19 \times 10^{-11} \approx 0$

12. a. 10 b. 0.000000954 c. 0.00018

CHAPTER 10

SECTION 10.1

1. $480

3. $21.26

5. $153.80

7. $4376.48

9. $14,073.09

11. $1474.81

13. $6162.51

15. $17,036.82

17. $6692.61

19. $1069.98

21. $1240.95

23. $2727.27

25. about $3\frac{1}{3}$ years

27. $328.19

29. $115.02

31. a. $264 b. $936 c. 5%

33. a. $311.72 b. $1,438.28 c. 14.45%

35. a. $405 b. $29,458.70

37. a. $8125 b. $146,250 c. $8125
    d. $67.71 e. $19,500.08
    f. $136,500 g. $15,600.08

39. a. $38,940 b. $311,520 c. $38,940
    d. $356.95 e. $95,013.60 f. $288,156.00
    g. $383,169.60

41. $P = M(1 - rt)$

43. $1875

45. $2448.63

SECTION 10.2

1. a. 0.03 b. 0.01 c. 0.000328767
   d. 0.0046153846 e. 0.005

3. a. 0.00775 b. 0.002583333 c. 0.000084932
   d. 0.001192307 e. 0.001291667

5. a. 0.02425 b. 0.008083333 c. 0.000265753
   d. 0.003730769 e. 0.004041667

7. a. 34 quarters b. 102 months c. 3103 days

9. a. 120 quarters b. 360 months c. 10,950 days

11. $7189.67

13. $9185.46

15. $3951.74

17. 8.30%

19. 4.34%

21. a. 10.38% b. 10.47% c. 10.52%

23. $583.49

25. $470.54

27. First National: 9.55%; Citywide: 9.45%

29. verifies

31. does not verify—true yield is 9.74%

33. a. $9664.97 b. $51.77

35. $19,741.51

37. a. $8108.07 b. $690.65

39. a. 5.116% b. $1025.26 c. $25.26 d. 2.526%
    e. Part (a) was a 12-month investment; part (d) is a 6-month investment.
    f. Interest is compounded 12 times.

41. $r = (1 + i)^n - 1$

43. 7.45%

45. a. 16.24%   **b.** 16.56%   **c.** 16.79%
d. 16.91%   **e.** 16.86%   **f.** 16.85%

57. a. 14.21 years   **b.** 13.89 years
c. 13.95 years   **d.** 13.86 years

59. 4.99 years

61. a. 2369 periods; 6.49 years   **b.** 9400 periods; 25.75 years

SECTION 10.3

1. $1478.56

3. $5422.51

5. a. $770.00   **b.** $750.00   **c.** $20.00

7. a. $1391.45   **b.** $1350.00   **c.** $41.45

9. a. $283,037.86   **b.** $54,600.00   **c.** $228,437.86

11. a. $251,035.03   **b.** $69,600.00   **c.** $181,435.03

13. $1396.14

15. $726.49

17. $18,680.03

19. $51.84

21. $33.93

23. $33.63

25. a. $537,986.93
b. 1 $537,986.93 $2734.77 $650 $540,071.70
    2    $540,071.70 $2745.36 $650 $542,167.06
    3    $542,167.06 $2756.02 $650 $544,273.08
    4    $544,273.08 $2766.72 $650 $546,389.80
    5    $546,389.80 $2777.48 $650 $548,517.28

27. a. $175,384.62   **b.** $111.84

29. $45.82

31. a. $595.83   **b.** $5367.01
c.   1   2   3   4   5
    $5367.01 $10,958.76 $16,784.67 $22,854.54 $29,178.58
    6   7   8   9   10
    $35,767.44 $42,632.21 $49,784.44 $57,236.17 $64,999.94

33. $P = \text{pymt}\dfrac{(1+i)^n - 1}{i(1+i)^n} = \text{pymt}\dfrac{1-(1+i)^{-n}}{i}$

35. $1396.14

37. $726.49

SECTION 10.4

1. a. $125.62   **b.** $1029.76

3. a. $193.91   **b.** $1634.60

5. a. $1303.32   **b.** $314,195.20

7. a. $378.35   **b.** $3658.36
c.

Payment Number	Principal Portion	Interest Portion	Total Payment	Balance Due
0	—	—	—	$14,502.44
1	$239.37	$138.98	$378.35	$14,263.07
2	$241.66	$136.69	$378.35	$14,021.41

9. a. $1602.91   **b.** $407,047.60
c.

Payment Number	Principal Portion	Interest Portion	Total Payment	Balance Due
0	—	—	—	$170,000.00
1	$62.28	$1540.63	$1602.91	$169,937.72
2	$62.85	$1540.06	$1602.91	$169,874.87

d. $4218.18/month

11. a. $18,950   **b.** $151,600
c. $18,950   **d.** $1501.28
e. $189.50   **f.** $271.61
g. $1962.39   **h.** $1501.28

13. a. $404.72; $353.68   **b.** $4597.09; $2730.25
c. simple interest loan (why?)

15. a. $354.29; $1754.44   **b.** $278.33; $2359.84
c. $233.04; $2982.40

17. a. $599.55; $115,838.00
b. $665.30; $139,508.00
c. $733.76; $164,153.60
d. $804.62; $189,663.20
e. $877.57; $215,925.20
f. $952.32; $242,835.20

19. Both verify, but not exactly. Their computations are actually more accurate than ours, since they are using more accurate round-off rules than we do.

21. a. $877.57; $215,925.20   **b.** $404.89; $215,814.20

23. a. $24,535.12

b.

Payment Number	Principal Portion	Interest Portion	Total Payment	Amount Due after Payment
0	—	—	—	$120,000.00
1	$23,647.62	$887.50	$24,535.12	$96,352.38
2	$23,822.51	$712.61	$24,535.12	$72,529.87
3	$23,998.70	$536.42	$24,535.12	$48,531.17
4	$24,176.19	$358.93	$24,535.12	$24,354.98
5	$24,354.98	$180.13	$24,535.11	$0.00

25. a. $12,232.14

b.

Payment Number	Principal Portion	Interest Portion	Total Payment	Amount Due after Payment
0	—	—	—	$48,000.00
1	$11,862.14	$370.00	$12,232.14	$36,137.86
2	$11,953.58	$278.56	$12,232.14	$24,184.28
3	$12,045.72	$186.42	$12,232.14	$12,138.56
4	$12,138.56	$93.57	$12,232.13	$0.00

27. add-on:

Payment Number	Principal Portion	Interest Portion	Total Payment	Amount Due after Payment
0	—	—	—	$14,829.32
1	$308.95	$95.77	$404.72	$14,520.37
2	$308.95	$95.77	$404.72	$14,211.42
3	$308.95	$95.77	$404.72	$13,902.47

simple interest:

Payment Number	Principal Portion	Interest Portion	Total Payment	Amount Due after Payment
0	—	—	—	$14,246.39
1	$248.32	$105.36	$353.68	$13,998.07
2	$250.15	$103.53	$353.68	$13,747.92
3	$252.00	$101.68	$353.68	$13,495.92

29. $3591.73

31. $160,234.64

33. a. $1735.74   **b.** $151,437.74   **c.** $1204.91
   d. $472,016.40   **e.** $343,404.24   **f.** yes (why?)

35. a. $1718.80   **b.** $172,157.40   **c.** $240,354.60
   d. $573,981.98   **e.** $1,200,543.86
   f. The decision is between saving more for retirement or
   increasing their standard of living now (why?).

37. a. $699.21   **b.** $559.97   **c.** $1670.88
   d. $98,621.73   **e.** 7.714%; $n = 348$
   f. $567.80   **g.** $3247.80

39. a. $474.54   **b.** $99,120.56   **c.** $602.07
   d. The monthly payments can change only on a yearly
   basis, so the difference in payment amount is added to
   the loan amount. The unpaid balance will increase.
   e. The loan is increasing, not decreasing, so the
   amortization is called negative.

43. b. $1511.42   **c.** $4269.84   **e.** $270.65

45. b. $18,449.10   **c.** $18,156.12   **e.** $1087.63

47. b. $50,702.84   **c.** $47,587.25   **e.** $3818.17

49. b. $735.76   **c.** $2353.63   **e.** $101.60

SECTION 10.5

1. 14.6%

3. 11.2%

5. $126.50

7. $228.77

9. Either one could be less expensive, depending on the APR.

11. Really Friendly S and L will have lower payments but higher fees and/or more points.

CHAPTER 10 REVIEW

1. $8730.15

2. $15,547.72

3. $4067.71

4. $4104.26

5. about 1.9 years

6. a. $308.25   b. $70,692.00   c. $235,092.00

7. Bank of the South at 7.88% is better than Extremely Trustworthy Savings at 7.63%

8. a. $4580.78   b. $2109.38

9. $37,546.62

10. $19,173.84

11. $37,738.26

12. a. $189,090.91   b. $88.22

13. a. $263.17   b. $3651.62

c.

Period Number	Principal	Interest	Total	Amount Outstanding
0	—	—	—	$12,138.58
1	$153.16	$110.01	$263.17	$11,985.42
2	$154.55	$108.62	$263.17	$11,830.87

d. does not verify   e. $6885.51

14. a. $174.50   b. does not verify

CREDITS

This page constitutes an extension of the copyright page. We have made every effort to trace the ownership of all copyrighted material and to secure permission from copyright holders. In the event of any questions arising as to the use of any material, we will be pleased to make the necessary corrections in future printings. Thanks are due to the following authors, publishers, and agents for permission to use the material indicated.

Photos   **p. 2:** Tony Freeman/PhotoEdit. **p. 8 (center and right):** Brown University Library. **p. 8 (left):** The Granger Collection, New York. **p. 9 (left):** North Wind Archive Pictures. **p. 9 (right):** Brown University Library. **p. 19:** AP Photo/Joe Marquette. **p. 104 (top):** Edward W. Souza/News Service, Stanford. **p. 104 (bottom):** Rick McClain/Gamma Liaison. **p. 112:** Cindy Charles/PhotoEdit. **p. 116:** Ed Lailo/Gamma Liaison. **p. 122:** AP/Wide World Photos. **p. 128:** AT&T Archives. **p. 143:** Kathleen Olson. **p. 152:** Cindy Charles/PhotoEdit. **p. 175:** Steve Raymer/Corbis. **p. 188 (left):** Novosti Information Agency. **p. 188 (right):** The Granger Collection, New York. **p. 211:** North Wind Archive Pictures. **p. 215 (left):** UPI/Corbis-Bettmann. **p. 215 (right):** Oxford University Press/Brown University Library. **p. 221:** James Schneph/Gamma Liaison. **p. 223 (left):** Brown University Library. **p. 223 (right):** Brown University Library. **p. 236:** Stock Montage. **p. 237:** Columbia University. **p. 240:** AP/Wide World Photos. **p. 260:** AP/Wide World Photos. **p. 265:** Stock Montage. **p. 266:** Far Side, Gary Larson/Universal Press Syndicate. **p. 270:** Tony Freeman/PhotoEdit. **p. 274:** The Granger Collection, New York. **p. 275:** The Granger Collection, New York. **p. 279:** AP/Wide World Photos. **p. 280:** Courtesy, Nancy Wexler. **p. 294:** Kathleen Olson. **p. 296:** Pictorial History Research. **p. 299:** Brooks/Cole. **p. 300:** Renato Retolo/Gamma Liaison. **p. 305:** The Kobal Collection. **p. 333:** AP/Wide World Photos. **p. 342 (left):** Novosti Information Agency. **p. 342 (right):** Brown University Library. **p. 356:** Robert Brenner. **p. 367:** Kathleen Olson. **p. 373 (left):** Los Alamos National Laboratory. **p. 373 (right):** Princeton University Press-Brown University Library. **p. 393:** Gary Larson/Universal Press Syndicate. **p. 416:** Charles Trotta/Gamma Liaison. **p. 417 (top):** AP/Wide World Photos. **p. 417 (bottom):** Far Side, Gary Larson/Universal Press Syndicate. **p. 442:** Stock Montage. **p. 443:** Stock Montage. **p. 462:** Bob Western/Brooks/Cole. **p. 475:** Bob Daemmirich/The Image Works. **p. 487:** Courtesy, Great Western Bank. **p. 490:** James Schneph/Gamma Liaison. **p. 509:** Michael Newman/PhotoEdit. **p. 516:** The Granger Collection, New York. **p. 518:** Tony Freeman/PhotoEdit. **p. 519:** Continental Savings of America. **p. 530:** The Granger Collection.

Quotations   **p. 281:** Excerpt from NEWSWEEK, February 12, 1973, Newsweek, Inc. All rights reserved. Reprinted by permission. **Fig. 9.27, p. 422:** Reprinted by permission of the Associated Press.

Grandma's Two, is the second book in the Ella Book series, about Ella, a 3 yr old bi-racial girl who has a very diverse family background. In this book, Ella visits her two grandmothers who live in two very different cities, with two different lifestyles. One grandmother lives in a small mountain town and the other lives in a large inner city. Ella learns valuable life lessons on her visits about respect, love, family and beliefs. Although both grandmothers are culturally and racially different, they both share positive goals for Ella. As grandmothers, they both love her and provide guidance, regardless of skin color and lifestyle.

Daneace Terry Jeffery was born in Detroit, Michigan, and currently resides in Baltimore, Maryland. Grandma's Two, is the second book in a series of five books. In her first book, Ella and Her Bubbles, readers were introduced to Ella, the main character. Daneace earned a B. A. in English from the University of Detroit Mercy and a M.S. in Literacy from Walden University. She is a Special Education teacher in a suburban high school near Baltimore, Maryland. Daneace is the grandmother of four granddaughters from the union of her two married sons. All four granddaughters are of mixed heritage and culturally different backgrounds. Two granddaughters are black and white mixed; while two other Muslim granddaughters are black and Filipino mixed. Ella, the main character is loosely based on the life of her oldest granddaughter.

ISBN 978-1-4990-5121-6

Xlibris

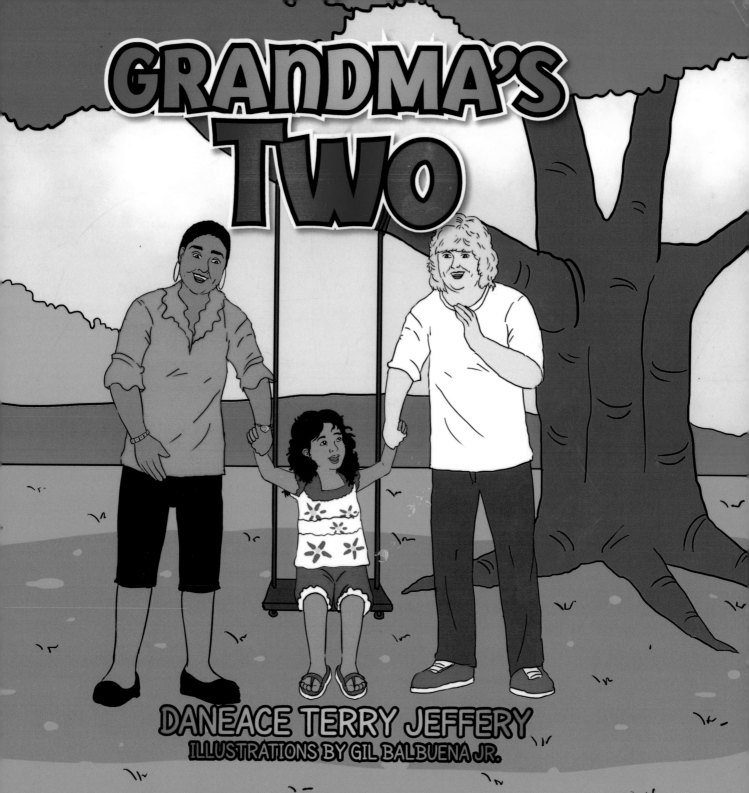

GRANDMA'S TWO

DANEACE TERRY JEFFERY
ILLUSTRATIONS BY GIL BALBUENA JR.

May all your dreams come
true.

Daneace